International Workshop on Membrane Transport in Plants, Jülich, 1974

Membrane Transport in Plants

Edited by
Ulrich Zimmermann and Jack Dainty

With 252 Figures and 49 Tables

Springer-Verlag New York Heidelberg Berlin 1974

Privat-Dozent Dr. Ulrich Zimmermann
Institut für Physikalische Chemie
der Kernforschungsanlage Jülich
5170 Jülich, W. Germany
and
Botanisches Institut
der Rheinisch-Westfälischen
Technischen Hochschule Aachen
5100 Aachen, W. Germany

Professor Dr. Jack Dainty
Department of Botany
University of Toronto
Toronto 181, Canada

ISBN 0-387-06989-5 Springer-Verlag New York Heidelberg Berlin
ISBN 3-540-06989-5 Springer-Verlag Berlin Heidelberg New York

Preface

In February, 1974, an 'International Workshop on Membrane Transport in Plants' was held at the Nuclear Research Centre, Jülich, West Germany. More than two hundred and fifty people, from fourteen countries, took part in this highly successful meeting. A somewhat similar meeting took place in Liverpool, England, two years ago and it became clear there that progress in the field of membrane transport in plants was now so marked that a second, and wider, meeting in Germany was more than fully justified. The members of our programme committee (U. Zimmermann, Chairman, Jülich (FRG); J. Dainty, Toronto (Canada); F. Führ, Jülich (FRG); N. Higinbotham, Pullman, Wa. (USA); A.B. Hope, Adelaide (Australia); A. Lev, Leningrad (USSR); U. Lüttge, Darmstadt (FRG); H.W. Nürnberg, Jülich (FRG); E.A.C. MacRobbie, Cambridge (UK); H. Stieve, Jülich (FRG); M. Thellier, Rouen (France); K. Wagener, Jülich (FRG)) decided on a broad spectrum of topics including thermodynamics of transport processes, water relations, primary reactions of photosynthesis, as well as the more conventional aspects of membrane transport. They were thus, for instance, particularly concerned to bring advanced thermodynamical concepts to the attention of biologists and to show physical chemists what the more complex biological systems were like.

We have taken considerable editorial liberty with the manuscripts submitted for this book, for we were very conscious of the necessity to make all papers understandable by as wide a circle of readers as possible. In effect we have obeyed the spirit of the meeting, which was to bring all works in plant membrane transport and relevant fields together for mutual understanding and clarification of the problems involved.

The meeting could not have taken place without the generous financial support of the Nuclear Research Centre, Jülich, of the Deutsche Forschungsgemeinschaft and the Sonderforschungsbereich 160 'Eigenschaften biologischer Membranen' and the German Industry (AEG-Telefunken AG, Ulm; Bayer AG, Leverkusen; Biotronik GmbH, Frankfurt/M.; Boehringer Mannheim GmbH, Mannheim; B. Braun Melsungen AG, Melsungen; Carl Zeiss, Oberkochen; Colora Meßtechnik GmbH, Lorch; E. Merck, Darmstadt; Farbwerke Hoechst AG, Frankfurt/M.; Kali und Salz AG, Hannover; Kontron GmbH, Eching; LKB Instrument GmbH, Düsseldorf; Riedel-de-Haen AG, Seelze; W. Pabisch KG, München). We gratefully thank all these organisations.

We are also indebted to Prof. Dr. K. Wagener, Director of the Institute of Physical Chemistry, for his support of the conference and to Dr. Borsch and his stuff, and to the coworkers of one of us (U.Z.) H.-J. Buers, H. Jäckel, H. Koch, Dr. G. Pilwat, Dipl.-Phys. F. Riemann, Dr. E. Steudle, and J. Zillikens for their day-to-day help in all the 'administrative' problems of the meeting: places to stay, slide projection, tape recording, meals, coffee at all times, and so on.

So far as the book is concerned, these same colleagues of Ulrich Zimmermann and his wife have rendered completely essential service in checking manuscripts, proof reading and redrawing figures. There is no doubt that the book could not have been ready in this relatively short time without their invaluable help. We are extremely grateful to them and to H.W. Arras, head of the printer shop, and H.-P. Pelzer for their support of the printing.

We also record our appreciation to Springer-Verlag, Berlin—Heidelberg—New York, for expediting the publication of these proceedings.

Finally we must single out the quite outstanding contribution made by Dr. Pilwat and Mrs. Pilwat. Dr. Pilwat produced the final copy of each manuscript and his wife really did nearly all the printing. They worked fantastically hard and effectively. Without them, the book wouldn't exist. So though we say thank you to the Pilwats, no words can express our deep gratitude to them.

Jülich, July 1974

J. Dainty
U. Zimmermann

Contents

Contributors

M. AVRON, Weizmann Institute, Rehovot, Israel.

J. BARBER, Department of Botany, Imperial College, University of London, England.

C.E. BARR, Department of Biological Sciences, State University of New York, Brockport, New York 14420, USA.

A. BEN-AMOTZ, Department of Biochemistry, Weizmann Institute, Rehovot, Israel.

F.W. BENTRUP, Institut für Biologie, Universität Tübingen, 74 Tübingen, Germany.

L. BERGMANN, Institut für Botanik, Universität Köln, 5 Köln, Germany.

T. BORNEFELD, Institut für Botanik, Universität Würzburg, 87 Würzburg, Germany.

D.J.F. BOWLING, Department of Botany, University of Aberdeen, Aberdeen, Scotland.

J.C. COLLINS, Department of Botany, University of Liverpool, Liverpool, England.

H.G.L. COSTER, School of Physics, University of New South Wales, Kensington, New South Wales, Australia.

W.J. CRAM, School of Biological Sciences, University of Sydney, Sydney 2006, New South Wales, Australia.

J. DAINTY, Department of Botany, University of Toronto, Toronto, Ontario, Canada.

R.F. DAVIS, Department of Botany, Rutgers University, Newark, New Jersey 07102, USA.

G. DUCET, Laboratoire de Physiologie Cellulaire, Centre Universitaire Marseille-Luming, 13 Marseille, France.

G. FINDENEGG, Institut für Botanik, Technische Hochschule Darmstadt, 61 Darmstadt, Germany.

F. FÜHR, Arbeitsgruppe Radioagronomie, Kernforschungsanlage Jülich, 517 Jülich, Germany.

C. GILLET, Laboratoire de Biologie Vegetale, Facultes Universitaires de Namur, B-500 Namur, Belgium.

B.Z. GINZBURG, Department of Botany, Hebrew University, Jerusalem, Israel.

D. GRADMANN, Institut für Biologie, Universität Tübingen, 74 Tübingen, Germany.

U.P. HANSEN, Institut für Angewandte Physik, Universität Kiel, 23 Kiel, Germany.

J.B. HANSON, Department of Botany, University of Illinois, Urbana, Illinois 61801, USA.

D.F. HASTINGS, Duke University Marine Laboratory, Beaufort, North Carolina 28516, USA.

W. HAUPT, Institut für Botanik, Universität Erlangen, 852 Erlangen, Germany.

R. HELLER, Laboratoire de Physiologie Vegetale, Şorbonne, Paris, France.

D.L. HENDRIX, Department of Botany, Washington State University, Pullman, Washington 99163, USA.

R. HERTEL, Institut für Biologie III, Universität Freiburg, 78 Freiburg, Germany.

N. HIGINBOTHAM, Department of Botany, Washington State University, Pullman, Washington 99163, USA.

T.K. HODGES, Department of Botany and Plant Pathology, Purdue University, Lafayette, Indiana 47906, USA.

M. HÖFER, Botanisches Institut, Universität Bonn, 53 Bonn, Germany.

A.B. HOPE, The Flinders University of South Australia, Bedfort Park, South Australia.

L.F. JAFFE, Department of Biological Sciences, Purdue University, Lafayette, Indiana, USA.

W.D. JESCHKE, Institut für Botanik, Universität Würzburg, 87 Würzburg, Germany.

W. JUNGE, Max-Volmer-Institut für Physikalische Chemie und Molekularbiologie, Technische Universität Berlin, 1 Berlin, Germany.
H.F. KAUSS, Fachbereich Biologie, Universität Trier-Kaiserslautern, 675 Kaiserslautern, Germany.
O. KEDEM, Weizmann Institute, Rehovot, Isreal.
M. KLUGE, Institut für Botanik, Technische Hochschule Darmstadt, 61 Darmstadt, Germany.
E. KOMOR, Fachbereich Biologie, Universität Regensburg, 84 Regensburg, Germany.
G.H. KRAUSE, Institut für Botanik, Universität Düsseldorf, 4 Düsseldorf, Germany.
C. KRISCHER, Institut für Neurobiologie, Kernforschungsanlage Jülich, 517 Jülich, Germany.
P.J.C. KUIPER, Laboratory for Plant Physiological Research, Wageningen, Netherlands.
A. KYLIN, Department of Plant Physiology and Anatomy, Royal Veterinary and Agricultural University, Copenhagen, Danmark.
A. LÄUCHLI, Institut für Botanik, Technische Hochschule Darmstadt, 61 Darmstadt, Germany.
T. LORENZ, Physik-Department, Technische Universität München, 8 München, Germany.
B.C. LOUGHMAN, Department of Agricultural Sciences, University of Oxford, Oxford, England.
U. LÜTTGE, Institut für Botanik, Technische Hochschule Darmstadt, 61 Darmstadt, Germany.
E. MARRE, Institute of Plant Sciences, University of Milan, Italy.
P. MEARES, Department of Chemistry, University of Aberdeen, Aberdeen, Scotland.
S.M. MERTZ, Jr., Department of Botany, University of Illinois, Urbana, Illinois 61801, USA.
J.I. MEYER, Max-Planck-Institut für Biophysik, 6 Frankfurt, Germany.
P. NISSEN, Botanical Laboratory, University of Bergen, Bergen, Norway.
P.S. NOBEL, Department of Biology, University of California, Los Angeles, California 90024, USA.
H.W. NÜRNBERG, Zentralinstitut für Analytische Chemie, Kernforschungsanlage Jülich, 517 Jülich, Germany.
H. OKAMOTO, Biological Institute, Faculty of Science, Nagoya University, Chikusa-ku, Nagoya, Japan.
G. PILWAT, Institut für Physikalische Chemie, Kernforschungsanlage Jülich, 517 Jülich, Germany.
M.G. PITMAN, School of Biological Sciences, University of Sydney, Sydney 2006, New South Wales, Australia.
J.A. RAVEN, Department of Biological Sciences, University of Dundee, Scotland.
L. REINHOLD, Department of Botany, Hebrew University, Jerusalem, Israel.
F. SAUER, Max-Planck-Institut für Biophysik, 6 Frankfurt/M., Germany.
E. SCHÄFER, Institut für Biologie II, Universität Freiburg, 78 Freiburg, Germany.
H. SCHÖNERT, Abteilung für Physikalische Chemie der Biopolymeren, Rheinisch-Westfälische Technische Hochschule Aachen, 51 Aachen, Germany.
M.S. SESHADRI, Max-Planck-Institut für Biophysik, 6 Frankfurt/M., Germany.
W. SIMONIS, Institut für Botanik I, Universität Würzburg, 87 Würzburg, Germany.
C.L. SLAYMAN, Department of Physiology, Yale University School of Medicine, New Haven, Connecticut, USA.
F.A. SMITH, Department of Botany, University of Adelaide, Adelaide, South Australia 5001.
E. STEUDLE, Institut für Physikalische Chemie, Kernforschungsanlage Jülich, 517 Jülich, Germany.
H. STIEVE, Institut für Neurobiologie, Kernforschungsanlage Jülich, 517 Jülich, Germany.
W. TANNER, Fachbereich Biologie, Universität Regensburg, 84 Regensburg, Germany.
M. THELLIER, Laboratoire de Nutrition Minerale, Faculte des Sciences de Rouen, 76 Mont-Saint-Aignan, France.

M.T. TYREE, Department of Botany, University of Toronto, Toronto, Ontario, Canada.
R.F.M. VAN STEVENINCK, Department of Botany, University of Queensland, Brisbane, Queensland 4067, Australia.
W.J. VREDENBERG, Centre for Plant Physiological Research, Wageningen, Netherlands.
G. WAGNER, Institut für Botanik, Universität Erlangen-Nürnberg, 852 Erlangen, Germany.
N.A. WALKER, School of Biological Sciences, University of Sydney, Sydney 2006, New South Wales, Australia.
J. WILLENBRINK, Institut für Botanik, Universität Köln, 5 Köln, Germany.
D. WOERMANN, Institut für Physikalische Chemie, Universität Köln, 5 Köln, Germany.
R.G. WYN JONES, Department of Biochemistry and Soil Science, University College of North Wales, Bangor, Wales, England.
H. ZIEGLER, Institut für Botanik, Technische Universität München, 8 München, Germany.
U. ZIMMERMANN, Institut für Physikalische Chemie, Kernforschungsanlage Jülich, 517 Jülich, Germany.

Session 1

Thermodynamics and Electrochemistry of Membrane Transport

Chairman: P. Meares

Permeation through Ion-Exchange Membranes

P. Meares

A. Introduction

It has been recognized for at least forty years that the selective permeabilities of some membranes towards ions of differing charges was due in many cases to the attachment, either by chemical bonds or by physical adsorption, of ionic groups on to the membrane material. The deliberate development of membranes with such properties and which have also high overall permeability and strength is of more recent origin. Such membranes have become available as a result of advances in the chemistry of synthetic ion-exchange resins. They have been developed and studied because of their technological value in electrically-driven membrane separation processes, such as electro-dialysis.

Most research on these membranes has been directed towards improving the efficiencies of separation processes. As a result of precise measurements on their permeabilities and selectivities and of substantial advances in the formulation of theoretical flux equations, a reasonably satisfactory picture of the functioning of such membranes has been built up.

Because biologists were among the first to realize the importance of and to study the effects of bound charges on membrane permeability, they have, very properly, followed with interest these developments in theory and practice made with synthetic ion-exchange membranes. It is essential to realize however that synthetic membranes differ completely in chemical and physical structure, in dimensions and in many other properties from any biological membranes. Their value to biology lies in providing stable model systems upon which to test ideas and from which to derive inspiration. It must be remembered that in the more complex transport phenomena that are now being discussed, for example all processes involving the coupling of fluxes, several competing processes are involved simultaneously. In such circumstances the net outcome is critically dependent on the relative importances of the various processes. These may be very different in the real and in the model systems and false conclusions might easily be drawn unless care is exercised in the choice of the model and in the interpretation of results.

Synthetic ion-exchange membranes consist of semi-rigid gels which contain polyelectrolytes. They may be produced in a variety of ways. In the earliest and simplest, the polyelectrolyte molecules are interconnected into a network by chemical crosslinks. Membranes of this kind are often fragile and swell or shrink considerably when the concentrations of the solutions in contact with them are changed. For commercial use in electrodialysis more complex structures, which are more stable dimensionally, have been preferred. Examples of these are polyelectrolyte chains grafted on to a nonpolar, inert and sometimes micro-crystalline polymer matrix, such as polyethylene, and mixtures of a polyelectrolyte with a non-electrolytic polymer. These may be cast into membranes in which the polyelectrolyte is concentrated into interconnecting micro-vesicles distributed through a nonpolar matrix.

B. Results and Discussion

I. Homogeneous Polyelectrolyte Gel Membranes

Studies of the permeation of ions and water through ion-exchange membranes have shown that permeation occurs only in the regions occupied by polyelectrolyte. Consequently, the more complex membranes behave as heterogeneous media. Transport processes in heterogeneous systems are not well understood even where the structures of the media can be described in detail. The polyelectrolyte network membranes are easier to treat theoretically and it has proved possible to develop a fairly satisfactory understanding of multi-component transport phenomena in these. This paper is restricted to the consideration of such essentially homogeneous gel membranes.

When such a membrane is brought into contact with an aqueous solution some of the ionogenic groups dissociate. The presence of these groups enables the membrane to imbibe water from the solutions and the amount of water taken up is restricted by the network crosslinks. The water enables the small counterions released by the ionogenic groups to diffuse within the membrane and to exchange with small ions of like charge in the contacting solutions. The bound charges restrict the entry into the swollen network of co-ions, i.e. ions of similar charge to the bound ions. These restrictions on the behaviour of the counterions and co-ions are brought about by the membrane taking up a uniform electric potential different from that of the solution. This potential is a function of the nature and concentration of the solution and also of the temperature and pressure which will be assumed constant throughout this paper.

In such an idealized membrane with a uniform internal electric potential the distribution of ions at equilibrium between the solution and the membrane may be described by the Donnan equation. This can be written in the form

$$\bar{a}_i = a_i \exp(-z_i F \psi_D / RT) \tag{1}$$

where z_i is the valency, including charge sign, of the ionic species, a_i its activity in the solution and $\bar{a}_i$ its activity in the membrane. F, R, and T are the Faraday number, gas constant and absolute temperature respectively. The Donnan potential ψ_D is the potential of the membrane interior relative to that in the external solution.

Ion-exchange membranes are thick in comparison with the dimensions of the electric double layers and the number of ionic charges required to establish the potential difference between the membrane and solution is a negligible fraction of the total number of ions in the system. Consequently both phases may be regarded as electrically neutral and the molalities $\bar{m}_i$ of all mobile ionic species i in the membrane are related to the concentration of ionized groups bound to the membrane, M equiv kg^{-1} of absorbed water, by

$$\Sigma z_i \bar{m}_i \pm M = 0 \tag{2}$$

where the positive sign applies when the bound charges are positive and *vice versa*.

The Gibbs-Donnan distribution of the ions between membrane and solution is obtained by combining Eqs. (1) and (2). Restricting attention to a single salt which dissociates into ν_g counterions and ν_c co-ions, the distribution equation takes the form

$$\bar{m}^{\nu_c} (\bar{m} + M/\nu_g)^{\nu_g} = (\alpha m)^{\nu} \tag{3}$$

where $\nu = (\nu_c + \nu_g)$, m is the molality of the salt in solution and $\nu_c \bar{m}$ the molality of co-ions in the membrane. The ratio of activity coefficients α should be relatively insensitive to m and, in a highly hydrated membrane, should be close to unity (GLUECKAUF, 1962).

Eq. (3) shows that the concentration of co-ions will be less and of counterions greater than their concentration in the solution in contact with the membrane. Furthermore, the concentration of co-ions increases with m raised to a power greater than unity (i.e. ν/ν_c in dilute solutions). Although these consequences of Eq. (3) are observed qualitatively, in practice there are systematic quantitative deviations which will be discussed below.

In principle the Donnan equation and the concept of electrically neutral phases can be used to describe also the distribution of a mixture of salts between a solution and a fixed-charge membrane. An important case is a mixture of two salts with a common ion, which is a co-ion in the membrane, and different counterions. In its simplest form, in which all mixtures are assumed ideal, this approach to the study of ion-exchange equilibrium is not very instructive. It predicts that when the pair of counterions have the same valency, e.g. Na^+ and K^+ or Mg^{2+} and Ca^{2+}, the mole fraction of an ion in the membrane will always equal its mole fraction in solution. This, as is well known, is not the case in practice and non-ideality must be taken into account. The Donnan treatment is more informative in the case of a pair of counterions with unequal valencies. It shows that the counterion of higher valence will be relatively more concentrated in the membrane than that of lower valence, in comparison with their concentrations in the solution, and the disparity is greatest in dilute solutions i.e. when $m/\bar{m}$ is large. A satisfactory treatment of ion-exchange equilibrium in such systems requires a considerably more complex treatment however (MEARES and THAIN, 1968). Permeation in such mixed systems is very complicated at the quantitative level (MACKAY and MEARES, 1960, 1961) and in this paper only single salts will be dealt with.

When a membrane separates two different solutions and distribution equilibria are set up at the opposite faces, the ion concentrations and the Donnan potential difference will be unequal at the opposite sides of the membrane. The ions and molecules within the membrane are thus subjected to a set of forces determined by the gradients of the intensive variables across the membrane. The fluxes generated by a given force per mole are as much influenced by the concentrations of the transported particles as by their mobilities. It will always be important to study both the distribution, i.e. thermodynamic properties, and mobility, i.e. kinetic properties, if a proper understanding of membrane permeability is to be obtained. In this paper some experiments along these lines on a well-characterized system are described and analysed in terms of the simple flux equations which are often used by biologists.

II. A Synthetic Homogeneous Membrane

The results in this paper were obtained with a synthetic membrane bearing bound anions and which was selectively permeable to cations. The membrane, known as Permutit Zeo-Karb 315, was prepared by reacting together phenol and formaldehyde in the presence of sodium metabisulphite and in an aqueous alkaline medium. It was prepared so as to be as nearly homogeneous as possible. The polymerization reaction is spatially random and ensures that the membrane charges are fixed to the membrane and not merely constrained within its phase as in a liquid ion-exchange membrane. Crosslinks were introduced into the membrane structure during its formation so that the molecular network encompassed the water and other substances present. On being transferred to a dilute salt solution the membrane swelled by 15–20 %. During this swelling process the more loosely crosslinked volume elements extended more than the more tightly crosslinked ones. Inevitably this led to some microscopic dispersity in the distribution of the fixed charges throughout the volume of the finished membrane.

Although some natural membranes may have a greater regularity in the distribution of their charges this does not necessarily lead to a more uniform electrical potential on the microscale because the ions and molecules whose diffusion is of interest are smaller than the structural elements of the membrane. It is believed therefore that results obtained with this synthetic membrane should give a guide as to how closely any membrane permeable to water, ions and other solutes may approach ideality in its behaviour.

Table 1

Properties of Zeo-Karb 315 in 0.1 M NaBr at 25°C.

$\bar{C}_{Na^+}$	0.58 M
$\bar{C}_{Br^-}$	0.031 M
Water content	$0.74\ g\ cm^{-3}$
Specific conductance	$12.3 \cdot 10^{-3}$ mho cm^{-1}
Electro-osmotic flux	$7.5 \cdot 10^{-3}\ cm^3\ C^{-1}$
Transport number of Br^-	$44.5 \cdot 10^{-3}$

Some of the more important parameters of Zeo-Karb 315 membranes are given in Table 1; others may be found in the references cited throughout this article.

III. The Donnan Equation and Co-Ion Sorption

It has always been found that the uptake of co-ions from dilute solutions is larger than predicted by the simple Donnan equation. As the concentration is increased the equation is increasingly well obeyed. This behaviour has been observed in Zeo-Karb 315 with several electrolytes of various valence types. Figs. 1 and 2 are based on the results of MACKIE and MEARES (1955a). They show data on NaCl and $MgCl_2$. Although the agreement between ideal theory and experiment was close above 0.5 m, at 0.01 m the observed uptake was more than twice the ideal value.

Experimental artefacts, such as cracks and voids in the resin matrix of the membrane (DAVIES and YEOMAN, 1953) or inability to separate completely the membrane and solution for analysis (FREEMAN, 1960) may contribute to these deviations but are insufficient to explain all of them. One must look instead to the oversimplifications inherent in the Donnan equation.

The only measurable parameter of the membrane which appears in this equation is the macroscopic average concentration of fixed charges. These charges are however discrete and, where the concentration of fixed charges is less than about 10 m, they are separated by mean distances which exceed 1 nm i.e. which are large compared with the thickness of the ionic atmospheres in the membrane which surround them. As a result there are local variations in the electric potential within the membrane and these will always lead to an ion uptake greater than the ideal prediction.

Three structural features may be described which could lead to local variations in the potential. The bound charges may be separated from one another along the macromolecular segments by distances large in the present context. The macromolecules in the swollen gel membrane may be separated by large distances. There may be irregularities in the micro-structure of these macromolecules or in their distribution in the gel which produce local fluctuations in the fixed charge concentration. These structural features may lead to fluctuations on the scale 10–100 nm as well as on the molecular scale.

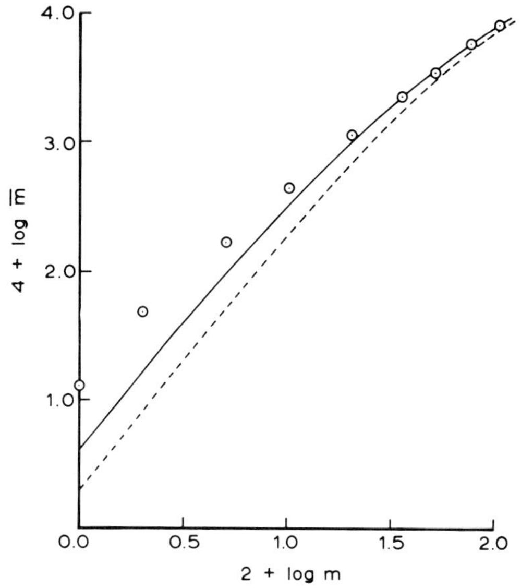

Fig. 1

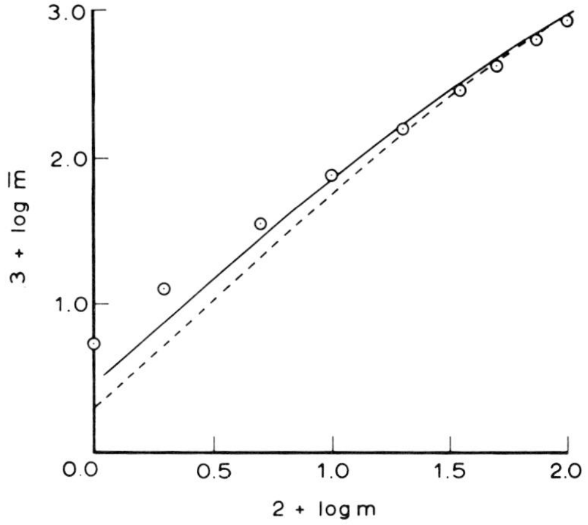

Fig. 2

Fig. 1 Sorption of NaCl by Zeo-Karb 315. m is molality in solution and $\overline{m}$ in the membrane. The points are experimental, the dotted curve was calculated from the ideal Donnan equation (3) with $\alpha = 1$ and the full line from the theory of KATCHALSKY (1954) and MACKIE and MEARES (1955a).

Fig. 2 Sorption of $MgCl_2$ by Zeo-Karb 315. m is molality in solution and $\overline{m}$ in the membrane. The points are experimental, the dotted curve was calculated from the ideal Donnan equation (3) with $\alpha = 1$ and the full line from the theory of KATCHALSKY (1954) and MACKIE and MEARES (1955a).

Several attempts have been made to treat this problem theoretically. In one group of theories the uniform charge of the ideal model is replaced by charges smeared along the macromolecules which may then be treated as parallel plates (LAZARE et al., 1956) or as rods (MANNING, 1969). Such theories predict effects qualitatively similar to those observed. Smeared charge models do not however take into account the discrete nature of the fixed ions.

The mean separation of the fixed charges along the polymer chains in Zeo-Karb 315 is about 2 nm. The separation between neighbouring polymer chains in the membrane when swollen is similar and this distance is several times the thickness of the ionic atmospheres. In such circumstances a smeared charge model is inappropriate and the theory of flexible polyelectrolytes, which has been extended to crosslinked gels by KATCHALSKY (1954), is more applicable. This takes into account the additional electrostatic and elastic contributions to the free energy which result from the interconnection of the fixed charges along the polymer chains and through the crosslinks.

MACKIE and MEARES (1955a) used KATCHALSKY's theory to analyse their data on Zeo-Karb 315. They showed that the observed co-ion uptakes fitted the theory well down to 0.1 m solutions and at lower concentrations the data deviated from KATCHALSKY's theory far less than from the Donnan equation. Figs.1 and 2 compare the observed data with those calculated on the basis of known structural parameters of the membrane. It is clear that the discreteness of the ionic charges and the polymeric nature of the real membrane lead to considerable deviations from the ideal Donnan treatment.

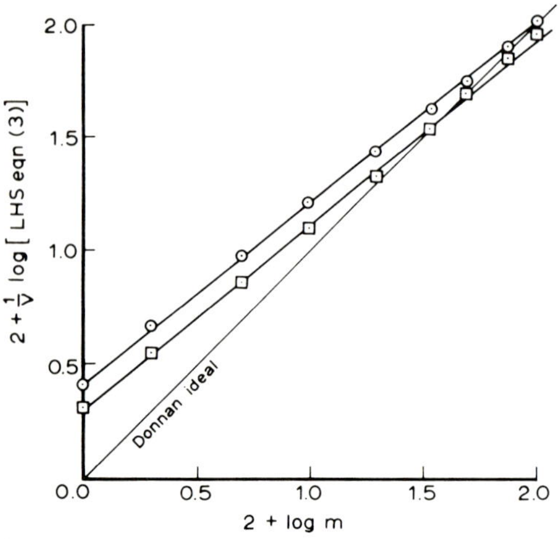

Fig. 3 Comparison of Eq. (3), i.e. ideal $a = 1$ Donnan line, with Eq. (4) for NaCl and $MgCl_2$. For NaCl $a = 0.991$ and $\beta = 0.83$; for $MgCl_2$ $a = 0.807$ and $\beta = 0.084$. ○ NaCl; □ $MgCl_2$.

The observed uptakes at the lowest concentrations were 50–60 % larger than could be accounted for on KATCHALSKY's theory. It is believed that this is due to local irregularities in the distribution of the fixed charges along the polymer chains and of the crosslinks between the chains. This belief is supported by observations on ionic mobilities and friction coefficients in the membrane (McHARDY et al., 1969b).

GLUECKAUF (1962) showed that ion uptake data could be analysed in terms of the Donnan equation so as to give information on the micro-heterogeneity in the distribution of the fixed charges. GLUECKAUF made use of an empirical distribution function to describe the structure of the systems he studied. This particular function is not suitable in all cases and the development of a more generally applicable distribution function might make possible the determination of important parameters of the membrane structure.

If $1/\nu$ times the left side of Eq. (3) is plotted logarithmically against log m a straight line of slope unity should be obtained with an intercept log a at m = 1. Such plots for NaCl and $MgCl_2$ in Zeo-Karb 315 are shown in Fig. 3. The plots are seen to be linear but the slopes are less than unity although almost equal for the two salts. Log$a$ turns out, as predicted, to be almost zero in each case. This observation shows that for Zeo-Karb 315 and perhaps for other homogeneous membranes the ideal Eq. (3) may be replaced by

$$\bar{m}^{\nu_c} (\bar{m} + M/\nu_g)^{\nu_g} = (am)^{\beta\nu} \tag{4}$$

Here a is a characteristic of the electrolyte and does not differ greatly from unity. β is a parameter which characterizes the dispersity of the charge distribution in the membrane. The greater $(1 - \beta)$ the greater must be the dispersity.

IV. Membrane Potentials and the Donnan Potential

When a homogeneous membrane separates two solutions of different concentrations the observable potential difference between the solutions can be regarded as the algebraic sum of the diffusion potential within the membrane and the interfacial potentials at the two membrane/solution boundaries. In the fixed charge theory of TEORELL, and MEYER and SIEVERS (see e.g. TEORELL (1953)) the idealized form of the Donnan equation is used to express the interfacial potentials.

The sorbed co-ions (valency z_c) are not bound by the fixed charges and are freely mobile in the membrane. The Donnan potential ψ_D can therefore be represented by

$$\psi_D = (RT/z_c F)\ln(\alpha m/\bar{m}) \tag{5}$$

In order to use this equation $\bar{m}$ must be obtained from Eq. (4) making use of the values of α and β determined experimentally. The sorption results show that $\bar{m}$ is greater than the ideal value especially in dilute solutions and the Donnan potential at the dilute face of the membrane will fall further below the ideal value than will that at the concentrated face. Unless this effect is taken into account membrane potentials calculated from the fixed charge theory may deviate by as much as 20 mV from the observed values.

V. Simplified Flux Equations

In modern adaptations of the TEORELL—MEYER—SIEVERS theory of fixed charge membranes the ion fluxes are usually described by the Nernst-Planck equation. Many particular cases have been considered and expressions obtained for ion fluxes and membrane potentials (SCHLÖGL, 1964). Despite the need for idealizations in their derivation, these equations have been widely used by biologists to interpret fluxes and potentials. It is useful and important to assess the validity of such equations in experiments on a synthetic ion-exchange membrane.

In the most commonly used version of the Nernst-Planck equation the ion fluxes ϕ_i are regarded as being driven by the gradients of concentration and potential in the membrane. The flux equation is

$$\phi_i = -\,\bar{u}_i\bar{c}_i\,(RT\,d\ln\bar{c}_i + z_i F\,d\psi)/dx \tag{6}$$

where $\bar{c}_i$ is the local concentration per unit volume in the membrane of ions i and x measures distance through the membrane. $\bar{u}_i$ is the absolute mobility of the ions in the membrane. In a homogeneous membrane $\bar{u}_i$ is intrinsically independent of x but may depend on $\bar{c}_i$ and on the concentrations of other substances present. Because there are concentration gradients across the membrane $\bar{u}_i$ may therefore be a function of x in a non-equilibrium situation. Nevertheless the integration of Eq. (6) is usually carried out while assuming $\bar{u}_i$ to be constant. Sometimes the potential gradient $d\psi/dx$ is also assumed independent of x (GOLDMAN, 1943).

VI. Non-Equilibrium Thermodynamic Flux Equations

More general flux equations may be developed by using non-equilibrium thermodynamics (KATCHALSKY and CURRAN, 1965). These equations include in a formal way all the forces acting on the system and the fluxes of all mobile components. They allow for the possibility of interactions between fluxes. For practical reasons these equations are usually restricted to linear relations between fluxes and forces and they are applicable only to systems that are close to equilibrium. Often the forces are chosen to be the differences between the intensive variables characteristic of the solutions on opposite sides of the membrane, the coefficients which connect these forces with the fluxes are average values across the membrane. In this 'discontinuous' form the equations, as usually written, hold over such narrow ranges of concentration that they are unlikely to be valid in many biological situations. This limitation has recently been greatly eased by a reformulation of the expressions for a discontinuous system which permits the linear equations to be retained so long as linear relations between fluxes and forces continue to hold locally within the membrane (SAUER, 1973).

Unfortunately this reformulation does not reduce the amount of experimental information that is required to determine the large number of empirical coefficients which describe each system of membrane and ambient solutions. For this reason the application of non-equilibrium thermodynamics to biological transport processes is likely to prove of greater value in discussions of competing theories than in the quantitative correlation of fluxes and forces in complex systems. It is likely that biologists will continue to rely heavily on the simpler flux equations mentioned above in order to examine the connections between passive fluxes and forces.

In the case of an aqueous solution of a single salt permeating a membrane, linear equations for the molar flux densities ϕ_i take the form

$$\phi_i = \sum_{k=1}^{3} L_{ik} \, \Delta\mu_k \tag{7}$$

where subscript 1 refers to cations, 2 to anions and 3 to water. The convention here is that fluxes are positive in the direction opposite from that chosen to express increases in the chemical potentials. $\Delta\mu_1$ and $\Delta\mu_2$ are differences in the electrochemical potentials of the cations and anions between the opposite sides of the membrane and $\Delta\mu_3$ is the difference in the chemical potential of water.

For a homogeneous membrane experiments can be devised in which the differential fluxes which would arise under vanishingly small forces can be evaluated (KRÄMER and MEARES, 1969). In such circumstances there is no problem regarding the validity of the linear relations because the system departs infinitesimally from equilibrium. When the flux equations are reduced to this differential form the phenomenological conductance coefficients L_{ik} are replaced by a set of differential coefficients $\mathcal{L}_{ik}$ which are defined by

$$\mathcal{L}_{ik} = \lim_{\text{all } \Delta\mu_i \to 0} (\partial\phi_i/\partial\mu_k)_{\mu_{j \neq k}} \tag{8}$$

The Onsager reciprocal relations $\mathcal{L}_{ik} = \mathcal{L}_{ki}$ hold for these differential coefficients. They may be evaluated as functions of known membrane composition from sufficient experimental data.

The matrix of Eq. (7) may be inverted to give another well known set of relations between fluxes and forces. This is

$$\Delta\mu_i = \sum_{k=1}^{3} R_{ik} \, \phi_k \tag{9}$$

These equations also can be expressed for vanishing fluxes and forces, and differential resistance coefficients $\mathcal{R}_{ik}$ are defined by

$$\mathcal{R}_{ik} = \lim_{\text{all } \phi_i \to 0} (\partial\mu_i/\partial\phi_k)_{\phi_{j \neq k}} \tag{10}$$

VII. Comparison of Nernst-Planck and Non-Equilibrium Thermodynamic Formulations

The Nernst-Planck equation (6) is a local differential equation. It can be compared with the non-equilibrium thermodynamic equation (7) provided this also is written in the following differential form

$$\phi_i = \delta[L_{i1}(\partial\mu_1/\partial x) + L_{i2}(\partial\mu_2/\partial x) + L_{i3}(\partial\mu_3/\partial x)] \qquad (11)$$

Here δ is the membrane thickness. Eqs. (6) and (11) are consistent with one another only if all L_{ik} ($i \neq k$) are zero. It may be seen from Eq. (8) that the coefficient L_{ik} connects the flux of i with the force on and, consequently, with the flux of k. In cases where the fluxes of different species are coupled the Nernst-Planck equation is inaccurate and can be used only if the coupling is weak. This restriction is likely to be important when there are ionic fluxes in a membrane in which there is also an osmotic flux of solvent. If an electric current passed through the membrane generates an electro-osmotic flux of water then there is clear evidence that the fluxes of the ions and water are coupled.

VIII. Extension of the Nernst-Planck Equation

The L_{ik} coefficients relate the fluxes to thermodynamic forces. A non-zero L_{ik} means that the force on k affects the flux of i. This does not necessarily mean that species i and k interact directly. The physical meaning of the R_{ik} coefficients is clearer. From Eq. (9) it is seen that R_{ik} expresses the drag force on i caused by unit flux of k when all other fluxes, including that of i itself, are held at zero. R_{ik} therefore measures the direct interaction between i and k. A non-zero R_{ik} implies that if k is a mobile species the flux of k will directly influence the flux of i. The straight coefficient R_{ii} represents the force required to produce unit flux of i while all other fluxes are held at zero.

The extent to which a flow of i is created by a flux of k when there is no external force on i and all flows other than ϕ_i and ϕ_k are zero can be expressed in terms of the degree of coupling q_{ik}. This parameter, defined by CAPLAN (1965, 1966),

$$q_{ik} = -R_{ik}/\sqrt{R_{ii}R_{kk}}, \qquad (12)$$

may take any value between zero, in the absence of coupling, and unity when the particles of i and k move with the same velocity.

Measurements on the degree of coupling are particularly useful because they show which coupling processes are likely to be most significant. From this information one may assess the consequences of omitting to take such interactions into account in the Nernst-Planck equation.

The ion/water and ion/ion degrees of coupling for NaBr in Zeo-Karb 315 are shown as functions of concentration in Fig. 4. Table 2 lists some values obtained with CsBr and $SrBr_2$. It can be seen that the cation flux is more tightly coupled to that of the water than is the anion flux. This is partly a result of the greater degree of hydration of cations compared with anions. The coupling of each of the ionic fluxes to water is considerable whereas the coupling between the counterion and co-ion fluxes is quite small. These results suggest that the Nernst-Planck equation would be substantially improved if an allowance were made for the coupling of the ion fluxes to that of the water.

An additional term has been introduced into the Nernst-Planck equation (SCHLÖGL and SCHÖDEL, 1955; MACKIE and MEARES, 1955b) in order to take account of the influence of volume flow on the ion fluxes. Since the major component of the volume flow is usually due to the solvent flux this extension of the Nernst-Planck equation has the effect of taking

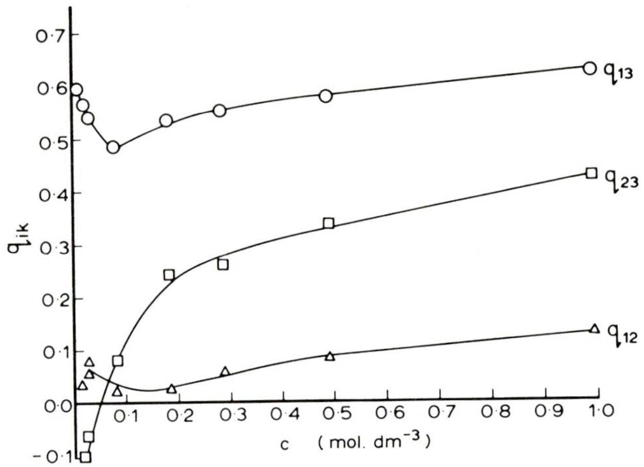

Fig. 4 Degrees of coupling q_{ik} against concentration of NaBr in the external solution c (M). 1 = Na^+, 2 = Br^-, 3 = water.

Table 2

Degrees of coupling of ion/ion and ion/water fluxes in Zeo-Karb 315.

Solution concentration (M)	q_{12}	q_{13}	q_{23}
CsBr			
0.085	−0.13	0.49	0.27
0.055	−0.03	0.45	0.12
0.035	0.02	0.46	0.00
0.020	0.03	0.47	−0.01
0.010	0.04	0.49	−0.02
$SrBr_2$ *			
0.499	−0.02	0.75	0.45
0.250	−0.02	0.64	0.42
0.104	−0.02	0.50	0.30
0.051	−0.01	0.45	0.20
0.015	0.00	0.41	0.08

*These values supersede in accuracy those used previously (MEARES, 1973).

into account the most important source of coupling. The modified Nernst-Planck equation can be written in the form

$$\phi_i = -\bar{u}_i\bar{c}_i\,(RT\,d\ln\bar{c}_i + z_iF\,d\psi)/dx + \gamma\phi_3\bar{c}_i/\bar{c}_3 \qquad (13)$$

where γ is a factor determined by the average degree of coupling between the ion and water fluxes. In the original papers γ was taken implicitly as unity i.e. it was assumed that there was complete coupling between the ion and water fluxes.

Experimentally it was found that when γ in Eq. (13) was set at unity the modified equation over-corrected for the effects of solvent drag. McHARDY et al. (1969a) determined γ for Na^+ and Sr^{2+} in Zeo-Karb 315 by fitting Eq. (13) to their experimental results. They found values of γ between 0.4 and 0.6 which are in good agreement with the degree of coupling q_{13} between cations and water. In the case of Br^-, γ was somewhat variable and its value was influenced also by the ion/ion interaction.

IX. Ionic Mobilities and the Nernst-Planck Equation

The Nernst-Planck equation is usually integrated across the membrane by making the assumption that the ionic mobilities $\bar{u}_i$ inside the membrane are independent of the concentrations of the solutions in contact with it. This assumption can be tested by making use of the non-equilibrium thermodynamic expressions and data.

Eqs. (6) and (11) can be made to agree exactly, irrespective of the values of the $\boldsymbol{L}_{ik}$ cross-coefficients, provided all forces are zero except on the species whose flux is being considered. If the gradient of the activity coefficient of i may be neglected then equating Eqs. (6) and (11) gives

$$\boldsymbol{L}_{ii}\,\delta = \bar{u}_i\bar{c}_i \qquad (14)$$

The variations of the mobility $\bar{u}_i$ with the concentration of the external solution may be seen from the values of $\boldsymbol{L}_{ii}\delta/\bar{c}_i$.

It must be remembered that where flux coupling occurs the cross-coefficients are not all zero and $\boldsymbol{L}_{ii}\delta/\bar{c}_i$ defines a mobility under unit force on i in circumstances where i is transported relative to a moving background. This complexity arises because fluxes of species other than i are generated by the coupling of these species with the flux of i resulting from the force applied directly to i.

The reciprocal quantity $\delta/c_i\boldsymbol{R}_{ii}$ also defines a mobility, in this case under restrictions whereby all fluxes other than that of i are zero. A mobility may also be defined through the tracer diffusion coefficient of i, D_i^*, obtained by measuring the flux of an isotope of i in the membrane under conditions of uniform chemical composition and zero external forces. In order to compare D_i^* with the non-equilibrium thermodynamic mobilities they must be multiplied by RT to bring them to the dimensions of diffusion coefficients. It is expected that D_i^* would differ from $RT\delta/\bar{c}_i\boldsymbol{R}_{ii}$ only as a result of coupling between the flux of the tracer and the counter-flux of unlabelled i. Data on the tracer diffusion coefficients in Zeo-Karb 315 have already been published (McHARDY et al., 1969b). These three indexes of ion mobility in Zeo-Karb 315 are compared for Na^+ in Fig. 5 and for Br^- in Fig. 6. It can be seen that in the case of counterions and co-ions $RT\boldsymbol{L}_{ii}\delta/\bar{c}_i$ is far from independent of concentration. In each case this quantity increases by more than 250 % in going from the most dilute to the most concentrated solution examined. It also shows little resemblance to the tracer diffusion mobility which is shown as the full line in these figures.

$RT\delta/\bar{c}_i\boldsymbol{R}_{ii}$ also increases with increasing solution concentration for counterions and co-ions but by a much smaller factor. From Fig. 5 it can be seen that $RT\delta/\bar{c}_1\boldsymbol{R}_{11}$ agrees excellently with D_1^*; the agreement between $RT\delta/\bar{c}_2 R_{22}$ and D_2^* is seen from Fig. 6 to be less good but the experimental difficulties in obtaining accurate values of the co-ion resistance coefficients are considerable and may account for most of the scatter in the experimental points.

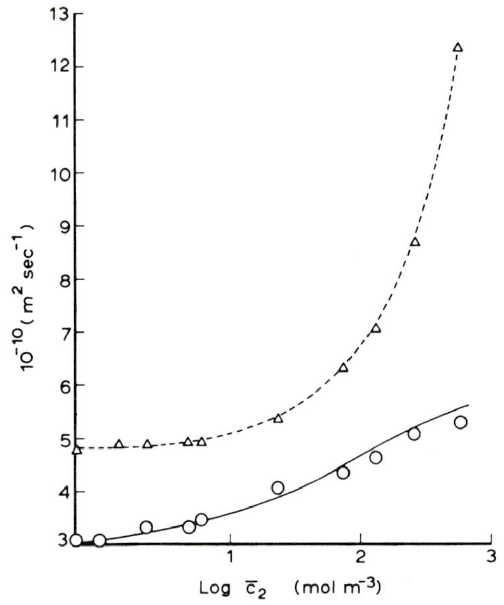

Fig. 5

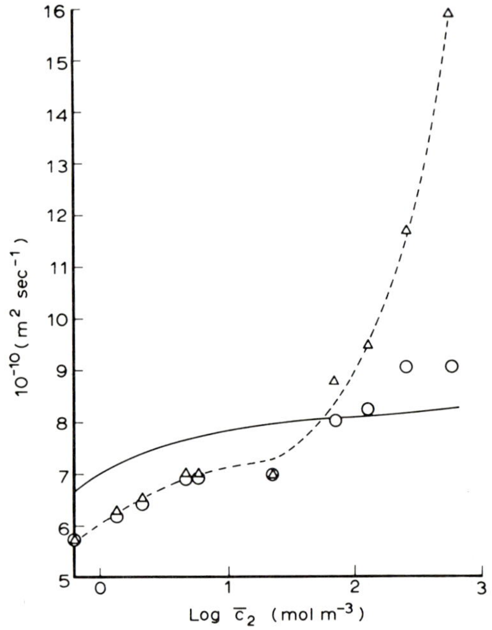

Fig. 6

Fig. 5 Mobility indexes of Na^+ in Zeo-Karb 315 against logarithm of sorbed Br^- concentration, log $\bar{c}_2$. The full line is the tracer diffusion coefficient D_1^* and the circles are $RT\delta/\bar{c}_1 R_{11}$. The triangles and dashed curve are $RTL_{11}\delta/\bar{c}_1$.

Fig. 6 Mobility indexes of Br^- in Zeo-Karb 315 against logarithm of sorbed Br^- concentration, log $\bar{c}_2$. The full line is the tracer diffusion coefficient D_2^* and the circles are $RT\delta/\bar{c}_2 R_{22}$. The triangles and dashed curve are $RTL_{22}\delta/\bar{c}_2$.

These observations suggest that $RT\delta/\bar{c}_i R_{ii}$ is the more appropriate mobility to use when applying the extended form of the Nernst-Planck equation. The variations of $RT\delta/\bar{c}_i R_{ii}$ with concentration are certainly not negligible in Zeo-Karb 315. When experiments over wide concentration intervals are under consideration such variations in mobility should be taken into account in the integration of the Nernst-Planck equation across the membrane if reliable conclusions are to be ensured.

C. Conclusions and Summary

The most readily applicable formulations of the fluxes and potentials across charged membranes which can be used to interpret data on biological systems are still those based on the TEORELL-MEYER-SIEVERS fixed charge model. Usually it is assumed that the ionic distributions defined by the ideal form of the Donnan equation hold at the membrane faces and that uncoupled ionic fluxes in the membrane may be represented by the Nernst-Planck equation. When these assumptions are tested by experiments on an almost homogeneous synthetic gel membrane selectively permeable to cations the following results are found.

1) The ideal Donnan equation is well-obeyed at high concentrations but at low concentrations the co-ion uptake by the membrane is several times larger than the ideal value. This behaviour is a result of the polymeric character of the membrane matrix and of fluctuations in the local charge density on the micro-scale.

2) Ion and osmotic water fluxes are strongly coupled in the membrane except for the co-ions in very dilute solutions. Omission of this coupling process from the Nernst-Planck equation can lead to considerable error in the estimation of ion fluxes when the flux of water is large.

3) Ion mobilities evaluated from non-equilibrium thermodynamic considerations approximately double as the solution concentration is increased from 0.01 to 1.00 m. Provided an appropriate average mobility is used in the integration of the Nernst-Planck equation this problem may be overcome but it must be understood that this average mobility is an empirical function of the concentrations of the solutions on both sides of the membrane. It can be satisfactorily determined from experiments with isotopic tracers in many membranes.

D. References

CAPLAN, S.R., J. Phys. Chem. **69**, 3801–3804 (1965)

CAPLAN, S.R., J. Theor. Biol. **10**, 209–235, 346–347 (1966)

DAVIES, C.W., YEOMAN, G.D., Trans. Faraday Soc. **49**, 968–974 (1953)

FREEMAN, D.H., J. Phys. Chem. **64**, 1048–1051 (1960)

GLUECKAUF, E., Proc. Royal Soc. (Lond) **A268**, 350–370 (1962)

GOLDMAN, D.E., J. Gen. Physiol. **27**, 37–60 (1943)

KATCHALSKY, A., Progr. Biophys. **4**, 1–59 (1954)

KATCHALSKY, A., CURRAN, P.F., Non-equilibrium Thermodynamics in Biophysics, Harvard University Press, Cambridge, 1965

KRÄMER, H., MEARES, P., Biophys. J. **9**, 1006–1028 (1969)

LAZARE, L., SUNDHEIM, B.R., GREGOR, H.P., J. Phys. Chem. **60**, 641–648 (1956)

McHARDY, W.J., MEARES, P., SUTTON, A.H., THAIN, J.F., J. Colloid and Interf. Sci. **29**, 116–128 (1969a)

McHARDY, W.J., MEARES, P., THAIN, J.F., J. Electrochem. Soc. **116**, 920–928 (1969b)

MACKAY, D., MEARES, P., Kolloid-Zeit. **171**, 139–149 (1960)

MACKAY, D., MEARES, P., Kolloid-Zeit. **176**, 23–29 (1961)

MACKIE, J.S., MEARES, P., Proc. Royal Soc. (Lond) **A232**, 485–498 (1955a)

MACKIE, J.S., MEARES, P., Proc. Royal Soc. (Lond) **A232**, 498–518 (1955b)

MANNING, G.S., J. Chem. Phys. **51**, 924–938 (1969)

MEARES, P., THAIN, J.F., J. Phys. Chem. **72**, 2789–2797 (1968)

MEARES, P., in Transport Mechanisms in Epithelia (eds. H.H. Ussing, N.A. Thom), pp. 51–67, Munksgaard, Copenhagen 1973

SAUER, F., Handbook of Physiology, Section 8, Renal Physiology (eds. J. Orloff, R.W. Berliner), pp. 399–414, Washington D.C. The American Physilogical Society 1973

SCHLÖGL, R., Stofftransport durch Membranen, Steinkopf, Darmstadt 1964

SCHLÖGL, R., SCHÖDEL, U., Z. Phys. Chem. N.F. **5**, 372–397 (1955)

TEORELL, T., Progr. Biophys. **3**, 305–315 (1953)

Transport Processes across Membranes with Narrow Pores

D. Woermann

A. Introduction

In the thermodynamic treatment of transport processes across membranes very little is postulated concerning the properties of the membrane. The membrane should have certain permeabilities and the ability to maintain a stationary state which is characterized by the vanishing of the time derivatives of all extensive quantities of the membrane. In order to gain insight into the physical basis of certain observed membrane transport properties it is necessary to construct a suitable model which is amenable to theoretical analysis. The real membrane is replaced by a fictitious model reflecting only the typical properties of the real membrane. SCHMID (1950, 1951a, 1951b, 1952) and SCHLÖGL (1964, 1966) have proposed a model which is applicable to membranes with narrow pores. The membrane is treated as a separate phase. It is assumed that the matrix of the membrane forming the boundaries of the pores can be a lattice-like structure or can consist of a statistically cross-linked high polymer network. It can carry fixed electric charges: ionic groups or adsorbed ions. The pores which are filled with a fluid medium are so narrow that the different components of the fluid are homogeneously distributed over the cross section of the pore by the thermal motion of the molecules. This requires pore diameters in the range from 0.1 to 10 nm. The topology of the pores can be complicated. The pores may branch out statistically and their cross sections can have any form. It is further assumed that the resistance of the membrane to transport processes is evenly distributed over the membrane phase and not located at the phase boundary membrane/solution. Ion exchange membranes and certain cellulose acetate membranes belong to this type.

B. Theory

The model of the membrane with narrow pores is further characterized by the following simplifying assumptions.

1) The solution contained in the pores is so dilute that it can be treated as ideal.
2) The concentrations of all components of the medium within the pores are only a function of the coordinate x perpendicular to the phase boundary membrane/solution.
3) If the matrix of the membrane carries electric charges (fixed ionic groups, adsorbed ions) the solution within the pores contains a surplus of mobile ions of sign opposite to that of the fixed charges. The density of space charge ρ_{el} formed by these mobile ions is only a function of the coordinate x.
4) Coupling between the motion of dissolved particle species is neglected. Not neglected are coupling effects between the dissolved particle species and the solvent, between the dissolved particle species and the membrane matrix, and between the solvent and the membrane matrix.

The density of the space charge within the pores is given by:

$$\rho_{el} = - F\omega X \qquad (1)$$

X is the concentration of the fixed ion groups in equivalents per unit volume of the solution within the pores; F is the Faraday constant and ω the sign of the charges of the fixed ionic groups.

For the whole membrane the condition of electroneutrality has to be fulfilled. This means

$$\sum_i z_i C_i + \omega X = 0 \tag{2}$$

C_i is the concentration of the dissolved particle species i per unit volume of the solution within the pores; z_i is its valence including the sign.

For the description of the flow of the dissolved particle species i across membranes with narrow pores SCHLÖGL has introduced the following expression:

$$\phi_i = C_i v - D_i \, (dC_i/dx + z_i C_i \, F/RT \cdot d\bar{\psi}/dx) \tag{3}$$

ϕ_i represents the molar flow density (moles cm^{-2} s^{-1}) and v the volume flow density (cm s^{-1}) across the membrane. D_i has the meaning of an effective diffusion coefficient inside the membrane. $\bar{\psi}$ is the electric potential inside the membrane.

The first term ($C_i v$) on the right-hand side of Eq. (3) describes a particle transport by convection caused by a streaming medium within the pores in the direction of the x axis. The second term ($-D_i dC_i/dx$) describes a diffusional flow caused by a concentration gradient inside the membrane phase. The third term takes into account the transport of charged particles by the electric potential gradient ($-d\bar{\psi}/dx$) inside the membrane. This gradient can be caused by a diffusion potential or by a resistance if an electric current is passed across the membrane.

According to SCHLÖGL (1964) the volume flow density (streaming velocity) across the membrane is given by

$$v = d_h \, (-\, d\bar{P}/dx - \rho_{el} d\bar{\psi}/dx) \tag{4}$$

The streaming velocity v of the solution across the membrane is proportional to the driving forces acting on unit volume of the medium within the pores. The proportionality factor d_h is called the specific mechanical permeability. The driving forces are a gradient of hydrostatic pressure inside the membrane ($-d\bar{P}/dx$) and an electrical force ($-\rho_{el} d\bar{\psi}/dx$) acting on the surplus charge of the solution within the pores.

Within the framework of this model the properties of the membrane are characterized by the parameters ωX, C_i, D_i, and d_h. In general they are functions of the electrolyte concentrations in the bulk phases outside the membrane. The Eqs. (3) and (4) can be established on the basis of the thermodynamics of irreversible processes in continuous systems (SCHLÖGL, 1964).

The model of the membrane with narrow pores will now be used to discuss two different types of experiments with ion exchange membranes. At first we will deal with osmotic properties and reflection coefficients of electrolytes. Thereafter we will be concerned with streaming potentials. Both experiments may be of interest from the biological point of view. For example recently HINGSON and DIAMOND (1972) have used measurements of streaming potentials in biological membranes to obtain values of reflection coefficients for a number of nonelectrolytes.

C. Results and Discussion

I. Osmotic Properties of Ion Exchange Membranes

We consider an isothermal-isobaric system in which an anion exchange membrane with strong basic groups ($-N(CH_3)_3^+$) separates two aqueous solutions of the same electrolyte. The electrolyte concentrations are different. The membrane is permeable to both components of the solution. The outer phases of the membrane are stirred mechanically to keep their composition homogeneous. They are connected to two horizontally located capillaries to measure the osmotic volume flow.

The model of the membrane with narrow pores can be used to predict the experimental conditions under which positive osmosis (volume flow in the direction of the gradient of the chemical potential of water) or negative osmosis (volume flow against the direction of the gradient of the chemical potential of water) will be observed. The volume flow density across the anion exchange membrane is given by Eq. (5) which follows from Eqs. (1) and (4) by integration.

$$v = d_h/\delta \cdot (\Delta\overline{P} - F\,X\Delta\overline{\psi}) \qquad (5)$$

$$\text{with } \overline{P} = \overline{P}' - \overline{P}'' \text{ and } \Delta\overline{\psi} = \overline{\psi}' - \overline{\psi}''$$

δ = thickness of the membrane.

The direction of the volume flow is determined by the resultant of the two driving forces of the volume flow $\Delta\overline{P}$ and $FX\Delta\overline{\psi}$. The potential gradient inside the membrane is a diffusion potential caused by concentration gradients and different mobilities of the mobile ion species.

Although the outer phases of the membrane are under equal pressure a pressure gradient is formed inside the membrane. At the phase boundaries membrane/solution there exist pressure jumps because the concentrations of osmotically active ions inside the membrane phase are always greater than those in the outer phase. Both pressure jumps can be interpreted as osmotic equilibrium pressure (SCHLÖGL, 1955). The electrolyte concentration in the left outer phase is higher than that in the right, the pressure jump at the left phase boundary membrane/solution is smaller than at the right. Since both outer phases are under the same hydrostatic pressure a pressure gradient is formed inside the membrane. It is directed from the dilute to the concentrated outer phase and has always the tendency to move the solution within the pores in this direction.

We will now discuss the phenomena caused by the superposition of the two driving forces of the volume flow (Fig. 1):
1) Both driving forces of the volume flow have the same direction (Fig. 1A). They are directed from the dilute to the concentrated outer phase. Consequently positive osmosis is observed. This situation is realized if $D_+/D_- < 1$.
2) The driving forces for the volume flow have opposite directions (Fig. 1B, C). Now two situations have to be distinguished. Positive osmosis is observed as long as the direction of the volume flow is determined by the pressure gradient inside the membrane. This will be the case if D_+/D_- is larger than 1 by only a small amount. Negative osmosis takes place if the electric driving force overcompensates the mechanical driving force of the volume flow. This is the case if $D_+/D_- \gg 1$.

For the quantitative treatment of these osmotic phenomena it is convenient to introduce the reflection coefficient, σ, defined by Eq. (6).

$$\sigma = \left(\frac{P}{\pi}\right)_{J_V=0} \qquad (6)$$

P is the hydrostatic pressure difference which has to be applied across the membrane to suppress the volume flow J_V caused by the osmotic pressure difference π between the outer phases of the membrane. Positive reflection coefficients correspond to positive osmosis, negative reflection coefficients to negative osmosis.

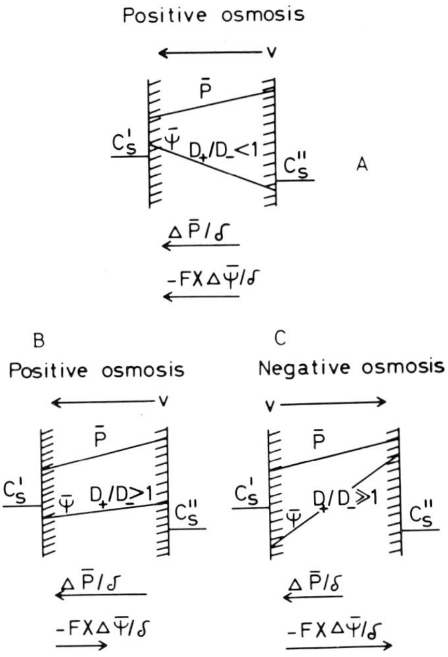

Fig. 1

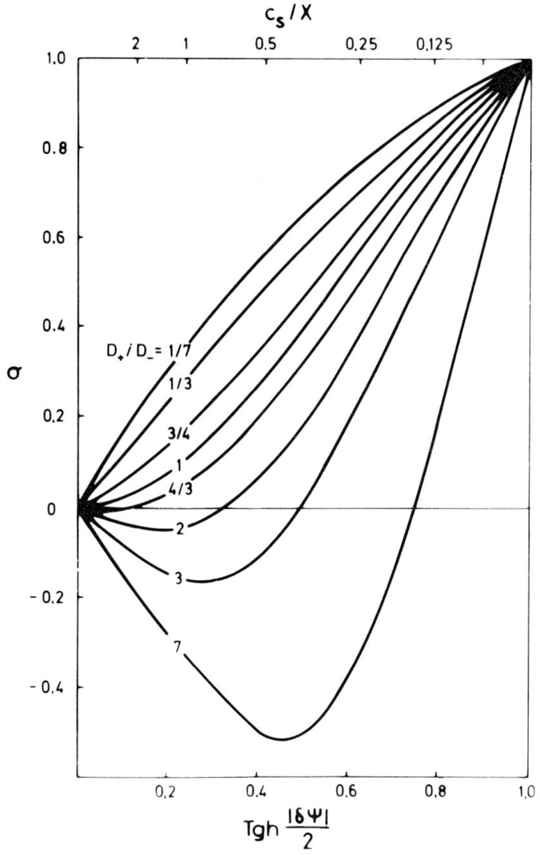

Fig. 2

Fig. 1 Driving forces of volume flow in an anion exchange membrane.

Fig. 2 Theoretically calculated reflection coefficients of an anion exchange membrane for a 1–1 valent electrolyte as function of c_s/X. Parameter D_+/D_-; c_s = mean value of electrolyte concentration in the outer phase of the membrane.

SCHLÖGL (1955) has shown that within the frame of the model of the membrane with narrow pores the reflection coefficient is a function of the parameters c_s/X and D_+/D_-. Fig. 2 shows a plot of theoretically calculated reflection coefficients as a function of c_s/X for different values of D_+/D_-. The reflection coefficient of the anion exchange membrane is positive only if $D_+/D_- < 1$. If on the other hand $D_+/D_- > 1$ the reflection coefficients will be positive for small values of c_s/X and negative for large values of c_s/X. This behaviour is in complete agreement with the qualitative arguments presented above.

We turn now to experimental results obtained by SCHÖNBORN (SCHÖNBORN and WOERMANN, 1967; WOERMANN, 1969; PUSCH and WOERMANN, 1970). The system anion exchange membrane/NaCl solution shows only positive reflection coefficients (Fig. 3). But with HCl solutions and the same membrane, positive and negative reflection coefficients are observed, depending on the mean value of the acid concentration in the bulk phases.

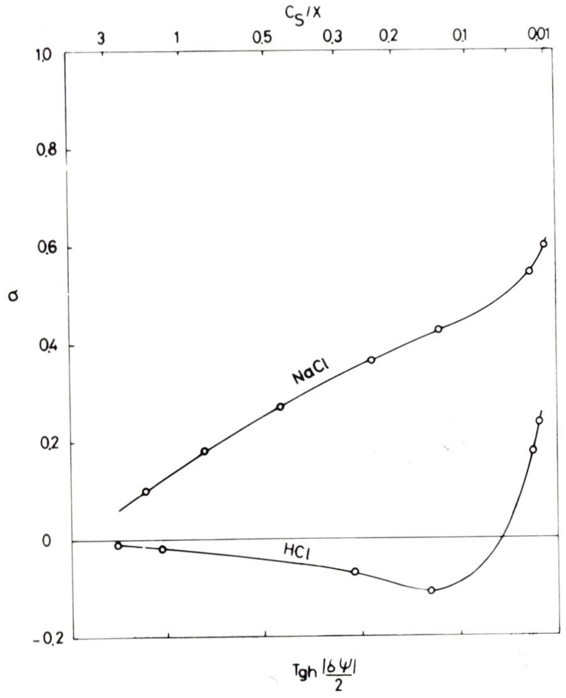

Fig. 3 Experimentally determined reflection coefficient of an anion exchange membrane as function of c_S/X.

The shapes of the experimental curves are similar to those of the calculated curves. The differences are due to the fact that the theoretical curves are calculated on the assumption that the parameter D_+/D_- is constant. This condition is not fulfilled in the experiments.

We will now discuss experiments concerning streaming potentials of cation exchange membranes.

II. Streaming Potentials in Ion Exchange Membranes

We consider an isothermal system in which a cation exchange membrane with strong acid groups ($-SO_3^-$) (condensation product of phenolsulphonic acid and formaldehyde) separates two aqueous solutions of KCl of equal concentration (bulk phases). The solutions are vigorously stirred during the experiment to keep their composition homogeneous. Two identical electrodes reversible to Cl^-, calomel electrodes, are immersed in the bulk phases. If a constant pressure difference is applied across the membrane a stationary electric potential difference between the bulk phases is observed after a certain transition time (Fig. 4). The membrane is not supported during the experiment. An increase of the stirring rate did not influence the value of the stationary electric potential difference. An analysis of the data shown in Fig. 4 shows that for small hydrostatic pressure differences there exists a linear relation between the stationary electric potential difference E $(= \psi'_{el} - \psi''_{el})$ and the hydrostatic pressure difference P (=P'—P''). It is characterized by the ratio $(E/P)_{I,\pi}$. The subscripts indicate the parameters which are kept zero during the experiment (I = electric current; π = osmotic difference). Fig. 5 summarizes results obtained from experiments carried out at different salt concentrations. The measurements of the streaming potentials were carried out by BUNTHOFF (1972) using a technique developed by HANEBECK (1968).

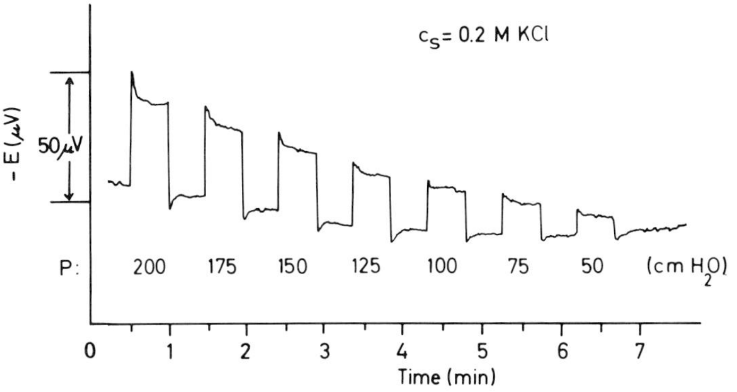

Fig. 4 Electric potential difference versus time curve. Parameters: pressure difference P (=P'—P''). E = $\psi'_{el.} - \psi''_{el.}$ (BUNTHOFF, 1972). The spikes in the E versus time curves are interpreted as transients caused by the interaction of processes connected with the sudden application and the sudden removal of the hydrostatic pressure difference and the stirring of the bulk phases.

The phenomena just described can be understood on the basis of the model of the membrane with narrow pores. The pressure difference causes a volume flow across the membrane. The fluid within the pores of the cation exchange membrane carrying a positive surplus charge is displaced by the volume flow. Thereby an electric potential difference between the bulk phases of the membrane, the so-called streaming potential, is generated; the electric potential of the high pressure phase is negative with respect to that of the low pressure phase.

Due to the presence of the negatively charged fixed ion groups within the membrane the concentration of Cl^- in the membrane phase is smaller than that in the bulk phases (Donnan distribution). Thereby the flow of salt across the membrane caused by the volume flow remains small. This results in a filtration of the salt solution; the salt concentration in the unstirred layer at the interface membrane/high pressure phase increases and at the interface membrane/low pressure phase decreases during the experiment. The concentration difference between the unstirred layers of the membrane generates an additional electric potential difference (concentration potential) which is superimposed on the streaming potential. Both potential differences have the same sign.

In the thermodynamic treatment of transport processes across membranes it is possible to take as the membrane the whole region of inhomogeneity including the unstirred layers. On the basis of this premise on can identify the stationary electric potential difference with the streaming potential.

The pressure and concentration dependence of the streaming potential can be described quantitatively on the basis of the model of the membrane with narrow pores. The simplifying assumption is made that the salt concentrations in both bulk phases are constant up to the phase boundary membrane/solution. This implies that the equations discussed below are valid only for the earliest time following the application of the hydrostatic pressure difference.

For cation exchange membranes the ratio $(E/P)_{I,\pi}$ is given by Eq. (7) (SCHMID and SCHWARZ, 1952; SCHLÖGL, 1964), which describes the dependence of this quantity on the mechanical permeability L_p, the fixed ion concentration X and the conductivity of the membrane L_E.

$$\left(\frac{E}{P}\right)_{I,\pi} = -\frac{F\,L_p\,X}{L_E} \tag{7}$$

$$L_p = \left(\frac{J_V}{P}\right)_{E,\pi} = \frac{A}{\delta}\,d_h \tag{8}$$

$$L_E = \left(\frac{I}{E}\right)_{P,\pi} = \frac{AF^2}{\delta}\left(\frac{1}{RT}\sum_i (D_i Z_i^2 C_i) + X^2 d_h\right), \quad i = K^+, Cl^- \tag{9}$$

J_V = volume flow ($cm^3\ s^{-1}$)
A = membrane area (cm^2)
I = electric current (amp)
δ = membrane thickness (cm)

For a given membrane the concentration dependence of $(E/P)_{I,\pi}$ is mainly determined by the concentration dependence of L_E. L_E increases with increasing electrolyte concentration in the bulk phases. This is pronounced in the concentration range $c_s/X \simeq 1$ (c_s = electrolyte concentration in the bulk phase), because of the concentration dependence of the Donnan distribution of the mobile ions between the membrane phase and the bulk phases. With this knowledge of the concentration dependence of L_E it can be concluded from Eq. (7) that the ratio $(E/P)_{I,\pi}$ will become smaller with increasing values of c_s and that the decrease of $(E/P)_{I,\pi}$ will be clearly indicated by the concentration range $c_s/X \simeq 1$.

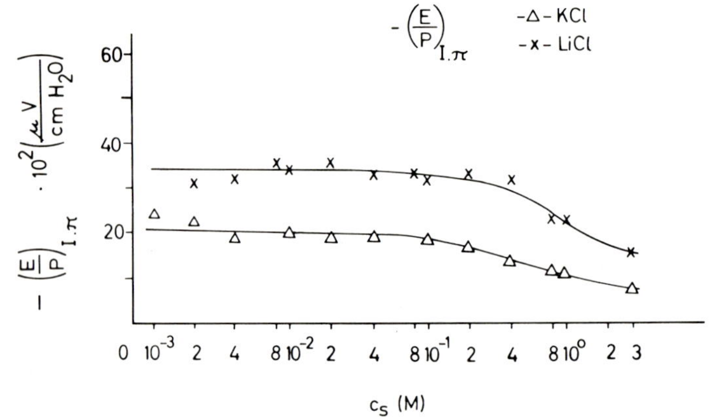

Fig. 5 — $(E/P)_{I,\pi}$ as function of salt concentration c_s (logarithmic scale) in the bulk phases of the membrane (BUNTHOFF, 1972).

In general the quantities L_p and X are also functions of concentration. But it turns out that for organic resin ion exchange material the product $L_p \cdot X$ is practically independent of c_s (HELFFERICH, 1959). The concentration dependence of $(E/P)_{I,\pi}$ as shown by Fig. 5 is in accord with these considerations.

In order to see how well the experimentally determined values of $(E/P)_{I,\pi}$ agree with the values of this quantity calculated from Eq. (9) the parameters L_p, L_E and X have to be determined by independent experiments. The results of these experiments carried out at a salt concentration of 0.5 M are compiled in Table 1. The calculated and the experimentally determined values of $(E/P)_{I,\pi}$ are of the same order of magnitude. But the calculated value is larger than the experimentally determined one. This difference is outside the margin of experimental error and cannot be attributed to the effect of concentration changes in the unstirred layers.

Table 1

Comparison of measured and calculated streaming potentials. See text. ($\delta = 3 \cdot 10^{-2}$ cm; A = 0.79 cm^2; c_s = 0.5 M; $L_p = L'_p\,(1 - FX(\frac{J_V}{I})_{\pi,P})$; $L'_p = (J_V/P)_{I,\pi}$).

X	L_E	$L_p \cdot 10^6$	$-(\frac{E}{P})^{calc}_{I,\pi} \cdot 10^3$	$-(\frac{E}{P})^{exp}_{I,\pi} \cdot 10^3$
(M)	(Ω^{-1})	(cm^6 J^{-1} s^{-1})	(cm^3 amp^{-1} s^{-1})	
2.5	0.76	8.4	2.7	1.5

The difference between the calculated and the experimentally determined value of the ratio $(E/P)_{I,\pi}$ can be rationalized by assuming that only a fraction of the analytically determined counterions is displaced by the volume flow (TOYOSHIMA and NOZAKI, 1969; FRANCK, 1963). One possible reason for such a behaviour may be related to the structure of the membrane used in this investigation. It has been reported (KOSCHEL and SCHLÖGL, 1957; SCHLÖGL and SCHURICH, 1961) that this type of membrane is homogeneous for dimensions large compared with the wave lenght of visible light but is inhomogeneous in smaller dimensions. According to these reports the membrane is composed of small tightly crosslinked regions separated by wider channels which determine the mechanical permeability. In these channels the density of the surplus charge formed by the counterions is smaller than in the strongly crosslinked regions.

III. Symmetry Relation of ONSAGER

From the linear laws of the thermodynamics of irreversible processes together with the symmetry relations of ONSAGER the following relation can be derived (MAZUR and OVERBEEK, 1951):

$$\left(\frac{J_V}{I}\right)_{P,\pi} = -\left(\frac{E}{P}\right)_{I,\pi} \qquad (10)$$

This equation can be used to check the reliability of the method we have used to measure streaming potentials. In independent experiments the electroosmotic volume flow was determined as a function of electric current using the same membrane. The bulk phases were stirred vigorously (WIEDNER, 1971). Also during these experiments concentration changes take place in the unstirred layers because the transference number of the cations within the membrane is larger than in the bulk phases. The results of the electroosmotic measurements together with the measurements of the streaming potential are shown in Fig. 6. The agreement between the two types of measurement is satisfactory.

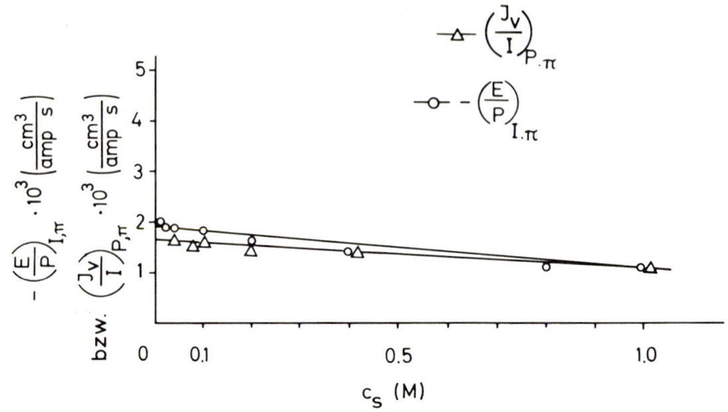

Fig. 6 $(J_V/I)_{P,\pi}$ and $-(E/P)_{I,\pi}$ as function of salt concentration c_S in the bulk phases of the membrane (BUNTHOFF, 1972).

IV. Effect of Unstirred Layers on Streaming Potentials

Fig. 7 shows a typical electric potential difference versus time curve which is observed when the bulk phases of the membrane are not stirred. Immediately after applying the hydrostatic pressure an electric potential difference is observed which corresponds to the streaming potential. Thereafter the filtration of the salt solution begins. This leads to concentration changes at the phase boundaries membrane/solution and to the generation of a concentration potential across the membrane. After a certain transition time the concentration changes at the phase boundaries membrane/solution and inside the membrane come to a standstill and the electric potential difference between the bulk phases reaches a steady state. After removing the hydrostatic pressure difference only the streaming potential

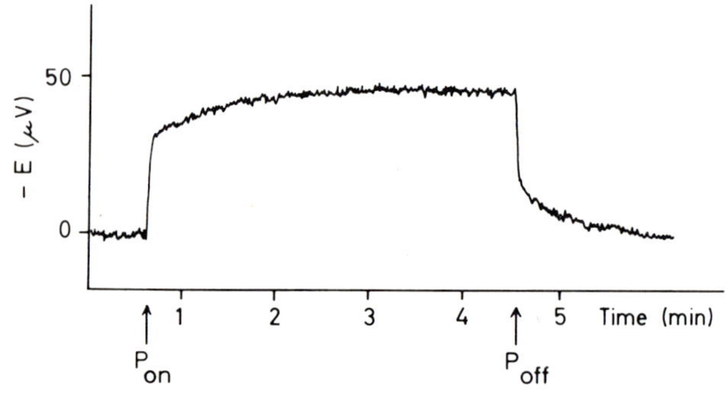

Fig. 7 Electric potential difference versus time curve. No stirring of the outer phases. P = 200 cm H_2O; c_S = 0.1 M (BUNTHOFF, 1972).

disappears. The concentration potential is maintained until the concentration gradients inside the membrane and in the layers adjacent to it have vanished as a result of diffusion. BRUN and VAULA (1967) have proposed a method of obtaining streaming potentials from curves such as shown in Fig. 7. The method is based on the assumption that the membrane is impermeable to salt. The streaming potentials obtained by this method are 5–10 % smaller than the stationary electric potential difference observed when both bulk phases are stirred vigorously.

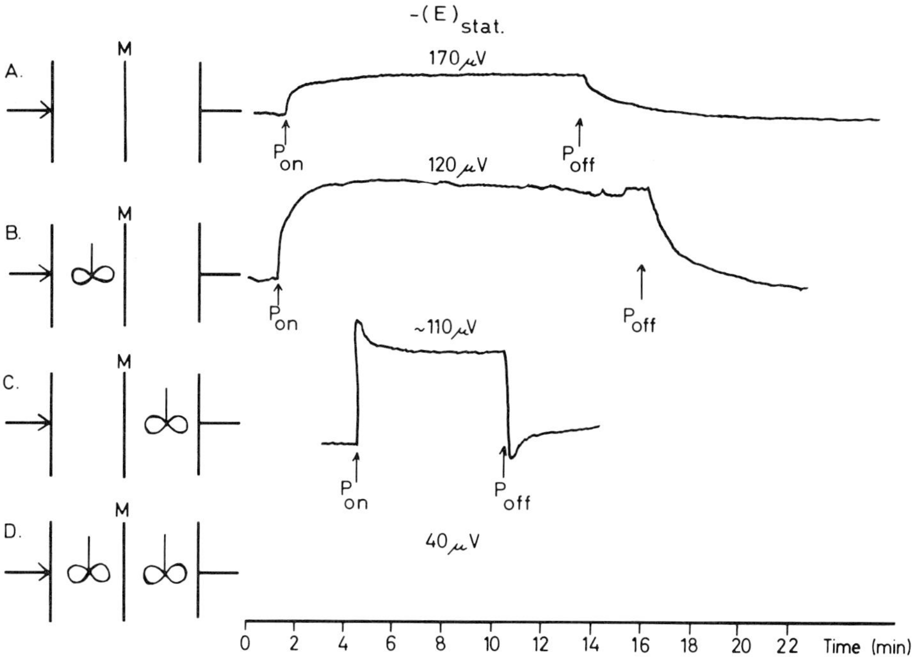

Fig. 8 Electric potential difference versus time curves obtained under different stirring conditions of the bulk phases of the membrane (BUNTHOFF, 1972) (for further explanation, see text).

In Fig. 8 three electric potential difference versus time curves are shown which were obtained with the same membrane under different stirring conditions of the bulk phases. The curve shown in Fig. 8A refers to a situation with no mechanical stirring. The stationary electric potential difference has a value of about 170 μV. If one of the bulk phases of the membrane is stirred it becomes smaller by about 60 μV. Fig. 8B and Fig. 8C show that it does not matter whether the high or low pressure phase is stirred. If both bulk phases are stirred the stationary electric potential difference is diminished by about 2 · 60 μV = 120 μV. The stirring destroys the concentration polarization at the phase boundaries of the membrane solution to a large degree. Therefore the contribution of the concentration potential to the stationary electric potential difference is almost eliminated. These observations indicate that the concentration profiles in the solution layers adjacent to the interphase membrane/solution are antisymmetric with respect to the membrane if the outer phases are not stirred. This is only true for small pressure differences.

Measurements of this type can be used to investigate concentration changes in the unstirred layers of a membrane. A concentration potential of 60 μV corresponds to a relative concentration difference of $\Delta C_s/C_s \simeq 3 \cdot 10^{-3}$. Here it is assumed that the membrane is impermeable to the salt. If the membrane is permeable to salt and water the value of $\Delta c_s/c_s$ corresponding to a given value of the concentration potential increases.

The model of the membrane with narrow pores can be used to calculate the flow of ions and the volume flow across the membrane at a given pressure difference and a given salt concentration in the bulk phases provided the concentration changes in the unstirred layers of the membrane are neglected (SCHLÖGL, 1964; HELFFERICH, 1959). That is:

$$\phi_i = \frac{d_h}{\delta} P \left(C_i - X \frac{\bar{t}_i}{z_i}\right), \quad i = K^+, Cl^- \tag{11}$$

with

$$\frac{\bar{t}_i}{z_i} = \frac{C_i \left(\frac{1}{RT} D_i z_i + d_h X\right)}{\frac{1}{RT} \sum_i (D_i z_i^2 C_i) + d_h X^2}$$

$$v = \frac{d_h}{\delta} P \left(1 - \frac{d_h X^2}{\frac{1}{RT} \sum_i (D_i z_i^2 C_i) + d_h X^2}\right) \tag{12}$$

$\bar{t}_i$ = transference number of ion species i in the membrane phase. These equations refer to a cation exchange membrane. As long as $c_s/X \ll 1$ the concentration of chloride ions (coions) in the membrane phase will be small due to the Donnan distribution. Consequently the salt flow across the membrane will be small (Eq. (11)); the filtrating properties of the membrane will be effective and the concentration-polarization in the unstirred layers will be pronounced under these conditions. If on the other hand $c_s/X \gg 1$ the concentration of chloride ions in the membrane phase will be comparable to that of the bulk phases; the

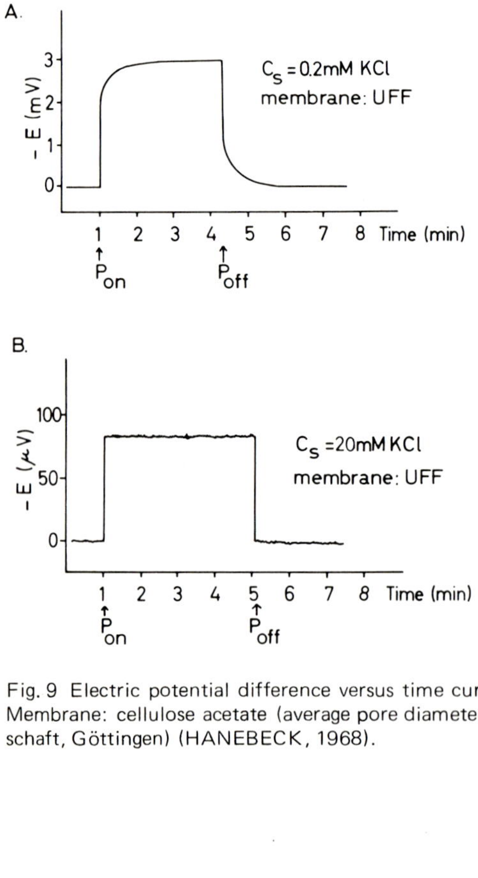

Fig. 9 Electric potential difference versus time curves. No stirring of the outer phases. P = 100 cm H_2O. Membrane: cellulose acetate (average pore diameter < 50 Å) (UFF Ultrafilter mittel, Membranfiltergesellschaft, Göttingen) (HANEBECK, 1968).

filtration properties of the membrane will be very much reduced and the concentration polarization in the unstirred layers will disappear. These predictions are borne out by the experimental results obtained by HANEBECK (1968) and shown in Fig. 9 where the membrane had a fixed ion concentration of the order of 10^{-3} M. (Fig. 9A: $c_s/X \ll 1$, concentration potential present; Fig. 9B: $c_s/X \gg 1$, concentration potential practically absent).

D. References

BRUN, T.S., VAULA, D., Ber. Bunsenges. Phys. Chem. **71**, 824–829 (1967)
BUNTHOFF, K., Zur Messung der Elektroosmose und des Strömungspotentials an synthetischen Membranen. Diplomarbeit, Köln 1972
FRANCK, U.F., Ber. Bunsenges. Phys. Chem. **67**, 657–671 (1963)
HANEBECK, N., Zur Messung des Strömungspotentials an feinporigen Membranen. Diplomarbeit, Köln 1968
HELFFERICH, F., Ionenaustauscher, Vol. 1, Verlag Chemie, Weinheim 1959
HINGSON, D.J., DIAMOND, J.M., J. Membrane Biol. **10**, 93–135 (1972)
KOSCHEL, D., SCHLÖGL, R., Z. Phys. Chem. **11**, 137–149 (1957)
MAZUR, P., OVERBEEK, J.T., Rec. Trav. Chim. **70**, 83–91 (1951)
PUSCH, W., WOERMANN, D., Ber. Bunsenges. Phys. Chem. **74**, 444–449 (1970)
SCHMID, G., Z. Elektrochem. **54**, 424–430 (1950)
SCHMID, G., Z. Elektrochem. **55**, 229–237 (1951a)
SCHMID, G., Z. Elektrochem. **55**, 295–307 (1951b)
SCHMID, G., SCHWARZ, H., Z. Elektrochem. **56**, 35–44 (1952a)
SCHMID, G., Z. Elektrochem. **56**, 181–193 (1952)
SCHLÖGL, R., Z. Phys. Chem. **3**, 73–102 (1955)
SCHLÖGL, R., SCHURICH, H., Z. Elektrochem. **65**, 863–870 (1961)
SCHLÖGL, R., Stofftransport durch Membranen. Steinkopff-Verlag, Darmstadt 1964
SCHLÖGL, R., Ber. Bunsenges. Phys. Chem. **70**, 400–414 (1966)
SCHÖNBORN, M., WOERMANN, D., Ber. Bunsenges. Phys. Chem. **71**, 843–855 (1967)
TOYOSHIMA, Y., NOZAKI, H., J. Phys. Chem. **73**, 2134–2141 (1969)
WIEDNER, G., Bestimmung der Transportkoeffizienten und des Transportverhaltens, auch im nichtlinearen Bereich, an einem speziellen Membransystem. Doktorarbeit, Frankfurt 1971
WOERMANN, D., Ber. Dtsch. Bot Ges. **82**, 431–443 (1969)

Coupling of Mass Transfer and Chemical Reaction across an Asymmetric Sandwich Membrane

J. Meyer, F. Sauer, and D. Woermann

A. Introduction and Theory[1])

It has been very difficult to incorporate active transport within the framework of irreversible thermodynamics. At first sight, it is hard to understand how it is possible that a chemical reaction forces a flow of an inert substance in a certain direction. A first thermodynamic approach was made by KEDEM (1961) and generalized by SAUER (1973), who was able to show that active transport is not a local but a global property of the membrane. This will be illustrated by a simple membrane model system, which is able to show active transport in a clear way (MEYER, 1973).

Two membranes α and β enclose a liquid phase $*$, in which an enzyme is dissolved (Fig. 1). The enzyme in the $*$ phase catalyses a chemical reaction inhibited elsewhere. The reaction follows the formal Eq. (1)

$$\sum_{i=0}^{n} \nu_i \, (B_i) = 0 \qquad (1)$$

where ν_i is the stoichiometric coefficient and B_i symbolises the chemical species, i, involved.

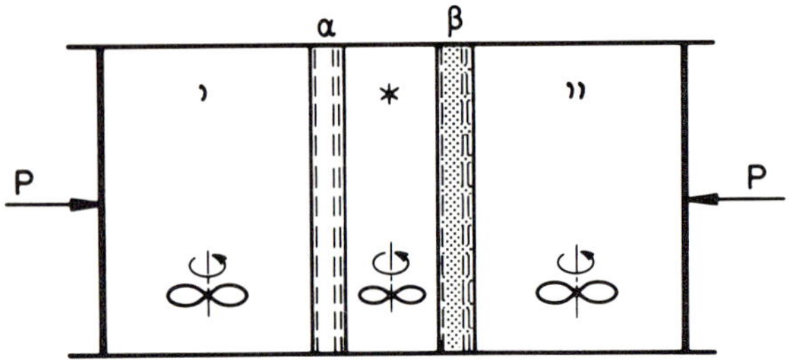

Fig. 1 The sandwich membrane. An enzyme trapped in the $*$ phase catalyses a chemical reaction inhibited elsewhere.

In the bulk phases, $'$ and $''$, are solutions of n permeable components in water. One solute may be an electrolyte but we do not allow electric current through our system in order to avoid the complications which arise when treating active transport of ions. To simplify the model we will assume highly dilute ideal solutions. Both phases $'$ and $''$ have identical

[1]) Systems which at first sight look very much like ours have been discussed. BLUMENTHAL et al. (1967) describe the generation of electrical energy by a chemical reaction in a sandwich membrane. An electric current is produced by components involved in the chemical reaction. Their system is therefore a special type of fuel cell and they have not shown active transport. CURRAN (KATCHALSKY and CURRAN, 1967) starts with a hypostatic pump for a component n in a membrane α and describes the influence of active transport on the volume flow if he builds up a sandwhich membrane in a stationary state with the chemical reactive membrane α and an inactive membrane β.

composition and pressure, so there are no passive forces across the membrane. We assume linear relations across each of the membranes α and β simplified by the fact that the dissolved species 1....n are so highly diluted that there is no coupling between them except via the solvent 0.

Under these assumptions the linear laws across the membranes α and β are (KATCHALSKY and CURRAN, 1967):

$$J_v^{\kappa''} = L_p^{\kappa} (P^{\kappa} - \sum_{j=1}^{n} \sigma_j^{\kappa} \pi_j^{\kappa}) \tag{2}$$

$$J_i^{\kappa''} = (1 - \sigma_i^{\kappa}) c_i J_v^{\kappa''} + \omega_i^{\kappa} \pi_i^{\kappa}, \tag{3}$$

where $J_v^{\kappa''}$ means the volume flux across membrane κ $(= \alpha$ or $\beta)$ into phase $''$; L_p^{κ} is the hydraulic conductivity of membrane κ (i.e. α or β); P^{κ} and π_i^{κ} are defined below; σ_i^{κ} is the reflection coefficient of membrane κ (i.e. α or β) for species i; $J_i^{\kappa''}$ is flux of species i across membrane κ (α or β) into phase $''$; c_i is an average concentration of species i; ω_i^{κ} is a permeability coefficient for species i in membrane κ (α or β).

All transport coefficients have to be measured at zero electric current. We do not have a chemical reaction term because there is no chemical reaction inside the membranes. The forces are always defined as differences as follows:

$$P^{\alpha} = P' - P^{*}; \; P^{\beta} = P^{*} - P'' \tag{4}$$

and correspondingly for the osmotic differences π_j^{κ}.

Because of the identical thermodynamic state of the bulk phases, the passive driving forces across the whole membrane system must vanish.

$$P^{\alpha} + P^{\beta} = 0 \text{ and } \pi_i^{\alpha} + \pi_i^{\beta} = 0 \tag{5}$$

Conservation of matter gives us the missing equations for an unique description of the membrane system in a stationary state:

$$J_v^{\alpha'} + J_v^{\beta''} = 0 \tag{6}$$

$$J_i^{\alpha'} + J_i^{\beta''} = \nu_i J_c ; \; i = 1,....,n \tag{7}$$

$$J_i^{\kappa'} + J_i^{\kappa''} = 0 ; \; \kappa = \alpha, \beta \tag{8}$$

These symbols have been explained in respect of Eqs. (2) and (3), except that J_c is the rate of production in phase * by chemical reaction. The first equation expresses the assumption that there is no volume change in the chemical reaction. We can now evaluate the flows and forces depending on the rate of the chemical reaction and on the transport coefficients of the membrane. We restrict ourselves to the volume flow and to a component n not involved in the chemical reaction ($\nu_n = 0$). Solution of the previous equations leads to:

$$P^{\beta} = \left(\frac{1}{L_p^{\beta}} - \sum_{j=1}^{n} \sigma_j^{\beta} \frac{\sigma_j^{\alpha} - \sigma_j^{\beta}}{\omega_j^{\alpha} + \omega_j^{\beta}} c_j\right) J_v^{\beta''} + \sum_{j=1}^{n} \sigma_j^{\beta} \frac{\nu_j \omega_j}{\omega_j^{\alpha} \omega_j^{\beta}} J_c \tag{9}$$

$$J_v^{\beta''} = L_p \sum_{j=1}^{n} \nu_j \frac{\sigma_j^{\alpha} - \sigma_j^{\beta}}{\omega_j^{\alpha} + \omega_j^{\beta}} J_c \tag{10}$$

$$\pi_n^\beta = - \frac{\sigma_n^\alpha - \sigma_n^\beta}{\omega_n^\alpha + \omega_n^\beta} c_n J_v^{\beta''} \tag{11}$$

$$J_n^{\beta''} = (1 - \sigma_n) c_n J_v^{\beta''}, \; \nu_n = 0. \tag{12}$$

Transport coefficients without Greek superscripts correspond to the whole membrane system (KEDEM and KATCHALSKY, 1963). From Eqs. (9) – (12) we see that all measurable effects are linearly dependent on the chemical flow, and nothing happens if there is no chemical reaction. Eq. (9) shows that the membrane can exist in a stationary state at a prescribed pressure difference between the bulk phases and the phase *. The value of this pressure difference may exceed one atmosphere. Thus, sealing the inner compartment * and applying atmospheric pressure to the bulk phases might be an unrealizable procedure to get the membrane into a stationary state. We have to apply additional pressure to the bulk phases.

The chemical reaction flow may give rise to an active volume flow $J^{\beta''}$, depending on the difference between the reflection coefficients of the membranes α and β.

The active volume flow may in turn drag along a component not involved in the chemical reaction $J_n^{\beta''}$, resulting in a net flow of this component relative to the membrane. A concentration difference π_n^β between the bulk phases and the * phase may be built up, if the two membranes have different reflection coefficients for n.

If component n is an electrolyte, the concentration difference may cause a potential difference between two electrodes in the outer phases. We call this potential difference an 'active EMF' because the electrolyte is not involved in the chemical reaction. Let us restrict our considerations to a binary electrolyte, say KCl or NaCl. We measure the electric potential difference E with identical, reversible electrodes, selective to the negative ion (Ag/AgCl electrodes), which record the electrochemical potential difference of the anion. Within the scope of our model, we neglect all coupling between dissolved components. The linear laws give the potential difference E^κ over the membrane κ at zero electric current as follows:

$$E^\kappa = -a^\kappa J_v^{\kappa''} - \frac{t_+^\kappa}{e_+ c_n} \pi_n^\kappa ; \; \kappa = \alpha, \beta \tag{13}$$

a^κ is the coefficient of the streaming potential, t_+^κ the transference number of the cation and e_+ the charge of one mole of the cations. The overall difference of the electrochemical potential is given additively by the difference over the membranes α and β:

$$e_- E = \bar{\mu}_-' - \bar{\mu}_-'' = \bar{\mu}_-' - \bar{\mu}_-^* + \bar{\mu}_-^* - \bar{\mu}_-'' = e_- \; (E^\alpha + E^\beta) \tag{14}$$

Coupling Eqs. (13) and (14) and eliminating the osmotic difference π_n^κ by Eq. (11) yields:

$$E = -\left(a^\alpha + a^\beta + \frac{(\sigma_n^\alpha - \sigma_n^\beta)\,(t_+^\alpha - t_+^\beta)}{e_+ \,(\omega_n^\alpha + \omega_n^\beta)}\right) J_v^{\beta''} \tag{15}$$

For the type of membranes we have used, we can neglect the contribution of the streaming potential $(a^\alpha + a^\beta)$ to the active EMF. The transference numbers have a value between 0 and 1, thus their difference has at the most the value 1, if we use an ideal cation and an ideal anion exchange membrane to compose our sandwich membrane.

We must modify the simple membrane system in order to fit the theory to the experiments. One of the membranes used (AMICON UMO 5) is built up of two layers: a very thin, narrow-pored and permselective layer, β_o, and a wide pored ($\sigma_i^{\beta\delta} = 0$) diffusion layer, β_δ, into which the enzyme may penetrate, if the diffusion layer is exposed to the compartment * (Fig. 2). The index β is assigned to the composite membrane ($\beta = \beta_o + \beta_\delta$).

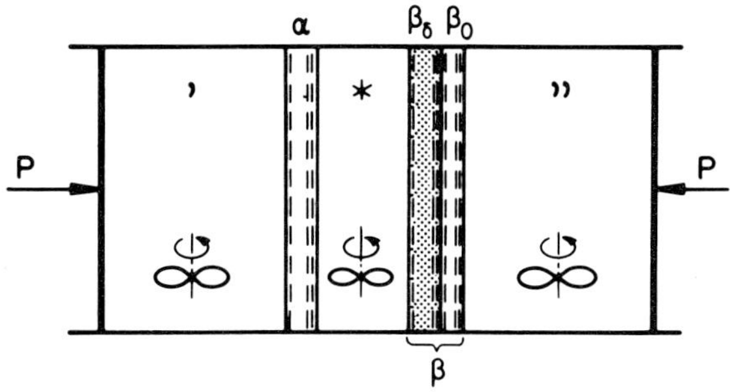

Fig. 2 The modified membrane system. The enzyme may penetrate into the layer β_δ.

The linear Eqs. (2) and (3) are unaltered for $\kappa = \alpha, \beta_o$ but applying the linear laws to the diffusion layer, β_δ, we have to account for the chemical flow inside the membrane. Assuming the diffusion layer to behave like a symmetric membrane, we get

$$J^{\beta\delta''} = (1 - \sigma_i^{\beta\delta}) c_i \, J_v^{\beta\delta''} + \omega_i^{\beta\delta} \, \pi_i^{\beta\delta} + \frac{\nu_i}{2} J_c^{\beta\delta} \tag{16}$$

From considerations analogous to those of the single membrane model, we get a modified active volume flow:

$$J_v^{\beta''} = L_p \left(\sum_{j=1}^{n} \nu_j \frac{\sigma_j^{\alpha} - \sigma_j^{\beta}}{\omega_j^{\alpha} + \omega_j^{\beta}} J_c - \sum_{j=1}^{n} \nu_j \frac{\sigma_j}{\omega_j^{\beta\delta}} \frac{J_c^{\beta\delta}}{2} \right) \tag{17}$$

where the chemical production rate of the whole membrane system (J_c) and that of the diffusion layer ($J_c^{\beta\delta}$) appear separately. If the contribution of the second term in the volume flow exceeds that of the first one, the volume flow may change sign when the membrane β is inverted and the enzyme cannot penetrate into the membrane any longer. ($J_c^{\beta\delta} = 0$). The Eqs. (11) and (12) for the concentration difference π_n^{β} of an inert component and for the active flow $J_n^{\beta''}$ remain unaltered. We regain the equations of the simple model without a diffusion layer if the restraints for transport are entirely given by the membranes α and β_o.

B. Materials and Methods

Two membrane systems have been successfully investigated for active transport phenomena. In either case the same type of cation exchange membrane (PSA) has been used on one side of the sandwich membrane. The PSA membrane is a condensation product of phenol sulphonic acid and formaldehyde (SCHLÖGL and SCHÖDEL, 1955). The second membrane was an anion exchange membrane (ASAHI CAI) or an ultrafiltration membrane (AMICON UMO 5), both commercially available. The AMICON UMO 5 consists of a thin layer (0.1 μm) of extremely narrow pores on a thick spongy layer (0.3 mm) of the same polymer with increasing pore size. The two membrane systems are denoted by PSA/*/ASAHI CAI and PSA/*/AMICON UMO 5.

The chemical reaction used was the enzymic splitting of sucrose into glucose and fructose, i.e.:

$$H_2O + \text{sucrose} \xrightarrow{E} \text{glucose} + \text{fructose} \tag{18}$$

where E = β-fructosidase (invertase).

A lucite cell was constructed, which consists essentially of three parts. Two outer cells in three different sizes (58 ml, 105 ml and 157 ml), and an inner cell with a volume of about 10 ml. In all three compartments Teflon-coated magnetic stirrer are inserted, which are driven by magnets rotating outside the cell. Electrodes or capillaries may be inserted into the outer cells through the inlet openings. The whole cell is held at 25°C in a water thermostat. The experimental set-up is totally symmetric. The stock solutions, which are under atmospheric pressure, are pumped into the pressurized compartments ' and ". The pumps are separated from the power line with the help of two transformers so as to exclude an electric short-circuit of the solutions in the outer cells. The pumps eject corresponding volumes of liquid out of the cell compartments, which drop into tanks connected to the pressure reservoir. (For more information see MEYER, 1973.)

The pressure difference between the liquid membrane phase and each bulk phase is obtained in two ways. Either we insert a capillary into the middle compartment and regulate the pressure on the bulk phases until the meniscus in the capillary does not change or we seal the middle compartment and apply a pressure to the bulk phases which exceeds the value necessary to reach the stationary state. The additional pressure will press liquid into the middle cell and build up a counter pressure until the stationary pressure difference is obtained. The first procedure is used with the system PSA/*/ASAHI CAI while the method of sealing the middle compartment is applied to the system PSA/*/AMICON UMO 5.

The solutions flowing out of the cell are analysed polarimetrically. The speed of the pumps is adjusted until the solutions have equal concentrations. Two Ag/AgCl electrodes in the bulk phases indicate the steady-state condition of the membrane. The potential difference is measured by a compensation method using a high precision voltage supply and a μV-meter, both commercially available.

When the membrane is in a stationary state, the active volume flow is measured. The various tubes are disconnected from the outer cells and the openings tightly closed. The capillary holders with the capillaries are screwed to the outer cells and the free openings of the capillaries are connected to the pressure reservoir. The volume flow is recorded until the change of volume observed is at least 0.5 ml (2–3 days). Finally, the concentration changes of Cl^- in the bulk phases are determined conductometrically.

C. Results and Discussion

I. The Dependence of Active Transport and Active EMF on Concentration of Sucrose in the Bulk Phases

1. The System PSA/*/AMICON UMO 5

The flow of the reactive components through the AMICON UMO 5 membrane is about five times the flow through the PSA membrane. That is why we use a big outer cell (158 ml) on the AMICON UMO 5 side and a small one on the other side. Thus only a small concentration difference will be built up when we measure the volume flow. The concentration of NaCl is 0.01 M and the concentration of enzyme in the middle compartment is 2 mg/ml. The hydrolysis does not obey first order kinetics but the chemical reaction takes place mainly within the diffusion layer of the AMICON UMO 5. This can be seen from the fact that the flow of reacting components through this membrane is much higher than through the other membrane, while on the other hand the permeability of AMICON UMO5 is lower than that of the PSA membrane. Thus we have:

$$J_c = J_c^* + J_c^{\beta\delta} \simeq J_c^{\beta\delta} \tag{19}$$

Indeed, the active EMF (Fig. 3) and the active volume flow (Fig. 4) show a linear dependence on reaction flow. The same must be true for the active transport of NaCl, which is proportional to the volume flow according to Eq. (12).

Volume flow and EMF have opposite signs. This means that the contribution of the cation exchange membrane PSA to the potential difference is, as expected, greater than that of the

AMICON UMO 5 (Eq. 15). The active salt flow relative to the membrane was found to be:

$$J^{\beta''}_{NaCl} = 3 \cdot 10^{-3}\ M \cdot J^{\beta''}_{V} \qquad (20)$$

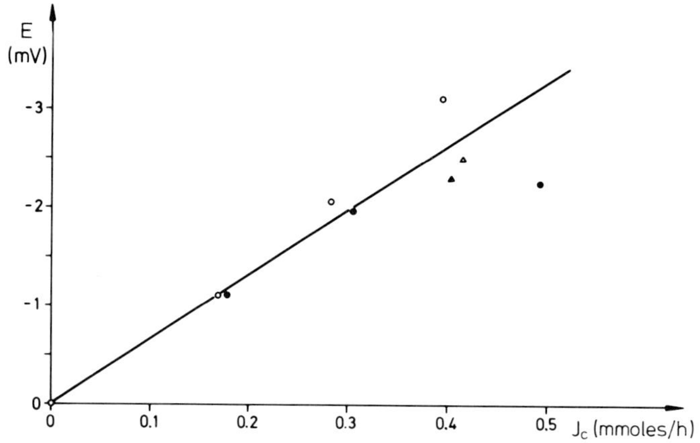

Fig. 3 Active EMF versus reaction flow as measured with three different pairs of membrane of the same batches. (PSA/*/AMICON UMO 5, c_{NaCl} = 0.01 M)

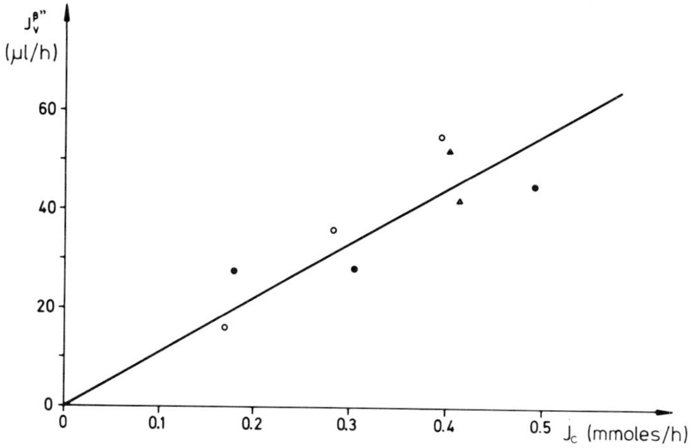

Fig. 4 Active volume flow versus reaction flow as measured with three different pairs of membranes of the same batches. (PSA/*/AMICON UMO 5).

2. The System PSA/*/ASAHI CAI

The hydraulic conductivity L_P of the anion exchange membrane is about ten times smaller than that of the cation exchange membrane PSA. It was therefore not possible to measure the volume flow. The experiment was restricted to the measurement of the active EMF. In addition, the pressure difference in the stationary state and the concentration of KCl in the fluid membrane phase were determined. The concentration of KCl was 0.1 M. All quantities measured are linearly dependent on the concentration of sucrose in the bulk phases (Figs. 5 and 6). The flow of sucrose through the membranes is so small that the enzymic reaction, being far from saturation, approximately obeys first order kinetics.

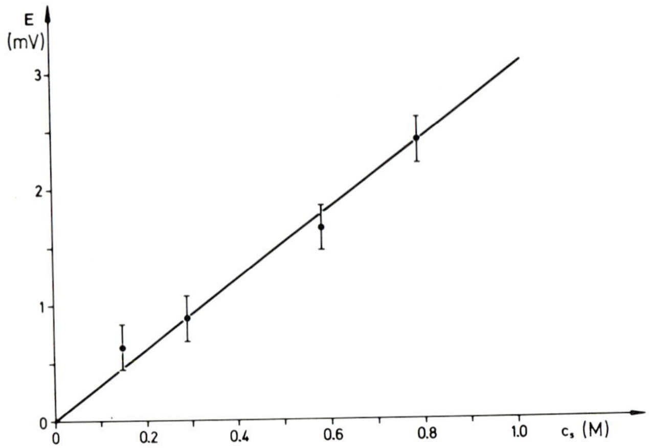

Fig. 5 Active EMF versus concentration of sucrose in the bulk phases. (PSA/*/ASAHI CAI, c_{KCl} = 0.10 M)

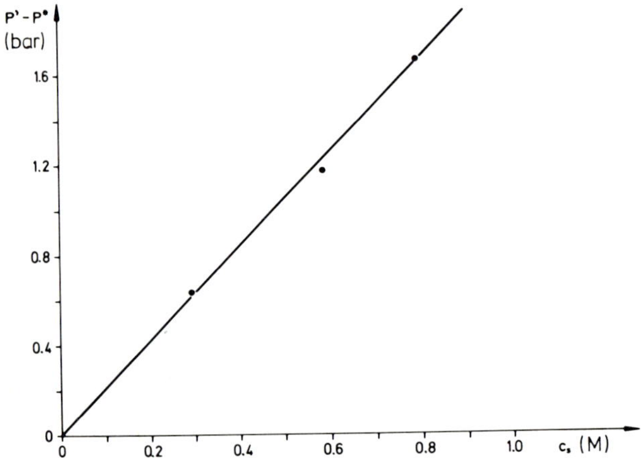

Fig. 6 Stationary state pressure difference P' – P* between the bulk phases and the inner compartment *, versus concentration of sucrose in the bulk phases. (PSA/*/ASAHI CAI).

II. Different Positioning of the Membrane AMICON UMO 5

In this experiment the diffusion layer of the AMICON UMO 5 points towards the outer phases. The concentration of the components was the same as in the experiment described before. The result is

$$E = +\,2.23\ \text{mV};\ J_V^{\beta''} = -\,27.5\ \mu l\ h^{-1}$$

The signs of the active EMF and of the active volume flow are opposite to those in the other experiments.

III. The Dependence of Active EMF and of Active Salt Transport on the Salt Concentration (PSA/*/AMICON UMO 5)

The concentration of sucrose was kept constant at 0.146 M. The concentration of NaCl was varied from 0.01 to 0.10 M. The active EMF (Fig. 7) decreases rapidly with increasing salt concentration and shows a concentration dependence similar to the reflection coefficients of KCl or NaCl for a PSA membrane. The reflection coefficient of the AMICON UMO 5 can be ignored compared with that of the PSA membrane. Thus the experimental result is in good qualitative agreement with theory (Eq. (15) and $\sigma^{\alpha}_{NaCl} \gg \sigma^{\beta}_{NaCl}$ and $a^{\alpha} = a^{\beta} = 0$). The active transport of NaCl increases with increasing salt concentration because of the factor c_{NaCl} in Eq. (12). The reflection coefficient of NaCl across the sandwich membrane was determined by the change of salt in the outer compartments by active transport. σ_{NaCl} decreases with increasing salt concentration, thus the active salt flow is further increased for it is proportional to $(1 - \sigma_{NaCl})$.

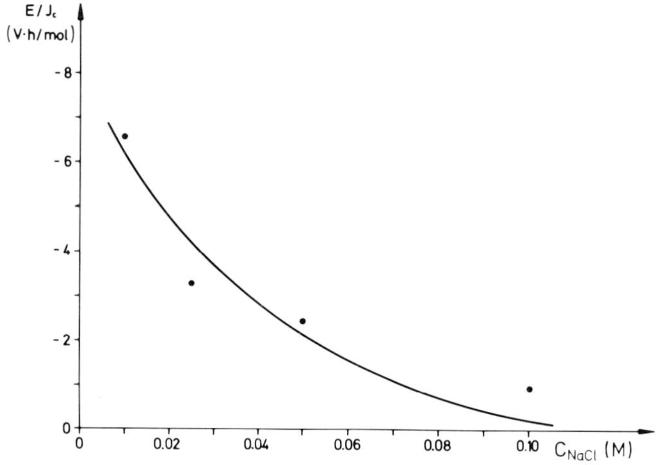

Fig. 7 Active EMF relative to chemical flow versus salt concentration. (PSA/*/AMICON UMO 5 c_S = 0.146 M).

From this we learn that the active transport of a salt and an active EMF are not necessarily coupled.

D. References

BLUMENTHAL, R., CAPLAN, S.R., KEDEM, O., Biophys. J. **7**, 735–757 (1967)

KATCHALSKY, A., CURRAN, P.F., Nonequilibrium Thermodynamics in Biophysics. Harvard University Press, Cambridge, Mass., 1967

KEDEM, O., in Proc. Symp. Transport and Metabolism, Academic Press, London and New York 1961

KEDEM, O., KATCHALSKY, A., Trans. Faraday Soc. **59**, 1918–1953 (1963)

MEYER, J., Kopplung von Stofftransport und chemischer Reaktion an einer zusammengesetzten, asymmetrischen Membran. Ein Beispiel für aktiven Transport. Dissertation, Frankfurt 1973

SAUER, F., in Handbook of Physiology Sec. 8 Renal Physiology (eds. J. Orloff and R.W. Berliner), pp. 399–414, The American Physiological Society, Washington D.C. 1973

SCHLÖGL, R., SCHÖDEL, U., Z. Phys. Chem. N.F. **5**, 372–397 (1955)

Multiple Steady States in a Reaction-Diffusion Coupled System

M.S. Seshadri

A. Introduction

The methods of irreversible thermodynamics have proved very fruitful in describing biological transport processes. We are here concerned with one aspect of this description, namely the treatment of the so-called steady state coupling between chemical reactions and diffusive flows. Almost two decades ago, HEARON (1950) recognized that such a coupling might play a special role in biological systems. More recently, KATCHALSKY and OSTER (1969) used the formalism of linear irreversible thermodynamics in their treatment of facilitated diffusion. It turns out that the flux of a species participating in a chemical reaction may be enhanced and that the enhancement is proportional to the rate of the chemical reaction. Similar in spirit is the treatment of carrier transport by BLUMENTHAL and KATCHALSKY (1969).

Chemical reactions have also been analysed from a slightly different point of view. The detailed investigations by PRIGOGINE and NICOLIS (1967) and PRIGOGINE and LEFEVER (1968) of the autocatalytic chemical reaction schemes, first proposed by TURING (1952), show the onset of instabilities and the consequent emergence of the so-called 'dissipative structures' in situations far from equilibrium. The significant findings of this work are summarized in a monograph by GLANSDORFF and PRIGOGINE (1971) and further reviewed by PRIGOGINE and NICOLIS (1971). These results may eventually pave the way to our understanding of spatial organization in biological systems.

In this paper we attempt a thermodynamic description of a reaction-diffusion coupled system when the constitutive relation between the chemical reaction rate and the chemical affinity is explicitly non-linear. In principle, the treatment could apply to biological systems characterized by carrier transport mechanisms. The problem can be treated in two parts. The first consists of writing down the local phenomenological relations between the entities that describe the two irreversible processes with which we are here concerned, namely diffusion and chemical reaction. Steady state coupling, in the local description, is essentially a statement of the conservation of mass. The second part is the global formalism, which results from integration of the local equations across the reaction column or the membrane. We shall here be concerned with the first of these.

B. Theory

I. Entropy Production

In the terminology of irreversible thermodynamics, a chemical reaction is characterized by the reaction rate J_r and its conjugate variable A, the affinity. The (local) entropy production is given by (DE GROOT and MAZUR, 1963):

$$\sigma = J_r A \geqslant 0 \qquad (1)$$

A phenomenological relation or a constitutive equation describes how J_r and A are related to each other. Invoking the law of mass action, it could be shown that the A-dependent part of $J_r(A)$ is of the form $(1 - \exp(-A/RT))$. Clearly, a linear relation between J_r and A presupposes that $|A|/RT \ll 1$. More generally, by expanding the exponential function in powers of A/RT and truncating the expansion at some point, it can be seen that $J_r(A)$ can be approximated by a polynomial in A. Thus,

$$J_r(A) = M_r(A)\, A \tag{2}$$

where $M_r(A)$ is some, as yet unspecified, polynomial in A. To explore the consequences of such a polynomial approximation is the principal objective of this paper. Since, in general, a polynomial does not have a unique sign, the inequality in Eq. (1) may be violated. This should be interpreted to mean the breakdown of the polynomial approximation itself.

II. Formulation of the Problem

Following KATCHALSKY and OSTER (1969), we consider unidimensional diffusion across a membrane or a reaction column delimited by the boundaries x = 0 and x = d. The local phenomenological relation for diffusion reads

$$-\vec{\mu}' = \hat{R}\,\vec{J} \tag{3}$$

and that for chemical reaction is given by Eq. (2). The prime means differentiation with respect to x, the spatial variable. In Eq. (3), $\vec{\mu}'$ is a column vector whose elements $(\vec{\mu}'_i)$ are the derivatives of the chemical potentials of the diffusing species, $\hat{R}$ a phenomenological matrix whose elements $(\hat{R}_{ik})$ are the frictional coefficients and $\vec{J}$ a vector whose elements (J_i) are the diffusive fluxes. At steady state, the equation of continuity can be written as

$$\vec{J}' = \vec{\nu}\, J_r \tag{4}$$

where $\vec{\nu}$ is the stoichiometric vector. The species that do not participate in the chemical reaction contribute each a null element to this vector. Using Eqs. (2), (3), and (4), the following differential equation for A(x) may be derived:

$$A'' = (\vec{\nu}, \hat{R}\,\vec{\nu})\, M_r(A)\, A \tag{5}$$

In deriving Eq. (5), the definition of the affinity

$$A(x) = -(\vec{\nu}, \vec{\mu}) \tag{6}$$

has been used and the constancy of $\hat{R}$ assumed. The notation $(\vec{a}, \vec{b})$ in the foregoing equations means the scalar product of the vectors $\vec{a}$ and $\vec{b}$. In principle, Eq. (5) can be solved for A(x) and then Eq. (2) determines the reaction rate.

III. The Affinity Profiles

In the domain of linear irreversible thermodynamics, $M_r(A) = M_r(0)$ is a positive constant. Integration of Eq. (5) leads to

$$A(x) = a \cosh(p_2^{1/2}\, x + \gamma) \tag{7}$$

with $p_2 = (\vec{\nu}, \hat{R}\vec{\nu})\, M_r(0)$. The boundary conditions determine the integration constants a and γ. It turns out that Eq. (5) is exactly solvable in terms of elliptic functions

(MILNE-THOMSON, 1965), when $M_r(A)$ has either the form $M_r(0) + bA$ or $M_r(0) + bA + CA^2$. In the sequel, we consider, for purposes of illustration, the latter situation and further set $b \equiv 0$ to render the treatment simple. Thus

$$A(x) = B \, \mathrm{sd}\,(\eta x + \delta) \tag{8}$$

where

$$B = \frac{p_0^{\,1/2}}{(p_2^2 + 4\,p_0\,|p_4|)^{1/4}} \tag{9}$$

$$\eta = (p_2^{\,2} + 4\,p_0\,|p_4|)^{1/4} \tag{10}$$

and

$$k^2 = \frac{p_2 + (p_2^{\,2} + 4\,p_0\,|p_4|)^{1/2}}{2(p_2^{\,2} + 4\,p_0\,|p_4|)^{1/2}} \tag{11}$$

In Eq. (8), $\mathrm{sd}\,u \equiv \mathrm{sn}\,u/\mathrm{dn}\,u$ is an elliptic function of modulus k, p_0 and δ are (real) integration constants and p_4 is proportional to the coefficient of A^2 in the expression for $M_r(A)$. In deriving Eq. (8), it has been assumed that $p_4 < 0$ and that $p_0 > 0$. It only remains to evaluate the integration constants from the prescribed boundary conditions.

We prescribe the following boundary condition:

$$A(0) = A(d) \tag{12}$$

Recalling that sd u has the period 4K(k), where

$$K(k) = \int_0^{\pi/2} \frac{d\phi}{(1 - k^2 \sin^2 \phi)^{1/2}} \tag{13}$$

is the complete elliptic integral of the first kind, it follows from Eq. (8) that

$$\eta d = 4mK(k) \quad (m = \text{any integer}) \tag{14}$$

Using Eqs. (10) and (11), Eq. (14) can be cast in the form

$$\frac{p_2^{\,1/2}\,d}{(2\,k^2 - 1)^{1/2}} = 4mK(k) \tag{15}$$

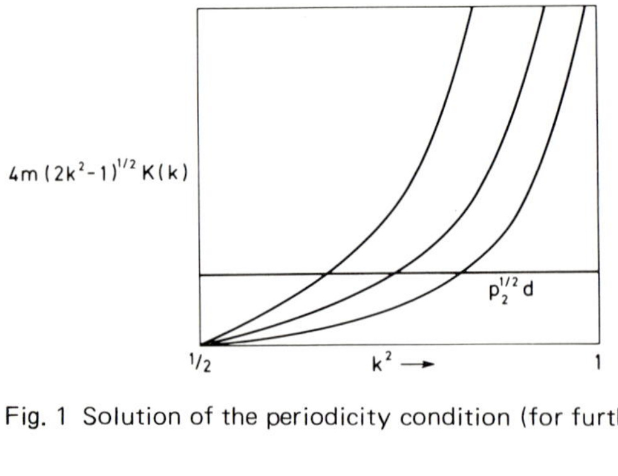

Fig. 1 Solution of the periodicity condition (for further explanation see text).

For given p_2 and p_4, k^2 varies from 1 to 1/2 as p_0 varies from 0 to $+\infty$. As k^2 varies from 1/2 to 1, the left hand side of Eq. (15) decreases monotonically from $+\infty$ to $p_2{}^{1/2}\,d$, while the right hand side increases monotonically from $4mK(1/\sqrt{2})$ to $+\infty$. Therefore, for each m, there exists a unique solution of Eq. (15). The family of curves in Fig. 1 is for different values of m. It thus follows that there exists a hierarchy of steady state solutions (Eq. 8) with amplitudes $B_m \equiv B(p_0{}^m)$ and frequencies $\eta_m \equiv \eta(p_0{}^m)$, where the $p_0{}^m$'s are the solutions of Eq. (15).

C. Discussion

The nonlinearity in Eq. (2) clearly has profound consequences. While the solution of the linear problem is unique and aperiodic, we now find a hierarchy of reaction rate profiles, each of which exhibits spatial oscillations as well. A physical criterion like the notion of stability of steady state solutions might enable us to single out a subset of the hierarchy as 'physically realizable solutions'. Presumably, there then exist nonoverlapping regions of stability, each of which contains precisely one stable solution. In the present context, the required stability analysis is rendered difficult by the fact that it requires a knowledge of the steady state concentration profiles, while we only know the steady state affinity profiles.

A polynomial, no matter how well it represents an exponential function, is still an approximation of the latter. One ought to make sure that the features observed above do not spuriously arise on account of the truncation procedure. In the limit of large negative affinities, it is possible to retain the exponential function as such and perform the calculations. Both the features, spatial oscillations and multiplicity of solutions, persist in this limit.

The conjugate thermodynamic variables in the global description can be recognized from the form of the global entropy production, which can be written in either of two equivalent forms:

$$\Phi = \sum_i J_i^0\, \Delta\mu_i + J_r^{tot} A(d) \tag{16}$$

$$= \sum_i J_i^d\, \Delta\mu_i + J_r^{tot} A(0) \tag{17}$$

Here

$$\Delta\mu_i = \mu_i(0) - \mu_i(d), \quad J_i^0 = J_i(0),\ J_i^d = J_i(d) \quad \text{and}$$

$$J_r^{tot} = \int_0^d dx\, J_r(x) \tag{18}$$

The evaluation of J_r^{tot} is straightforward since the local reaction rate profiles are known. The phenomenological relations between the global forces and fluxes may be written in the form

$$J_i^0 = J_i^0\, [\Delta\mu_i, A_d] \tag{19}$$

$$J_r^{tot} = J_r^{tot}\, [\Delta\mu_i, A_d] \tag{20}$$

The right hand sides of Eqs. (19) and (20) need not necessarily be linear in the $\Delta\mu_i$'s and A_d (SAUER, 1973). In the linear domain, the value of $p_2{}^{1/2}d$ determines the magnitude of the 'facilitated component' of the flux. In nonlinear situations, however, the additivity property of passive and facilitated components of the flux itself breaks down. A more detailed account of these will be published elsewhere.

D. Conclusion

The emergence of 'structures' in an initially structureless (homogeneous) medium appears to be significant both in physical and biological terms. From the purely physical point of view, this would be interpreted to mean a lowering of the entropy. It is attractive to view this as bringing about spatial organization in biological systems. Current interest in reaction-diffusion coupled systems stems largely from the hope that such a correlation can in fact be established. Moreover, important biochemical processes indeed seem to occur well outside the domain that lies in the neighbourhood of thermodynamic equilibrium. Indeed spatial oscillations and multiple steady states, as found in the present treatment, are typically far from equilibrium phenomena. One observes that the affinity profiles become more and more complex as one moves away from equilibrium.

E. References

BLUMENTHAL, R., KATCHALSKY, A., Biochim. Biophys. Acta **173**, 357–369 (1969)

DE GROOT, S.R., MAZUR, P., Non-Equilibrium Thermodynamics. North Holland, Amsterdam 1963

GLANSDORFF, P., PRIGOGINE, I., Thermodynamic Theory of Structure, Stability and Fluctuations. Wiley, New York 1971

HEARON, J.Z., Bull. Math. Biophys. **12**, 57–106 (1950)

KATCHALSKY, A., OSTER, G., The Molecular Basis of Membrane Function (ed. D.C. Tosteson), pp. 1–44, Prentice-Hall, New Jersey 1969

MILNE-THOMSON, L.M., Handbook of Mathematical Functions (eds. M. Abramowitz, I.E. Stegun), pp. 567–683, National Bureau of Standards, Washington D.C. 1965

PRIGOGINE, I., LEFEVER, R., J. Chem. Phys. **48**, 1695–1700 (1968)

PRIGOGINE, I., NICOLIS, G., J. Chem. Phys. **46**, 3542–3550 (1967)

PRIGOGINE, I., NICOLIS, G., Quart. Rev. Biophys. **4**, 107–148 (1971)

SAUER, F., in Handbook of Physiology (eds. J. Orloff, R.W. Berliner), Sec. 8, pp. 399–414, The American Physiological Society, Washington D.C. 1973

TURING, A.M., Phil. Trans. Roy. Soc. Lond. B **237**, 37–72 (1952)

On the Problem of Nonstationary Ion Fluxes

H. Schönert

A. Introduction

The Nernst-Planck equations describe the flux of ions in solutions which exhibit gradients of chemical and electrical potential. They are well known to electrochemists and also to those who study the movement of ions in exchange resins, in membranes and in some other electrolytic systems. These equations have been useful in successfully explaining a variety of phenomena.

Yet we could not find in the pertinent literature solutions of these equations dealing explicitly with time-dependent processes (except LONGSWORTH, 1943; KOBOTAKE and FUJITA, 1964). Instead additional assumptions are usually made, i.e. steady state or diffusion contraints etc., which help in integrating these equations. The solutions so arrived at give a good understanding of the processes but, in view of the problems of excitable membranes, electrophoresis, etc. it would be desirable to know how fast one state of a system changes to another state when one of the variables, for example the electrical current, is changed. Furthermore it would be interesting to know if this change is accompanied by a capacitive or inductive voltage response, as has been pointed out by TEORELL (1953).

B. Theory

In trying to find solutions to these problems we concluded that we should not just be looking for an appropriate mathematical method for we found that the Nernst-Planck equations are not well suited as a starting point; they should be reformulated on the basis of the equations of the thermodynamics of irreversible processes.

As a simple example take a solution of a binary electrolyte confined in a rectangular cell which is closed by reversible electrodes, for example a solution of $CaCl_2$ in water and Ag/AgCl electrodes. This simple example has the advantage that all parameters such as diffusion coefficients, transference numbers, electrical conductivity and activity coefficients are known so that any theoretical result can be compared with experiment.

The system is depicted in Fig. 1 which should be self-explanatory. For this system the Nernst-Planck equations are:

$$\frac{\partial c_+}{\partial t} = \frac{\partial}{\partial x}\left[D_+\left(\frac{\partial c_+}{\partial x} + z_+ c_+ \frac{F}{RT}\frac{\partial \psi}{\partial x}\right)\right], \tag{1a}$$

$$\frac{\partial c_-}{\partial t} = \frac{\partial}{\partial x}\left[D_-\left(\frac{\partial c_-}{\partial x} + z_- c_- \frac{F}{RT}\frac{\partial \psi}{\partial x}\right)\right], \tag{1b}$$

$$\frac{\partial^2 \psi}{\partial x^2} = 0, \quad z_+ c_+ + z_- c_- = 0 \quad . \tag{1c}$$

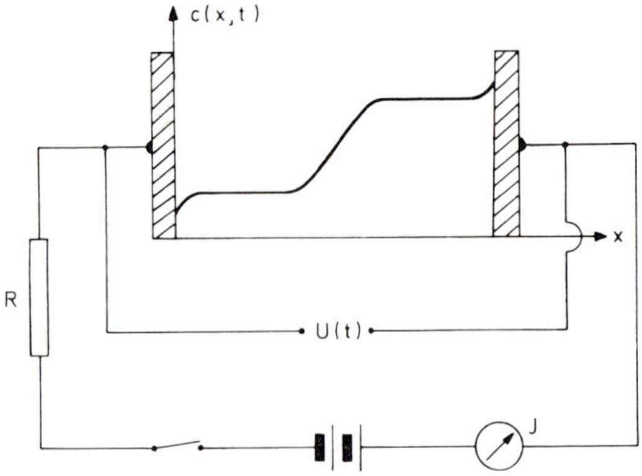

Fig. 1 Schematic diagram of an electrolytic cell showing the concentration c(x,t) as a function of the cell coordinate x at a time t.

Here the symbols have their usual meanings: D_i, diffusion coefficient of ion i; c_i, concentration of ion i; z_i, electrochemical valency of ion i; t, time; x, coordinate along the cell; ψ, electrical potential; R, gas constant; T, absolute temperature; F, Faraday constant. Eq. (1c) is the condition of local electroneutrality.

Analysing the structure of these equations, we find two variables, c(x,t) and ψ(x,t), to be solved simultaneously. Two transport coefficients, D_+ and D_-, characterise the electrolyte.

	System I Conductivity $\varkappa$	System II Diffusion D and EMF	System III Transference number t_i
Concentration in the cell	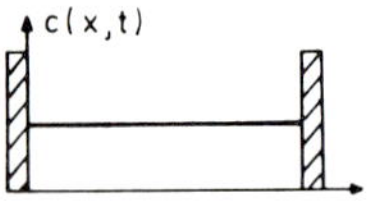	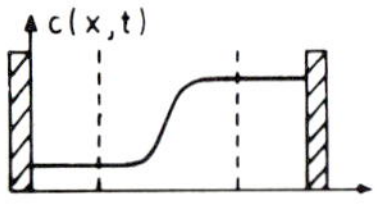	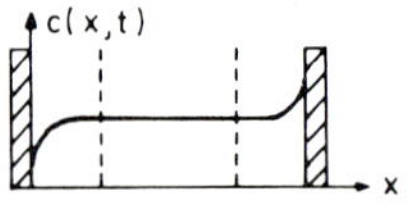
Boundary conditions	$I \neq 0$ $\frac{\partial c}{\partial x} = 0$	$I = 0$ $\frac{\partial c}{\partial x} \neq 0$, middle compartment $\frac{\partial c}{\partial x} = 0$, electrode compartments	$I \neq 0$ $\frac{\partial c}{\partial x} = 0$, middle compartment $\frac{\partial c}{\partial x} \neq 0$, electrode compartments
Transport equations	$-\int \frac{I}{\varkappa} dx = \int d\psi = U$ (2)	$\Phi = -D \frac{\partial c}{\partial x}$, or $\Phi_i = -\nu_i D \frac{\partial c}{\partial x}$ (3) $EMF = \pm \frac{RT}{F} \frac{1}{\nu_+ z_+} \int (1 - t_i) \, d \ln a$ (4)	$t_i = \frac{z_i F \Phi_i}{I}$, or $\Phi_i = \frac{t_i}{z_i} \frac{I}{F}$ (5)
Superposition		$\Phi_i = -\nu_i D \frac{\partial c}{\partial x} + \frac{t_i}{z_i} \frac{I}{F}$, or $\frac{\partial c}{\partial t} = \frac{\partial}{\partial x} \left(D \frac{\partial c}{\partial x} - \frac{t_i}{\nu_i z_i} \frac{I}{F} \right)$ (6) Transport equation, describing c(x,t)	
	$-\int \frac{I}{\varkappa} dx = U - EMF$ (7) Ohm's law, describing U(t)		

Fig. 2 Electrolytic cell of Fig. 1 divided into three subsystems.

Looking at the same system from the point of view of an electrochemist studying the transport properties along classical lines, we note that he decomposes the system into three subsystems by choosing different boundary conditions. These three subsystems are depicted in Fig. 2.

In the first subsystem the solution is homogeneous throughout and the electrochemist determines the specific conductivity κ, using Ohm's law as a transport equation:

$$I = -\kappa \frac{\partial \psi}{\partial x}, \quad \text{or} \quad -\int \frac{I}{\kappa} dx = \int d\psi = U \quad \text{for} \quad \frac{\partial c}{\partial x} = 0 . \tag{2}$$

Here I denotes the electrical current density and U the voltage. In the second subsystem the current density is zero and there is a concentration gradient in the middle compartment. Two measurements can be made.
1) Observing the concentration c as a function of time t and coordinate x gives the diffusion coefficient D of the electrolyte component (not of the single ions), according to Fick's equation:

$$\phi = -D\frac{\partial c}{\partial x}, \quad I = 0 \tag{3a}$$

where ϕ is the diffusion current density (flux) of the component. This equation may be written for the cation flux ϕ_+ and the anion flux ϕ_- ($\phi_+ = \nu_+ \phi$, $\phi_- = \nu_- \phi$):

$$\phi_i = -\nu_i D \frac{\partial c}{\partial x}, \quad I = 0 \tag{3b}$$

where ν_i is a stoichiometric coefficient.

2) The EMF of this galvanic cell is

$$\text{EMF} = \pm \frac{RT}{F} \frac{1}{\nu_+ z_+} \int (1-t_i)\, d \ln a . \tag{4}$$

Index i refers to the ion to which the electrode is reversible, $a = c\,\gamma_\pm$ is the activity of the electrolyte component ($\gamma_\pm$ = activity coefficient), and t_i the transference number of the ion i.

In the third subsystem electrical current is flowing (Hittorff experiment) and from the concentration changes in the electrode compartments the transference numbers t_i are determined, using the definition of t_i:

$$t_i = \frac{z_i F \phi_i}{I}, \quad \frac{\partial c}{\partial x} = 0,$$

which may be written as a transport equation for a homogeneous solution:

$$\phi_i = \frac{t_i I}{z_i F}, \quad \frac{\partial c}{\partial x} = 0; \quad i = +,- \tag{5}$$

These three types of experiments have given consistent results for κ, D, t_i for all binary electrolyte solutions so far investigated. Therefore we may take them for granted experimentally and use them as a starting point for superimposing the three subsystems to regain

the original system (Fig. 1) with simultaneous diffusion and electrical migration. The justification for these superposition procedures resides in the following arguments:
1) because the 'driving forces' $\partial c/\partial x$ and I are independent of each other and because we have linear processes they may be superimposed;
2) the result of such a superposition is equivalent to the phenomenological equations of the thermodynamics of irreversible processes as will be shown later[1]).
Superimposing subsystems II and III produces the original concentration gradient of the cell in Fig. 1. Likewise superimposing the relevant transport equations (3b) and (5) gives the transport equation

$$\phi_i = -\nu_i D\frac{\partial c}{\partial x} + \frac{t_i I}{z_i F}$$

or since $c_i = \nu_i c$

$$\frac{\partial c}{\partial t} = \frac{\partial}{\partial x}\left(D\frac{\partial c}{\partial x} - \frac{t_i I}{\nu_i z_i F}\right), \qquad (6)$$

which describes simultaneous diffusion and electrical migration. This equation should be compared with the Nernst-Planck equation; one it written in observables whereas the other is expressed in unobservables. Combining subsystems I and II and Eqs. (2) and (4) gives the current-voltage-relationship

$$-\int \frac{I}{\kappa} dx = U - EMF, \qquad (7)$$

in which the left hand side is the ohmic drop in the cell and U is the applied voltage. If U balances the EMF, I=0.

Eqs. (4), (6), and (7) are straightforward to use. From the geometry of the cell, the current density I is known as a function of x; for a rectangular cell I is independent of x. Thus Eq. (6) has only one variable, c(x,t), which can be found by integration. Knowing this, the EMF can be calculated from Eq. (4) and therefore, by Eq. (7), the current-voltage relationship. Thus we have all quantities of interest.

This is quite different from the Nernst-Planck procedure where c(x,t) and ψ(x,t) have to be solved simultaneously.

There are other differences between the two points of view. The three Eqs. (4), (6), and (7) have three transport coefficients t_+, D and κ. At this point it is well to remember that D is independent of the other two (ROBINSON and STOKES, 1959); only for infinite dilution is D linked to these two by the well-known Nernst relation. The Nernst-Planck equations have two transport coefficients, in contrast to three, here.

All the quantities in the Eqs. (4), (6), (7), are observables, unlike the D_i and ψ. In tracer diffusion experiments it seems that the D_i's are directly measurable; but one cannot move just one ion species in the solution from one point to another, at I=0, without destroying electroneutrality. Thus the tracer experiment gives a quantity which is in general different from the D_i in the Nernst-Planck equations. This point has been clarified by LIUKKONEN (1973).

---

[1])Actually the separation into subsystems and the subsequent superposition is given here mainly to demonstrate the simple interpretation of the equations of the thermodynamics of irreversible processes if they are reformulated on the basis of conventional transport coefficients. The same procedure may be used to treat more complicated systems than the one given here.

The main difference between the Nernst-Planck equations and the set of Eqs. (4), (6), (7) is that the former are not formulated on a phenomenological basis using only observable coefficients and functions, but rather on some model of transport kinetics in solution.

So far we have assumed that the subsystems and their transport equations can be superimposed. This seems plausible but it is far from certain. It is therefore gratifying that the set of Eqs. (4), (6), and (7) is equivalent to the phenomenological equations of the thermodynamics of irreversible processes (MILLER, 1967; SCHÖNERT, 1967). These equations are:

$$\phi_+ = - a_{++} \left(\frac{\partial \mu_+}{\partial x} + z_+ F \frac{\partial \psi}{\partial x}\right) - a_{+-} \left(\frac{\partial \mu_-}{\partial x} + z_- F \frac{\partial \psi}{\partial x}\right) \tag{8a}$$

$$\phi_- = - a_{-+} \left(\frac{\partial \mu_+}{\partial x} + z_+ F \frac{\partial \psi}{\partial x}\right) - a_{--} \left(\frac{\partial \mu_-}{\partial x} + z_- F \frac{\partial \psi}{\partial x}\right) \tag{8b}$$

The μ_i's are the chemical potentials.

At first sight it looks as if things have been made worse. The unobservable quantities have been increased; besides ψ we now have the single ion activity coefficients γ_i (in μ_i) which are not measurable. Furthermore, well-behaved transport coefficients like D_i, or D, t_i and κ are replaced by three coefficients a_{ik} of unknown properties (with $a_{ik} = a_{ki}$, ONSAGER's reciprocity relation). These equations do not look easy to integrate, and probably this is why they are so seldom used. However, by taking the Eqs. (8a), (8b) and inserting them into the defining equations of the subsystems I, II, and III, Eqs. (2), (3a), and (5), we can solve the resulting equations for the conventional transport equations (MILLER, 1960) and obtain three relations:

$$\kappa = \kappa\,(a_{ik}),\ t_i = t_i(a_{ik}),\ D = D\,(a_{ik}\,, \frac{\partial \mu}{\partial c}),$$

which give the conventional transport coefficients κ, t_i and D as functions of the phenomenological coefficients a_{ik} and of the thermodynamic factor

$$\frac{\partial \mu}{\partial c} = (\nu_+ + \nu_-)\,\frac{RT}{c}\,\left(1 + \frac{\partial \ln \gamma_\pm}{\partial \ln c}\right).$$

These relations apply to any concentration and are not restricted to dilute solutions, because Eqs. (8a) and (8b) are general. The three relations may be solved for the a_{ik}'s:

$$a_{++} = a_{++}\,(t_i, D, \kappa),\ a_{--} = a_{--}\,(t_i, D, \kappa)$$

$$a_{+-} = a_{-+} = a_{+-}\,(t_i, D, \kappa)$$

and when they are substituted in Eqs. (8a) and (8b), Eqs. (4), (6), and (7) result.

Thus the superposition of the subsystems is consistent with the thermodynamics of irreversible processes. Also all non-observable quantities can be eliminated and the resulting transport equations take into account any non-ideal solution behaviour through the thermodynamic factor ($\partial\mu/\partial c$) in D and any kinetic coupling of the species through the cross coefficient a_{+-}, by taking t_+, D and κ as three independent coefficients. In contrast the Nernst-Planck equations neglect both the thermodynamic and the kinetic interactions and they use unobservable variables.

How well do the equations check with experiments? Eq. (4) for the EMF is well known to be valid. Eq. (6) can be applied to a finite or to an infinitely long cell. The electrical current density is an experimental parameter. LONGSWORTH (1943) has experimentally found, with the help of Eq. (6), the following cases.
1) Solutions with binary electrolytes where the transference number $t_i = u_i/(u_+ + u_-)$ is independent of concentration. This leads to

$$\frac{\partial}{\partial x}\left(\frac{t_i}{\nu_i z_i}\frac{I}{F}\right) = 0 \tag{9}$$

and therefore the transport Eq. (6) reduces to

$$\frac{\partial c}{\partial t} = \frac{\partial}{\partial x}\left(D\frac{\partial c}{\partial x}\right) \tag{10}$$

Thus the concentration gradient changes with time according to Fick's law of diffusion and the electrical current has no influence at all. This is the case for the system H_2O + KCl. It is interesting to note that this conclusion had already been drawn by PLANCK (1890), using an equivalent of Eq. (6). But he overlooked the cases where t_i is concentration-dependent.

2) The transference number changes linearly with concentration:

$$\frac{\partial}{\partial x}\left(\frac{t_i I}{\nu_i z_i F}\right) = \frac{I}{z_i \nu_i F}\frac{\partial t_i}{\partial c}\frac{\partial c}{\partial x} = k\,\frac{\partial c}{\partial x}.$$

This implies that the concentration gradient changes its shape according to Fick's law and that the Gaussian peak travels along the x-axis with a constant velocity. H_2O + NaCl is such a system.

3) The transference number varies in a non-linear fashion with concentration. In this case the peak departs from a Gaussian form and becomes skewed. This effect depends on the direction of the current. The system $H_2O + CdI_2$ exhibits such behaviour.

A general solution of Eq. (6) has been found (SCHÖNERT, 1974). The agreement between observed and calculated c(x,t) is within the limits of experimental error.

In a cell of finite length the concentration changes in the electrode compartments and the middle compartment overlap after some time and the resulting steady state is reached with a relaxation time which is, according to Eq. (10), independent of the electric current density. This surprising effect can be verified experimentally by observing the concentration changes in a Schlieren optical cell (SCHÖNERT, 1974). If t_i varies with concentration its influence on the relaxation time can be calculated.

The Eqs. (4), (6), and (7) can be generalized to multicomponent systems (MILLER, 1967; SCHÖNERT, 1967) and also to membrane systems. Any model feature can be built into these equations by taking the appropriate concentration dependence of the transport coefficients.

C. Acknowledgements

This work was supported by the Deutsche Forschungsgemeinschaft (Sonderforschungsbereich 160).

D. References

KOBOTAKE, Y., FUJITA, H., J. Chem. Phys. **40**, 2219–2222 (1964)
LIUKKONEN, S., Acta Polytechn. Scand., Chem. Ser., **113** (1973)
LONGSWORTH, L.G., J. Am. Chem. Soc. **65**, 1755–1765 (1943)
MILLER, D.G., J. Phys. Chem. **64**, 1598–1599 (1960)
MILLER, D.G., J. Phys. Chem. **71**, 3588–3592 (1967)
PLANCK, M., Ann. Phys. Chem. **39**, 161–185 (1890)
ROBINSON, R.A., STOKES, R.H., Electrolyte Solutions, Butterworth, London 1959
SCHÖNERT, H., Z. Phys. Chem. **54**, 245–265 (1967)
SCHÖNERT, H., Forschungsb. Nordrhein-Westf., **2396**, Westd. Verlag Opladen 1974
TEORELL, T., Progr. Biophys. Biophys. Chem. **3**, 305–369 (1953)

The Influence of Double Layer Effects on Chemical Reactions at Charged Interfaces

H.W. Nürnberg

A. Introduction

In both biology and electrochemistry the immediate environment of electrically charged interfaces forms a region where significant influences on important types of chemical reactions are exerted. This is particularly true for reactions of ions, for the dissociation of a neutral or charged reactant into oppositely charged products and for reactions of substances specifically adsorbed at the interface. Among these reaction types proton transfer reactions such as the dissociation and recombination of carboxylic acids, the dissociation and formation of metal complexes, the dissociation and association of ion pairs and the behaviour of biopolymers specifically adsorbed at charged interfaces are very significant in biological systems. Although there are indeed substantial differences in composition, structure and properties between such different interfaces as the membrane of a living cell and the comparatively smooth surface of a metal electrode, various general physico-chemical properties and parameters are relevant and significant for both types of charged phase boundaries.

B. Theory and Discussion

I. The Double Layer and its Significant Parameters

If interfaces carry a charge excess there will be an electrical double layer which has the properties of a capacity. One plate of this condenser is the electrode or membrane surface while the other plate on the solution side is represented by a distribution of a corresponding excess of oppositely charged ions (PARSONS, 1961; NÜRNBERG and v.STACKELBERG, 1962; DELAHAY, 1965; MAIRANOVSKII, 1972). The distribution of these counter ions can be described by a Boltzmann distribution and therefore the whole double layer forms in principle a very complicated 3-dimensional problem; but it can be surprisingly successfully treated by the rather approximate and simple 1-dimensional Gouy-Chapman theory (NÜRNBERG, 1967b; NÜRNBERG and WOLFF, 1969).

In the absence of specific adsorption the electrode surface is hydrated by a monolayer of adsorbed water dipoles while the membrane has a similar water layer formed by the hydration shells of its hydrophilic constituents. The centres of the counterions, having an effective diameter which includes their primary hydration shells, can approach the interface up to a plane of closest approach. This plane is termed the outer Helmholtz plane (o.H.p.). An even closer approach of a particle to the interface is only possible in the case of specific adsorption and then surface water molecules are displaced from the electrode surface and a partial dismantling of the primary hydration shell of the adsorbed particle also occurs. The o.H.p. divides the total double layer into two parts, a discrete part between the electrode or membrane surface and the o.H.p. and the diffuse part over which the excess of counterions decreases exponentially to their bulk concentration level. The thickness of this diffuse double layer increases with decreasing ionic strength.

The potential ψ drops roughly exponentially from the o.H.p. towards the bulk solution. The potential at the o.H.p., ψ_H, depends on the charge density of the interface, Q_d, the dielectric constant, ϵ, and the ionic strength, J, of the solution. The formal thickness, ρ_d, of the double layer is defined as that distance, roughly speaking, at which the potential ψ has fallen to 1/e of its value at the o.H.p., i.e. 1/e of ψ_H. This thickness is inversely proportional to the square root of the ionic strength, J, of the solution. One can say that at a distance $x = 4\rho_d$, ψ becomes negligible and all double layer effects have disappeared. A most important double layer parameter with respect to reaction kinetics is the field strength, E, which of course is given by $(\partial\psi/\partial x)$. For details the reader is referred to the references cited above.

II. Effects on Chemical Reactions in the Diffuse Double Layer

In order to discuss the effects of the electric field we shall assume at first that the interface is a mercury electrode which carries a negative excess charge. For a metal electrode this is easily achieved by adjusting the electrode potential. This situation is the one we have studied extensively both theoretically and experimentally for the important homogeneous chemical reaction type represented by the dissociation and recombination of carboxylic acids (NÜRNBERG, 1967a, b; NÜRNBERG and WOLFF, 1969).

1. The Dielectric Constant

Since ψ gets more negative as the o.H.p. is approached from the bulk solution, there is an exponential increase in the concentration of positive ions, say e.g. Li^+. This accumulation of counterions and the corresponding decrease of the ions carrying a charge of the same sign as the interface, i.e. the anions in this case, is called the static ψ-effect. As the interface is approached an increasing fraction of the total available water will clearly be bound in the primary hydration shells of Li^+; an average of 2.3 water dipoles will be in a state of dielectric saturation. Furthermore the progressively increasing field strength E leads to a corresponding polarization and orientation of the remaining 'free' water dipoles outside the hydration shells of the cations. The net effect is a corresponding decrease in the dielectric constant ϵ of the solvent which turns into a drastic decrease of ϵ near the o.H.p..The magnitude of the effect depends on the charge density Q_d of the double layer and on its thickness ρ_d which is a function of the ionic strength, because Q_d and ρ_d determine the field strength E_x at any distance x within the diffuse double layer (cf. Table 1).

2. The Dissociation Field Effect

In his general theoretical treatment of the dissociation field effect ONSAGER (1934) has shown that the degree of dissociation of ion pairs is enhanced as a function of the parameter $(E/\epsilon)_x$. We have studied the dissociation mechanism of weak acids where a latent ion pair occurs as an intermediate stage

$$HA \underset{k_r}{\overset{k_d}{\rightleftharpoons}} H^+ + A^-$$

or in more detail:

$$HA\cdot(aq)_m + n\ aq \rightleftharpoons [H^+A^-]\cdot(aq)_m + n\ aq \rightleftharpoons [H^+(aq)_n A^-\cdot(aq)_m] \rightleftharpoons \ldots H_9O_4{}^+ + A^-(aq)_{m+(n-4)}$$

with aq (water molecules) involved in hydration shells of HA and its dissociation products. The first step is the breaking of the covalent bond between O and H in the carboxylic group (–COOH) of the acid and the localisation of the positive and negative charges at the H^+ and the carboxylate anion ($-COO^-$). This leads to the formation of an ion pair $[H^+A^-]$. The dissociation is then completed by the hydration of the products H^+ and A^- and their separation by a critical distance of 7.5 to 15 Å depending on the ionic strength of the solution. The electric field acts in the intermediate ion pair stage where only coulombic interactions between H^+ and A^- still exist. The net effect is the enhancement of the separation rate between the dissociation products in the final stages of the dissociation process. Thus, the dissociation field effect enhances not only the dissociation degree but also the dissociation rate (NÜRNBERG, 1967a, b; NÜRNBERG and WOLFF, 1969).

It should be noted, incidentally, that in principle a dissociation mechanism passing through an intermediate ion pair stage is generally to be expected if a neutral or charged particle dissociates into two oppositely charged products. Thus, a dissociation field effect will usually also operate if ligands dissociate from a metal complex.

With the above mechanism in mind we have, for the first time, applied ONSAGER's theory to compute the enhancement of the dissociation rate constants of neutral and charged carboxylic acids as a function of the field strength and the dielectric constant in the double layer. For charged acids their concentration profile due to the static ψ-effect had also to be taken into account. The results of our theoretical treatment were confirmed experimentally for numerous carboxylic acids of different structure in aqueous LiCl solution by determining their dissociation and recombination rate constants, k_d and k_r, with an advanced electrochemical relaxation technique of polarographic origin (NÜRNBERG, 1967a, b; NÜRNBERG and WOLFF, 1969).

3. Field Influences on Recombination

The electric field acts also on recombination in several ways (NÜRNBERG and WOLFF, 1969). Thus, in principle, recombination is favoured by the decrease of the dielectric constant ϵ and generally by the disintegration of the ion clouds around the recombining reactants H^+ and A^-. The latter effect leads to some increase in the rate of approach of H^+ and A^- and thus in their recombination rate. However, the most important effect of the electric field on recombination is a retarding effect. Due to the static ψ-effect there is, over the diffuse double layer, an exponentially increasing drop in the concentration of one of the recombining reactants, H^+ or A^-, of which that having a charge of the same sign as the interface is electrostatically repelled. The resulting decrease in the concentration of the respective reactant for recombination, H^+ or A^-, leads to a corresponding decrease in recombination rate and the net result is a further additional enhancement in the degree of dissociation in the diffuse double layer.

4. General Conclusion

We conclude that the rather approximate and simplified evaluation of the parameters of the diffuse double layer based on the Gouy-Chapman theory (including some modifications introduced by us (NÜRNBERG, 1967b; NÜRNBERG and WOLFF, 1969) leads, in conjunction with our treatment of the dissociation field effect based on ONSAGER's theory (1934), to a rather satisfactory comprehensive description of all major influences in the diffuse double layer on homogeneous dissociation reactions. These double layer effects are significant when $|\psi_H| \geqslant 50$ mV, for bulk ionic strengths $J \geqslant 0.1$.

5. Complications due to Associative Field Effects

However, it must added that the situation can become more complicated. As JENARD and HURWITZ (1968) and we (NÜRNBERG and WOLFF, 1969) have shown an effect of high fields on dissociation, compensating partially the dissociation field effect, may become significant for certain dissociation reactions. This field effect on dissociation arises from the substantial polarization of the solvent by high fields ($E > 3 \cdot 10^6$ V cm^{-1}) and the corresponding decrease of the dielectric constant which causes a stabilization of the ion pair stage. While the plane of closest approach remains, for larger particles such as carboxylic acids, too far from the interface to reach the region of sufficiently high field strength the situation may be different for ion pairs of small ions or for substances specifically adsorbed. Indeed experimental evidence for the stabilisation of ion pairs formed by small ions in high field regions of the double layer has been reported (GIERST et al., 1966). Theory shows that finally, when the field strength becomes large enough to cause dielectric saturation of the solvent, the dissociation field effect again breaks through and operates alone.

III. The Situation at Charged Biological Interfaces

The previous discussion of the double layer effects on chemical reactions was based on experimental evidence obtained with metal electrodes in aqueous electrolyte solutions. However, all the important conclusions remain equally relevant if the charged interface is the membrane of a living cell. This membrane consists of proteins and lipids which are combined in a given structure by weak interactions such as hydrogen bonds, hydrophobic binding and dispersion forces. Under given conditions the membrane will carry on its protein and lipid moieties a certain excess of COO^--groups or of protonated amino groups-NR_3H^+. Furthermore other ions may be incorporated in the membrane (LÄUGER, 1970). Thus, the membrane is usually an interface with an excess charge of positive or negative sign and all the physicochemical consequences discussed before become operative. A simple calculation shows that concentrations of singly charged monomeric units of the membrane $\Gamma \geqslant 1 \cdot 10^{-10}$ moles cm^{-2} are sufficient to make mandatory the allowance for double layer effects and especially for the influences of the electric field on chemical reactions in the environment of the membrane (Table 1). For instance it should be emphasized that the formation and dissociation of intermediate ion pair stages is almost certainly involved in the active transport of ions through membranes via carrier-complexes. Furthermore proton transfer reactions play a significant role in several important reactions at this charged biological interface.

Table 1

Computed typical values of double layer parameters for biological membranes for various concentrations Γ of charged monomeric units ($|z| = 1$), and corresponding interfacial excess charge densities $Q_d = zF\Gamma$, at different bulk ionic strength J adjusted with LiCl at 20°C. $\bar{\epsilon}_{eff}$ = average dielectric constant over diffuse double layer. Values for formal thickness ρ_d of diffuse double layer are average values for the given range ot $\bar{\epsilon}_{eff}$.

$\|z\| = 1$		J = 0,1; $\rho_d \approx 9{,}5$ Å			J = 1,0; $\rho_d \approx 2{,}6$ Å		
Γ (moles cm^{-2})	Q_d ($\mu C\ cm^{-2}$)	ψ_H (mV)	E_H ($V cm^{-1}$)	$\bar{\epsilon}_{eff}$	ψ_H (mV)	E_H ($V\ cm^{-1}$)	$\bar{\epsilon}_{eff}$
$1 \cdot 10^{-10}$	10	90	$14.5 \cdot 10^5$	75	42	$18 \cdot 10^5$	64
$2 \cdot 10^{-10}$	20	123	$27 \cdot 10^5$	75	75	$40 \cdot 10^5$	61
$3 \cdot 10^{-10}$	30	150	$48 \cdot 10^5$	74	100	$65 \cdot 10^5$	57

IV. Effects in the Adsorbed State

The discussion has been focussed hitherto on the effects of the electric field on chemical reactions in the diffuse double layer of charged interfaces. However, very significant influences on substances on the interface must also be expected. Due to their specific adsorption they populate the discrete part of the double layer. Again some results from the extensive studies of many authors with metal electrodes illustrate this point.

Many organic substances show, within a certain potential range, i.e. within a certain range of the interfacial charge density, a tendency to become specifically adsorbed (NÜRNBERG and v.STACKELBERG, 1962; FRUMKIN and DAMASKIN, 1964; KASTENING and HOLLECK, 1965; GILEADI, 1967). They usually form a monomolecular adsorption layer where the hydrophobic part of the molecule interacts with the interface, displacing the water molecules present at the adsorption site, and the hydrophilic part of the adsorbed substance remains hydrated and is oriented towards the solution. If the adsorbant contains aromatic or heterocyclic rings with easily movable π-electrons, at positively charged interfaces a planar orientation of the adsorbed particles with respect to the surface is usually observed due to the interactions of the π-electrons with the positively charged surface. The adsorption layer remains rather penetrable for smaller ions or molecules and the interactions between the adsorbed particles remain feeble. However, with increasing negativity of the interfacial charge one observes, beyond a certain threshold value, a substantial change in the orientation and properties of the adsorption layer to a position more or less normal to the interface. At the same time a rather compact adsorption film is formed in which strong interactions between the adsorbed particles prevail. At very negative charges a very sharp desorption of the whole film may occur. Observations of this kind have been made for instance with coumarines (PARTRIDGE et al., 1966, 1969), pyridine derivatives (GIERST et al., 1959; NÜRNBERG and WOLFF, 1965), and quinolines (VANLAETHEM-MEUREE et al., 1972).

A very pronounced adsorption over an extended range of interfacial charge has been recently estabilished by us (VALENTA et al., 1974a, b) for denatured and native DNA at the mercury electrode. The adsorption occurs via the base moieties of DNA and the particular orientation of the film is a function of the sign and value of the interfacial charge.

However, the most important and fundamental finding in this study was the observation that the electric field induces a partial opening of the double helix of native DNA adsorbed at the interface. A striking correlation between field strength and the degree of induced helix-coil-transformation of native DNA from various origins and of different molecular weights was obtained (VALENTA et al., 1974b). The possibility of a field induced helix-coil-transition was predicted theoretically, for fields in which $E > 10^4\ V\ cm^{-1}$, by HILL (1958) many years ago already.

C. Conclusion

In general one may state that double layer effects play a very significant if not a key role for all chemical steps occurring at biological membranes as these usually carry an excess charge. Thus, despite all the differences in structure and composition, the physicochemical conclusions emerging from electrochemical studies at the more easily adjustable interface, metal electrode/solution, have a fundamental relevance for a deeper and better understanding of the biophysics of processes at the membranes of living cells.

D. References

DELAHAY, P., 'Double Layer and Electrode Kinetics', Interscience Publ., New York 1965

FRUMKIN, A.N., DAMASKIN, B.B., in Modern Aspects in Electrochemistry (eds. J.O'M. Bockris, B.E. Conway), Vol. 3, pp. 149–223, Butterworth, London 1964

GIERST, L., BERMANE, D., CORBUSIER, L., Ric. Sci. **29**, 75–107 (1959)

GIERST, L., VANDENBERGHEN, L., NICOLAS, E., FRABONI, A., J. Electrochem. Soc. **113**, 1025–1036 (1966)

GILEADI, E., 'Electrosorption', Plenum Press, New York 1967

HILL, T.L., J. Am. Chem. Soc. **80**, 2142–2147 (1958)

JENARD, A., HURWITZ, H.D., J. Electroanal. Chem. Interfac. Electrochem. **19**, 441—445 (1968)
KASTENING, B., HOLLECK, L., Talanta **12**, 1259—1288 (1965)
LÄUGER, P., Naturwissenschaften **57**, 474—480 (1970)
MAIRANOVSKII, S.G., in Progress in Polarography (eds. P. Zuman, L. Meites, I.M. Kolthoff), Vol. 3, pp. 287—369, Interscience Publ., New York 1972
NÜRNBERG, H.W., Fortschr. chem. Forschg. **8**, 241—308 (1967a)
NÜRNBERG, H.W., Ber. Kernforschungsanlage Jülich, Jül—475—CA, pp. 1—402 (1967b)
NÜRNBERG, H.W., v.STACKELBERG, M., J. Electroanal. Chem. **4**, 1—47 (1962)
NÜRNBERG, H.W., WOLFF, G., Coll. Czechoslov. Chem. Commun. **30**, 3997—4015 (1965)
NÜRNBERG, H.W., WOLFF, G., J. Electroanal. Chem. Interfac. Electrochem. **21**, 99—122 (1969)
ONSAGER, L., J. Chem. Phys. **2**, 599—615 (1934)
PARSONS, R., in Advances in Electrochemistry and Electrochemical Engineering (eds. P. Delahay, C.W. Tobias), Vol. 1, pp. 1—64, Interscience Publ., New York 1961
PARTRIDGE, L.K., TANSLEY, A.C., PORTER, A.S., Electrochim. Acta **11**, 517 (1966)
PARTRIDGE, L.K., TANSLEY, A.C., PORTER, A.S., Electrochim. Acta **14**, 223 (1969)
VALENTA, P., GRAHMANN, P., NÜRNBERG, H.W., J. Electroanal. Chem. Interfac. Electrochem. **49**, 41—54; 55—75 (1974a)
VALENTA, P., NÜRNBERG, H.W., KLAHRE, P., Bioelectrochemistry and Bioenergetics **1**, in press (1974b)
VANLAETHEM-MEUREE, N., LAMBERT, J.P., GIERST, L., Chem. Ing. Techn. **44**, 219—220 (1972)

Round Table Discussion 1

Chairman: O. Kedem

DAINTY initiated the discussion by asking what a membrane, or a molecule in a membrane 'sees'; does it see the electrochemical potential of, say, an approaching ion, or what? In other words, if the electrochemical potential is accurately assessed by counting every potential, etc., what determines the rate of reaction? KEDEM commented that there is no general answer to this question; you must specify the two or more molecules involved and ask what parts of the electric field, for example, each are in; i.e. you must specify the scale you are talking about, for instance the gradient of the field with respect to the molecules. NÜRNBERG cited the example of the diffuse double layer, which is certainly inhomogeneous, with a very strong electric field gradient over a scale approaching the molecular.

KEDEM asked when do 'things become complicated' with respect to the above discussion of the relevance of thermodynamic parameters to reactions at or near membranes. SAUER explained that complications arise in strongly inhomogeneous systems, for instance when an external field strength becomes comparable with molecular interaction forces. In such a situation, the concept of chemical potential becomes questionable; it can be introduced but it is no longer a simple function of the local state, but depends on the state variables of the whole system. By using the example of a chemical reaction in a centrifugal or gravitational field, he showed that providing the field strength is weak its potential can be superposed on the other thermodynamic potentials in, say, the free energy and the usual procedures can then be carried out, according to Gibbs and others, with no loss of validity. In response to a question by DAINTY as to whether the dubiety of concepts such as chemical potential arose from the smallness of the regions involved, SAUER repeated that it is not the smallness but the 'strength of the inhomogeneities' in the appropriate region. He pointed out that one can show also from statistical mechanics that the superposition principle breaks down for strong fields, except for the case of an ideal gas in which there are no intermolecular interactions.

KEDEM next directed the discussion towards the question of the sizes of the pores involved in studies on some artificial membrane systems. They are usually appreciably wider than the supposed narrow pores in biological membranes and thus any osmotic effects largely arise from charges on the walls of the pore. She referred to the work of MEARES on ion-exchange membranes, where quite a good fit between theory and experiment obtains. However she warned that such a good fit could not be expected in biological membranes; for instance one could easily find a situation where the charge has one sign and the streaming potential the opposite sign. WOERMANN pointed out that this was the whole point of his paper; physiologists should not be surprised if the charge density, say, that they calculate from measuring potential differences and streaming potentials across biological membranes does not agree with the charge density calculated from analytically determined charge densities of the COO^- groups on the proteins, etc.

The discussion then moved to questions of linearity and non-linearity, stability and instability, and oscillations. KEDEM said that there was no reason to associate non-linearity with instability, nor to associate linearity (of, say, a current-voltage curve) with simplicity (of a membrane). After SESHADRI had explained that non-linearity in his chemical reaction rates arises because the rate depends on $(1-\exp(-A/RT))$ and the linear approximation, of the rate being proportional to the affinity (A), breaks down when A/RT becomes sufficiently large, KEDEM retorted that in most chemical reactions, A/RT can be quite large and yet the systems behave beautifully with no tendency to have more than one stationary state. SESHADRI replied that there may be a multiplicity of steady states, only one of which is stable; he had not yet examined the question of the stability of his predicted set of steady states. At this point SCHÖNERT spoke of the inconsistency between, on the one hand, classical reaction kinetics and, on the other hand, irreversible thermodynamics applied to chemical reactions: from classical reaction kinetics one does not get the mass action law in the proper form, with activity coefficients and so on, and from irreversible thermodynamics you cannot get the classical reaction laws far from equilibrium. KEDEM agreed that the range of validity of linear irreversible thermodynamics, applied to chemical reactions, is too narrow to be of much use; outside this range you get the quiet, well-behaved, laws of kinetics; much further out there are situations where more than one stationary state occurs, but this is much further away from equilibrium than indicated by SESHADRI's non-linear affinity. SAUER then introduced the idea of looking at a system from the point of view of the precision with which the experimental parameters can be measured, for this effects how many terms in the polynomial expansion of any of the thermodynamic functions involved must be used. Clearly the less precise the measurements, the fewer the polynomial terms and hence the smaller the departure from linearity resulting in instability. MEARES did not like this idea at all. He could not see how imprecise measurements would lead to, say, an oscillatory situation, unless the amplitude of the oscillation were less than the precision of the measurements! However KEDEM did stress, partly in reference to a point made by DAINTY about action potentials, that there are predictable instabilities which are real and have a functional role in biology; the nerve action potential does not arise from imprecision! Finally in respect of this long discussion of stability and multiplicity of states, LORENZ raised the question of fluctuations of the parameters of a system leading to instability. He mentioned the possibility of fluctuations in local potential functions reaching a critical range and thus leading to instability, though KEDEM pointed out that double layers, in which there are very large gradients, are quite stable. She added, with reference to non-linearity, strong inhomogeneities, etc., that 'things may go crazy' but not in such a way as to bring about instability; for this you need some positive feed-back.

The final topic of this discussion was the relevance of the kinds of coupling between chemical reaction and diffusion referred to in the papers of this session and active transport as biologists understand it. KEDEM remarked that to get the kind of high stoichiometry found in nature you need a very different kind of coupling, in fact something like a carrier.

SLAYMAN then described Mitchell's latest model for the transport ATPases; this is an asymmetric sandwich membrane, as was the membrane system described in Meyer's paper. The interior of the membrane (or of the molecule of ATPase) is supposed to possess the catalytic site. The interior also has access to the outer surface of the membrane (extracellular space) via channels permeable to Na^+ and K^+, and has access to the inner surface of the membrane (cytoplasm) via channels permeable to Na_3HATP, to K_2HADP, and to $PO_4{}^{3-}$ The ATPase acts as a pump, therefore, in the following manner: Na_3HATP diffuses to the ATPase interior, where it is hydrolyzed to Na^+, ADP, and P_i; Na^+ then diffuses outward through the Na^+ or K^+ channels, while K^+ diffuses inward. Then K_2HADP and P_i diffuse to the cytoplasm through the K_2HADP and P_i channels. The model gets around the requirement for transmembrane conformational changes, and relies on fixed but highly asymmetric permeabilities of the membrane surfaces. KEDEM pointed out that this nice simple model depended on there being highly specific channels and SAUER said that it was of course not in disagreement with general thermodynamic considerations, and he stressed the importance of making models within the framework of thermodynamics. KEDEM closed the discussion by emphasizing the need for very high coupling between chemical reactions and diffusion in order to get the highly efficient active transport observed in nature.

Session 2

Water Transport and Osmotic Processes

Chairman: A. B. Hope

A Study of Transcellular Osmosis and the Kinetics of Swelling and Shrinking in Cells of *Chara corallina*

J. Dainty, H. Vinters, and M.T. Tyree

A. Introduction

The theory of water transport across plant cell membranes is now reasonably satisfactory and straightforward, especially since the coupling of solute and water flows has been correctly introduced by the theory of irreversible thermodynamics. Adequate treatments of the theory can be found in the reviews of DAINTY (1963, 1969) and the book of SLATYER (1967). For a single plant cell the basic flow equation is:

$$J_V = L_p\,(\Delta P - \sigma\Delta\pi) \qquad (1)$$

where J_V is the volume flux in cm s^{-1}, L_p is the hydraulic conductivity of the membrane in cm s^{-1} bar^{-1}, ΔP and $\Delta\pi$ are the hydrostatic and osmotic pressure differences, respectively, across the membrane in bar and σ is the reflection coefficient of the solute, a concept well explained in the thermodynamic section of this book. Clearly L_p and σ are important membrane parameters and we need as much information about them for as many different cells (and solutes) as possible. There is, however, another very important parameter hidden in Eq. (1); this is the volume elastic modulus ϵ of the plant cell wall. For it can easily be shown that if, for instance, one changes the external osmotic pressure, i.e. changes $\Delta\pi$, then water will flow in or out of the cell; such flows swell or shrink the cell volume, i.e. increase or relax the tensions in the cell wall, and thus markedly alter ΔP; the change in P_i, the internal or turgor pressure, is related to the fractional change in cell volume, dV/V, through the volume elastic modulus, ϵ, i.e.

$$dP_i = \epsilon(dV/V) \qquad (2)$$

If the mathematics of the consequences of, say, changing the external osmotic pressure by a relatively small amount, $\Delta\pi_e$, are worked out, it is not difficult to show that the rate of change of volume (rate of swelling or shrinking) is given by the following equation:

$$\frac{1}{V_o}\frac{dV}{dt} \approx \frac{A_oL_p}{V_o} \cdot \Delta\pi_e \cdot \exp\left[-(\epsilon + \pi_{io})\frac{A_oL_p}{V_o}\,t\right] \qquad (3)$$

In this formula, which is actually written for a shrinking process, V(cm^3) is the cell volume at time t(s). The assumption has been made, hence the 'approximately equals' sign in the equation, that neither the volume nor the area of a plant cell change very much between full turgor and incipient plasmolysis and hence we can use the values of area ($A_o cm^2$), volume ($V_o cm^3$) and internal osmotic pressure (π_{io}bar) at zero turgor pressure, i.e. incipient plasmolysis, as constant, average representatives of these quantities over the swelling or shrinking process. It has also been assumed that the solutes involved are non-permeating, i.e. their σ is unity.

Such an analysis shows that the swelling or shrinking process, i.e. the entry or exit of water, is not only rate-controlled by L_p but also equally by ϵ. In fact the half-time for the process is obviously given by:

$$T_{1/2} = 0.693\, V_o/A_oL_p\,(\epsilon + \pi_{io}) \qquad (4)$$

and, usually, $\epsilon \gg \pi_{ic}$. Thus the volume elastic modulus, ϵ, is an extremely important parameter for a quantitative understanding of the water relations of plant cells.

Confining our attention to non-permeating solutes, $\sigma = 1$, it is now clear that the key information we need is a knowledge of the values of L_p and ϵ for plant cells. We have surprisingly little of such information. We know L_p for a few giant algal cells only; for higher plant cells we have no unambiguous values, for all such values in the literature have either been obtained on plasmolysed protoplasts and the unstirred layer corrections have been ignored.

However in this paper we are more concerned with the volume elastic modulus, ϵ, for which there is much the same paucity of information. Until ZIMMERMANN (see ZIMMERMANN et al., 1969; STEUDLE and ZIMMERMANN, 1974) introduced his new technique of direct measurement of ϵ (and L_p), only a very few indirect measurements had been made (KELLY et al., 1963; BARRY, 1970). These indirect measurements involved comparing fractional length changes of *Nitella* or *Chara* species produced by changes in turgor pressure. It was thus necessary to convert fractional length changes (dl/l) to fractional volume changes (dV/V) by introducing a factor k, defined by:

$$(dV/V) = k(dl/l) \tag{5}$$

k is a complex function of the elastic moduli and Poisson's ratios of the cell wall and will be discussed later in this paper. In the previous work of KELLY et al. (1963), k was estimated (incorrectly as it happened) from some independent work on isolated cell walls. In this paper we use the indirect approach, *via* length changes, to measuring ϵ for *Chara corallina*, but we are able to get a direct measurement of k and are thus able to make a reasonably satisfactory conversion of our (dl/l)'s to (dV/V)'s and hence get ϵ from Eq. (2). The parameter k is itself very interesting and is a kind of bonus from this approach.

As will be explained in the next section, in order to get k (and to get ϵ by an alternative method) we need to measure L_p for our cells; this is done by the now classical method of transcellular osmosis (KAMIYA and TAZAWA, 1956; DAINTY and HOPE, 1959; DAINTY and GINZBURG, 1964).

B. Materials and Methods

Measurements of cell shrinkage (and swelling) were quite simple. An internodal cell of *Chara corallina* bearing a fine silver marker was fastened into position in a capillary glass tube (internal diameter = 2.5 mm), using a fine thread. Solutions bathing the cell could be quickly exchanged (1–2 s) by means of a two-way stopcock. The absolute length change of the cell was measured using a micrometer eyepiece fitted to a stereomicroscope. The osmoticum used was polyethylene glycol, M.W. 300 to 400. The normal bathing medium of the cell was artificial pond water (A.P.W. = 1.0 mM NaCl, 0.1 mM KCl, 0.1 mM $CaCl_2$). Each cell was exposed to as many as seven different concentrations of osmoticum, with repeated (4 to 10) exposures to any one concentration. Experiments (static and kinetic) were carried out both at room temperature (23°–30°C) and in a thermostatically controlled cold room (7°C). The cells used varied in length (40–120 mm) and had diameters of 0.7 to 0.9 mm.

The kinetics of cell volume changes were studied in the same apparatus. A special timing device, using a chart recorder, was employed to ensure reasonable accuracy in the kinetic measurements. This will be described in detail elsewhere. The result of a 'kinetics run' is given in Fig. 1. In several cases a series of static and kinetic measurements were made on the same cell, both at room temperature and at 7°C.

The apparatus used for L_p measurements was essentially the same as that of DAINTY and GINZBURG (1964). Osmoticum was again polyethylene glycol of various concentrations.

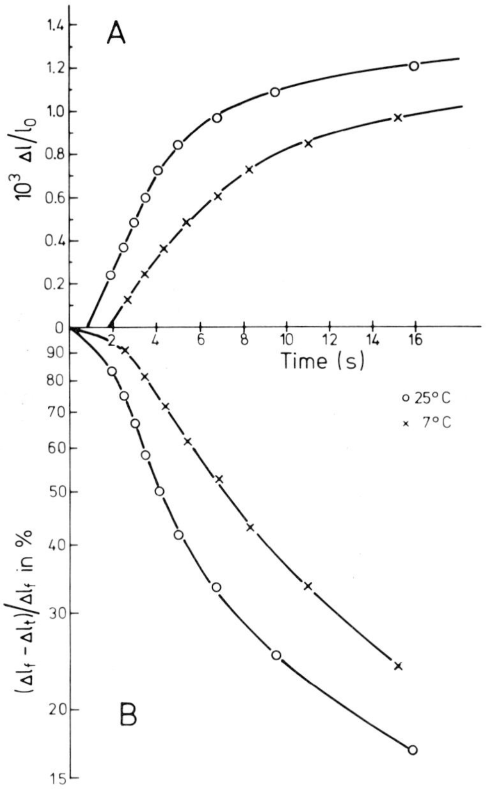

Fig. 1 Two sets of graphs derived from chart recordings taken during a kinetic study of *Chara* cell shrinkage at room temperature ($\approx$ 25°C) and at 7°C. Stopcock opened at t = 0, to exchange solutions. 120 mosmolal osmoticum.

A. Graph of relative length change with time. The initial slope provides an estimate of $\frac{1}{l_o}\left(\frac{dl}{dt}\right)_{t=o}$.

B. Semi-log plot of % length change with time. The linear part of the graph is used to estimate $T_{1/2}$. $(T_{1/2})_{7°C}$ = 5.3 s and $(T_{1/2})_{25°C}$ = 3.0 s. Δl_f is the change in length after a long time; Δl_t is change at time t.

C. Results and Discussion

The mean value of L_p, obtained from the transcellular osmosis experiments, from 76 determinations on 13 different cells, was $6.8 \pm 0.3 \cdot 10^{-6}$ cm s^{-1} bar^{-1}; this includes data on cells placed symmetrically and asymmetrically in the transcellular osmosis chambers, and using polyethylene glycol solutions of various osmolalities. This value is some 30 % lower than that of DAINTY and GINZBURG (1964) ($1.01 \pm 0.08 \cdot 10^{-5}$ cm s^{-1} atm^{-1} for ***C. australis* (= *C. corallina*)**, osmoticum 0.1 ml sucrose, 25°C). This discrepancy may not be too important, considering that we have included, in computing the mean, measurements taken under a broader set of conditions. DAINTY and HOPE (1959) had previously reported a value of $0.95 \cdot 10^{-5}$ cm s^{-1} atm^{-1} for the same species.

For a number of cells (L_p/k) was measured from the initial rates of swelling or shrinking, determined from a graph such as Fig. 1A. It will be seen from Eq. (3) that the initial rate of change in the fractional length is given by:

$$\left(\frac{1}{l_o}\frac{dl}{dt}\right)_{t=o} = \frac{1}{k}\left(\frac{1}{V_o}\frac{dV}{dt}\right)_{t=o} = \frac{A_o L_p}{V_o k} \cdot \Delta\pi_e \qquad (6)$$

For a series of room temperature measurements, from 12 sets of determinations on 4 different cells, (L_p/k) was $2.2 \pm 0.2 \cdot 10^{-6}$ cm s^{-1} bar^{-1}, yielding $k = 3.1 \pm 0.4$ (S.E.'s).

Fig. 1B is a semi-log plot of the same data as those contained in Fig. 1A, and it is used to find the half-time of cell volume change. It will be noticed from Fig. 1B that there is an initial lag of 1 or 2 s, arising from the finite time need to change solutions and the existence of unstirred layers. After that the semi-log plot is reasonably linear for 50–60 % of the volume change, the subsequent departure from linearity being almost certainly due to the fact that ϵ is pressure-dependent to some extent. The half-times were determined over the linear portions of such semi-log plots as Fig. 1B. For each of the 12 sets of determinations of (L_p/k) for the 4 cells mentioned above, $T_{1/2}$ was measured. Using the appropriate value of L_p in Eq. (4), $(\epsilon + \pi_{io})$ was calculated. The values of $(\epsilon + \pi_{io})$ ranged from 630 to 990 bar, with the mean value being 770 ± 30 bar. Since $\pi_{io} \approx 7$ bar, the mean value of ϵ estimated by these kinetic experiments, was 763 ± 30 bar. All these measurements were made over a range of turgor pressures within 2 bar of the maximum turgor; over this range ϵ should be reasonably independent of turgor pressure.

The other way of estimating ϵ is from the 'static' experiments, using Eq. (2), which can be more pertinently written as

$$dP_i = \epsilon\ (dV/V) = \epsilon k\ (dl/l) = d\pi_e \qquad (2a)$$

The value of k (ratio of fractional volume change to fractional length change) is obtained from L_p and the 'kinetic' experiments. For the same 4 cells as used above, the value of ϵ was 620 bar. The error in this value is essentially that of k (= 3.1 ± 0.4) and is therefore about ± 80 bar. In a larger sample of cells, the mean value of ϵ was 590 ± 80 bar. Thus the discrepancy between the two methods of measuring ϵ is not too large; we can say that ϵ is of the order of 700 bar. This agrees with the earlier value of KELLY et al. (1963), when their value is corrected for the fact that they assumed an incorrect value for k (7 instead of 3), and it also agrees with the recent, directly determined, values of ZIMMERMANN and STEUDLE (1974).

The study of the temperature coefficient of ϵ is very preliminary and all we wish to say at this stage is that the temperature variation of the elastic properties of the cell wall of ***C. corallina*** is surprisingly small and is being further investigated.

The parameter k is a complex function of the cell wall elastic moduli which has not been correctly discussed previously. The theory of the dependence of k on the elastic moduli will be fully given elsewhere. Here we give the final formula for k:

$$k = (dV/V)/(dl/l) = [1-2\sigma_1 + 2(2-\sigma_2)/\beta]/(1-2\sigma_1) \qquad (7)$$

In this formula β is the ratio of the Young's moduli, (γ_r/γ_l), for circumferential strain (γ_r) and longitudinal strain (γ_l); σ_1 and σ_2 are Poisson's ratios, σ_1 referring to the fractional change in length produced by a circumferential stress and σ_2 to the fractional change in circumference produced by a longitudinal stress. It can be plausibly argued that $\sigma_2 < \sigma_1$ and inspection of Eq. (7) in any case indicates that σ_2 has less influence on k than has σ_1. Thus we can tentatively simplify Eq. (7) to:

$$k = 1 + 4/\beta\,(1-2\sigma_1) \qquad (8)$$

Since k = 3.1, the value of $\beta\,(1-2\sigma_1) \approx 2$; thus $\beta \geqslant 2$. If the value of $\beta = 3.3$, as referred to in BARRY (1970), then $\sigma_1 \approx 0.2$.

D. Acknowledgements

We are very grateful to Drs. U. Zimmermann and E. Steudle of the KFA, Jülich, for stimulating discussions.

E. References

BARRY, P.H., J. Membrane Biol. **3**, 335–371 (1970)
DAINTY, J., Advan. Bot. Res. **1**, 279–326 (1963)
DAINTY, J., in The Physiology of Plant Growth and Development (ed. M.B. Wilkins), pp. 421–452, McGraw-Hill, London 1969
DAINTY, J., GINZBURG, B.Z., Biochim. Biophys. Acta **79**, 102–111 (1964)
DAINTY, J., HOPE, A.B., Aust. J. Biol. Sci. **12**, 136–145 (1959)
KAMIYA, N., TAZAWA, M., Protoplasma **46**, 394–422 (1956)
KELLY, R.B., KOHN, P.G., DAINTY, J., Trans. Bot. Soc. (Edin.) **39**, 373–391 (1963)
SLATYER, R.O., Plant-Water Relationships, Academic Press, London and New York 1967
STEUDLE, E., ZIMMERMANN, U., Biochim. Biophys. Acta **332**, 399–412 (1974)
ZIMMERMANN, U., RAEDE, H., STEUDLE, E., Naturwissenschaften **56**, 634 (1969)
ZIMMERMANN, U., STEUDLE, E., Aust. J. Plant Physiol., in press (1974)

Hydraulic Conductivity and Volumetric Elastic Modulus in Giant Algal Cells: Pressure- and Volume-Dependence

U. Zimmermann and E. Steudle

A. Introduction

An adequate treatment of water transport in algal cells and higher plants can be achieved by applying the phenomenological equations of non-equilibrium thermodynamics, since this theory allows for the coupling between water flow and solute flows (KATCHALSKI and CURRAN, 1965). Analysis of time-dependent water transport shows that the rate constant of water exchange between the cell and its surrounding medium is determined by the geometric parameters of the cell, by the hydraulic conductivity, L_p, and the volumetric elastic modulus ϵ of the cell wall (cf. DAINTY, 1963). With a few exceptions L_p and ϵ have not yet been accessible to direct measurement, although knowledge of their values would be of great importance in studying osmoregulation (STEUDLE and ZIMMERMANN, 1974a), water relations of higher plants and cell growth. We lack such data partly because of the difficulties in measuring the driving forces directly, particularly the hydrostatic pressure inside the cell. As described previously (ZIMMERMANN et al., 1969), we have developed a pressure probe for directly measuring the pressure inside giant algal cells. This method can also be applied to higher plants. Using the pressure probe, L_p and ϵ can be determined from the cell volume changes and the water flows induced both by osmotic and hydrostatic pressure gradients. The dependence of L_p and ϵ on the driving forces can also be measured. Such measurements are reported in this paper on ***Nitella flexilis*** and ***Valonia utricularis***, and on isolated ***Nitella*** cell walls. The measurements were performed with younger and older cells to investigate changes of L_p and ϵ due to growth.

B. Materials and Methods

Nitella flexilis was collected from ponds near Jülich and cultivated in artificial pond water containing (in mM): 1 NaCl, 0.1 KCl, 0.1 $CaCl_2$ and 0.1 $MgCl_2$. The cultivation of *Valonia utricularis* originally obtained from Naples (Italy) is described elsewhere (ZIMMERMANN and STEUDLE, 1974a).

A schematic diagramm of the pressure probe is given in Fig. 1. The hydrostatic pressure inside the cells is measured by a silicon miniature pressure transducer mounted in a closed plexiglass chamber. The pressure transducer (CQS 125–200 Kulite Semiconductor Products, Ridgefields, N.Y.) transforms the applied pressure into a proportional voltage which is measured with a Keithley 602 voltmeter and recorded. A microcapillary is introduced into the plexiglass chamber through a rubber seal. Both microcapillary and plexiglass chamber are filled with silicone oil of low viscosity, which can be taken as nearly incompressible. After inserting the microcapillary into the cell under the microscope, the oil transmits the pressure in the cell to the pressure transducer. As the reference point for measuring the pressure the boundary layer oil/cell sap is chosen at the tip of the capillary. Displacements of this boundary layer during the pressure-measurement due to leakages in the pressure probe can be corrected for by a metal rod, which is inserted into the plexiglass chamber through a rubber seal and which can be moved by turning a micrometer screw. Capillary forces in the tip of the microcapillary, which could produce an apparent additional pressure, can be neglected because of the tip diameter used (ca. 50 μm).

Volume flows and hence changes in pressure inside the cell can be induced by changes in the osmotic pressure of the external solution (e.g. by addition of glucose). These experiments will be called 'osmotic experiments' in the following. With our device, however, volume flows can also be generated by increasing or decreasing the hydrostatic pressure P in the cell. Moving the metal rod forwards or backwards causes a displacement of the boundary layer oil/cell-sap, the new position of which is then taken as the new reference point. These experiments will be called 'hydrostatic experiments' in the following. From the initial changes of pressure and cell volume, V, which can be calculated from the diameter of the metallic rod and the turn of the micrometer screw, the important parameter

$$\epsilon = V\frac{dP}{dV}, \qquad (1)$$

the volumetric elastic modulus, can be obtained, provided that the change in volume takes place so rapidly, that it is not falsificated by the initial volume flow. This requirement is fulfilled with the algae under investigation.

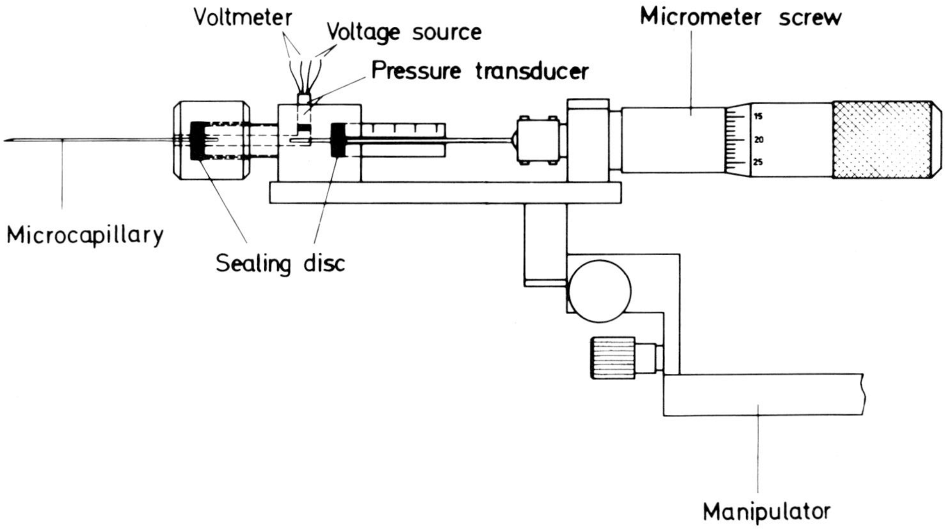

Fig. 1 Probe for direct turgor pressure and volume flow measurements (for explanation see text).

From the exponential time course of the pressure change caused by osmotically and hydrostatically induced volume flows the hydraulic conductivity L_p can be calculated using the phenomenological equations of non-equilibrium thermodynamics. The calculation yields the following two equations for L_p under osmotic (Eq. (2)) and hydrostatic (Eq. (3)) conditions, respectively (ZIMMERMANN and STEUDLE, 1974a; STEUDLE and ZIMMERMANN, 1971):

$$L_p = \frac{k \cdot V\,(P_o - P_E)}{\epsilon \cdot A \cdot \sigma_s \cdot \pi_s^o} \qquad (2)$$

and

$$L_p = \frac{k \cdot V\,(P_A - P_E)}{\epsilon \cdot A(P_A - P_o)} \qquad (3)$$

Before the start of an experiment the cell, of surface area A, is in water flux equilibrium with an equilibrium pressure P_o. In osmotic experiments change of the external osmotic pressure, π_s^o (osmoticum s with reflection coefficient σ_s) causes a water flow and an exponential change in pressure with rate constant k and a final equilibrium pressure P_E. In hydrostatic experiments water flow is induced by changing the original pressure from P_o to P_A. Measurements of the reflection coefficients using the pressure probe have been previously described (ZIMMERMANN and STEUDLE, 1970).

In order to determine L_p in isolated *Nitella* cell walls, the pressure measuring device was inserted into one end of the cylindrical cell wall tube, while the other end was connected to a 12 step Braun-Melsungen pump producing pumping rates between 0.125 and 8.33 μl/min. A stationary pressure is obtained at different pumping rates, whenever the volume flow produced by the pump equals the volume flow across the cell wall. The hydraulic conductivity is then easily calculated from the pumping rate, the pressure inside the cell wall measured with the pressure probe and the surface area of the cell wall. The volumetric elastic modulus of the isolated cell wall tubes was determined by a modification of the method described by KAMIYA et al. (1963).

C. Results and Discussion

I. Pressure- and Volume-Dependence of the Volumetric Elastic Modulus

In Fig. 2 the pressure-dependence of ϵ is plotted for *N. flexilis* and *V. utricularis*. For large volumes, the hyperbolic shape of the pressure-dependence of ϵ is typical, as shown in Fig. 2. For smaller cells, varying for each cell and species, ϵ is nearly independent of pressure over considerable pressure ranges. Simultaneously the absolute ϵ-value decreases with decreasing volume. This result, which was also verified with cells of *Chara* sp. (ZIMMERMANN and STEUDLE, 1974b) indicates a volume-dependence of ϵ, a surprising finding, as ϵ like Young's modulus, should be an intrinsic cell wall property. This volume-dependence is also confirmed, when we compare the ϵ-values of different volumes in the pressure range where ϵ is pressure-independent, i.e. at higher pressures in the saturation range. Thus the influence of the parameter 'pressure' is eliminated. Fig. 3 shows $\epsilon = f(V)$ for *Valonia*, a similar result is obtained for *Nitella*, although the slope of the curve is much steeper in the latter case. In order to determine exactly the relation between volume V and ϵ, we measured ϵ with mercury-filled *Nitella* cell wall tubes, reducing the tube length of each one stepwise. These measurements were carried out in the range of 4 to 8 bar where ϵ is independent of pressure. The results plotted in Fig. 3 show that ϵ is strongly dependent on the volume (or the length, for the diameter remains nearly constant to a first approximation). The upper limit equals the value for ϵ obtained when the cell was still alive. Artefacts can be ruled out, especially since L_p is not volume-dependent. One possible explanation for this volume-dependence of ϵ might be found in the increase of cell wall thickness, as observed in older cells, i.e. at larger volumes. The explanation might be true for *Valonia* but not for *Nitella*, since we proved by measurements on isolated cell wall tubes that changes in the thickness of the wall are not responsible for the volume-dependence of ϵ. We assume (ZIMMERMANN and STEUDLE, 1974b) that ϵ as determined with our pressure-measuring device is a complex quantity, made up of two ϵ-values, representing two regions of the cell wall with different elastic properties. It seems reasonable to assign one modulus, ϵ_1, to the cylindrical part of the cell with volume V_1, and the other one, ϵ_2, to the small node or end regions with volume V_2. We assume that both moduli are intrinsic properties of the cell wall and hence volume-independent. It can easily be shown, that for cell wall tubes, where one has to take into consideration only a single node region, the following equations must hold.

$$V(P) = V_1(P) + V_2(P) \tag{4}$$

Introducing Eq. (1) in Eq. (4) yields, provided $V \approx V_1$:

$$\epsilon = \frac{\epsilon_1 \cdot \epsilon_2 \cdot V}{V_1 \cdot \epsilon_2 + V_2 \cdot \epsilon_1} \approx \frac{\epsilon_1 \cdot \epsilon_2 \cdot V}{V \cdot \epsilon_2 + V_2 \cdot \epsilon_1} \qquad (5)$$

or

$$\frac{1}{\epsilon} = \frac{V_1}{V} \cdot \frac{1}{\epsilon_1} + \frac{V_2}{\epsilon_2} \cdot \frac{1}{V} \approx \frac{1}{\epsilon_1} + \frac{V_2}{\epsilon_2} \cdot \frac{1}{V} \qquad (6)$$

Eqs. (5) and (6) indicate that, with these assumptions, ϵ is indeed a function of the volume. For small volumes, i.e. for young cells, ϵ is linearly dependent on the volume, if $\frac{V_2}{V_1} \gg \frac{\epsilon_2}{\epsilon_1}$, or if $\epsilon_2 \ll \epsilon_1$, i.e.

$$\epsilon \approx \frac{\epsilon_2}{V_2} \quad V + \epsilon_2 \qquad (7)$$

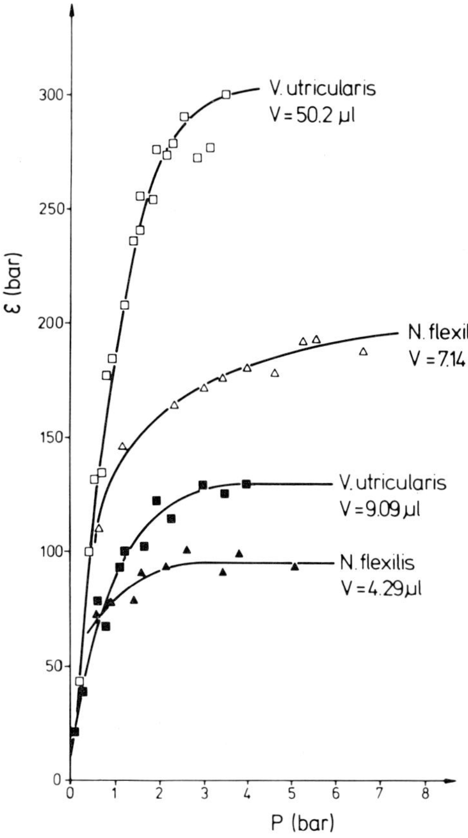

Fig. 2 Typical pressure-dependence of the volumetric elastic modulus, ϵ, of cells of ***Valonia utricularis*** and ***Nitella flexilis*** for different cell volumes. ϵ was obtained using the pressure probe for measuring small changes in cell-turgor pressure, ΔP, and producing changes in cell volume, ΔV, according to Eq. (1). ϵ-values are plotted against the midpoints of the pressure steps, ΔP.

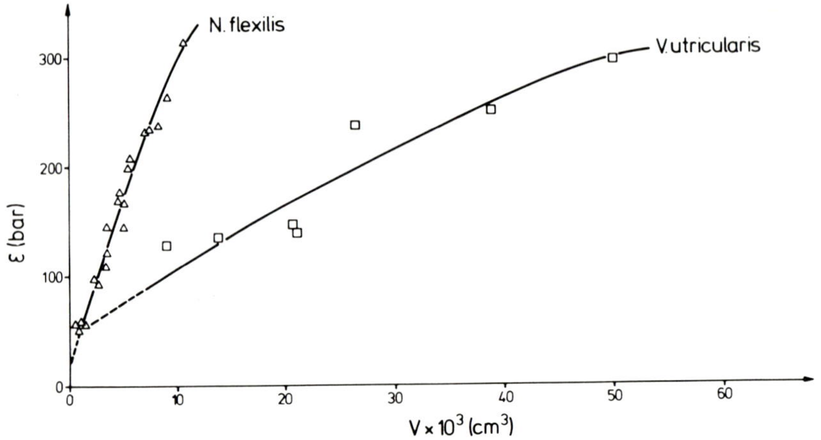

Fig. 3 Volumetric elastic modulus, ϵ, plotted as a function of cell volume, V, for 7 living cells of ***Valonia utricularis*** and for mercury-filled cell wall tubes of ***Nitella flexilis*** (ZIMMERMANN and STEUDLE, 1974b). Determinations of ϵ were performed at high cell turgor pressures (3, 5 and 4–8 bar, respectively), where ϵ is independent of P (Fig. 2).

Eq. (7) holds for small volumes, if $V \approx V_1$ is still valid. This is confirmed by experiment (Fig. 3). If, however $V_1 \cdot \epsilon_2 \gg V_2 \cdot \epsilon_1$, i.e. if $\Delta V_1 \gg \Delta V_2$ for a given pressure change, ϵ becomes equal to ϵ_1. This condition should be satisfied for large volumes, i.e. for older cells. From the intercept of the double reciprocal plot in Eq. (6) ϵ_1 is calculated to be about 730 bar (ZIMMERMANN and STEUDLE, 1974b). ϵ_2 can be estimated to be about 10 bar from the intercept of the linear relation between ϵ and V, which is valid for small volumes (Eq. (7), Fig. 3). This value is much smaller than ϵ_1. From Eq. (7) V_2 is calculated to be 0.2 μl; the whole volume of the cells under investigation being 5 to 10 μl; this means that V_2 is about 1/25 to 1/50 of the total cell volume.

The value of ϵ_1 is in good agreement with the value of 700 bar, which has been calculated for the cell walls of ***Chara corallina*** and ***Nitella translucens*** by DAINTY et al. (1974). As these authors determined ϵ from the longitudinal extension of the cylindrical part of the cell, their value for ϵ should reflect ϵ_1, while the ϵ-value given by BARRY (1970) for ***N. flexilis*** might partly incorporate ϵ_2. Any discussion of ϵ-values obtained from extension measurements must therefore be cautious. The present theoretical analysis suggests that probably all ϵ-values for the cylindrical part of ***Nitella*** and ***Chara*** cells are about the same and that there are no species differences.

This interpretation of the volume-dependence may also be valid for ***V. utricularis***, since for the determination of ϵ the cells used had some daughter-cells still connected.

The importance of this analysis of the ϵ-values is mainly that it now seems possible to follow the changes in elastic properties of the nodal region during growth processes.

II. The Pressure-Dependence of L_p

In Fig. 4 the pressure-dependence of L_p is plotted for ***V. utricularis*** and ***N. flexilis***. L_p was calculated from volume flows, induced by increasing or decreasing the internal hydrostatic pressure. The turgor pressures of these cells were about 1 and 7 bar and the hydraulic conductivities $6 \cdot 10^{-7}$ and $2.6 \cdot 10^{-5}$ cm s^{-1} bar^{-1}, respectively.

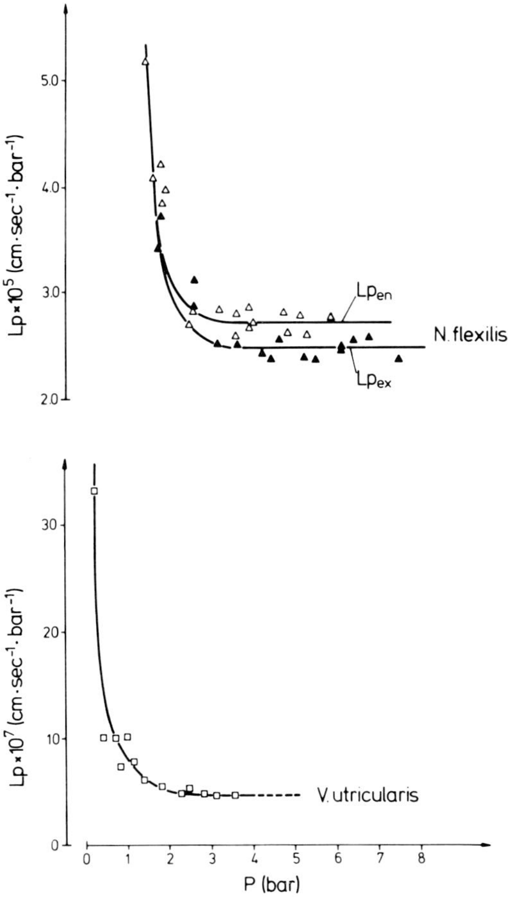

Fig. 4 Pressure-dependence of the hydraulic conductivity, L_p, of ***Valonia utricularis*** and ***Nitella flexilis***. The L_p values were obtained from hydrostatic experiments (Eq. (3)); endosmotic and exosmotic water flows show a polarity of water movement for ***Nitella***. The hydraulic conductivities were determined using small pressure steps, over which ϵ was constant, and are plotted against the midpoints of these pressure steps.

For both species L_p is not dependent on the pressure over a wide pressure range. At smaller pressures (for ***Valonia*** below 0.5 bar, for ***Nitella*** below 2 bar), the hydraulic conductivity L_p increases drastically. The polarity in the water-flow, which can at high pressures be observed with ***Nitella***, (ratio of endosmosis to exosmosis = 1.1), but not with ***Valonia***, disappears on approaching the plasmolytic point. This characteristic dependence of L_p on the hydrostatic pressure within the cell can also be observed with ***Chara intermedia*** and ***Chara fragilis*** (ZIMMERMANN and STEUDLE, 1974b). A volume-dependence of L_p is not discernible. A change in the value of L_p could also not be detected when performing the same experiments in a pressure chamber where the external pressure was varied from 0 to 20 bar indicating that the absolute pressure has no influence on L_p, e.g. by changing the water structure of the membranes. This was also confirmed by keeping the cells in the pressure chamber under an argon atmosphere, for argon is one of the best water structure formers. In spite of increasing the argon pressure to 20 bar, no change in L_p was observed.

The rise in the hydraulic conductivity should in reality be even steeper than indicated in Fig. 4, for the experimentally obtained values for L_p represent the hydraulic conductivity for the whole barrier including the cell wall. For ***N. flexilis*** cell walls the hydraulic conductivity was examined over the total pressure range of 0 to 8 bar, and L_p did not depend on pressure (ZIMMERMANN and STEUDLE, 1974b). Its value of $6.9 \cdot 10^{-5}$ cm s^{-1} bar^{-1} is in good agreement with the value measured by TYREE (1968). The lower value measured by TAZAWA and KAMIYA (1965) was probably due to incomplete removal of fragments of the cell membrane and cytoplasm.

At about 0.5 bar, the L_p-value of the total barrier approaches the value of the hydraulic conductivity of the cell wall. Thus, at approximately this pressure, the cell wall becomes the rate-limiting barrier for water flow and therefore the increase in the hydraulic conductivity of the cell membrane is in reality even steeper. Similar considerations are valid also for ***V. utricularis***.

A decrease in L_p is observed in osmotic experiments, particularly for ***Nitella*** internodes (STEUDLE and ZIMMERMANN, 1974b). On increasing the osmotic pressure of the external solution by adding sucrose, the hydraulic conductivity decreases. This observation which agrees with the findings of DAINTY and GINZBURG (1964), can be explained by a dehydration theory. A decrease in the chemical potential of water in the vacuole and the outer phase causes a decrease in the water content of the membranes. The dehydration theory can also explain the constancy of the hydraulic conductivity at higher pressures in the hydrostatic experiments, for under our experimental conditions the chemical potential of water remains constant. That theory does not explain, however, the increase in the hydraulic conductivity near the plasmolytic point; this increase is probably directly caused by the pressure or by processes, which are triggered by the decreased pressure.

The decrease in L_p at higher external concentrations of sucrose can also be regarded as evidence that the increase of L_p in the hydrostatic experiment is not due to leakages around the microcapillary tip. Preliminary measurements of L_p in ***Nitella*** in the neighbourhood of the plasmolytic point in the presence of higher concentrations of sucrose demonstrated a reduced increase of L_p.

For the time being we do not have, for higher plants, L_p values obtained with our pressure probe. Nevertheless, we feel sure that based on our results with giant algal cells, one must be very careful in interpretating L_p values, which have been measured by the plasmolysis-method (cf. STADELMANN, 1966). The L_p values of plasmolysed and turgid cells can clearly be quite different judging by the increase in L_p which we have observed as the plasmolytic point is approached. On the other hand, applying osmotica for plasmolysis may compensate for this effect and reduce the hydraulic conductivity. In any case, a method which incorporates a number of unknown influences is not very satisfactory. With our method we hope to overcome these difficulties.

D. Acknowledgements

We are very grateful to Professor J. Dainty (Department of Botany, University of Toronto) for helpful discussions. We also wish to thank Mr. H. Jaeckel and Mr. J. Zillikens for skillful technical assistance. This work was supported by a grant from the Deutsche Forschungsgemeinschaft, Sonderforschungsbereich 160.

E. References

BARRY, P.H., J. Membrane Biol. **3**, 335–371 (1970)

DAINTY, J., Adv. Bot. Res. **1**, 279–326 (1963)

DAINTY, J., GINZBURG, B.Z., Biochim. Biophys. Acta **79**, 102–111 (1964)

DAINTY, J., VINTERS, H., TYREE, M.T., in Membrane Transport in Plants (eds. U. Zimmermann, J. Dainty), Springer-Verlag, New York–Heidelberg–Berlin, 1974

KAMIYA, N., TAZAWA, M., TAKATA, T., Protoplasma **57**, 501–521 (1963)

KATCHALSKY, A., CURRAN, P.F., Nonequilibrium Thermodynamics in Biophysics, Harvard University Press, Cambridge, Mass., 1965

STADELMANN, E., in Methods in Cell Physiology (ed. D.M. Prescott), Vol. 2, pp. 143–216, Academic Press, London and New York 1966

STEUDLE, E., ZIMMERMANN, U., Z. Naturforsch. **26b**, 1302–1311 (1971)

STEUDLE, E., ZIMMERMANN, U., in Membrane Transport in Plants (eds. U. Zimmermann, J. Dainty), Springer-Verlag, Berlin–Heidelberg–New York 1974a

STEUDLE, E., ZIMMERMANN, U., Biochim. Biophys. Acta **332**, 399–412 (1974b)

TAZAWA, M., KAMIYA, N., Ann. Rep. Scient. Works, Fac. Sci., Osaka University **13**, 123–157 (1965)

TYREE, M.T., Can. J. Bot. **46**, 317–327 (1968)

ZIMMERMANN, U., RÄDE, H., STEUDLE, E., Naturwissenschaften **56**, 634 (1969)

ZIMMERMANN, U., STEUDLE, E., Z. Naturforsch. **25b**, 500–504 (1970)

ZIMMERMANN, U., STEUDLE, E., J. Membrane Biol. **16**, 331–352 (1974a)

ZIMMERMANN, U., STEUDLE, E., Aust. J. Plant Physiol., in press (1974b)

Turgor Pressure Regulation in Algal Cells: Pressure-Dependence of Electrical Parameters of the Membrane in Large Pressure Ranges

E. Steudle and U. Zimmermann

A. Introduction

It has been often postulated (KESSELER, 1964; GUTKNECHT, 1968; ZIMMERMANN and STEUDLE, 1974a) that the regulation of the salt content and the osmotic balance in algal cells is governed by cell turgor pressure, i.e. that the hydrostatic pressure directly influences the membrane permeability by changing its structure or by stimulating ion pumps. Another type of osmoregulation, where osmotic balance is achieved by the synthesis of osmotically active substances, has been described for the fresh water alga ***Ochromonas*** (cf. KAUSS, 1973) and the halophilic alga ***Dunaliella*** (BEN-AMOTZ and AVRON, 1973). In either case the molecular mechanisms of the postulated pressure-sensitive receptors, which transduce the regulating signal 'pressure' into a physiological event, are still unknown.

The investigation of turgor pressure regulation needs direct and continuous measurements of the hydrostatic pressure in the plant cell. A pressure probe, which permits such measurements was previously described by us (cf. ZIMMERMANN and STEUDLE, 1974b). Combining this technique with measurements of the electrical parameters of the membrane should yield more information about the several steps probably involved in pressure control of ionic regulation of algal cells like ***Chaetomorpha*** and ***Valonia***, which — as will be shown — can both regulate their ionic content in response to changed external osmolarity.

B. Materials and Methods

Experimental procedures and culture conditions have been already described elsewhere (STEUDLE and ZIMMERMANN, 1971; ZIMMERMANN and STEUDLE, 1974a). Briefly, cells of ***V. utricularis*** from Naples and ***Ch. linum*** from List/Sylt (North Sea) were cultured at external osmolarities of 1300 and 1050 mosM, respectively. Prior to measurement the algae were pre-incubated in isotonic artificial sea water (A.S.W.) for 3 to 4 days under continuous illumination (10 000 lux; HQL-lamps). To induce turgor pressure regulation the osmolarity of the A.S.W. was changed by varying the NaCl concentration. Additionally at different osmolarities the external K^+ concentration was varied between 0.6 and 50 mM maintaining the osmotic pressure of the A.S.W. constant by a corresponding variation of the external NaCl concentration. At different time intervals cell sap was obtained by a syringe ***(V. utricularis)*** or by centrifugation ***(Ch. linum)*** and analysed flame-photometrically and cryoscopically.

The set-up for measuring the pressure-dependence of the membrane PD and membrane resistance in ***Valonia*** employed the pressure probe (ZIMMERMANN and STEUDLE, 1974b) and two 3M KCl-filled glass-microelectrodes (tip diameter about 15 μm), which were introduced into the centre of the vacuole of the spherical cells and tightly connected to Ag/AgCl electrodes, to prevent leakages even at pressures up to 5 bar. The membrane PD was amplified by a Grass P 16 microelectrode amplifier with 10^{11} Ω input impedance, which fed a chart recorder with 100 mV s^{-1} response time. To determine the membrane resistance, current pulses of 20 s duration were applied and the corresponding changes of membrane PD were measured. The hydrostatic pressure inside the cells was increased stepwise by adding small amounts of 12 mM KCl solution to the external medium to maintain constant $[K^+]_o$, and decreased by stepwise addition of NaCl. The membrane resistance was measured at each step, after the establishment of water flux equilibrium.

C. Results and Discussion

I. Regulation of the Ionic Content in *Valonia* and *Chaetomorpha*

Figs. 1 and 2 show the results of turgor pressure regulation experiments performed on ***V. utricularis*** and ***Ch. linum*** at various hypotonic stresses. The internal osmolarity of both algae is essentially regulated via a net loss of KCl within 10 and 20 h, respectively. Sodium is scarcely involved at normal values of $[K^+]_o$ but its influence seems to be important at extremely low $[K^+]_o$ (STEUDLE and ZIMMERMANN, 1971). Cryoscopic measurements prove that changes in internal osmolarity are due to changes in $[K^+]_i$ and $[Na^+]_i$.

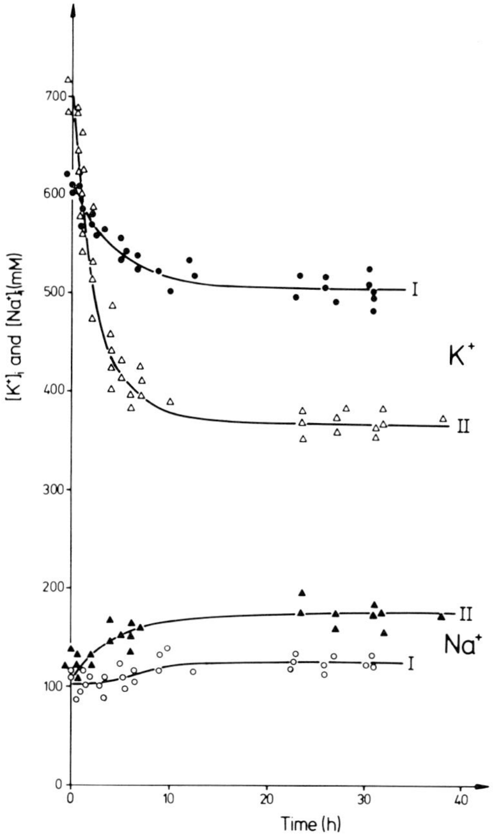

Fig. 1 Time course of internal K^+ and Na^+ concentrations during turgor pressure regulation of ***Valonia utricularis*** in hypotonic media. Curves I and II denote regulations in response to external osmotic pressure decreases of 3.8 (I) and 13.5 (II) bar, respectively, at external K^+ concentrations ranging from 0.6 to 50 mM. For regulations II the algae were pre-incubated in hypertonic sea water (1600 mosM). For clarity not all the data points are given (taken from ZIMMERMANN and STEUDLE, 1974a).

For ***Valonia*** the change of the internal osmolarity exactly corresponds to the change of the osmolarity of the external medium at $[K^+]_o$ above 0.6 mM. The same is true for experiments in isotonic and hypertonic media at $[K^+]_o$ values above 1 and 5 mM, respectively (ZIMMERMANN and STEUDLE, 1974a). Thus ***Valonia*** is capable of carrying out an exact turgor pressure regulation over a large range of values of $[K^+]_o$. However, the osmolarity range for regulation is still restricted compared with other osmoregulating plants. At osmotic stresses larger than ±6 bar (= 250 mosM) the cells died. A larger hypotonic stress could be applied, if the cells were pre-incubated in sea water of higher tonicity (curves II in Fig. 1).

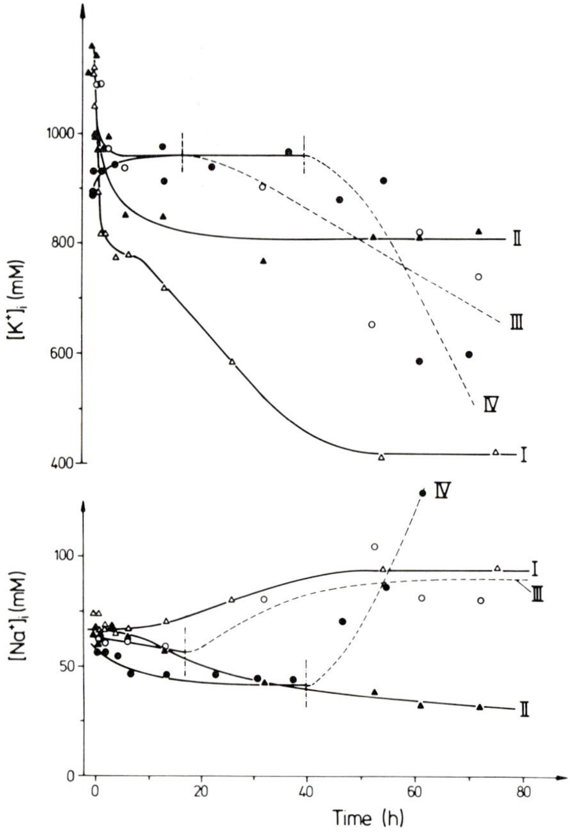

Fig. 2 Time-course of internal K^+ and Na^+ concentrations during turgor pressure regulation of ***Chaetomorpha linum*** at different values of $[K^+]_o$. The osmotic pressure of the sea water (27.4 bar = 1050 mosM; $[K^+]_o$ = 10 mM) was reduced for all experiments by 13.2 bar at t=0 to induce regulation. The $[K^+]_o$ in the different media were I 4.6, II 11.0, III 34.2, and IV 52.6 mM (redrawn from ZIMMERMANN and STEUDLE, 1971).

In contrast to the findings on ***V. utricularis***, turgor pressure regulation in ***Ch. linum*** is strongly dependent on both external osmolarity and $[K^+]_o$. As can be seen in Fig. 2 stepwise increase of $[K^+]_o$ suppresses the normal course of turgor regulation as found in experiments in diluted sea water (curve I in Fig. 2 ; KESSELER, 1964) and even reverses it at concentrations above 50 mM. The resulting high internal pressure at high $[K^+]_o$ is tolerated for only a short time (curves III and IV) and leads to irreversible damage of the cells.

These results for ***Valonia*** and ***Chaetomorpha*** suggest changes in the K^+ transport stimulated by turgor pressure, as in the studies of GUTKNECHT (1968) of pressure-dependent K^+ fluxes in the pressure range between 0 and 1 bar in ***V. ventricosa***. We have, therefore, extended our investigations to measurements of the pressure-dependence of electrical parameters of the membranes of *V. utricularis* over a large pressure range; this seems to be a useful way of getting more insight into the mechanism of pressure-induced ion regulation.

II. Pressure-Dependence of Membrane PD and Resistance in *Valonia*

Fig. 3 shows the pressure-dependence of the membrane PD and membrane resistance for 4 cells of ***Valonia utricularis*** in the pressure range between 0 and 4 bar. The membrane resistance, which is of the order of 100 to 500 $\Omega \cdot cm^2$ at normal cell turgor pressure (1 to 1.5 bar), increases with increasing pressure, reaches a maximum value usually at about 2 bar 2 to 3 times greater than the normal resistance and then decreases again. In some cases the maximum occured at higher pressures (curves III and IV) and the absolute height of the maximum was different. In the pressure range of maximum resistance the potential (normally between 0 and +15 mV) becomes more negative by 5 to 40 mV.

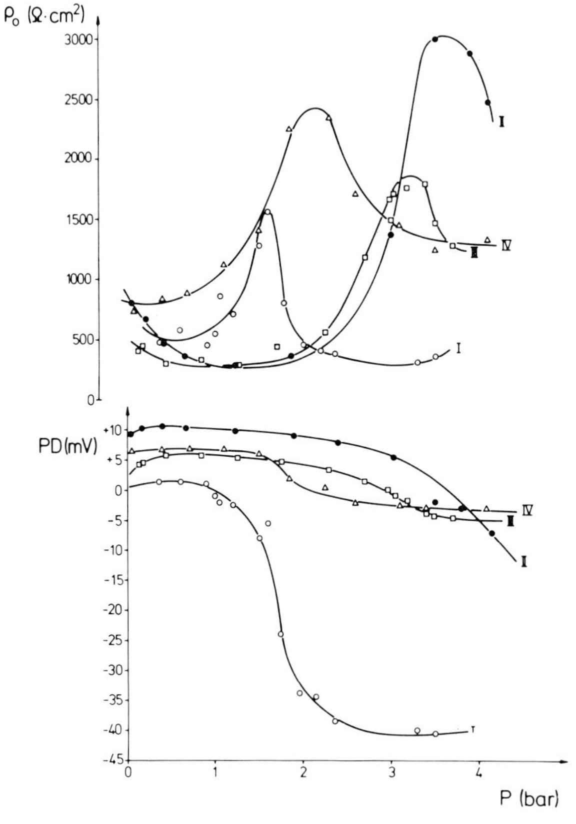

Fig. 3 Pressure-dependence of membrane resistance, ρ_0, and of membrane PD for 4 different cells of ***Valonia utricularis***. Cell turgor pressure was increased stepwise by adding a 12 mM KCl solution to the sea water and lowered by the addition of NaCl.

We feel that these changes in membrane PD and resistance reflect a direct change of ion permeabilities and do not result from the procedures of measurement. As we have shown before (ZIMMERMANN and STEUDLE, 1974a), the changes of potential and resistance are reproducible for a single cell and independent of the rate at which the pressure was changed, of the osmotic agent used and the value of $[K^+]_o$ in the range of 6 to 15 mM K^+.

The effect also seems to be independent of the internal K^+ concentration as shown in Fig. 4. In this figure the time course of cell turgor pressure and of membrane PD and resistance during turgor pressure regulation is given. The pressure was rapidly increased by the addition of distilled water and it then slowly decreases due to the extrusion of K^+ from the cell according to Fig. 1 until the original pressure is re-attained after about 12 h. During the rapid increase and slow decrease of pressure the same pressure-dependence as in Fig. 3 is obtained. A new increase of pressure at the end of the turgor pressure regulation causes the same changes in resistance and potential as at the beginning of the experiment, although the internal K^+ concentration has been changed.

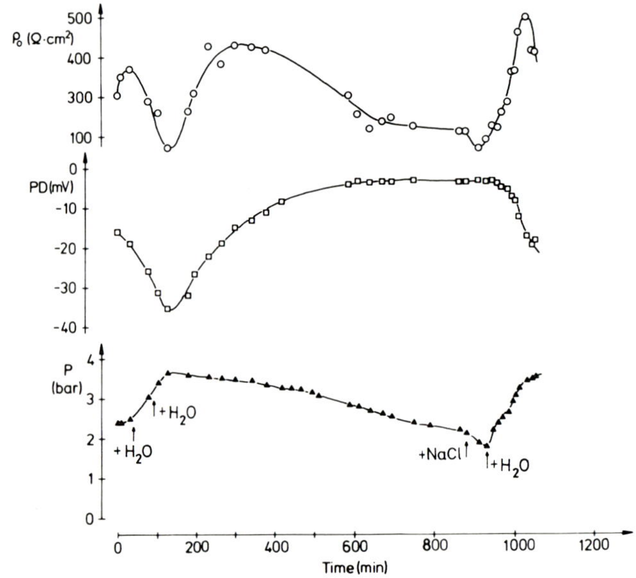

Fig. 4 Changes of turgor pressure, P, of membrane PD and of membrane resistance, ρ_o, during turgor pressure regulation by ***Valonia utricularis***. Cell turgor was increased rapidly by adding water to the medium and it then declined and reached the original level after about 12 h. Further addition of water at the end of regulation caused changes of PD and resistance similar to those at beginning and during the slow decline of pressure.

The finding of a maximum resistance points to two pressure-dependent processes involved in the regulatory mechanism of ***Valonia*** and to large differences in the ion transport at low and high pressures. This latter is confirmed by the light-dependence of the membrane PD at different pressures (Fig. 5). At normal, or lower, cell turgor pressures (upper curve in Fig. 5) the membrane PD drops in the dark and reaches a stable value after about 60 min, about 8 mV more negative than the resting potential in the light. This drop is accompanied by fast and slow hyperpolarizing (more positve) and depolarizing (more negative) transitions. 'Light on' normally causes only one small fast transient depolarization and a rather steady hyper-polarization until the original resting potential is reached after 1 h. At high pressures, beyond those which produce the membrane resistance maximum, reverse changes take place in the membrane PD and more impressive transient potentials occur (lower curve in Fig. 5).

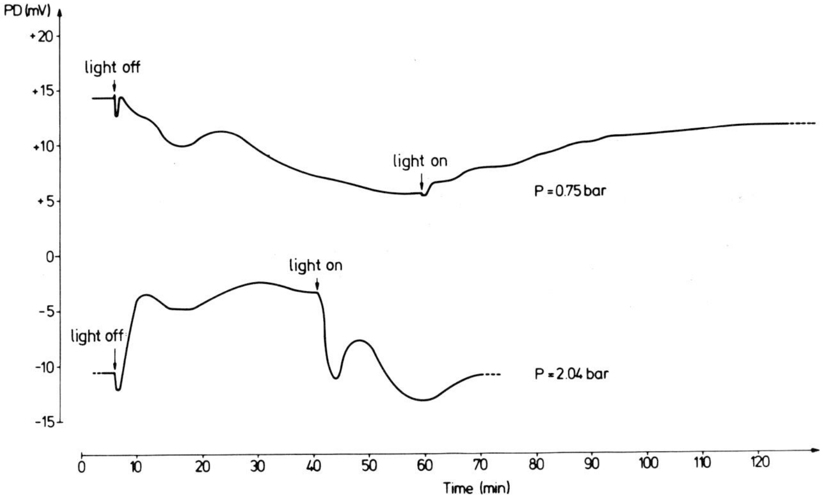

Fig. 5 Membrane PD of ***Valonia utricularis*** as a function of time in light/dark regimes at two different cell turgor pressures, smaller (P = 0.75 bar) and larger (P = 2.04 bar) than the pressure at maximum membrane resistance (see Fig. 3).

We suggest that the increase of membrane resistance at lower pressures is due to a decrease of the K^+ influx (GUTKNECHT, 1968). At higher pressures the decrease of resistance may arise from a K^+ efflux, which increases with increasing pressure. And the reversal of the change in the resting potential under light/dark regimes at different pressures may be due to the effect of pressure on the electrogenic K^+ pump postulated for ***Valonia*** (GUTKNECHT, 1967).

The pressure-dependence of the electrical membrane parameters may arise from structural changes in the cell membranes due to the extension of the cell with pressure. Variations in the critical pressure for the maximum resistance and in the absolute values of resistance and potential as indicated in Fig. 3 for different cells might be then explainable in terms of different cell wall extensibilities. We hope to follow up these ideas combining extension measurements in the different cell wall regions (ZIMMERMANN and STEUDLE, 1974b) with electrical measurements. Such studies are highly relevant to the question of whether turgor pressure regulation is directly influenced by cell wall extensibility as was found for cell growth regulation in ***Nitella*** (GREEN, 1971).

D. Acknowledgements

We should like to thank Professor J. Dainty (Department of Botany, University of Toronto) for reading and discussing the manuscript and Mr. J. Zillikens and Mr. H. Jaeckel for expert technical assistance. This work was supported by a grant from the Deutsche Forschungsgemeinschaft, Sonderforschungsbereich 160.

E. References

BEN-AMOTZ, A., AVRON, M., Plant Physiol. **51**, 875–878 (1973)
GREEN, P.B., ERICKSON, R.O., BUGGY, J., Plant Physiol. **47**, 423–430 (1971)
GUTKNECHT, J., J. Gen. Physiol. **50**, 1821–1834 (1967)
GUTKNECHT, J., Science **160**, 68–71 (1968)
KAUSS, H., Plant Physiol. **52**, 613–615 (1973)
KESSELER, H., Helgoländer Wiss. Meeresunters. **10**, 73–90 (1964)
STEUDLE, E., ZIMMERMANN, U., Z. Naturforsch. **26b**, 1276–1282 (1971)
ZIMMERMANN, U., STEUDLE, E., Marine Biol. **11**, 132–137 (1971)
ZIMMERMANN, U., STEUDLE, E., J. Membrane Biol. **16**, 331–352 (1974a)
ZIMMERMANN, U., STEUDLE, E., in Membrane Transport in Plants (eds. U. Zimmermann, J. Dainty), Springer-Verlag, Berlin–Heidelberg–New York, 1974b

Turgor Pressure Regulation: Modulation of Active Potassium Transport by Hydrostatic Pressure Gradients

D.F. Hastings and J. Gutknecht

A. Introduction

The giant celled marine alga, ***Valonia macrophysa***, maintains a 1.5 atm turgor pressure by holding the osmolarity of its intracellular fluid about 65 mosM above the osmolarity of the external sea water. This difference is maintained approximately constant over a wide range of sea water salinities. The osmotically active species in the vacuolar sap are the inorganic ions K^+, Na^+, and Cl^- whose concentrations are, in mM: 480, 150, and 630, respectively. Our earlier observations on internally perfused cells of both ***V. ventricosa*** and ***V. macrophysa*** showed that inward K^+ transport is stimulated by abolishing the normal turgor pressure of 1.5 atm (GUTKNECHT, 1968, and unpublished observation). The observations suggest that a negative feedback loop is used to regulate the turgor pressure. An increase in the turgor pressure is sensed by the cell and this leads to a reduction of K^+ transport. An important part of the feed back loop is the pressure sensitive transducer, which links the turgor pressure (a physical parameter) to the K^+ pump (a biochemical process). In this investigation we further studied the pressure transducer.

B. Methods

Valonia macrophysa was obtained from the Florida Keys and cultured in 1100 mosM (36.4 ‰) sea water enriched with Alga-Grow (Carolina Biological Supply Co.). Cells weighing 0.2–0.4 g and having a surface area of 1.5–2.5 cm^2 were used in the perfusion experiments. The perfusate was a 1100 mosM artificial cell sap composed of 50 ml of 1100 mosM sea water + 7.00 g NaCl and 34.00 g KCl made up to one litre; the major ion concentrations (mM) were; Na^+, 145; K^+, 454; Cl^-, 602 at pH 5.5. The external solution was 1100 mosM seawater. Illumination was constant and sufficient to maintain an active K^+ influx. The bath temperature was held in the range 23–25°C.

The perfusion system (Fig. 1) was a modified version of that reported by GUTKNECHT (1968). The perfusate was driven through the cell by gravity at a rate of 0.2 ml/min and collected at 10 min intervals through the outflow port for counting in a liquid scintillation counter. The perfusion pipettes were connected via agar salt bridges to potential measuring calomel electrodes and current supplying Ag/AgCl electrodes. The vacuolar potential and d.c. membrane resistance measurements were used to detect the gross damage to the cell which sometimes occurs during perfusion. The closed perfusion reservoirs were connected to a supply of mixed gases through a common pressure regulator. The O_2 + N_2 gas mixtures were chosen to maintain a constant O_2 tension of 0.2 atm.

The outside bath containing $^{42}K^+$ was stirred continuously by blowing a stream of water-saturated gas across the bath surface. The bath and micromanipulators were enclosed in a plastic bag to prevent contamination of the gas mixture by room air. Evaporation losses were replaced at 0.5 h intervals with distilled water.

The pressurized perfusion system was placed in a room-sized hyperbaric chamber at the Hall Laboratory, Duke University Medical Center. The external pressure was adjusted to 1.0 or 2.0 atm as required, while the turgor pressure of the cell was adjusted independently by the pressurized perfusion system to be 0.0 or 1.0 atm above the hyperbaric chamber pressure.

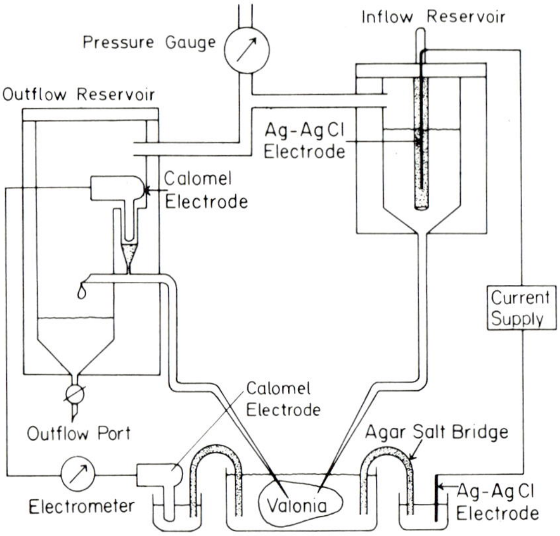

Fig. 1 The pressurized perfusion system consists of two closed reservoirs. The inflow reservoir is situated higher than the outflow reservoir, causing the perfusion solution to flow into the vacuole through the right-hand pipette (200 μm tip diameter) and from the vacuole to the outflow reservoir through the second pipette. The sample is collected at 10 min intervals through the outflow port. The height of the inflow reservoir is used to adjust the flow rate to 0.2–0.3 ml/min, a rate sufficient to exchange the vacuolar volume once a minute. The reservoirs are pressurized via a common gas supply without affecting the perfusion rate. The electrodes make electrical contact with the vacuole and the artificial cell sap in the perfusion tubing via salt bridges.

Active K^+ transport was estimated by measuring the unidirectional influx of $^{42}K^+$. The inward movement of K^+ is against a net 454 mM to 10.6 mM concentration gradient and a net 10 mV potential gradient. The effective electrochemical potential gradient is 106 mV. Using the Goldman equation (KATZ, 1966), and efflux data we estimate the passive influx to be less than 2 % of the total K^+ influx. The data presented here has not been corrected for passive K^+ influx.

The experiments presented were conducted after the isotopic flux had reached a stable value. K^+ influx, in pmoles cm^{-2} s^{-1}, was calculated by the formula

$$\phi K = \frac{\text{Sample activity} \cdot 10^{12}}{\text{Bath specific activity} \cdot \text{area} \cdot \text{sampling time}}$$

From the activity of a sample in counts per min (cpm) we subtracted a background and a correction for the naturally occurring radioactive isotope $^{40}K^+$ in the sample. Counting efficiency was good and constant. The $^{42}K^+$ activity of the sample was corrected for isotopic decay to the calibration time of the bath specific activity. To measure bath specific activity an aliquot of the $^{42}K^+$ bath solution was diluted 1000 fold and the radioactivity was measured in the same manner as the samples. The area of the cell was measured prior to the experiment. In the above formula 'sampling time' represents the duration of sample collection, i.e. 600 s.

C. Results

In a series of four experiments we applied four temporal sequences of the two stimulae, i.e. absolute pressure and the pressure difference (turgor pressure). An increase in K^+ influx occurred in presponse to a reduction in the turgor pressure and not to a reduction of the absolute pressure. Conversely, the high K^+ influx of a zero turgor cell decreased in response to an increase in turgor pressure and not to an increase in absolute pressure.

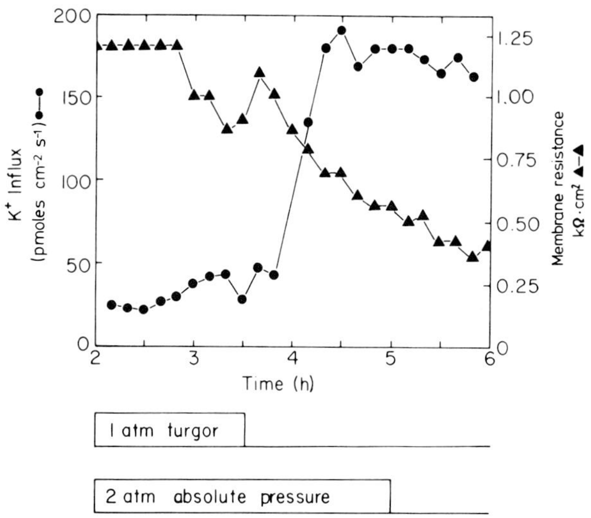

Fig. 2 The removal of 1.0 atm turgor pressure increases the K^+ influx (solid dots) 4 fold. The removal of 1 atm turgor pressure tends to decrease the d.c. membrane resistance (solid triangles) measured across the tonoplast, cytoplasm, and plasmalemma.

Fig. 2 shows the K^+ influx for a cell exposed to both stimulae prior to beginning the experiment. 5 min after implanting the pipettes the turgor pressure was raised to 1 atm. The vacuolar potential was allowed to stabilize at + 11 mV prior to raising the pressure of the vacuole and the outside chamber each by 1 atm. The oxygen tension in the bath and the perfusion reservoir was held constant by adjusting the composition of the appropriate O_2 + N_2 gas mixture.

After allowing an additional half hour to stabilize the system, K^+ influx measurements were begun. The K^+ influx during the first 2 h control period was low and stable. Soon after removing the 1 atm turgor pressure the K^+ influx increased 4 fold. The subsequent reduction of the absolute pressure caused no additional increase in the K^+ influx.

In a second experiment the absolute pressure was reduced before the turgor pressure was reduced. In all other respects the experimental conditions were identical to the preceding experiment. The reduction of the absolute pressure caused no increase in the K^+ influx.

We have shown that a reduction of cell turgor can stimulate K^+ influx. In the following experiments we will show the converse statement to be true, i.e. the K^+ influx is reduced by an increase of cell turgor from 0 to 1.0 atm.

A cell was punctured and held at zero turgor and 1 atm absolute pressure for 2 h prior to adding the $^{42}K^+$. The isotope influx stabilized within 1.5 h, at which time the absolute pressure was raised from 1 to 2 atm. During the subsequent 1.5 h the K^+ influx did not decrease (Fig. 3). At 5.2 h the turgor pressure was increased from 0 to 1.0 atm. In response to the increased turgor pressure, the K^+ influx dropped gradually over the final hour to half the maximum rate. We repeated this experiment and obtained the same result; i.e. K^+ influx decreases in response to an increase in turgor pressure from 0 to 1.0 atm.

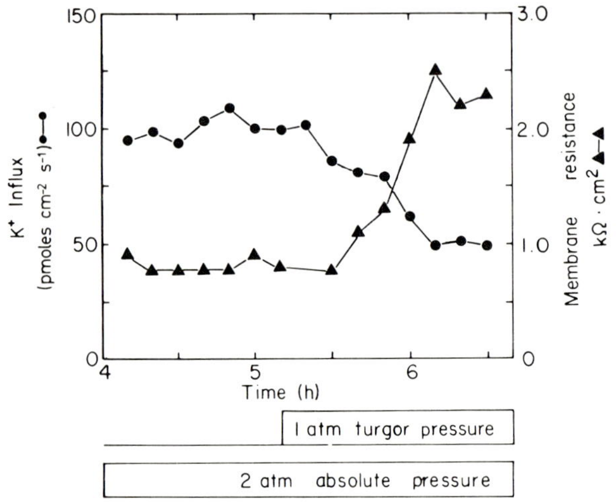

Fig. 3 The high K^+ flux of a zero turgor cell does not decrease in response to 2.0 atm absolute pressure. The application of an additional atm of turgor pressure causes the K^+ influx (solid dots) to decrease by 50 % and the membrane resistance (solid triangles) to increase 2.5 fold.

D. Discussion

In principle, numerous physical forces could be used by the cell to measure turgor. We have controlled four physical forces related to pressure: the osmotic pressure difference, oxygen partial pressure, absolute pressure, and hydrostatic pressure difference (turgor pressure). The pressurized perfusion technique provides the advantage that osmotic pressure and turgor pressure can be controlled independently. In these experiments the osmotic pressure difference was held constant at zero atm. We had previously found that changes in the osmotic pressure have little effect on the K^+ uptake by perfused cells (GUTKNECHT, 1968).

Oxygen tension may influence the production of usable energy by chloroplasts and mitochondria. In illuminated cells, the oxygen tension in the chloroplast is a balance between production rate and metabolic and diffusional losses. We feel that the magnitude of the turgor pressure will have little effect on the diffusion path length and the diffusion coefficient of oxygen across the cell wall (2–10 μm) and the external unstirred layer. To avoid inadvertently changing the oxygen diffusion gradients, we held the oxygen tension in both the seawater and the perfusate at 0.2 atm.

Our results indicate that a turgor of 1.0 atm pressure is sufficient to reduce K^+ influx by at least 50 %. The fact that a pressure difference is required indicates that the pressure transducing mechanism must reside along a pressure gradient. The vacuole is essentially structureless and is incapable of sustaining a pressure gradient. Similarly the cytoplasm has a semifluid structure that is unable to maintain a significant pressure gradient. Thus, by elimination, we conclude that the entire pressure gradient must occur across the plasmalemma and cell wall.

If one could separate the functions of the plasmalemma and the cell wall, then the plasmalemma may be viewed as a semipermeable membrane across which osmotic pressure differences and hydrostatic pressure differences may come to equilibrium. The cell wall may be viewed as a structural support. However, this simple perspective overlooks some possibly significant interactions between the plasmalemma and the cell wall. Some that may prove important in ***Valonia*** include, first, intercalation of the plasmalemma between the large cellulose microfibrils and second, interaction of the negatively charged cell wall and charged membrane components (ROELOFSEN, 1958). A third interaction may be the influence of compression and/or tension forces applied to the 100 Å plasmalemma upon membrane structure and function. At this point we can only state with certainty that one component of the plasmalemma-cell wall complex, i.e. the pressure transducing mechanism, is modified by the pressure gradient.

E. Acknowledgements

We wish to thank W.L. Greeman for his expert assistance in modifying the experimental equipment and operating the hyperbaric chamber. We wish to thank H.A. Saltzman, M.D., Director of the Hall Laboratory, for the use of the hyperbaric chamber. This work was supported by the U.S. Public Health Service Grant HE 12157.

F. References

GUTKNECHT, J., Science **160**, 68–70 (1968)
KATZ, B., Nerve, Muscle, and Synapse, pp. 60–61, McGraw-Hill 1966
ROELOFSEN, P.A., The Plant Cell Wall, G. Borntraeger, Berlin 1959

The Location and Measurement of the Resistances to Bulk Water Movement in the Shoots of Hemlock *(Tsuga canadensis)*

M.T. Tyree, C. Caldwell, and J. Dainty

A. Introduction

Using the Scholander-Hammel pressure-bomb it is possible to measure the kinetics of water exchange between the shoot and a suitable collecting vial in response to a rapid water potential change (= pressure increment), and by measuring the kinetics under different experimental conditions, it is possible to build up a quantitative picture of the magnitude and location of the major resistances to bulk water flow within woody shoots. This paper reports our continuing efforts to measure and locate the resistances to bulk water flow in the shoots of eastern hemlock *(Tsuga canadensis)*.

B. Materials and Methods

The details for measuring the kinetics of water exchange in hemlock have been reported elsewhere (TYREE and DAINTY, 1973b). A small hemlock shoot is enclosed in a pressure-bomb with the cut end protruding through an air tight seal to the outside. The cut end is connected via a water column in a plastic capillary to a small beaker of water on an electronic balance (Fig. 1). Small gas pressure changes from the balance pressure will change the shoot water potential causing measurable amounts of water to exchange. Other procedural details appear below.

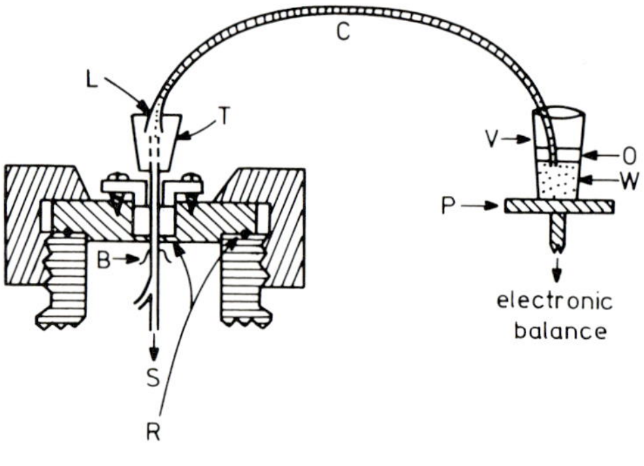

Fig. 1 A diagram showing the experimental setup. On the left, the top part of the pressure bomb is shown in cross section. The symbols have the following meaning: B, bark of stem; C, capillary tube (micro-catheter); L, slip joint end of microcatheter; O, oil layer on top of water; P, pan of electronic balance; R, rubber gasket; S, hemlock stem; T, translucent silicone-rubber stopper; V, collection vial; W, water in vial and capillary shown by dots.

C. Results and Discussion

I. The Preliminary Characterization of the System

In the summer of 1973 we were able to reproduce the kinetic experiments obtained previously (TYREE et al., 1973a, b). A typical recent experiment is reproduced in Fig. 2. Sample C12 had an initial fresh weight of 35 g; it was equilibrated at 21°C at 0.36 bar pressure. At time zero the bomb pressure was rapidly increased by 0.434 bar and volume flow rate, dV_e/dt, was measured. Fig. 2 is a semilogarithmic plot of dV_e/dt versus time (open circles). The relation is curvilinear until about 4000 s whereupon dV_e/dt appears to decrease exponentially with a large half time ($7 \cdot 10^3$ s). This slow exponential decline of dV_e/dt can be subtracted from the total for times less than 4000 s, and when the difference is replotted another curve is obtained (pluses). Again the relation is curvilinear until about 500 s whereupon the residual dV_e/dt declines exponentially with a half time equal to $1.2 \cdot 10^3$ s. Extrapolating this fast exponential back to time zero and subtracting from the pluses for times less than 500 s, yields a very fast exponential with a half time equal to 105 s.

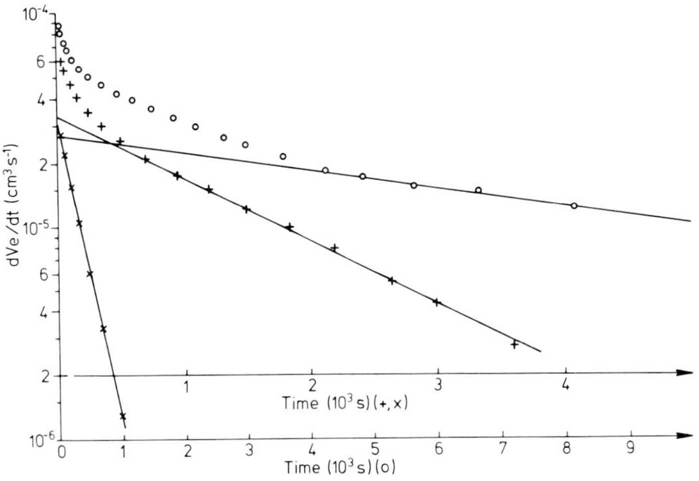

Fig. 2 A plot of the logarithm of the efflux (dV_e/dt) of a hemlock shoot versus time (on the linear scale). The open circles correspond to the lower time scale and the pluses and x's correspond to the upper time scale.

Thus hemlock shoots appear to approach equilibrium as the sum of three exponential processes with distinct half times. TYREE et al. (1973a, b) demonstrated that this behaviour is consistent with a model system containing three distinct cell types that are rate-controlled in their approach to equilibrium by a population cell constant, K (bar cm^{-3}), and a population membrane hydraulic conductivity, AL_P ($cm^3\ s^{-1}\ bar^{-1}$). The volume flow in this model is given by,

$$\frac{dV_e}{dt} = \sum_{n=1}^{3} (AL_P)_n\, \Delta P \exp(-K_n (AL_P)_n\, t), \tag{1}$$

where ΔP (bar) is the pressure increment applied at time zero and t (s) is the time. The K_n is a measure of the combined rate of change of turgor pressure and osmotic pressure for the nth population.

TYREE et al. (1973c) measured the activation energy for the temperature dependence of the tempo of water exchange in hemlock shoots in the temperature range from 2°C to 40°C. The measured value is 25.9 ± 0.6 · 10^3 J $mole^{-1}$. This figure is higher than the activation energy for laminar flow of water in a pipe (17 · 10^3) and is also higher than the activation energy for self diffusion of water (20.7 · 10^3). By comparing the shoot activation energy to the activation energy for water permeation across other plant membranes, TYREE et al. (1973c) concluded that the membrane resistance makes up no less than 1/4 of the total shoot resistance.

II. In Search of the Main Resistance to Water Exchange

1. The Experimental Approach

A sequence of experiments was performed on the same shoot in order to minimize variability between shoots. The following procedures were carried out on several shoots. a) A shoot was allowed to equilibrate at atmospheric pressure (zero bomb pressure) overnight at about 20°C. The following morning the bomb pressure was increased by 0.345 bar and the tempo of water exchange measured. The shoot was returned to zero bars overnight. b) On the second morning the shoot was removed from the bomb and inverted; the leaves were in open air and the cut end of the shoot protruded into the bomb into a weighed beaker of water. When the bomb pressure was increased, infiltration of the air spaces in the spongy mesophyll of the leaves began. Infiltration was allowed to proceed at about 14 bar until guttation was observed on most of the under-surfaces of the leaves. After infiltration the shoot was reinverted and the tempo of water exchange again measured for a pressure increment of 0.345 bar. The difference in the kinetics of water exchange between infiltrated and normal shoots yields information on the normal pathway of water movement in hemlock shoots, provided we have a detailed knowledge of the anatomy of hemlock leaves (GAMBLES, 1973). c) On the third morning the shoot was removed from the pressure-bomb and the leaves carefully plucked from the stem. The defoliated shoot was used to measure the resistance offered by the xylem pathway (less the vascular bundles of the leaves) to water flow. The xylem resistance was measured by submerging the defoliated stem in a column of water except for the cut end which protruded above the water surface; the column of water and shoot were inserted into the bomb and the efflux of water as a function of pressure measured. No resin blockage occurs at the petiole stumps.

2. Results and Discussion of these Experiments

The results of our experiments are summarized in Table 1.

The shoot resistance of uninfiltrated samples was calculated from the volume flow rate at time zero, $J_o = (dV_e/dt)_{t=0}$, after a pressure increment of ΔP; if R_s is the shoot resistance, then $R_s = \Delta P/J_o$.

The most surprising result of this study is that the xylem resistance is a large fraction of the shoot resistance. In Table 1 we report values of R_s, R_{xy} (the xylem resistance), and the ratio of R_{xy}/R_s. A crude value of the leaf resistance can be obtained by subtracting R_{xy} from R_s. The mean value (± S.E.) of R_{xy}/R_s is 0.66 ± 0.07, i.e. two thirds of the shoot resistance resides in the xylem. This result was contrary to our predictions for in 1969 TYREE (unpublished) measured the hydraulic conductivity of hemlock trunk xylem to be 11.4 cm^2 s^{-1} bar^{-1}. It can be shown that if our shoots had this hydraulic conductivity then R_s would be roughly 10 bar s cm^{-3}. The discrepancy was later explained by measurements that indicate that the hydraulic conductivity of minor branches of hemlock is much lower (= 3.4 · 10^{-2} cm^2 s^{-1} bar^{-1}) than the value for trunk xylem. The difference in hydraulic conductivities can be explained in part by anatomical differences between branchlet and trunk xylem.

Table 1

Values of resistances to bulk water movement in ***Tsuga canadensis***. R_s = shoot resistance, R_{xy} = xylem resistance (less the xylem in the leaves), R_e = the external resistance to water movement from the air spaces. The value of R_e is estimated in two ways (see text) using values of dV_e/dt and dV'_e/dt at a time equal to 6000 s.

Sample No.	Temp. (°C)	R_s	R_{xy}	Estimates of R_e I	Estimates of R_e II	R_{xy}/R_s
		(10^{-3} bar s cm^{-3})				
C–21	17.5	5.35	–	1.80	4.09	–
C–22	19.5	5.85	4.11	10.2	15.6	0.702
C–23	19.7	5.32	4.32	11.6	17.5	0.812
C–24	19.6	3.45	2.60	3.45	1.13	0.754
C–25	19.7	4.82	2.91	3.10	4.78	0.603
C–26	19.7	6.00	2.51	2.03	3.20	0.418

Mean ± S.E. of R_{xy}/R_s = 0.66 ± 0.07
Mean ± S.E. of R_e/R_s = 1.0 ± 0.3 for estimate I
= 1.4 ± 0.5 for estimate II

Yet another surprising result was that the resistance for water efflux from infiltrated air spaces appears to be about twice the resistance for water efflux from the living cells.

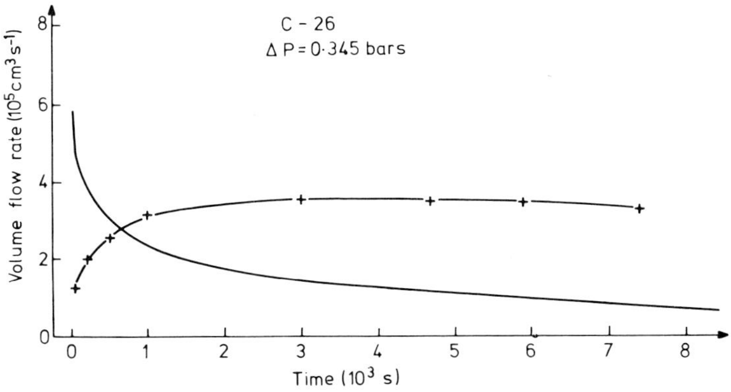

Fig. 3 Plots of dV_e/dt (continuous line) and $dV'_e/dt - dV_e/dt$ (pluses connected by a broken line) versus time. See text for details.

In Fig. 3 we present a plot of a representative experiment on hemlock shoot C26. The solid line gives the water efflux, dV_e/dt, vs. time for a normal (uninfiltrated) shoot after applying a small pressure increment at time zero. Later the air spaces of the leaves were infiltrated with water and the water efflux was again measured after an identical pressure increment. In every experiment the water efflux from the infiltrated shoot, dV'_e/dt, exceeded the water efflux from the normal shoot, dV_e/dt, at all corresponding times. In Fig. 3 the pluses connected by a broken line represent a plot of $dV'_e/dt - dV_e/dt$.

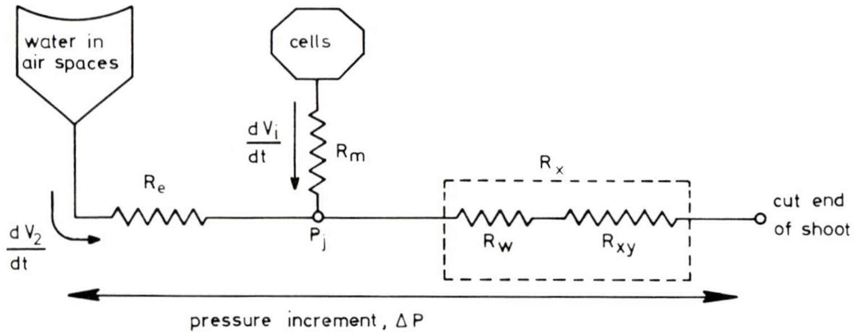

Fig. 4 A simplified resistance diagram for a hemlock shoot. See text for meaning of symbols.

The theoretical analysis of these experiments is very complicated because the overall xylem resistance of the shoot is of the same order as the leaf resistance; the actual xylem resistance experienced by each leaf depends on how far the leaf is from the cut end of the shoot. In Fig. 4 we present a resistance diagram which will aid in a better qualitative understanding of what we believe to be the situation in hemlock shoots. In some experiments the air spaces are filled with water and when the pressure is increased water flows from the air spaces at the rate dV_2/dt across an external resistance, R_e, into an extracellular resistance, R_x, (which can be further divided into the xylem resistance, R_{xy}, and an extracellular leaf resistance due to the cell walls and leaf xylem, R_w). The cells also deliver water at the rate dV_1/dt to the extracellular resistance, R_x, across the membrane resistance, R_m.

The initial pressure increment, ΔP, can be related to the flows and resistances by analogy with an electrical circuit,

$$\Delta P = \frac{dV_2}{dt} R_e + \left(\frac{dV_1}{dt} + \frac{dV_2}{dt}\right) R_x , \quad (2)$$

from which we obtain,

$$R_e = \frac{\Delta P}{dV_2/dt} - \left(1 + \frac{dV_1/dt}{dV_2/dt}\right) R_x > \frac{\Delta P}{dV_2/dt} - \left(1 + \frac{dV_1/dt}{dV_2/dt}\right) R_s , \quad (3)$$

because $R_s > R_x$. R_x cannot be measured directly.

We can use Eq. (3), together with some rather complicated arguments about the magnitudes of the pressure drops in the system, to make some reasonable estimates of the external resistance, R_e. In Table 1 we present two such estimates of R_e. In estimate I, dV_1/dt is taken as zero and dV_2/dt is equated to dV'_e/dt; in estimate II, dV_1/dt is equated to dV_e/dt and dV_2/dt is equated to $dV'_e/dt - dV_e/dt$. The values of R_e are comparable to R_s, i.e. the mean values of R_e/R_s from the two estimates of R_e are about equal to one.

We feel that the non-zero value of R_e indicates a suberized layer at the air/water interfaces in hemlock.

It is important to know the activation energy associated with R_{xy}. This activation energy has been roughly measured to be 17 to 19 $\cdot$ 10^3 J $mole^{-1}$, but this measurement is attended by all the problems inherent in multiple determinations of R_{xy} (see TYREE and ZIMMERMANN (1971) and ZIMMERMANN and BROWN (1971) for a lengthy discussion of the sources of error).

Our tentative conclusion is that the higher temperature dependence of R_s over that of R_{xy} indicates some rate controlling influence of living membranes. This confirms previous predictions that the extracellular resistance in hemlock could be no more than 3 times the membrane resistance (TYREE et al., 1973c).

We always find that the tempo of influx and efflux can be described by the sum of three exponential functions of time. Previously we thought that this fact was best described by a model shoot containing three populations of cells which exchange water with the apoplast independently of each other. Since we now know that the xylem contributes about 2/3 of the total shoot resistance, this simple model must be wrong. Due to the complex geometry of the series resistances in the shoot, little will be gained by trying to create a more complicated mathematical model.

D. References

GAMBLES, R.L., The Leaf Anatomy of ***Tsuga canadensis***. M.Sc. Thesis, Botany Department, University of Toronto, 1973

TYREE, M.T., DAINTY, J., BENIS, M., Can. J. Bot. **51**, 1471–1480 (1973a)

TYREE, M.T., DAINTY, J., Can. J. Bot. **51**, 1481–1489 (1973b)

TYREE, M.T., BENIS, M., DAINTY, J., Can. J. Bot. **51**, 1537–1543 (1973c)

TYREE, M.T., ZIMMERMANN, M.H., J. Exp. Bot. **22**, 1–18 (1971)

ZIMMERMANN, M.H., BROWN, C.L., Trees, Structure and Function. Springer-Verlag, New York–Heidelberg–Berlin, 1971

Osmoregulation in *Ochromonas*

H. Kauss

A. Introduction

There are many examples in which the osmotic balance of cells is maintained constant under changing external osmotic stress, as well as other examples in which the osmotic pressure in cells is changed in order to perform work, e.g. to open and close stomata. In either case osmotic pressures have to be sensed by the cells and converted into information capable of regulating biochemical steps, presumably at the level of enzymes or enzyme-like proteins involved in the metabolic mechanisms. In general the osmotic agents are ions. However, at present the biochemistry of ion transport and especially regulation of ion transport is not well understood. In contrast, osmotic balance in the golden brown flagellate ***Ochromonas malhamensis*** is mediated by elements of carbohydrate metabolism (KAUSS, 1967; KAUSS, 1969; SCHOBERT et al., 1972), the general biochemistry of which has been well investigated. *Ochromonas* offers, therefore, a simple system for studying biophysical and biochemical steps of osmoregulation; and it will be seen that this particular system shows many similarities to other systems involving transport of solutes across membranes.

B. Results and Discussion

Fig. 1 shows the concentrations of some low molecular weight solutes in cells of *Ochromonas* which have been shown to act in cells of other organisms as osmotic agents (e.g. GUTKNECHT, 1968; KANESHIRO et al., 1969; STONER and DUNHAM, 1970). It can be seen (Fig. 1) that 3 h after an increase in external osmotic pressure, the levels of Na^+ and K^+

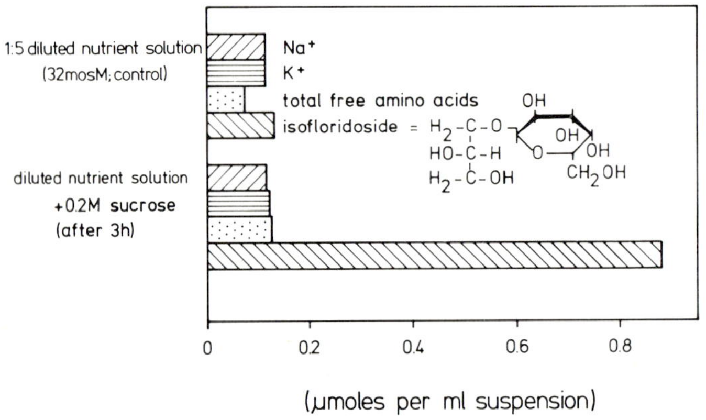

Fig. 1 Concentration in cells of ***Ochromonas*** of some low molecular weight solutes in relation to outside osmotic pressure. For methods see SCHOBERT et al. (1972).

have not changed. Although the sum of the free amino acids is somewhat higher this increase appears small when compared with the very great change shown in a specific low molecular weight carbohydrate. This carbohydrate has been identified as α-galactosyl-(1→1)-glycerol (isofloridoside, KAUSS, 1967). It has been shown that the fluctuations in isofloridoside content are indeed directly related to an osmotic effect.

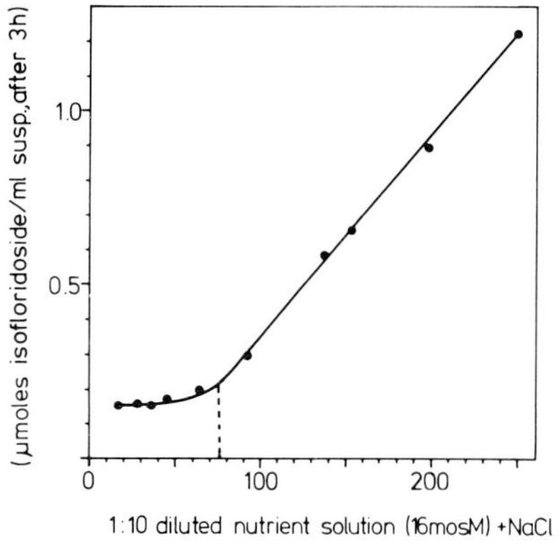

Fig. 2

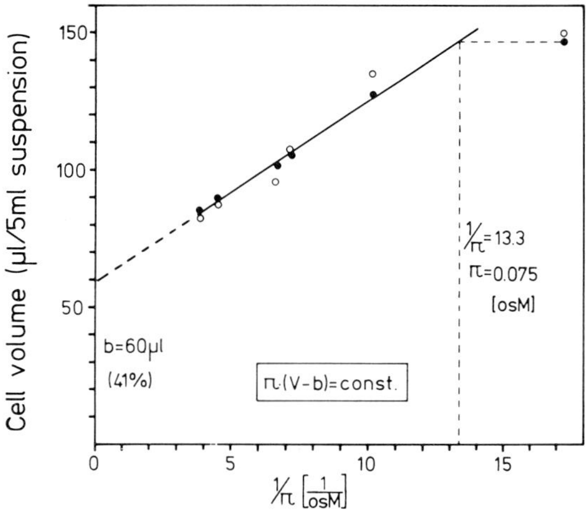

Fig. 3

Fig. 2 Isofloridoside content in cells of ***Ochromonas*** at different osmotic values. NaCl in dilute nutrient solution was used as the osmotic substance and its osmotic pressure determined by means of a micro-osmometer (Knauer & Co, Berlin). Other methods see SCHOBERT et al. (1972).

Fig. 3 Cell volume in ***Ochromonas*** 6 min after increase in osmotic pressure. NaCl in dilute nutrient solution was used as the osmotic substance. Sediment volume was corrected for intercellular space with ^{14}C-Inulin (SCHOBERT et al., 1972) and read either directly in a hematocrit-like calibrated centrifuge tube (○ —— ○) or determined by weight (● —— ●) according to NOBEL (1968) in the above tubes. Other methods see SCHOBERT et al. (1972).

The amount of isofloridoside stored in the cells is proportional to the outside osmotic pressure if it is above 75 mosM (Fig. 2); below 75 mosM osmotic balance is achieved by excretion of water by pulsating vacuoles. It has been observed in previous studies (SCHOBERT et al., 1972) that cell shrinkage is complete about 2–3 min after an increase in external osmotic pressure; the cells then regain their volume as isofloridoside is produced intracellularly. The time necessary to reach a new and higher isofloridoside concentration depends on the osmotic change; with an increase of 100 mosM the time is 1–2 h.

We have recently reinvestigated the extent of shrinkage (Fig. 3) and compared it with the amount of isofloridoside produced. The osmotic behavior of the cells of ***Ochromonas*** can be considered in terms of the Boyle-Van't Hoff relation where π is the osmotic pressure, V the total cell volume and b the non-osmotic volume as NOBEL (1969) has shown for chloroplasts. A plot of $1/\pi$ against cell volume V gives a straight line indicating that the cells follow the above law and behave as osmometers. The results shown in Fig. 3 allow two further conclusions to be made. First, a comparison with the original cell volume shows that shrinkage starts at 75 mosM. This is in agreement with the results shown in Fig. 2 where an increase in the isofloridoside pool also starts at an external osmotic pressure of 75 mosM.

Second, one can read on the ordinate the non-osmotic part b of the cell volume, which in this experiment was 41 % of the total volume. In other experiments it varied between 40 and 50 %.

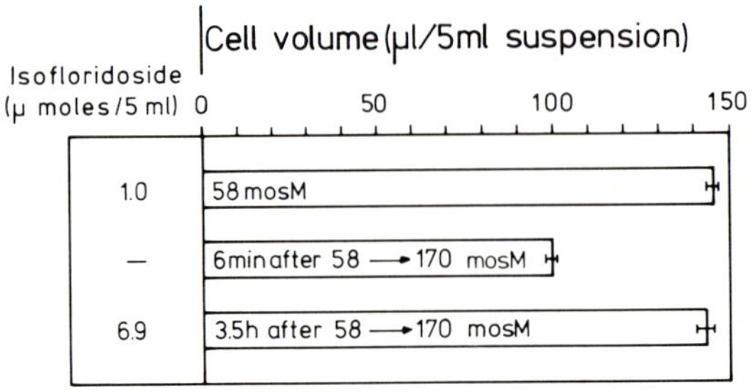

Fig. 4 Shrinkage, swelling and isofloridoside content in cells of ***Ochromonas***. Same algal suspension as in Fig. 3.

With the same algal suspension as used in Fig. 3 we have again determined the volume after 3.5 h (Fig. 4). The shrinkage caused by the osmotic stress has been fully cancelled by this time. The increase in isofloridoside pool size in this experiment is from 1.0 to 6.9, i.e. 5.9 µmoles/5 ml cell suspension. Using the non-osmotic volume derived from Fig. 3 one can calculate that this represents an internal increase of 70 mM. Roughly another 10 % increase is due to the increase of the free amino acids (Fig. 1) which, together with the isofloridoside, comes to about 77 mM. Assuming an activity coefficient of about 1.0 this would cause the internal osmotic pressure to increase by 77 mosM. The effective external increase was 95 mosM. Thus the measured increase in isofloridoside plus amino acids can account for more than 80 % of the observed volume increase. This justifies the statement that the osmotic balance in ***Ochromonas*** is mainly established by the isofloridoside mechanism.

Since the main aim of our work is to contribute towards the understanding of the biochemistry of osmoregulation, we have been mainly involved in the clarification of the

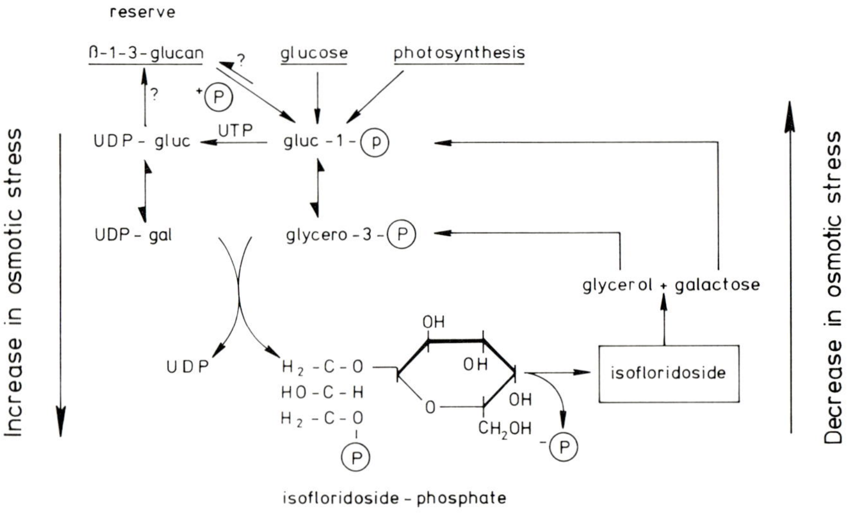

Fig. 5 The isofloridoside pathway in ***Ochromonas***.

biochemical reactions responsible for the isofloridoside metabolism. It was shown by a combination of tracer studies and enzymic methods (Fig. 5) that classical glycolytic reactions lead from any carbon source to UDP-galactose and glycero-3-phosphate. The key enzyme for isofloridoside formation is an UDP-gal: *sn*-glycero-3-phosphoric acid 1-α-galactosyl-transferase (KAUSS and SCHOBERT, 1971) forming phosphorylated isofloridoside which is then dephosphorylated to its free form. Breakdown of the galactosylglycerol most likely involves an α-galactosidase; the corresponding products glycerol and galactose are transferred back after gluconeogenesis to a reserve polysaccharide. The enzymes we have studied in detail are a specific phosphorylase responsible for the mobilization of the reserve β-1 → 3 glucan (ALBRECHT and KAUSS, 1971) and the galactose-transferase (KAUSS and SCHOBERT, 1972); the phosphatase and the α-galactosidase have also been demonstrated. It seems clear that an increase in outside osmotic pressure results in the promotion of reactions leading to isofloridoside formation, whereas a decrease in exterior osmotic pressure causes the degradation of isofloridoside. There is some evidence that at least one of the regulated enzymes is the galactose-transferase (KAUSS and SCHOBERT, 1971).

I will not discuss this in more detail but will document the physiological situation with results of chase-experiments with ^{14}C-glucose, performed in order to demonstrate the flow of material into and out of the isofloridoside pool under various osmotic conditions (Table 1). There is both formation and degradation of isofloridoside at any osmotic pressure and at any pool-size; however, the rates of production and degradation are higher when the pool-size is greater. There is a significant increase in formation of isofloridoside after an increase in external osmotic pressure, as well as an increased degradation of the carbohydrate after a decrease in external osmotic pressure. This suggests that regulation most probably occurs both at enzymic steps involved in formation as well as in degradation of isofloridoside. The situation in isofloridoside metabolism is therefore very similar to the results shown in the transport of ions or other small molecules across membranes, where both influx and efflux can be subject to alteration.

Table 1

Flow of material into and out of the isofloridoside pool under various osmotic conditions. Calculated with data from KAUSS (1973).

External osmotic condition (+ NaCl or dilution)	Isofloridoside pool size (µmoles/ml suspension)	formed (inflow) (µmoles/h per ml suspension)	degraded (outflow) (µmoles/h per ml suspension)
30 mosM, constant	0.12	0.07	0.07
10 min after 30 → 185 mosM	0.23	0.64	? (small)
185 mosM, constant	1.01	0.34	0.34
10 min after 185 → 30 mosM	0.70	0.04	0.85
70 min after 185 → 30 mosM	0.28	0.04	0.08

Another point which may connect isofloridoside-mediated osmoregulation with membrane properties is presently under investigation. First experiments indicate that the mechanism for its proper function requires constituents of the nutrient solution, presumably metal ions, to be present at the cell surface. The effect needs much more exploration. One may speculate as a working hypothesis that here the cell membrane is responsible and that during shrinkage and swelling of the cell not only the mechanical tension but also the ion-dependent electric properties change, thereby producing the signal which in turn may regulate the isofloridoside metabolism in the required direction.

C. Acknowledgements

I would like to thank Dr. W. Tanner (Regensburg) and Dr. U. Heber (Düsseldorf) for very helpful suggestions, Chr. Glaser for technical assistence, and the DFG for financial support.

D. References

ALBRECHT, G.J., KAUSS, H., Phytochemistry **10**, 1293–1298 (1971)
GUTKNECHT, J., Science **160**, 68–70 (1968)
KANESHIRO, E.S., HOLZ, G.G., JR., DUNHAM, P.B., Biol. Bull. **137**, 161–169 (1969)
KAUSS, H., Z. Pflanzenphysiol. **56**, 453–465 (1967)
KAUSS, H., Ber. Dtsch. Bot. Ges. **82**, 115–125 (1969)
KAUSS, H., Plant Physiol. **52**, 613–615 (1973)
KAUSS, H., SCHOBERT, B., FEBS Letters **19**, 131–135 (1971)
NOBEL, P.S., Plant Physiol. **43**, 781–787 (1968)
NOBEL, P.S., Biochim. Biophys. Acta **172**, 134–143 (1969)
SCHOBERT, B., UNTNER, E., KAUSS, H., Z. Pflanzenphysiol. **67**, 385–398 (1972)
STONER, L.C., DUNHAM, P.B., J. Exp. Biol. **53**, 391–399 (1970)

Osmoregulation Mechanism in the Halophilic Alga *Dunaliella parva*

A. Ben-Amotz

A. Introduction

One of the most remarkable characteristics of certain marine algae is their ability to survive and to withstand changes in the osmotic concentration of seawater. Certain marine algae tolerate salt concentrations ranging from 0.1 to 3.0 times that of seawater (BIEBL, 1952; BIEBL, 1958; EPPLEY and CYRUS, 1960). In nature the phenomenon of marine algal osmotic tolerance toward large changes in salinity is often associated with intertidal habitats and salt lakes.

Algae tend to maintain an internal osmotic pressure somewhat above that of the medium. The mechanism responsible for the osmotic regulation and the basis for salt tolerance in certain marine algae have not been elucidated (GUILLARD, 1962). The osmotic balance of marine algae involves a high internal osmotic pressure, but the nature of the osmotically-active components is still obscure.

The green halophilic alga ***Dunaliella*** requires an unusually high salt concentration for growth (JOHNSON et al., 1968); and a high external osmotic pressure is required for photosynthetic oxygen evolution (BEN-AMOTZ and AVRON, 1972) and glycerol production (CRAIGIE and McLACHLAN, 1964; WEGMANN, 1971). ***Dunaliella*** cells are often subjected to widely fluctuating salt concentrations and can tolerate a broad range of salt from low salinity to almost saturated NaCl solutions. Recently it was demonstrated that free glycerol is produced and accumulated within ***Dunaliella parva*** cells to a final concentration (around 2.1 M) required to osmotically balance the medium concentration of 1.5 M NaCl (BEN-AMOTZ and AVRON, 1973b).

The purpose of this study is to demonstrate that the mechanism involved in the osmotic regulation in ***D. parva*** is mediated through two interrelated processes. Both the intracellular glycerol content and the water content change in response to the external salt concentration, resulting in a new glycerol concentration which is isoosmotic with the medium concentration of NaCl.

B. Materials and Methods

Dunaliella parva, a green halophilic flagellate was isolated from the Dead Sea and was maintained by periodic transfers in artificial media or on agar slants. The growth medium contained: 1.5 M NaCl, 24 mM $MgSO_4$, 20 mM $MgCl_2$, 10 mM $CaCl_2$, 4 mM $NaNO_3$, 1 mM KNO_3, 0.1 mM K_2HPO_4, 1.5 μM $FeCl_3$, 30 μM EDTA, 185 μM H_3BO_3, 7 μM $MnCl_2$, 0.8 μM $ZnCl_2$, 0.02 μM $CoCl_2$, 0.2 nM $CuCl_2$, and 20 mM Tris-Cl pH 7.4. For agar slants the above medium was supplemented with 2 % agar and 2 mM acetate. Cultures were grown under illumination with fluorescent lamps (light intensity of about 200 ft candles). The cultures were gassed with a slow stream of 5 % CO_2 in air and slowly shaken. Temperature was maintained at 25°C. Under these conditions the generation time was around 15 h.

Logarithmic phase algae were harvested and washed twice with 1.5 M NaCl, 10 mM $MgCl_2$, and 25 mM sodium-potassium phosphate buffer, pH 7.5, at room temperature. Samples of the concentrated algae containing about 400 μg chlorophyll were treated in two different ways. For non-metabolic conditions the algae were diluted in a series of tubes containing 10 mM $MgCl_2$, 25 mM sodium-potassium phosphate buffer, pH 7.5, the desired NaCl concentration, 5 mM sorbitol, 3 μCi/ml HTO, and 2 μCi/ml ^{14}C-sorbitol in a total volume of 1 ml. The algal suspensions were incubated for 2 min with the labelled compounds and then placed in microfuge tubes (about 0.4 ml per tube) and centrifuged (Beckman model 152) for 5 min. For metabolic conditions the algae were diluted first in a series of tubes containing 25 mM sodium-potassium phosphate buffer, pH 7.5, 10 mM $MgCl_2$, and the desired NaCl concentration in a total volume of 20 ml. All tubes were placed in a water bath for 90 min at 25°C. Where indicated light was supplied by lamps at about 45 000 lux. After 1.5 h the algae were quickly concentrated by centrifugation at 1500 X g for 3 min at room temperature, diluted with the same solution containing the labelled compounds and treated as described above for algae under non-metabolic conditions. After centrifugation in the microfuge a 50 μl sample from the supernatant and a slice from the bottom of the pellet were taken for counting and for glycerol determination. Each sample was mixed with 2 ml of a solution of 80 % acetone. The tubes containing the pellet slices were vigorously shaken until all the chlorophyll dissolved and the precipitated protein was centrifuged in a clinical centrifuge. Each sample was then mixed with 0.5 ml 50 mM buffer phthalate, pH 5.5, and 5 ml of diethyl ether was added to extract the chlorophyll. The amount of chlorophyll was determined in the ether. Samples of 0.2 ml were taken from the water phase for counting of ^3H and ^{14}C (ROTTENBERG et al., 1972) and for chemical determination of glycerol (BEN-AMOTZ and AVRON, 1973b).

C. Results

The effect of extracellular NaCl concentration upon the intracellular concentration of glycerol was tested under two different physiological conditions. 1) Under non-metabolic conditions, a suspension of ***D. parva*** cells (containing 400 μg chlorophyll per ml) was

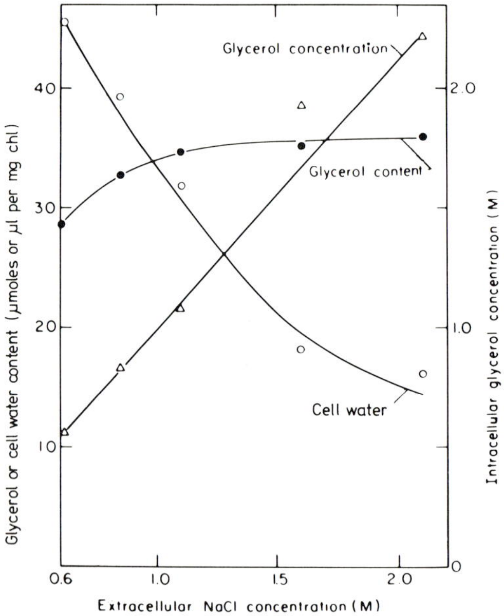

Fig. 1 Effect of NaCl concentration of the medium on the intracellular concentration of glycerol in ***D. parva*** under non-metabolic conditions. The reaction mixture contained: 10 mM $MgCl_2$, 25 mM sodium-potassium phosphate buffer, pH 7.5, the indicated concentration of NaCl, 5 mM sorbitol, 3 μCi/ml THO, 2 μCi ^{14}C-sorbitol and ***D. parva*** cells containing about 400 μg chl/ml. After 2 min of preincubation the algae were spun in a microfuge. For further details see 'Materials and Methods'.

exposed to the desired NaCl concentration for less than 5 min. 2) Under metabolic conditions, a diluted ***D. parva*** cell suspension (containing 20 μg chlorophyll per ml) was incubated in light or dark for 90 min at 25°C.

As illustrated in Fig. 1 upon addition of NaCl, under non-metabolic conditions, above the salt level of the culture medium a decrease in the cell water content is observed. When the osmolarity of the medium is decreased the cell water content increases. Since the intracellular glycerol content remained constant over the tested range of extracellular NaCl concentrations the internal concentration of glycerol remained in osmotic equilibrium with the medium concentration of NaCl. It seems clear that, under non-metabolic conditions, ***D. parva*** cells, lacking a rigid cell wall (TREZZI et al., 1965), behave like osmometers which shrink and swell under hypertonic and hypotonic conditions, respectively.

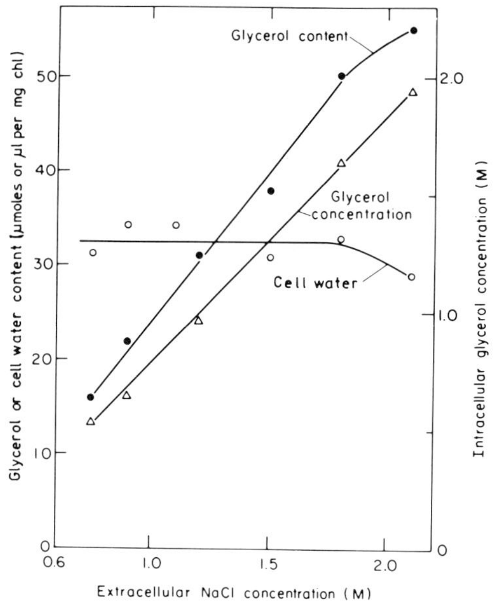

Fig. 2 Effect of NaCl concentration of the medium on the intracellular concentration of glycerol under metabolic conditions. The reaction mixture contained: 25 mM sodium-potassium phosphate buffer, pH 7.5, 10 mM $MgCl_2$, the indicated concentration of NaCl and ***D. parva*** cells containing 400 μg chl in a total volume of 20 ml. After 90 min at 25°C the algae were concentrated by centrifugation, resuspended in the same solution containing the labelled compounds and treated as described in Fig. 1. This metabolic behavior was observed in the light or in the dark.

Fig. 2 illustrates the effect of NaCl (0.6 M to 2.1 M) on the intracellular glycerol content under metabolic conditions. About 90 min were required for metabolic changes in the glycerol content of the cells to reach its final level. The cell water content returned then to a constant value equivalent to the original cell water in cells suspended in 1.5 M NaCl. As a consequence an equilibrium between internal and external osmotic conditions was established with the new final glycerol concentration required to balance the new external salt concentration. Thus osmoregulation in ***Dunaliella*** depends on the synthesis and degradation of intracellular glycerol (BEN-AMOTZ and AVRON, 1973b).

As shown in Fig. 3, below 0.6 M NaCl glycerol leaks out or is excreted into the medium in gradually increasing amounts. The glycerol distribution, i.e. ratio of amounts inside to outside, drops and the non-osmotic space (sorbitol space) increases. Thus at low osmotic pressures, the cells burst, the degradation and formation of glycerol stop and large amounts of glycerol are excreted into the medium (BEN-AMOTZ and AVRON, 1972).

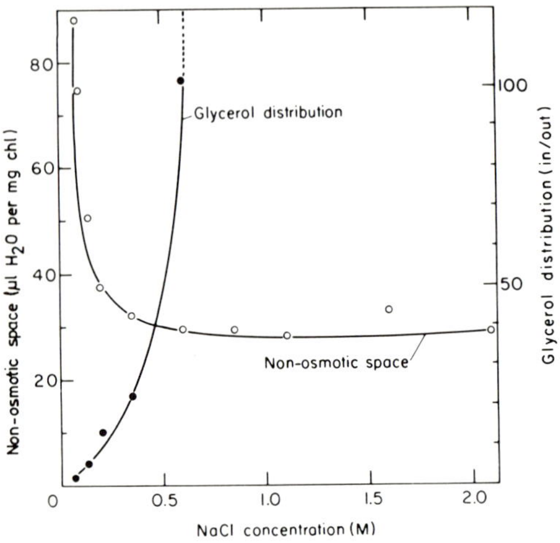

Fig. 3 The effect of NaCl concentration on the distribution of glycerol and on the non-osmotic (sorbitol) space in ***D. parva*** pellet. The reaction mixture and time were the same as described in Fig. 1, except for the addition of NaCl at the indicated concentration.

D. Discussion

Several reports have suggested that the osmotic balance of ***Dunaliella*** is mediated by high salt concentration inside the cells (MARRE and SERVATTAZ, 1959; TREZZI et al., 1965). However, various attempts to determine the intracellular concentration of NaCl did not yield a conclusive result. Direct measurements of the internal concentration of Na^+, K^+, and Cl^- in a halophilic strain of ***Chlamydomonas*** showed salt concentrations much lower than those of the external concentration (OKAMOTO and SUZUKI, 1964).

Our data indicate that ***D. parva*** produces and accumulates large amounts of glycerol within the cell. Through measurements of the distribution of glycerol between the cell water and the medium the internal concentration of glycerol was found to be isoosmotic with the medium concentration of NaCl. Our observation that above 0.6 M NaCl glycerol is not excreted into the medium suggests that the major function of glycerol is to maintain the osmotic balance in ***D. parva***.

Previous investigators (TREZZI et al., 1965) have concluded on the basis of rapid exchange of water and cell volume changes of ***D. salina*** that the reversible cell volume changes are mediated by fluxes of NaCl through the cytoplasmic membrane. As illustrated in Fig. 4 our data show that under non-metabolic conditions the reversible cell volume changes are due to rapid exchange of water so as to keep an osmotic equilibrium between the cells and the medium. Under metabolic conditions 90 min were required for the metabolic changes needed for glycerol to reach its new level. Since no leakage of intracellular glycerol was observed above 0.6 M NaCl, these alterations in glycerol are interpreted as due to metabolic formation and degradation of intracellular glycerol under hypertonic and hypotonic conditions, respectively. As illustrated in Fig. 4, cell volume changes followed the metabolic changes in glycerol content to return to the original volume of cells suspended at 1.5 M NaCl. Thus, glycerol formation and degradation should be regarded as an osmoregulatory mechanism necessary to maintain a suitable osmotic pressure within the cells.

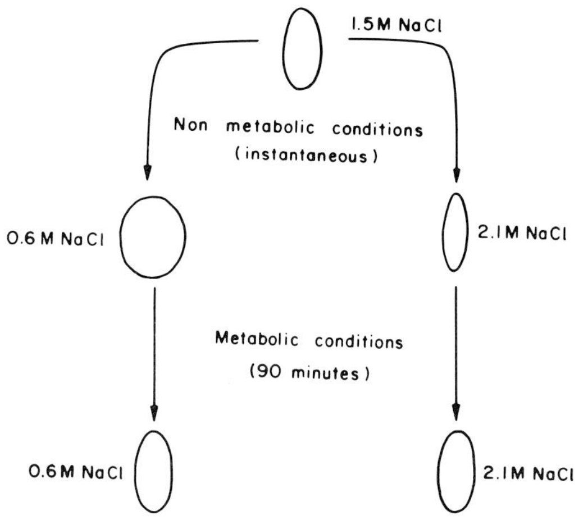

Fig. 4 Schematic illustration of volume changes of ***D. parva*** cells under different osmotic and metabolic conditions.

The results demonstrate that the formation and degradation of glycerol are not light-dependent. However, since it has been shown that glycerol is a photosynthetic product of ***Dunaliella*** (CRAIGIE and McLACHLAN, 1964; WEGMANN, 1971), it is reasonable to assume that two different metabolic pathways may be responsible for glycerol formation: one using photosynthetic products and the other via the metabolic degradation of starch, the storage product in ***Dunaliella***. A key enzyme in the pathway of glycerol formation was recently isolated from ***D. parva*** and partly purified (BEN-AMOTZ and AVRON, 1973a).

E. Acknowledgements

I wish to thank Professor M. Avron for his guidance and helpful suggestions during the progress of this work.

F. References

BEN-AMOTZ, A., AVRON, M., Plant Physiol. **42**, 240–243 (1972)
BEN-AMOTZ, A., AVRON, M., FEBS Letters **29**, 153–155 (1973a)
BEN-AMOTZ, A., AVRON, M., Plant Physiol. **51**, 875–878 (1973b)
BIEBL, R., J. Marine Biol. Ass. **31**, 307–315 (1952)
BIEBL, R., Protoplasma **50**, 217–242 (1958)
CRAIGIE, J.S., McLACHLAN, J., Can. J. Bot. **42**, 777–778 (1964)
EPPLEY, R.W., CYRUS, C.C., Biol. Bul. **118**, 55–65 (1960)
GUILLARD, R.R.L., in Physiology and Biochemistry of Algae (ed. R.A. Lewin), pp. 529–540, Academic Press, London and New York 1962
JOHNSON, M.K., JOHNSON, E.J., MACELORY, R.D., SPEER, H.L., BRUFF, B.S., J. Bacteriol. **95**, 1461–1468 (1968)

MARRE, E., SERVATTAZ, O., Atti Accad. Naz. Lincei, Rend. Cl. Sci. Fis. Mat. Natur. **26**, 272–277 (1959)
OKAMOTO, H., SUZUKI, V., Z. Allg. Microbiol. **4**, 350–357 (1964)
ROTTENBERG, H., GRUNWALD, T., AVRON, M., Eur. J. Biochem. **25**, 54–63 (1972)
TREZZI, M., GALLI, G., BELLINI, E., Gaz. Bot. Ital. **72**, 255–263 (1965)
WEGMANN, K., Biochim. Biophys. Acta **234**, 317–323 (1971)

The Role of Malic Acid Fluxes in Regulation of Crassulacean Acid Metabolism (CAM): Osmoregulation of Efflux at the Tonoplast

M. Kluge and U. Lüttge

Crassulacean Acid Metabolism (CAM) of photosynthesis is characterized by large diurnal variations of H^+ and malate, i.e. malic acid, levels in the leaves (RANSON and THOMAS, 1960). These diurnal oscillations which, under certain experimental conditions, can also be observed as endogenous rhythms (KLUGE, 1969; WARREN and WILKINS, 1961) appear to be largely regulated enzymatically (QUEIROZ, 1970). However, there is unequivocal evidence that the bulk of the malic acid involved in CAM oscillations is stored in and mobilised from the vacuoles (KLUGE and HEININGER, 1973; LÜTTGE and BALL, 1974a, b). Hence, mechanisms at the tonoplast transporting malic acid, H^+ and/or malate are probably involved in regulation of CAM rhythmicity. It can be shown that the loss ('efflux') of H^+ and malate from ***Bryophyllum*** (a CAM plant) leaf slices into an ambient medium is highly dependent on the osmotic gradient between the cells and the medium (KLUGE, 1969; LÜTTGE and BALL, 1974a, b). The osmotic gradient can be varied by adding osmotically active substances to the medium (e.g. salts, sorbitol, mannitol) and/or by using acidified and de-acidified leaves respectively, where acidified leaves contain high levels of malic acid (up to about 200 mM) at the end of an 8 h dark phase, and de-acidified leaves contain low malic acid levels (ca. 30 mM) at the end of a light phase.

The correlations between osmotic gradients and the loss of malic acid from the vacuoles, i.e. malic acid fluxes across the tonoplast, described by KLUGE and by LÜTTGE and BALL suggest that changes in turgor may play a role in the overall regulation of CAM. It can be shown that the turgor of ***Bryophyllum*** leaf cells is highly correlated with the amounts of malic acid present in the cells (LÜTTGE and BALL, 1974a). Thus it appears to be possible to draw analogies to other systems for which turgor dependent transport processes have been described (e.g. DAINTY and GINZBURG, 1964; HUMPHREYS, 1973; ZIMMERMANN and STEUDLE, 1974), and it looks promising to design experiments to investigate a model of CAM regulation in which turgor dependent malic acid transport at the tonoplast plays a certain role.

References

DAINTY, J., GINZBURG, B.Z., Biochim. Biophys. Acta **79**, 112–121 (1964)
HUMPHREYS, TH.E., Phytochem. **12**, 1211–1219 (1973)
KLUGE, M., Planta **88**, 113–129 (1969)
KLUGE, M., HEININGER, B., Planta **113**, 333–343 (1973)
LÜTTGE, U., BALL, E., Z. Pflanzenphysiol., in press (1974a)
LÜTTGE, U., BALL, E., Z. Pflanzenphysiol., in press (1974b)
QUEIROZ, O., Physiol. Veg. **8**, 75–110 (1970)
RANSON, S.C., THOMAS, M., Ann. Rev. Plant Physiol. **11**, 81–110 (1960)
WARREN, D.M., WILKINS, M.B., Nature **191**, 686–687 (1961)
ZIMMERMANN, U., STEUDLE, E., J. Membrane Biol. **16**, 331–352 (1974)

Round Table Discussion 2

Chairman: J. Dainty

NOBEL opened the discussion by commenting on glycerol as the internally produced or destroyed substance for osmotic regulation in ***Dunaliella***. He noted that the permeability of the cell membrane to glycerol must be low, hence its reflection coefficient should be near 1. He pointed out that for (pea) chloroplasts σ for glycerol is 0.6. He also drew attention to the large non-osmotic volume of ***Dunaliella***, around 60 %. With these background facts in mind, he suggested that experiments should be done using glycerol as an external osmoticum to determine σ for the cell membrane and that some studies be attempted to assess the compartmentation of internal glycerol between the chloroplasts and the cytoplasm of ***Dunaliella***. GINZBURG said that experiments on σ had been started, but the behaviour of the cells in external glycerol solutions was somewhat strange and more experiments were needed to reach any definite conclusions. A rather long discussion then started on the mechanism of osmotic regulation, particularly relating to how turgor pressure changes can be sensed by the cells. CRAM initiated the discussion by saying that he could envisage two types of turgor regulation, which he called the homeostatic type of control and the servo type of control. He said that ***Valonia*** showed the homeostatic type of regulation in which the signal is the difference between the actual turgor pressure and the 'desired' turgor pressure; in the servo type the signal is one which indicates changes in the difference between two values, inside and outside osmotic pressure, say, and the signal tells the cell to restore this desired difference. There was some inconclusive argument as to whether there was any difference between these two types. HASTINGS, for instance, commented that if CRAM meant by this servo type a mechanism which maintained a constant difference between the internal and external concentrations of some ionic species, it would have to be terribly sensitive to small fluctuations in this difference. He pointed out that in ***Valonia***, the normal $[K^+]_i$ is 10 mM and $[K^+]_o$ is 500 mM; a 1 bar turgor pressure change corresponds to a change, in this large K^+ concentration difference, of about 22 mM which is only a small fraction of the difference. He found it difficult to conceive of a biological system having that degree of precision. CRAM agreed that such a mechanism seemed unlikely to operate in a marine organism, but it might become quite appropriate for organisms living in low external concentrations.

At this point THORHAUG questioned the GUTKNECHT ***Valonia*** model which, she said, puts most of the potassium pumping activity at the tonoplast. She thought that his electrical potential, and therefore electrochemical potential, data were dubious because he had deduced the plasmalemma (and hence tonoplast) potential from the plasmalemma potential of a newly sporulated aplanospore, and the relation between this potential and that in the adult cell is unknown. Her studies indicate that the main K^+ pump is at the plasmalemma: externally-applied ouabain (10^{-3} M) causes the cell potential to drop from +10 to −60 mV and the short circuit current to drop to zero; ouabain does not affect the tonoplast if placed in the vacuole; temperature gradient studies on the short circuit current show that the plasmalemma is much more affected by such gradients than the tonoplast. HASTINGS replied that he was not aware that the GUTKNECHT group had really located the K^+ pump at the tonoplast so definitely and DAVIS said that his microelectrode studies on mature ***Valonia*** cells had confirmed the GUTKNECHT picture, i.e. the cytoplasm potential is about −70 mV. STEUDLE got down to the question of the location of the turgor pressure sensor in, say, ***Valonia***. He questioned HASTING's conclusion that it must be in the plasmalemma-cell wall complex, by pointing out that the smallest pressure gradient across the tonoplast will stretch it a great deal and we can hence imagine consequent structural changes taking place in this membrane. By appealing to the classical relation between pressure difference, ΔP, surface tension, T, and radius, R, i.e. $\Delta P = 2T/R$, HASTINGS remarked that the pressure gradient across the tonoplast was too negligible to bother about. KEDEM, however, strongly supported STEUDLE's point of view. JAFFE then drew attention to the so-called periplasmic space between the plasmalemma and the cell wall; he thought a sensor might be located there and it might also be involved in growth control. REINHOLD gave some support to this notion. She suggested that a small quantity of protein resides in this space, linked by ionic bonds to the cell wall and the plasmalemma. Such proteins are very sensitive to changes in osmotic pressure; for instance mild osmotic shock causes the release of these proteins and, at the same time, the cell loses its capacity for active transport. However KAUSS pointed out that such well-studied cells as ***Ochromonas*** and erythrocytes had no wall; he thought that the only possible signal, to which the cell could respond for osmoregulation purpose, would be the tension in the membrane.

In reference to the studies on osmoregulation in ***Dunaliella***, KEDEM asked whether the cell was trying to achieve a certain volume, or a certain osmotic pressure, or a certain difference in some appropiate parameter. No real answer was forthcoming to this question, although GINZBURG, from his experience, thought that the cells always tried to get back to whatever volume they had on the particular day of an osmoregulation experiment. (There was some disagreement at this point between BEN-AMOTZ and GINZBURG on the volume behaviour of cells in cultures of ***Dunaliella***.)

The discussion closed with a debate, largely between GINZBURG and ZIMMERMANN, on the errors involved in the use of Coulter Counters for measuring volume. Essentially ZIMMERMANN pointed out that the Coulter Counter must be so designed that all the cells take the same pathway through the orifice (achieved by hydrodynamic focusing) and that dielectric breakdown in the cell membranes would distort the volume determination.

Session 3

Electrical Properties of Membranes

Chairman: N. A. Walker

Proton Pumping and Generalized Energetics of Transport: A Review

C.L. Slayman

Introduction

Over the past 10 years our understanding of the nature of biological membrane potentials and the relationship of those potentials to metabolism, to the transport of ionic substances, and to the transport of uncharged substrates has undergone a major revolution. My purpose in this review is to sketch out the terrain on which that revolution has left us. I should begin by asserting that three distinctly different mechanisms are employed to supply energy to 'active' transport systems: I. redox transfer; II. direct enzymatic splitting of covalent bonds; and III. cotransport, in which the electrochemical gradient for one species is used to drive the uphill transport of another. Each of these will be taken up in turn.

I. Redox Transport

The most important contribution of MITCHELL (1961a) to our current understanding of transport has been to state, in testable form, the earlier hypothesis that the energy of biological oxidation-reduction reactions can appear directly as transmembrane potential differences. In doing this he drew our attention away from cell plasma membranes and focussed it properly on the membranes of chloroplasts and mitochondria, which contain well-developed redox reaction systems. Since 1961, a variety of evidence has accumulated to indicate that classical electron transfer in these organelles is a ***bona fide*** transmembrane transport process, one which leaves H^+ ions and equivalent OH^- ions separated across the thylakoid membrane or the mitochondrial inner membrane, with an attendant difference in electric potential.

The best demonstration that a potential difference exists across the mitochondrial inner membrane was provided by BAKEEVA et al. (1970), and GRINIUS et al. (1970), who showed that lipid-soluble cations, such as dibenzyl dimethyl ammonium ion (DDA^+) or triphenyl methyl phosphonium ion ($TPMP^+$) rapidly enter intact respiring mitochondria, but not morphologically inverted submitochondrial vesicles. The latter are penetrated instead by lipid soluble anions, such as phenyl dicarbaundecaborane (PCB^-), tetraphenyl boron (TPB^-), or picric acid. Direct evidence for the involvement of proton transport in the redox process comes from the observation that respiration-driven uptake of lipid-soluble cations, solubilized alkali metal ions, and Ca^{2+} ions is accompanied by a more or less stoichiometric release of H^+ ions (BAKEEVA et al., 1970; MOORE and PRESSMAN, 1964; SARIS, 1963). Furthermore, as was pointed out initially by MITCHELL (1961b), the action of the classical energy-uncoupling agent 2,4-dinitrophenol (DNP) in suppressing net H^+ efflux and all of the associated inward cation fluxes also argues for the existence of a respiration-dependent membrane potential and proton transport process, since DNP makes lipid membranes highly permeable to H^+ ions (HOPFER et al., 1968).

Studies on the chemical asymmetry of the mitochondrial inner membrane are also consistent with a transmembrane orientation of the redox reactions. Attacks with specific antibodies, proteolytic enzymes, and other non-penetrating chemical reagents, as well as saline extraction procedures, have been used to localize the functional proteins; appropriate to its participation in the TCA cycle, succinic dehydrogenase is localized at the matrix side of the membrane (FESSENDEN-RADEN, 1969); cytochrome c is at the normal outer side (RACKER et al., 1971); and cytochrome aa_3, accessible from both surfaces (SCHNEIDER et al., 1972), presumably spans the membrane. Relatively inaccessible components such as cytochrome c_1 and the b-type cytochromes are presumed buried within the membrane, though their position, and that of the non-heme iron elements, is not yet documented. Thus, as predicted by MITCHELL, the electron transfer chain does appear to be folded across the membrane.

Two novel lines of experiment on photophosphorylation in chloroplasts add further fuel to the argument. First, illumination of broken chloroplasts, suspended in unbuffered saline solution, results in rapid alkalinization of the solution (NEUMANN and JAGENDORF, 1964). When the effect is interpreted as H^+ accumulation inside the thylakoids, acidification by as much as 3–4 pH units can be calculated. The H^+ shift is suppressed by H^+-conducting uncoupling agents (JAGENDORF and NEUMANN, 1965). Second, WITT, JUNGE and their collaborators have demonstrated that a light-induced shift in the absorption spectrum of chlorophyll and carotenoid pigments behaves as if it were monitoring the kind of membrane potential postulated by MITCHELL (JUNGE et al., 1970). This so-called electrochromic change has a rise half-time of 15–20 ns (WOLFF et al., 1969) and a decay half-time of nearly 350 ms (JUNGE et al., 1970). Decay of the absorbance change is greatly accelerated by the presence of ADP plus inorganic phosphate, by cation-solubilizing antibiotics such as valinomycin and gramicidin, by ionic leaks resulting from osmotic shock (JUNGE and WITT, 1969) and by conventional uncoupling agents.

The recent demonstration by OESTERHELT and STOECKENIUS (1973) and by RACKER and STOECKENIUS (1974) that the purple membrane (rhodopsin) of ***Halobacterium halobium*** releases protons upon illumination would seem to be the ***coup de grace*** for all contrary theories on the action of light in photophosphorylation. The pigment, which is essential to photophosphorylation in ***Halobacterium***, behaves like an H^+ pump both ***in vivo*** and after extraction and incorporation into liposomes.

In the terminology of generalized membrane transport, redox transport can be described as electrogenic hydrogen ion pumping: the energy-dependent transfer of one ionic species (H^+) through a membrane without the obligatory movement of a neutralizing charge. Membrane potentials which can result from this electrogenic pump have been estimated to be between about 100 and 240 mV (SCHLIEPHAKE et al., 1968; MITCHELL and MOYLE, 1969; JACKSON and CROFTS, 1969; LIBERMAN and SKULACHEV, 1970; ROTTENBERG, 1973).

II. Transport Driven by Hydrolysis of 'High-Energy' Bonds

1. The Na^+/K^+ Pump in Animal Cells

Without question the best characterized transport system in this category is the K^+/Na^+ exchange system which is ubiquitously distributed in the plasma membranes of animal cells. The system's primary action is to extrude Na^+ from cells living in high-sodium environments, thereby making the cells functionally impermeable to Na^+ and establishing an osmotic equilibrium. Na^+ extrusion is normally coupled to K^+ entry, and because the initial studies of Na^+/K^+ exchange in squid nerve (HODGKIN and KEYNES, 1955) and also in erythrocytes (GLYNN, 1956) revealed little evidence of electrical asymmetry, the system was believed to be electrically neutral, with a 1:1 chemical coupling of Na^+ efflux and K^+ influx. For purposes of further discussion, I shall represent Na^+/K^+ exchange transport by the cyclic carrier model of SHAW (1954), slightly modified, as shown in Fig. 1.

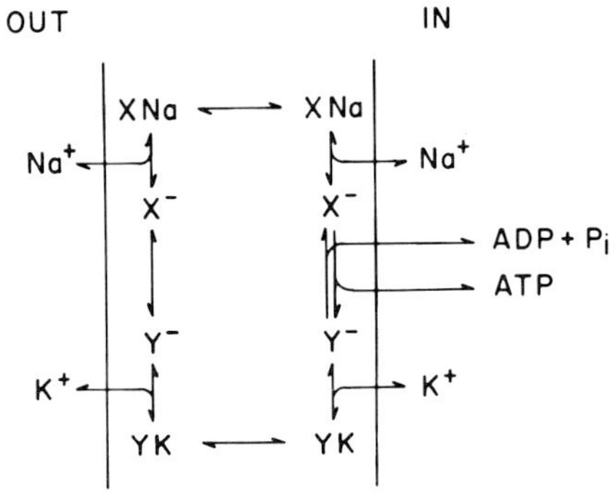

Fig. 1 Cyclic carrier model of the type first postulated by SHAW (1954) for the Na^+/K^+ exchange pump.

The carrier is assumed to exist in two uncombined forms, Y^- and X^-, each of which can 'face' either the inner surface of the membrane or the outer surface. At the membrane inner surface, the form Y^-, with a high affinity for K^+, is rapidly converted to form X^-, having a high affinity for Na^+, by reaction with ATP. X^- spontaneously reverts to Y^- at the outer surface. Binding of K^+ to Y^- and of Na^+ to X^- neutralizes the charges, and are the neutral forms YK and XNa which 'cross' the membrane.

Fueling of the system by ATP was demonstrated by injecting ATP into metabolically poisoned squid nerves (CALDWELL et al., 1960), by trapping ATP inside resealed erythrocyte ghosts (GARDOS, 1954), and – most importantly – by separating from a wide variety of cell membrane preparations an ATP-hydrolizing enzyme which is activated only in the presence of both Na^+ and K^+. Subsequent studies have shown that this Na^+- and K^+-dependent ATPase has many properties in common with the intact transport system: equivalent kinetic constants ($K_{1/2}$'s for Na^+ and K^+), equivalent inhibitory constants for a series of cardiac glycosides and aglycones, and equivalent distribution among a wide variety of tissues and organisms (reviewed by GLYNN, 1968).

In the model of Fig. 1, reaction with ATP drives the carrier in the counterclockwise direction, carrying 1 Na^+ ion outward and 1 K^+ inward with each turn. But there is no ***a priori*** reason why either a strictly 1:1 coupling between Na^+ and K^+ or a strictly clockwise cycling should be necessary. In fact, GARRAHAN and GLYNN (1967) and later GLYNN and LEW (1970) showed, on intact erythrocytes and on resealed ghosts, that clockwise operation does occur, if the system is primed with a large external Na^+ concentration, a large internal K^+ concentration, and a large value of the ratio $[ADP]_i \cdot [H_2PO_4^-]_i/[ATP]_i$. Furthermore, careful studies in a variety of systems have suggested that the Na^+/K^+ pump is indeed electrogenic rather than electroneutral (POST and JOLLY, 1957; MULLINS and BRINLEY, 1969; ADRIAN and SLAYMAN, 1966; THOMAS, 1969), with Na^+/K^+ stoichiometries of 3:1 down to 1.1 : 1.

From the point of view of the cyclic carrier model in Fig. 1, electrogenic operation of the pump can be obtained either by binding (e.g.) 3 Na^+ and 2 K^+ ions during each turn of the cycle; or by allowing the charged, low-energy form of the carrier (Y^-) to cross the membrane in addition to the neutral forms, as suggested by CROSS et al. (1965). The former device is more reasonable on an energetic basis (GLYNN, 1956), but the latter more readily admits the observed variable stoichiometry between Na^+ and K^+ fluxes. In the extreme, when binding of K^+ to Y^- or transit of YK is negligibly slow relative to transit of uncombined Y^-, the pump becomes purely electrogenic.

The essential feature of both electrogenic pump models is that one form or another of the carrier traverses the membrane bearing a charge, so that a very large membrane potential (superimposed by an independent source) should drive the system clockwise to synthesize ATP. Thus far, no one has attempted to demonstrate this effect by voltage-clamped hyperpolarization of nerve or muscle cells, but the experiment should be possible with the large mollusc neurons worked on by THOMAS (1969) and others.

2. Pumping in Non-Animal Cells

It is now clear that the animal-type Na^+/K^+ transport system is very rare among non-animal cells and tissues, if it exists there at all. Superficially, the process of Na^+/K^+ exchange transport in bacteria, fungi, and algae resembles that seen in nerve and muscle, but there are many differences in detail. Generally an excess of K^+ is pumped in, rather than an excess of Na^+ pumped out, so that Na^+/K^+ ratios of 0.9 and downward are observed, with the charge difference made up by H^+ ions extruded (SCHULTZ et al., 1963; ROTHSTEIN and ENNS, 1946; SLAYMAN and SLAYMAN, 1968; MacROBBIE, 1962). The specificity of the cation exchange in these organisms is low: Na^+ is concentrated from low-potassium media (SHIEH and BARBER, 1971; CONWAY and MOORE, 1954) and both Na^+ and K^+ are rapidly released by amino cations (SLAYMAN and SLAYMAN, 1968). In the few cases where cardiac glycosides inhibit Na^+/K^+ exchange at all, they do so only at concentrations (e.g. ouabain, 10^{-3}M) where non-specific effects may be expected (RAVEN, 1967; MacROBBIE, 1962). Finally, attempts to demonstrate non-animal Na^+- and K^+-dependent ATPase have failed (ATKINSON and POLYA, 1967; HAFKENSCHEID and BONTING, 1968; EVANS, 1969).

Cation transport processes in higher plant tissues show a similar lack of specificity, and balance charges in a variety of ways, depending on the metabolic condition of the tissue and the availability of particular intracellular or extracellular ions. NOBEL (1969), for example, found that light-dependent uptake of K^+ by pea leaves, in medium containing KCl and sucrose, is balanced 23 % by Cl^- uptake, 60 % by H^+ efflux, and the remainder by small losses of Na^+, Mg^{2+}, and Ca^{2+}. JACKSON and ADAMS (1963) and PITMAN (1970), comparing K^+ uptake by barley roots in solutions with different anions, concluded that influx could be coupled 100 % to H^+ efflux. Cation-stimulated ATPases from higher plant tissues (BROWN and ALTSCHUL, 1964; DODDS and ELLIS, 1966; GRUENER and NEUMANN, 1966; SEXTON and SUTCLIFFE, 1969; FISHER et al., 1970) again fail to show glycoside inhibition or synergistic action of Na^+ and K^+.

Although the point was obscured for many years, the most conspicuous transport property of fungal and plant cell membranes is not their ability to maintain the internal milieu by exchanging Na^+ and K^+, but their ability to maintain very large membrane potentials under a wide variety of circumstances. They will not do so following prolonged Ca^{2+} leaching (DAINTY, 1962), but in the presence of 1 mM Ca^{2+} the membrane potentials[1]) of ***Neurospora, Nitella***, and ***Avena*** coleoptile cells, for example, become quite insensitive to extracellular K^+ (SLAYMAN, 1965a; SPANSWICK et al., 1967; HIGINBOTHAM et al.,

---

[1]) The term membrane potential will be used for convenience throughout this discussion of plant tissues. Vacuolar potential would be more satisfactory, since microelectrode measurements usually record between the vacuole and the cell exterior, across both the plasma membrane and the tonoplast. The contribution of the tonoplast to the total voltage and resistance is generally known or assumed to be small. With non-vacuolated cells, such as ***Neurospora***, the problem does not arise at all.

1964). Attempts to accommodate the results to the diffusion equations, by addition of finite terms for Ca^{2+} and H^+, have been unsuccessful. Moreover, in all of these systems and in a number of other algae and higher plants as well, the membrane potential is extremely sensitive to the state of metabolism. Anoxia, respiratory inhibitors such as cyanide and carbon monoxide, lowered temperatures, uncouplers of phosphorylation, or simply light deprivation all depolarize the cells by significant amounts: 40–150 mV (NAGAI and TAZAWA, 1962; SLAYMAN, 1965b; KITASATO, 1968; SADDLER, 1970; HIGINBOTHAM et al., 1970; SPANSWICK, 1972; BENTRUP et al., 1973). In the case of ***Neurospora***, for example, cyanide depolarizes the cells by about 150 mV, with a half-time of 4 seconds and in the absence of appreciable resistance changes (a slight increase is observed) or ionic leakages (SLAYMAN, 1965b; SLAYMAN et al., 1970).

These results taken together argue for the existence of some kind of electrogenic ion pump, moving either anions inward or cations outward through the plasma membrane. Uptake of HCO_3^- ions may be electrogenic in the angiosperm ***Potamogeton*** (DENNY and WEEKS, 1970); and uptake of Cl^- ions, in the alga ***Acetabularia*** (SADDLER, 1970) and in the salt gland of ***Limonium*** (HILL and HILL, 1973a, 1973b). However, evidence now accumulating suggests that electrogenic extrusion of H^+ ions is the dominant active transport process carried out by the bacteria, fungi, algae, and higher plants. The most substantial electrophysiological data implicating H^+ transport in the generation of membrane potentials were presented by KITASATO (1968) using internodal cells of ***Nitella clavata***. With the potential difference between the vacuole and the external medium clamped at equilibrium for K^+ diffusion, changes in external pH generated large changes in clamping current 2 μamp/cm^2 inward, from pH 6 to pH 4), but no significant alteration of K^+ flux (less than 0.1 μamp / cm^2). Cl^- flux was also found to be nearly independent of external pH (in the absence of action potentials). Furthermore, under ambient illumination the resting potential was 65–70 mV negative to that expected for a diffusion potential involving K^+, Na^+, and H^+ ions; and SPANSWICK (1972) found the membrane resistance during illumination to be low and relatively insensitive to external pH, although it was very sensitive to changes of temperature. A number of plausible objections notwithstanding (WALKER and HOPE, 1969), the simplest interpretation of this ensemble of results is that during illumination a current of H^+ ions is driven across the plasma membrane against an electrochemical gradient. That amounts to an electrogenic pump for H^+ ions. Similar but less detailed observations have led to the same conclusion for ***Neurospora*** (SLAYMAN, 1970). It should be noted, too, that hyperpolarizing responses to illumination of certain higher plants ***(Vallisneria, Atriplex, Chenopodium)*** are accompanied by changes of net H^+ flux; but the magnitude and direction of the changes, as well as their dependence on passive permeability shifts are unsettled (LÜTTGE and PALLAGHY, 1969; PALLAGHY and LÜTTGE, 1970; BENTRUP et al., 1973).

Involvement of ATP in generating the putative H^+ pump voltages is suggested for green plant cells by the fact that uncoupling agents depress membrane potentials more effectively than do various manoeuvres which block photosynthesis (BENTRUP et al., 1973; KITASATO, 1968; LÜTTGE and PALLAGHY, 1969). But the argument is indirect and is complicated by the fact that uncoupling agents are likely to affect the plasma membrane directly. The situation is simplified in ***Neurospora***, where energy comes mainly from mitochondrial phosphorylation. Detailed time-course studies on intracellular ATP have defined a precise relationship between the membrane potential and ATP concentration (SLAYMAN et al., 1973). Both voltage and ATP decay with a half-time of 4 s (24°C) during the onset of cyanide inhibition; and both recover through a linear rise, overshoot, and exponential fall to the steady-state level, when inhibitor is removed. Plots of membrane potential versus ATP yield Michaelis curves with an apparent $K_{1/2}$ of 2 mM ATP and a maximal pump voltage of 311 mV (24°C).

Indirect evidence strongly supports the existence of an ATP-driven electrogenic H^+ pump in bacterial, mitochondrial, and chloroplast membranes as well, although attempts to record membrane potentials directly have not yet been successful. The most important finding is that, in the absence of respiration or photoelectron transfer, ATP generates essentially all of the H^+ ion flux phenomena that were listed previously as dependent upon electron transfer. Both rotenone-treated mitochondria and bacteria which lack a respiratory system, like ***Streptococcus faecalis***, release H^+ ions in exchange for K^+ (in the presence of valinomycin), (COCKRELL et al., 1966) or in exchange for lipid-soluble cations such as DDA^+ and $TPMP^+$ (BAKEEVA et al., 1970; HAROLD and PAPINEAU, 1972) when exogenous or glycolytic ATP is provided. ATP also supports uptake of lipid-soluble anions by inverted sub-mitochondrial particles (GRINIUS et al., 1970); and, in chloroplasts, it drives H^+ uptake when photoelectron transfer is inhibited by darkness (CARMELI, 1970).

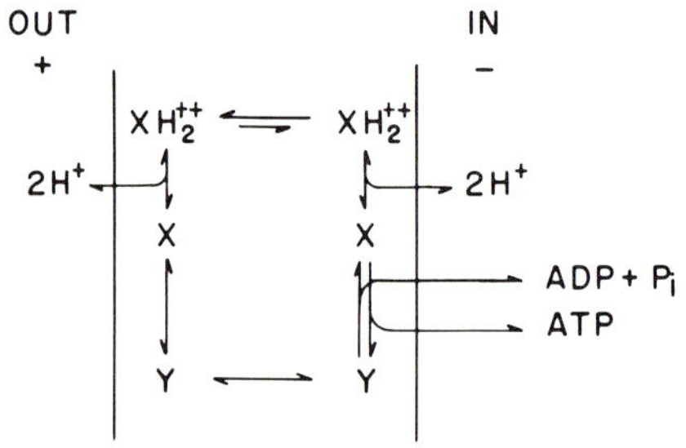

Fig. 2 Cyclic carrier model for a purely electrogenic H^+ ion pump wich uses energy from ATP hydrolysis to drive H^+ ions outward, with a stoichiometry of 2 H^+ : 1 ATP. Unequal arrows indicate that the normal function of the pump is to drive XH_2^{++} outward, polarizing the membrane negatively at the inside surface.

Following the scheme of Fig. 1, for ATP-dependent Na^+/K^+ pumping, it is possible to write a cyclic carrier scheme for electrogenic proton extrusion, as shown in Fig. 2. The stoichiometry of 2 H^+ transported for each ATP molecule hydrolyzed reflects the stoichiometry measured in mitochondrial material (MITCHELL and MOYLE, 1968; THAYER and HINKLE, 1973), and is consistent with the data on actual pump voltages obtained from ***Neurospora*** (SLAYMAN et al., 1973). The scheme allows a distinction between the H^+ carrier, ***per se***, and the enzyme catalyzing ATP hydrolysis, a distinction which is suggested by the peculiar architectural arrangement of the enzyme on the membrane (PENEFSKY and WARNER, 1965; KOPACZYK et al., 1968). It also incorporates the idea of reversibility, which is central to MITCHELL's 'chemiosmotic' mechanism of energy conservation.

Whereas reversal of the Na^+ pump in animal cell membranes is probably rare physiologically, clockwise running of the H^+ pump schematized in Fig. 2 must be regarded as the normal circumstance in mitochondria, chloroplasts, and respiring bacteria. That membrane potentials indeed drive ATP synthesis in mitochondria was most elegantly argued by GLYNN (1967), from the data of COCKRELL et al. (1967). The latter authors found that mitochondria which had been loaded with K^+ and then placed into a K^+-free medium would commence net ATP synthesis in the presence of added ADP and inorganic phosphate when valinomycin was added to the suspension. Driving of ATP synthesis by a large pH gradient was demonstrated by JAGENDORF and URIBE (1966a, b) in suspensions of broken chloroplasts which had been equilibrated at pH 3.8 and then were shifted to pH 8. And more recently URIBE and LI (1973) have used a gradient of K^+ in the presence of valinomycin to impose voltage-modulation upon the ATP synthesis occurring with a pH jump. The question of whether the electrogenic H^+ pump in the plasma membranes of plants, fungi, and non-respiring bacteria can be reversed by imposed voltages or ionic gradients is as yet untested, but should be feasible on voltage-clamped internodal segments of ***Nitella***, or on fungi and bacteria whose membrane potentials are poised by a large K^+ gradient in the presence of valinomycin.

The entire picture for the energy-conserving membranes is bolstered by the fact that bacteria, mitochondria, and chloroplasts have all yielded a soluble ATPase which is a required factor in reconstituting functional membranes (RACKER et al., 1964; KAGAWA and RACKER, 1966; RACKER and HORSTMAN, 1967); and when the (mitochondrial) ATPase is incorporated into liposomes made from purified soybean lipids, addition of ATP in the presence of valinomycin plus K^+ leads to H^+ release and K^+ uptake by the liposomes (HINKLE, personal communication). If the ATPase is incorporated along with rhodopsin from halophilic bacteria, illumination of the liposomes produces net ATP synthesis (RACKER, 1973).

III. Ion-Dependent Cotransport

1. Cotransport of Sodium with Amino Acids or Sugars

The prototype cotransport system is that which concentrates amino acids in ascites tumor cells and in avian erythrocytes, and which was first investigated by CHRISTENSEN and his colleagues (see review by CHRISTENSEN, 1970). The rate of glycine uptake by these tissues was found to increase not just with the external concentration of glycine, but also with the external concentration of Na^+, according to a typical saturation function (KROMPHARDT et al., 1963), and a detailed analysis of the relationship has been presented more recently by EDDY and his associates (EDDY et al., 1967; EDDY, 1968a, 1968b). Elevating external Na^+ has a dual kinetic effect on glycine transport, increasing the maximal velocity and decreasing the apparent Michaelis constant for glycine. Treatment of the cells with 2 mM cyanide, which reduces intracellular ATP by 85 % (LAWS and STICKLAND, 1967; EDDY and HOGG, 1969), has little effect on the initial rate of glycine uptake, although the maximal accumulation ratio is reduced 2-to-3-fold in the non-respiring cells. The significance of this latter observation is in dispute, but some direct involvement of ATP (POTASHNER and JOHNSTONE, 1971), perhaps via multiple transport systems for a given amino acid (CHRISTENSEN, 1970), rapid decay of the Na^+ gradient when energy metabolism is blocked, and a fractional dependence of glycine uptake on the countertransport of K^+ (RIED and EDDY, 1971) may all be involved. From data on epithelial membranes as well as on isolated cells, the Na^+-dependent cotransport system appears to be reversible (VIDAVER, 1964; HAJJAR et al., 1970), to be driven either by a gradient of the uncharged substrate or by a gradient of Na^+ (EDDY, 1969; CURRAN et al., 1970), and under normal conditions, to carry a depolarizing current across the membrane (ROSE and SCHULTZ, 1970).

2. Cotransport in Non-Animal Cells

Among plant, fungal, and bacterial cells, Na^+-dependent cotransport of sugars or amino acids is uncommon. This is no doubt an evolutionary consequence of the fact that most of these cells either are indifferent to Na^+ or do not normally exist in high-sodium media. H^+ ions, instead of Na^+, seem to function as the major driving species for cotransport systems among non-animal cells generally, though the most extensive evidence has come from studies on bacteria. PAVLASOVA and HAROLD (1969) observed that accumulation of thiomethyl-galactoside (TMG; a non-metabolizable substrate for the β-galactoside transport system) in

Escherichia coli is inhibited by H^+-conducting uncoupling agents, without any appreciable effect on the levels of ATP supplied by glycolysis. Subsequently, WEST (1970) and WEST and MITCHELL (1972, 1973) showed that influx of lactose or TMG is accompanied by influx of H^+ ions with an initial stoichiometry of nearly 1:1. Complementary results were obtained by HAROLD and BAARDA (1968) and ASGHAR et al. (1973) in studies on a group of four amino acids — glycine, alanine, threonine, and serine — accumulated by ***Streptococcus faecalis***. Under the conditions of these experiments, energy starvation essentially blocked accumulation of the amino acids, and the block was alleviated by an outward gradient of K^+, in the presence of valinomycin. KASHKET and WILSON (1973) systematically examined this phenomenon for TMG accumulation in non-metabolizing ***Streptococcus lactis***, and obtained the following linear equation: $59 \log ([TMG]_i/[TMG]_o) = -23 + 0.86 \cdot 59 (\log([K^+]_i/[K^+]_o) - pH_o + pH_i)$, from a least-squares fit of data obtained during the rise and decay of gradients (for pH, K^+, and TMG) produced by injection of valinomycin into a preincubated cell suspension. H^+-dependent cotransport systems for sugars and amino acids have also been described in the yeasts ***Saccharomyces cerevisiae*** and ***S. carlsbergensis*** (EDDY and NOWACKI, 1971; SEASTON et al., 1973), in ***Neurospora crassa*** (SLAYMAN and SLAYMAN, 1974), and in ***Chlorella vulgaris*** (KOMOR, 1973; KOMOR and TANNER, 1974); but in these systems the stoichiometry is thus far less clearly defined than in the bacterial systems.

A cyclic carrier model for H^+-dependent cotransport, simplified from those for various Na^+-dependent cotransport systems postulated to exist in animal cells (SCHULTZ and CURRAN, 1970), is shown in Fig. 3. The carrier (X) at the outer membrane surface is presumed to combine sequentially with the uncharged substrate and then with H^+, to be driven across the membrane by the electrochemical gradient for the ternary complex (XSH^+), to dissociate at the inner surface, and then to return to the outer surface as the uncomplexed carrier. For non-animal cell systems the membrane potential appears to be of paramount importance in driving the cycle.

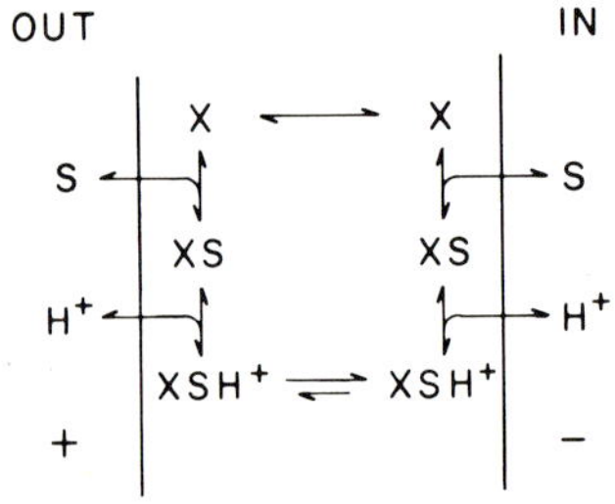

Fig. 3 Simplified cyclic carrier model for cotransport of H^+ ions and an uncharged substrate. Unequal arrows indicate that a preexisting membrane potential and the gradient of Na^+ ions normally act to drive XSH^+ inward.

A direct demonstration of the interaction between membrane potential and a cotransport system comes from studies on ***Neurospora*** (SLAYMAN and SLAYMAN, 1974). This organism is known (SCARBOROUGH, 1970; SCHNEIDER and WILEY, 1971) to possess an inducible active transport system for glucose and related sugars. Induced cells transport glucose at maximal velocities greater than 10 mM/min, with a $K_{1/2}$ of 10–40 μM, and maintain membrane potentials of 200 mV or more for hours, apparently by using endogenous carbohydrate reserves. Addition of glucose, or non-metabolized analogues (2-deoxyglucose, 3-O-methylglucose) produces a two-phase shift of potential, with a maximal depolarization of 90–150 mV and a steady depolarization of ca. 60 mV. The transient minimum is reached with a half-time of 0.4 s and occurs without any appreciable change in

ATP levels. A plot of maximal depolarization vs. glucose concentration yields a saturation curve with a $K_{1/2}$ value of ca. 40 μM. ETHERTON (personal communication) has observed similar electrical effects in oat coleoptiles, upon treating the tissue with a variety of amino acids; but no corresponding amino acid transport studies yet exist for the higher plants. It should be possible with cotransport systems of this type to hyperpolarize the membrane by driving H^+ ions outward along a large gradient of non-metabolizable sugars or amino acids. The proposition is presently being tested in ***Neurospora.***

It is apparent that ions, as well as neutral molecules, can be driven by cotransport with H^+, though in the simplest cases such cotransport should be electroneutral overall. Phosphate entry into mitochondria (CHAPPELL and CROFTS, 1966) and into ***Streptococcus faecalis*** (HAROLD and BAARDA, 1968) is probably mediated by an electrically neutral OH^-/phosphate countertransport system, which is operationally equivalent to an H^+/phosphate cotransport system. SMITH (1970) has postulated, on the basis of pH studies, that Cl^- transport into giant algal cells may be driven by countertransport with OH^-, which again is not distinct from cotransport with H^+ (SPEAR et al., 1969). The postulate would account for the observation (MacROBBIE, 1965) that agents uncoupling photophosphorylation have relatively little effect on Cl^- influx, since a pH gradient across the plasma membranes of these large cells might be expected to dissipate slowly, compared with an ATP-generated membrane potential, when the driving ATP supply is depleted.

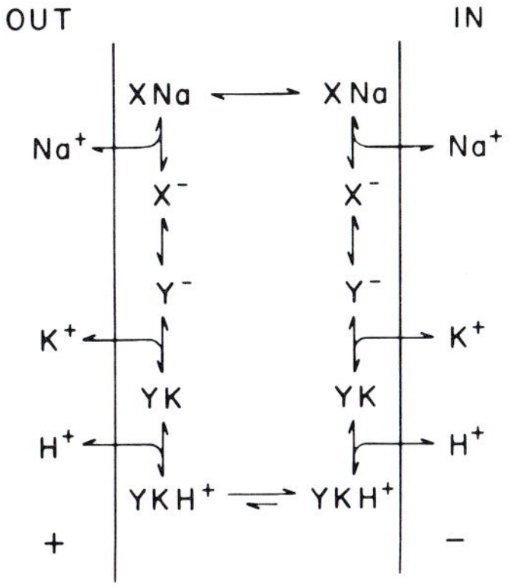

Fig. 4 Cyclic carrier model for Na^+/K^+ exchange, postulated to be driven by the electrochemical gradient of H^+ ions. Same conventions as in Fig. 3.

The fact that Na^+/K^+ pumping in the Characeae seems to depend strictly on ATP, coupled with the fact that no appropriate cation-stimulated ATPase has been demonstrated raises the question of whether the cation exchange process may also be driven by the electrochemical gradient for H^+ ions. A cyclic carrier model to accomplish this is diagrammed in Fig. 4. With respect to Na^+ and K^+ the system is electrically neutral; but H^+ ions (YKH^+) carry one net charge inward through the membrane for each turn of the cycle. The model resembles that in Fig. 1 for the Na^+/K^+ ATPase and differs from the model for cotransport of neutral molecules (Fig. 3) in that a cyclic change in the selectivity of the carrier for Na^+ and K^+ is required in order for the system to concentrate K^+ while excluding Na^+. The scheme would account for many cation transport phenomena in fungi – as well as in the algae – and in particular, for two curious observations on ***Neurospora***: 1) in a mutant strain of ***Neurospora*** which transports K^+ more slowly than the wild type the resting membrane potential is actually 10 % higher than that of the wild type (SLAYMAN and SLAYMAN, 1965), perhaps indicative of a higher membrane resistance; and 2) Na^+/K^+ exchange and steady-state K^+/K^+

exchange (K^+ substituting for Na^+ on the exiting carrier) take place with equal maximal velocities and are both energy-dependent, but switching from K^+ extrusion to Na^+ extrusion has little effect on the membrane potential (SLAYMAN and SLAYMAN, 1968; SLAYMAN, 1970).

Concluding Remarks

From the point of view of electrical equivalent circuits, it is obvious that much more must now be included than simply an EMF and non-linear resistance to describe 'leakage' pathways in biological membranes. Specific elements must be included for electrogenic pumps, for charge-carrying cotransport systems, and (in the case of energy-conserving membranes) for redox transport systems. In a naive way this can be done by arranging in parallel four batteries or EMF's, each with its series resistance, as shown in the diagram of Fig. 5. The circuit drives home the point that generation of membrane potentials either via an ATP-driven electrogenic pump or via a redox system affords a very general device for transfer of metabolic energy to transport systems and to other kinds of metabolic processes. The relative sizes of the four EMF's in any given circumstance will determine whether, for example, redox gradients, substrate gradients, or ionic gradients drive net ATP synthesis, or instead are driven at the expense of ATP. It is well known that standing ionic gradients in themselves can supply energy to other physiological processes, such as signal transmission through the nerve action potential. Standing ionic gradients, however, rarely exceed a ratio of about 100, whether for Na^+, for K^+ or for H^+, so that the use of ionic gradients alone to drive ATP synthesis or substrate transport would be very restrictive; but when the existence of 100 to 240 mV of membrane potential is added into the formulation, electrochemical activity ratios in the range of 10^4 to 10^6 become possible. ATP synthesis, substrate transport from very dilute media, and even reversal of partial reactions within the redox process can therefore follow.

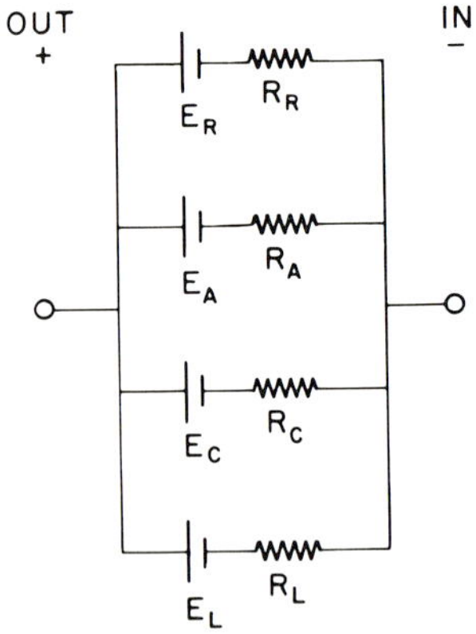

Fig. 5 Simplified equivalent circuit for a membrane containing four distinct 'electrogenic' mechanisms: L) ionic diffusion pathways; C) ion-dependent cotransport systems; A) an ATP-driven ion pump; R) a redox pump. E designates a battery (or EMF), and R designates a resistance in each of the four limbs. All eight elements may vary with conditions.

Acknowledgements

Research for this review was supported by Research Grant GM-15858 and by RCD Award GM-20164, from the National Institute for General Medical Sciences.

References

ADRIAN, R.H., SLAYMAN, C.L., J. Physiol. **184**, 970–1014 (1966)
ASGHAR, S.S., LEVIN, E., HAROLD, F.M., J. Biol. Chem. **248**, 5225–5233 (1973)
ATKINSON, M.R., POLYA, G.M., Aust. J. Biol. Sci. **20**, 1069–1086 (1967)
BAKEEVA, L.E., GRINIUS, L.L., JASAITIS, A.A., KULIENE, V.V., LEVITSKY, D.O., LIBERMAN, E.A., SEVERINA, I.I., SKULACHEV, V.P., Biochim. Biophys. Acta **216**, 13–21 (1970)
BENTRUP, F.W., GRATZ, H.J., UNBEHAUEN, H., in Ion Transport in Plants (ed. W.P. Anderson), pp. 171–182, Academic Press, London and New York 1973
BROWN, H.D., ALTSCHUL, A.J., Biochem. Biophys. Res. Commun. **15**, 479–483 (1964)
CALDWELL, P.C., HODGKIN, A.L., KEYNES, R.D., SHAW, T.I., J. Physiol. **152**, 561–590 (1960)
CARMELI, C., FEBS Letters **7**, 297–300 (1970)
CHAPPELL, J.B., CROFTS, A.R., in Regulation of Metabolic Processes in Mitochondria (eds. J.M. Tager, S. Papa, E. Quagliariello, E.C. Slater), pp. 293–314, Elsevier, Amsterdam 1966
CHRISTENSEN, H.N., in Membranes and Ion Transport (ed. E.E. Bittar) **1**, pp. 365–394, Wiley-Interscience, New York 1970
COCKRELL, R.S., HARRIS, E.J., PRESSMAN, B.C., Biochem. J. **5**, 2326–2335 (1966)
COCKRELL, R.S., HARRIS, E.J., PRESSMAN, B.C., Nature **215**, 1487–1488 (1967)
CONWAY, E.J., MOORE, P.T., Biochem. J. **57**, 523–528 (1954)
CROSS, S.B., KEYNES, R.D., RYBOVA, R., J. Physiol. **181**,865–880 (1965)
CURRAN, P.F., HAJJAR, J.J., GLYNN, I.M., J. Gen. Physiol. **55**, 297–308 (1970)
DAINTY, J., Ann. Rev. Plant Physiol. **13**, 379–402 (1962)
DENNY, P., WEEKS, D.C., Ann. Bot. **34**, 483–496 (1970)
DODDS, J.J.A., ELLIS, R.J., Biochem. J. **101**, 31P (1966)
EDDY, A.A., Biochem. J. **108**, 195–206 (1968a)
EDDY, A.A., Biochem. J. **108**, 489–498 (1968b)
EDDY, A.A., Biochem. J. **115**, 505–509 (1969)
EDDY, A.A., HOGG, M.C., Biochem. J. **114**, 807–814 (1969)
EDDY, A.A., MULCAHY, M.F., THOMSON, P.J., Biochem. J. **103**, 863–876 (1967)
EDDY, A.A., NOWACKI, J.A., Biochem. J. **122**, 701–711 (1971)
EVANS, D.J., J. Bacteriol. **100**, 914–922 (1969)
FESSENDEN-RADEN, J.M., Fed. Proc. **28**, 472 (1969)
FISHER, J.D., HANSEN, D., HODGES, T.K., Plant Physiol. **46**, 812–814 (1970)
GARDOS, G., Acta physiol. hung. **6**, 191–199 (1954)
GARRAHAN, P.J., GLYNN, I.M., J. Physiol. **192**, 237–256 (1967)
GLYNN, I.M., J. Physiol. **134**, 278–310 (1956)
GLYNN, I.M., Nature **216**, 16–17 (1967)
GLYNN, I.M., Brit. Med. Bull. **24**, 165–169 (1968)
GLYNN, I.M., LEW, V.L., J. Physiol. **207**, 393–402 (1970)
GRINIUS, L.L., JASAITIS, A.A., KADZIAUSKAS, Yu.P., LIBERMAN, E.A., SKULACHEV, V.P., TOPALI, V.P., TSOFINA, L.M., VLADIMIROVA, M.A., Biochim. Biophys. Acta **216**, 1–12 (1970)
GRUENER, N., NEUMANN, J., J. Physiol. Plantarum. **19**, 678–682 (1966)
HAJJAR, J.J., LAMONT, A.S., CURRAN, P.F., J. Gen. Physiol. **55**, 277–296 (1970)
HAFKENSCHEID, J.C.M., BONTING, S.L., Biochim. Biophys. Acta **151**, 204–211 (1968)
HAROLD, F.M., BAARDA, J.R., J. Bacteriol. **96**, 2025–2034 (1968)
HAROLD, F.M., PAPINEAU, D., J. Membrane Biol. **8**, 45–62 (1972)
HIGINBOTHAM, N., ETHERTON, B., FOSTER, R.J., Plant Physiol. **39**, 196–203 (1964)
HIGINBOTHAM, N., GRAVES, J.S., DAVIS, R.F., J. Membrane Biol. **3**, 210–222 (1970)
HILL, A.E., HILL, B.S., J. Membrane Biol. **12**, 129–144 (1973a)
HILL, B.S., HILL, A.E., J. Membrane Biol. **12**, 145–158 (1973b)
HODGKIN,A.L., KEYNES,R.D., J. Physiol. **128**, 28–60 (1955)
HOPFER, U., LEHNINGER, A.L., THOMPSON, T.E., Proc. Nat. Acad. Sci. **59**, 484–490 (1968)
JACKSON, J.B., CROFTS, A.R., FEBS Letters **4**, 185–189 (1969)
JACKSON, P.C., ADAMS, H.R., J. Gen. Physiol. **46**, 369–386 (1963)
JAGENDORF, A.T., NEUMANN, J., J. Biol. Chem. **240**, 3210–3214 (1965)
JAGENDORF, A.T., URIBE, E., Proc. Nat. Acad. Sci. **55**, 170–177 (1966a)
JAGENDORF, A.T., URIBE, E., Brookhaven Symp. Biol. **19**, 215–241 (1966b)
JUNGE, W., EMRICH, H.M., WITT, H.T., in Physical Principles of Biological Membranes (eds. F. Snell, J. Wolken, G. Iverson, J. Lam), pp. 383–394, Gordon and Breach, New York 1970
JUNGE, W., RUMBERG, B., SCHRÖDER, H., Eur. J. Biochem. **14**, 575–581 (1970)
JUNGE, W., WITT, H.T., Z. Naturforsch. **23b**, 244–254 (1969)

KAGAWA, Y., RACKER, E., J. Biol. Chem. **241**, 2475–2482 (1966)
KASHKET, E.R., WILSON, T.H., Proc. Nat. Acad. Sci. **70**, 2866–2869 (1973)
KITASATO, H., J. Gen. Physiol. **52**, 60–87 (1968)
KOMOR, E., FEBS Letters **38**,16–18 (1973)
KOMOR,E., TANNER, W., in Membrane Transport in Plants (eds. U. Zimmermann, J. Dainty), Springer-Verlag, Berlin, Heidelberg, New York, 1974
KOPACZYK, K., ASAI, J., ALLMANN, D.W., ODA, T., GREEN, D.E., Arch. Biochem. Biophys. **123**, 602–621 (1968)
KROMPHARDT, H., GROBECKER, H., RING, K., HEINZ, E., Biochim. Biophys. Acta **74**, 549–551 (1963)
LAWS, J.O., STICKLAND, L.H., Biochem. J. **104**, 158–164 (1967)
LIBERMAN, E.A., SKULACHEV, V.P., Biochim. Biophys. Acta **216**, 30–42 (1970)
LÜTTGE, U., PALLAGHY, C.K., Z. Pflanzenphysiol. **61**, 58–67 (1969)
MacROBBIE, E.A.C., J. Gen. Physiol. **45**, 861–878 (1962)
MacROBBIE, E.A.C., Biochim. Biophys. Acta **94**, 64–73 (1965)
MITCHELL, P., Nature **191**, 144–148 (1961a)
MITCHELL, P., Biochem. J. **81**, 24P (1961b)
MITCHELL, P., MOYLE, J., Eur. J. Biochem. **4**, 530–539 (1968)
MITCHELL, P., MOYLE, J., Eur. J. Biochem. **7**, 471–484 (1969)
MOORE, C., PRESSMAN, B.C., Biochem. Biophys. Res. Commun. **15**, 562–567 (1964)
MULLINS, L.J., BRINLEY, F.J., Jr., J. Gen. Physiol. **53**, 704–740 (1969)
NAGAI, R., TAZAWA, M., Plant Cell Physiol. **3**, 323–339 (1962)
NEUMANN, J., JAGENDORF, A.T., Arch. Biochem. Biophys. **107**, 109–119 (1964)
NOBEL, P.S., Plant Cell Physiol. **10**, 597–605 (1969)
OESTERHELT, D., STOECKENIUS, W., Proc. Nat. Acad. Sci. **70**, 2853–2857 (1973)
PALLAGHY, C.K., LÜTTGE, U., Z. Pflanzenphysiol. **62**, 417–425 (1970)
PAVLASOVA, E., HAROLD, F.M., J. Bacteriol. **98**, 198–204 (1969)
PENEFSKY, H.S., WARNER, R.C., J. Biol. Chem. **240**, 4694–4702 (1965)
PITMAN, M.G., Plant Physiol. **45**, 787–790 (1970)
POST, R.L., JOLLY, P.C., Biochim. Biophys. Acta **25**, 118–128 (1957)
POTASHNER, S.J., JOHNSTONE, R.M., Biochim. Biophys. Acta **233**, 91–103 (1971)
RACKER, E., Biochem. Biophys. Res. Commun. **55**, 224–230 (1973)
RACKER, E., HORSTMAN, L.L., J. Biol. Chem. **242**, 2547–2551 (1967)
RACKER, E., LOYTER, A., CHRISTIANSEN, R.O., in Probes of Structure and Function of Macromolecules and Membranes (eds. B. Chance, C.-P. Lee, J.K. Blasie), pp. 407–410, Academic Press, London and New York 1971
RACKER, E., STOECKENIUS, W., J. Biol. Chem., in press (1974)
RACKER, E., TYLER, D.D., ESTABROOK, R.W., CONOVER, T.E., PARSONS, D.F., CHANCE, B., in Oxidases and Related Redox Systems (eds. T. King, H. Mason, M. Morrison), pp. 1077–1094, Wiley-Interscience, New York 1964
RAVEN, J.A., J. Gen. Physiol. **50**, 1607–1625 (1967)
RIED, M., EDDY, A.A., Biochem. J. **124**, 951–952 (1971)
ROSE, R.C., SCHULTZ, S.G., Biochim. Biophys. Acta **211**, 376–378 (1970)
ROTHSTEIN, A., ENNS, L.H., J. Cell. Comp. Physiol. **28**, 231–252 (1946)
ROTTENBERG, H., J. Membrane Biol. **11**, 117–137 (1973)
SADDLER, H.D.W., J. Gen. Physiol. **55**, 802–821 (1970)
SARIS, N.-E., Comm.Physico-Math. (Soc. Scient. Fenn.) **28** (11), 3–59 (1963)
SCARBOROUGH, G.A., J. Biol. Chem. **245**, 3985–3987 (1970)
SCHLIEPHAKE, W., JUNGE, W., WITT, H.T., Z. Naturforsch. **23b**, 1571–1578 (1968)
SCHNEIDER, D.L., KAGAWA, Y., RACKER, E., J. Biol. Chem. **247**, 4074–4079 (1972)
SCHNEIDER, R.P., WILEY, W.R., J. Bacteriol. **106**, 479–486 (1971)
SCHULTZ, S.G., CURRAN, P.F., Physiol. Rev. **50**, 637–718 (1970)
SCHULTZ, S.G., EPSTEIN, W., SOLOMON, A.K., J. Gen. Physiol. **47**, 329–346 (1963)
SEASTON, A., INKSON, C., EDDY, A.A., Biochem. J. **134**, 1031–1043 (1973)
SEXTON, R., SUTCLIFFE, J.F., Ann. Bot. **33**, 683–694 (1969)
SHAW, T.I., Ph.D. Thesis, Cambridge University 1954
SHIEH, Y.J., BARBER, J., Biochim. Biophys. Acta **233**, 594–603 (1971)
SLAYMAN, C.L., J. Gen. Physiol. **49**, 69–92 (1965a)
SLAYMAN, C.L., J. Gen. Physiol. **49**, 93–116 (1965b)
SLAYMAN, C.L., Amer. Zool. **10**, 377–392 (1970)
SLAYMAN, C.L., LONG, W.S., LU, C.Y.-H., J. Membrane Biol. **14**, 305–338 (1973)

SLAYMAN, C.L., LU, C.Y.-H., SHANE, L., Nature **226**, 274–276 (1970)
SLAYMAN, C.L., SLAYMAN, C.W., Abstr. 23rd Intl. Cong. Physiol. Sci, (Tokyo), Item 129 (1965)
SLAYMAN, C.L., SLAYMAN, C.W., J. Gen. Physiol. **52**, 424–443 (1968)
SLAYMAN, C.L., SLAYMAN, C.W., Proc. Nat. Acad. Sci., in press (May, 1974)
SMITH, F.A., New Phytol. **69**, 903–917 (1970)
SPANSWICK, R.M., Biochim. Biophys. Acta **288**, 73–89 (1972)
SPANSWICK, R.M., STOLAREK, J., WILLIAMS, E.J., J. Exp. Bot. **18**, 1–16 (1967)
SPEAR, D.G., BARR, J.K., BARR, C.E., J. Gen. Physiol. **54**, 397–414 (1969)
THAYER, W.S., HINKLE, P.C., J. Biol. Chem. **248**, 5395–5402 (1973)
THOMAS, R.C., J. Physiol. **201**, 495–514 (1969)
URIBE, E.G., LI, B.C.Y., Bioenergetics **4**, 435–444 (1973)
VIDAVER, G.A., Biochem. **3**, 795–799 (1964)
WALKER, N.A., HOPE, A.B., Aust. J. Biol. Sci. **22**, 1179–1195 (1969)
WEST, I.C., Biochem. Biophys. Res. Commun. **41**, 655–661 (1970)
WEST, I.C., MITCHELL, P., Bioenergetics **3**, 445–462 (1972)
WEST, I.C., MITCHELL, P., Biochem. J. **132**, 587–592 (1973)
WOLFF, C., BUCHWALD, H.E., RÜPPEL, H., WITT, K., WITT, H.T., Z. Naturforsch. **24b**, 1038–1041 (1969)

Light-Dependent Changes of Membrane Potential and Conductance in *Riccia fluitans*

H. Felle and F.W. Bentrup

A. Introduction

Transport through plasma membranes of green plant cells has turned out to be linked intimately to photosynthetic reactions. This is particularly evident from higher plant cells which generally display large light-dependent changes in their membrane potentials (ANDERSON, 1973). Unfortunately, for experimental reasons, a comparably rigorous analysis of both ion fluxes and electrical membrane parameters still lags behind the studies on giant algal cells.

We have investigated the ionic relations of the aquatic liverwort ***Riccia fluitans***; its richly lobed thallus is good for flux measurements, its large cells facilitate microelectrode techniques, and its rhizoid cells protruding into the milieu lend themselves favourably to impedance measurements. Moreover, the basic electrophysiological phenomena strikingly resemble those of higher plants, notably ***Vallisneria*** (BENTRUP et al., 1973). The purpose of the present paper is to show that light not only powers an active, electrogenic ion pump, but also interferes with passive ion channels.

B. Materials and Methods

Thalli of *Riccia fluitans* were grown in natural pond water in a greenhouse, and incubated 24 h before the experiments with the following test solution (mM): KCl, 0.1; NaCl, 10; $CaCl_2$, 0.1; Na_2HPO_4, 5.9; NaH_2PO_4, 13.9. The solution permitted isoosmolar changes of $[K^+]_o$ between 0.01 and 10 mM by replacing KCl with NaCl, as well as changes in pH between 4.5 and 7.5 by varying the ratio of the buffer components. All experiments were carried out at room temperature. A quartz iodine lamp delivered about 10 W/m^2 of white light through a heat absorbing filter.

Thalli were incubated with $^{86}Rb^+$ and $^{36}Cl^-$, respectively, under the particular conditions of $[K^+]_o$ etc. For efflux studies the plants were loaded for 24 h. The wash-out of radioactivity was sampled at 30 or 60 s intervals for 180 min. Unidirectional membrane fluxes were calculated using compartment analysis (HIGINBOTHAM, 1973).

Electrophysiological measurements were carried out under the microscope in a plexiglass chamber which could be perfused with the test solution. The membrane (= vacuolar) potential was recorded from thallus or rhizoid cells by means of the conventional microelectrode technique, pen chart or tape recorder, and oscilloscope. Resistance was measured on the cylindrical transparent rhizoid cells which are 20–30 μm thick and 0.5–3.0 mm long. These cells are vacuolated except for their growing tip, so electrodes could be positioned deliberately into the vacuole or cytoplasm. Up to three electrodes were inserted into single rhizoid cells. Using currents (square pulses of 2–5 s), I_o, producing a current density of less than 0.3 μamp/cm^2, the space constant λ, was found to be between 200 and 750 (mean 370) μm, depending upon the experimental conditions and the cell's age. Current-voltage curves were obtained using one electrode each for current injection and for voltage recording at the cell's midpoint.

C. Results and Discussion

I. The Membrane Potential

The membrane potential, E_m, of ***Riccia*** thallus cells subjected to a light/dark regime and different $[K^+]_o$ is shown in Fig. 1. The steady state light potential, ${}_LE_m$, is less responsive to $[K^+]_o$ than the dark potential, ${}_DE_m$. On the other hand, only ${}_LE_m$ responds to changes in $[H^+]_o$. Following KITASATO (1968) the membrane potential of ***Riccia*** can be described by Eq. (1):

$$E_m = \frac{RT}{F} \ln \frac{P_H[H^+]_o + P_K[K^+]_o + P_{Na}[Na^+]_o + P_{Cl}[Cl^-]_i}{P_H[H^+]_i + P_K[K^+]_i + P_{Na}[Na^+]_i + P_{Cl}[Cl^-]_o} + \frac{F \cdot J}{g'_m} \qquad (1)$$

Here $[H^+]_o$ etc. denote the ion concentrations on the outside, o, and the inside, i, of the membrane, R, T, F have their usual meaning; P_H etc. denote the ion permeabilities, J is the electrogenic flux through the membrane, g'_m denotes the passive conductance due to the permeabilities in the first, diffusive term of the equation.

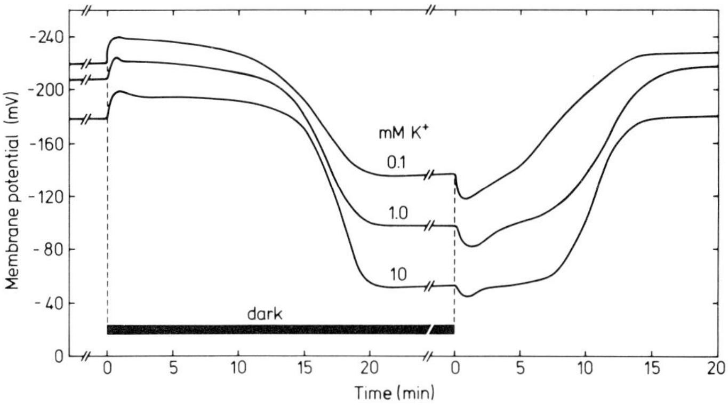

Fig. 1 Membrane potential, E_m, of ***Riccia fluitans*** thallus cells under a light/dark regime recorded in media (pH 5.5) of different K^+ concentration as indicated.

In the light Eq. (1) represents ${}_LE_m$ with $P_H \gg P_K$, P_{Na}, and P_{Cl}. In the dark ${}_DE_m$ is correctly given by Eq. (1) if the terms $P_H\,[H^+]_o$, $P_H\,[H^+]_i$, and the electrogenic term $F \cdot J/g'_m$ are taken as zero, and $P_{Na}/P_K = 0.012$. P_{Cl} hardly affects E_m, because $P_{Cl} < P_{Na}$.

II. Membrane Fluxes and Conductance

Table 1 summarizes the membrane potentials, E_m, the Nernst potentials, E_j, and the steady state influxes, ϕ_j of K^+ and Cl^-. From the given data individual conductances, g_j, have been calculated according to Eq. (2).

$$g_j = F \cdot \phi_j/(E_m - E_j) \qquad (2)$$

The resulting values of g_K and g_{Cl} have been plotted in Fig. 2 together with the electrically measured slope conductance, g_m. The data show that for $[K^+]_o$ = 1 mM in the dark

$$g_m = 5 \cdot 10^{-5}\ \text{mho cm}^{-2}, \text{ and } g_K + g_{Cl} = 4.4 \cdot 10^{-5}\ \text{mho cm}^{-2}$$

whereas in the light

$$g_m = 3.3 \cdot 10^{-5} \text{ mho cm}^{-2}, \text{ and } g_K + g_{Cl} = 0.45 \cdot 10^{-5} \text{ mho cm}^{-2}.$$

Table 1

Membrane potentials, E_m, ion equilibrium Nernst potentials, E_j, and light-dependent plasmalemma influxes, ϕ_j, of K^+ ($^{86}Rb^+$) and $^{36}Cl^-$ in *Riccia fluitans* in media of different K^+ concentration, $[K^+]_o$, at pH 5.5 and 22°C.

$[K^+]_o$ (mM)	E_m (mV) light	E_m (mV) dark	E_K (mV)	ϕ_{K^+} (pmoles cm^{-2} s^{-1}) light	ϕ_{K^+} dark	E_{Cl} (mV)	ϕ_{Cl^-} (pmoles cm^{-2} s^{-1}) light	ϕ_{Cl^-} dark
0.1	-215±10	-130±5	-165	0.56	0.47	+23	13.5	11.0
1.0	-200±10	-100±5	-106	1.83	2.7	+19.5	8.6	6.8
10.0	-185±10	- 55±5	- 50	10.2	5.8	+18.5	4.4	3.4

From Fig. 2 it is clear that g_K, but not g_{Cl}, is substantially light-dependent, and accounts almost completely for g_m in the dark. But in the light 86 % of g_m remains unidentified. We have not yet measured Na^+ fluxes; but even if ϕ_{Na^+} increases in the light as significantly as ϕ_{Cl^-} (Table 1), the resulting increase of g_{Na} according to Eq. (2) will at most yield only 50 % of g_m in the light, because E_m-E_{Na} only doubles in the light. Hence most of g_m in the light will tentatively be attributed to H^+ (see below).

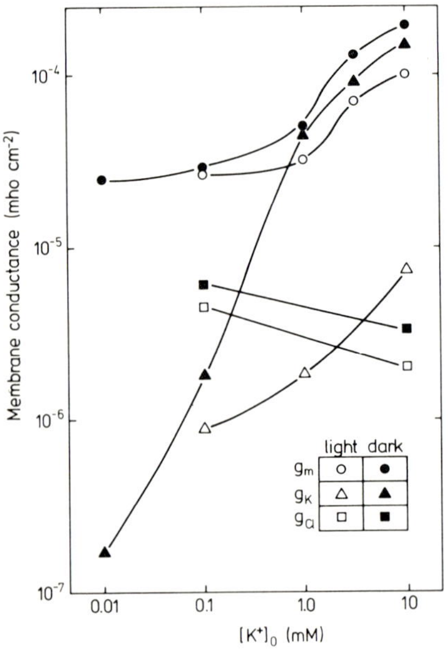

Fig. 2 Membrane conductances of *Riccia fluitans* cells in the light and dark as a function of the external K^+ concentration. g_m = electrically measured slope conductance. g_K and g_{Cl} were calculated from the flux data of Table 1.

If one calculates the K^+ permeability from the K^+ efflux according to HODGKIN and KATZ (1949), the g_K versus $[K^+]_o$ curve in the dark (Fig. 2) shows an essentially constant value of P_K of about $4 \cdot 10^{-7}$ cm s^{-1}.

III. Current-Voltage Relationships

The current-voltage (I—V) plot from a representative rhizoid cell is illustrated by Fig. 3. In the low current range there are obvious light-dependent differences; in the dark the membrane exhibits rectification, i.e. a high conductance range for depolarizing currents and a low conductance range for hyperpolarizing currents, with a transition zone around E_K. As we have argued earlier (GRADMANN and BENTRUP, 1970), this I—V characteristic might indicate purely passive diffusion channels, actually the voltage-dependent K^+ concentration profile within the membrane (GRADMANN and KLEMKE, 1974).

In terms of an equivalent circuit (Fig. 4) we conclude that the steady state dark potential, $_DE_m$, is fully represented by the passive channels for K^+, Na^+, and Cl^-.

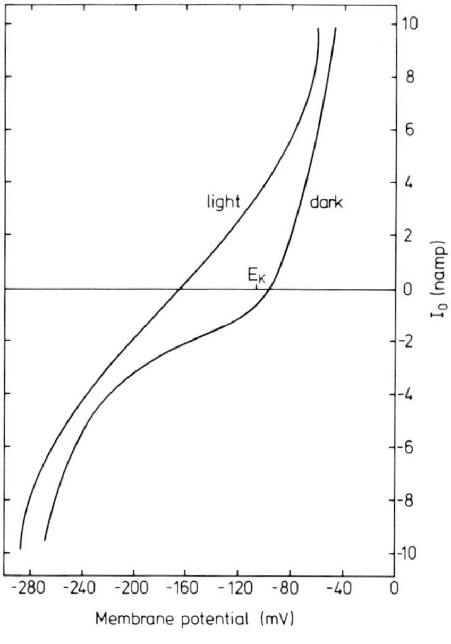

Fig. 3

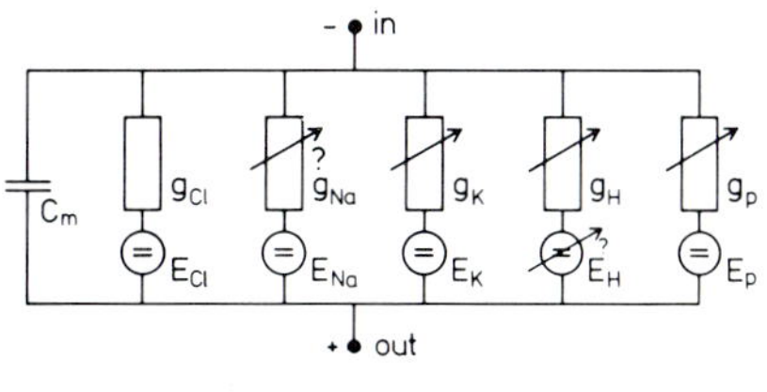

Fig. 4

Fig. 3 Current-voltage relationship of a single ***Riccia fluitans*** rhizoid cell in the light and dark in a medium containing 1 mM K^+ at pH 5.5. I_o is the current injected at the midpoint of the cylindrical cells.

Fig. 4 Equivalent circuit for the ***Riccia fluitans*** cell membrane. Arrows indicate presumably light-dependent elements. See text.

In the light, another passive, i.e. proton conducting channel, g_H, and the electrogenic pump, E_p, having a conductance, g_p, is added to the circuit. Putting passive and active electrogenic membrane elements in parallel has proved to be a satisfactory treatment of the membrane potential of many plant cells, for instance *Neurospora* (SLAYMAN, 1965), *Characeae* (KITASATO, 1968; SPANSWICK, 1973), and *Acetabularia* (GRADMANN, 1970; GRADMANN and KLEMKE, 1974; SADDLER, 1970).

FINKELSTEIN (1964) has worked out a carrier model for an electrogenic pump which features a linear conductance range at low applied currents and limiting membrane currents for large voltage displacements due to the finite number of carrier molecules. The I—V curve of *Riccia* in the light (Fig. 3) resembles FINKELSTEIN's model in having such a linear conductance range for low currents. This would infer that $g_p \gg g'_m$. Note that $g_m = g'_m + g_p$. The I—V characteristic for the pump alone cannot be derived from Fig. 3. Since g_H will contribute to g'_m in the light, g_p cannot be obtained by simply subtracting the I—V curve in the dark $[g_m=(g'_m)_{dark}]$ from that in the light $[g_m=(g'_m)_{light} + g_p]$. Although the current through the pump, I_p, is balanced by the passive current, I_H , this equality does not imply that $g_p = g_H$ (see Fig. 4). Hence the individual contributions to g_m of these conductances remain unknown. We can only suggest that they add up to, say, 80 % of g_m in the light, i.e. about $3 \cdot 10^{-5}$ mho cm^{-2} (Fig. 2). Similarly, the relative contributions of the passive and electrogenic terms of Eq. (1) to the steady state membrane potential in the light cannot be estimated precisely.

Finally, Fig. 3 suggests that the membrane breaks down at a sufficiently high displacement of E_m, which is lower in the light. This is to be expected from the fixed charge model of the *Chara corallina* membrane by COSTER (1969), if the H^+ concentration within the membrane is assumed to be higher in the light than in the dark.

IV. Multiple Light Effects

The arrows in the equivalent circuit denote those channels for which a light-dependence has been inferred. For g_{Na} this has been guessed at by analogy with g_K ; changes of E_H occur, if photosynthesis affects the cytoplasmic pH. Although the present paper excludes the kinetics of the light effects, it should be noted that membrane potential and the individual conductances and fluxes are different functions of time. The reported changes of g_K and g_H, for instance, occur within a minute, whereas g_p shows a lag, so that E_m needs about 20 min after light-on or -off to reach its new steady state (Fig. 1). Evidence from experiments with a sufficiently high $[K^+]_o$ which take $_D E_m$ to or below -40 mV suggests that right after light-on E_m swings toward $E_H \approx -40$ mV; this would indicate a cytoplasmic pH of 4.8. In *Nitella*, a dual, time-variant effect of photosynthetic energy conversion upon membrane conductance and the electrogenic pump has been reported by VREDENBERG and TONK (1973). HANSEN (pers. comm.) draws comparable conclusions from network analysis using frequency-modulated light intensity.

D. Acknowledgements

This work was supported by the Deutsche Forschungsgemeinschaft.

E. References

ANDERSON, W.P., Ion Transport in Plants, Academic Press, London and New York 1973

BENTRUP, F.W., GRATZ, H.J., UNBEHAUEN, H., in Ion Transport in Plants (ed., W.P. Anderson), pp. 171–181, Academic Press, London and New York 1973

COSTER, H.G.L., Aust. J. Biol. Sci. **22**, 365–374 (1969)

FINKELSTEIN, A., Biophys. J. **4**, 421–440 (1964)

GRADMANN, D., Planta **93**, 323–353 (1970)
GRADMANN, D., BENTRUP, F.W., Naturwissenschaften **57**, 46–47 (1970)
GRADMANN, D., KLEMKE, W., in Membrane Transport in Plants (eds. U. Zimmermann, J. Dainty), Springer-Verlag, Berlin–Heidelberg–New York 1974
HIGINBOTHAM, N., Bot. Rev. **39**, 15–69 (1973)
HODGKIN, A.L., KATZ, B., J. Physiol. **108**, 37–77 (1949)
KITASATO, H., J. Gen. Physiol. **52**, 60–87 (1968)
SADDLER, H.D.W., J. Gen. Physiol. **55**, 802–821 (1970)
SLAYMAN, C.L., J. Gen. Physiol. **49**, 93–116 (1965)
SPANSWICK, R.M., in Ion Transport in Plants (ed. W.P. Anderson), pp. 113–128, Academic Press, London and New York 1973
VREDENBERG, W.J., TONK, W.J.M., Biochim. Biophys. Acta **298**, 354–368 (1973)

Changes in Transport Determining Electrical Parameters of Cell and Chloroplast Membranes Associated with Primary and Associated Photosynthetic Reactions

W.J. Vredenberg

A. Introduction

Electrophysiological studies on membranes of metabolizing plant cells (HIGINBOTHAM, 1973) and of chloroplasts *in situ* (BULYCHEV et al., 1972) appear to be promising in contributing to knowledge about the bioenergetics of membrane transport processes. Ample evidence has accumulated from several approaches which suggests that the ion transport processes at the plasmalemma of several types of plant cells are controlled by an energy-dependent electrogenic pump.

It has been shown for cells of *Nitella translucens* (VREDENBERG and TONK, 1973) that the energy state of a cell is reversibly altered due to accumulation of energy products which trigger an electrogenic pump. The energy state and activity of the pump after energization could be characterized by changes in the current-voltage characteristics of the plasmalemma. The multiphasic kinetics of energy-dependent changes of plasmalemma potential and resistance in *Nitella* are shown in this paper. The kinetics indicate that these changes are mainly, if not exclusively, due to the activation and de-activation of an electrogenic pump.

Characteristics of the millisecond light-induced potential changes of the chloroplast enclosing membrane in mesophyll cells of *Peperomia metallica* leaves are discussed. The changes are interpreted in terms of an association with primary energy converting reactions at the thylakoid membrane.

B. Materials and Methods

Conditions and techniques for measuring the kinetics of light-induced changes in the potential and resistance of the plasmalemma of *Nitella* cells were similar to those described elsewhere (VREDENBERG, 1972).

Potential measurements on plant mesophyll cells and chloroplasts were performed with leaf sections of *Peperomia metallica* (VREDENBERG et al., 1973). Dark potentials of chloroplasts, relative to the outside medium, were usually between -100 and -130 mV, and did not differ much from the potential measured in the cytoplasm. Illumination of the cells (chloroplasts) was carried out by two light beams from two identical lamphouse assemblies, or, for one beam, from an electronic flash (rise time 5 μs, half-width 2 ms). Light from each beam reached the specimen via a common collecting light guide. The time resolution of the set up was usually set at 30 μs, but for illumination by light from the lamphouse assemblies was limited by the opening time (3 ms) of the mechanical shutters which were placed in these beams. Further details of the arrangement have been described previously (VREDENBERG et al., 1973).

Light-induced changes of absorption at 515 and 496 nm in intact leaves were measured in a single beam absorption difference apparatus. Low intensity monochromatic measuring light (transmitted through 515 or 496 nm interference filters) and high intensity actinic light transmitted through a red color filter (Schott RG665) was from the same illumination systems used in the potential measurements, except for the final transmission of the beams through a light guide.

C. Results and Discussion

The kinetics of changes in membrane potential and resistance in ***Nitella***, mainly of the plasmalemma (VREDENBERG, 1972), occurring upon illumination and darkening, or upon addition of 5 μM DCMU in the light, are shown in Fig. 1. The initial depolarization in the light, reflecting the decrease in potential generated by an electrogenic pump due to the light-driven increase in the membrane resistance (VREDENBERG and TONK, 1973), is followed, after 10 min, by a hyperpolarization. The hyperpolarization from -120 to -140 mV is accompanied by an increase in the resistance, and from -140 to -160 mV by a decrease in the resistance. Qualitatively the same changes, but in the reverse direction, occur after a few minutes in the dark, or in the light after the addition of DCMU. These kinetics indicate that an electrogenic pump is activated in the light, and de-activated upon cessation of photosynthetic energy production, either by darkening, or by inhibition of light-driven electron transport by DCMU. According to the specific non-linear current-voltage characteristics of the membrane at moderate and extreme, current-stimulated, hyperpolarization levels (KISHIMOTO, 1966; VREDENBERG and TONK, 1973), the resistance should increase and decrease in these respective hyperpolarization regions. It has conclusively been found that the normalized I—V curves computed (by integration) from the potential and corresponding resistance values of these kinetic recordings, are in approximate fit with the I—V curves measured by an adopted method (VREDENBERG and TONK, 1973). The quantitative deviations are assumed to be due to small alterations in the passive permeability properties of the membrane. These alterations are appreciable at external K^+ concentrations above 0.3 to 0.5 mM, probably due to induced permeability transitions of the membrane (SPANSWICK, 1972; SAITO and SENDA, 1973), and need further study. The slow rise in resistance which is observed in this particular experiment (Fig. 1) during the first minutes in the light after the initial decrease, is probably a reflection of this transition at low K^+ concentrations. I suggest that the resistance and potential transient upon darkening is due to a primary effect of darkness on the membrane resistance.

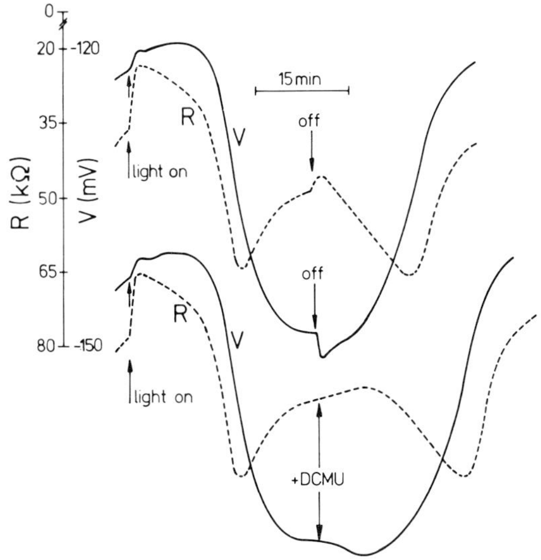

Fig. 1 Simultaneous recordings of changes in potential (V) (solid lines) and resistance (R) (dotted lines) of the plasmalemma in ***Nitella translucens*** upon illumination and darkening (upper two traces), and upon the addition of 5 μM DCMU, during illumination (lower traces). Changes in the resistance during the hyperpolarizing phase in the light and during the depolarization phase after darkening, or after the addition of DCMU appear to be associated with the potential changes. These potential changes are caused by changes in the activity of an electrogenic ion pump (see text).

The kinetics of light-induced changes in the potential across the plasmalemma of mesophyll cells of ***Peperomia*** have been found to be qualitatively similar to those found for the plasmalemma of ***Nitella*** cells. An initial depolarization of about 5 mV, completed in 1 to 2 min, is followed by a membrane hyperpolarization of about 15 to 20 mV completed in 5 to 15 min. These kinetics suggest an energy triggering of the membrane resistance (depolarization) and an electrogenic ion pump (hyperpolarization) in these leaf cells, in a similar way as was concluded for ***Nitella*** cells, and summarized in a bioenergetic model (VREDENBERG, 1973). However, definite conclusions await experiments in which changes in membrane resistance can be followed at the same time. The light-induced potential changes at the plasmalemma were found to be inhibited by 5 μM DCMU.

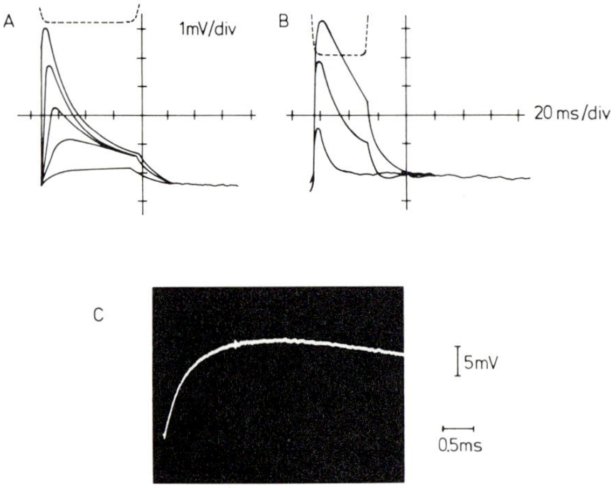

Fig. 2 A. Potential response of chloroplast enclosing membrane in ***Peperomia metallica*** in flashes of white light of different intensities. The highest response is of the highest intensity, I(max) (approx. 800 kerg cm^{-2} s^{-1}). The other intensities were 0.3, 0.06, 0.03 and 0.01, relative to I(max).
B. Potential response, for a different chloroplast, in absence (upper tracing) and presence of low intensity (approx. 10 kerg cm^{-2} s^{-1}) continuous background light. Middle and lower curve were measured 1 and 3 or 5 s, respectively, after the onset of background illumination. Traces of A and B are reproduced from oscilloscope pictures. Upward movement means a depolarization of membrane.
Opening and closing of the shutter, probed by a photodiode, can be read from the upper dotted curves.
C. Initial potential response of chloroplast enclosing membrane upon a short saturating white light flash (rise time 5 μs, 2 ms width at half intensity).

A typical potential response of the chloroplast enclosing membrane upon illumination with light flashes (rise time 3 ms) of variable intensity is shown in Fig. 2A. Three distinct phases can be distinguished at high intensities. A rapid depolarization (phase 1) is followed by a slower hyperpolarization (phase 2) towards a steady state (phase 3), which in general is slightly above the dark potential. Phase 3 appears to be saturated at much lower light intensities than phases 1 and 2. The magnitude of the phase 1 depolarization is variable for different cells (chloroplasts), usually between 10 and 40 mV. The effect of low intensity background light on the phase 1 and phase 2 potential changes in high intensity light is shown in Fig. 2B. It appears that the extent of the phase 1 change is decreased, whereas the rate of change in phases1 and 2 is unaffected or, at least for phase 2, is even enhanced.

The potential response upon a saturating short light flash (rise time 5 μs, half-width 2 ms) is shown in Fig. 2C. For a variety of cells (chloroplasts) the rise time of phase 1 was found to be between 0.5 and 1.5 ms. The half-time of the potential change in phase 2 was found to be variable (10 to 200 ms) for cells and chloroplasts from different leaves.

In contrast to the results reported by BULYCHEV et al. (1972), neither the electron transport inhibitor DCMU (5 μM) nor the uncoupler CCCP (4 μM) were found to affect any of the three phases of the chloroplast potential changes.

It has been suggested that the phase 1 potential change of the chloroplast membrane is associated with H^+ binding at the negative sites on the outer side of the thylakoid membrane (VREDENBERG et al., 1973). These sites are generated in the primary charge-separating photochemical acts within the membrane-bound photosynthetic reaction centres (WITT, 1971). Such a process would indeed be expected to be insensitive to DCMU and CCCP. Preliminary experiments have suggested that the kinetics and extent of the phase 1 potential changes in saturating light flashes are quantitatively dependent on the actual number of negative binding sites at the thylakoid membrane. The number of active binding sites will be different in the absence and presence of weak background light (Fig. 2B). A detailed analysis of these experimental results will be published elsewhere.

It has been suggested that the permeability of the enclosing chloroplast membrane is reversibly altered as a consequence of secondary reactions (ion diffusion, ATP formation) at the thylakoid membrane, following the primary H^+ binding (VREDENBERG et al., 1973). The phase 2 potential change has been interpreted in terms of this conformational transition. However, membrane modifying agents like CCCP, valinomycin and nigericin were not found to affect the phase 2 transition.

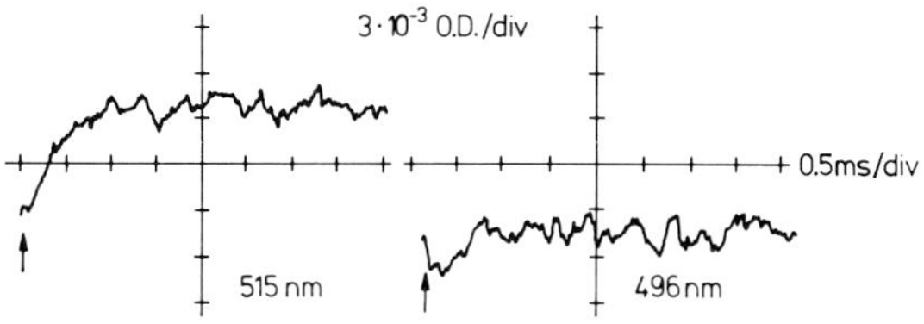

Fig. 3 Kinetics of changes in absorption at 515 and 496 nm with saturating light flashes (red light transmitted by Schott RG 665) from electronic flash. The pointing arrows mark the start of the flash.

The absorption changes at 515 and 496 nm in intact leaves with saturating actinic flashes (rise time 5 μs, 2 ms width at half intensity) are shown in Fig. 3, and indicate a photoreaction of P 515. The rise time of the absorption increase at 515 nm was found to be of the order of 1 ms. The kinetics of these changes suggest an association of the 515 nm change with the phase 1 potential change of the chloroplast enclosing membrane. In our concept this would mean that the 515 nm change occurs in association with the H^+ binding at the thylakoid membrane. The 515 nm change has been proposed by several authors (see WITT, 1971) to be associated with the primary energy conserving events at the thylakoid, e.g. charge separation, secondary ion fluxes and ATP formation. The fast change associated with the charge separation has probably been small in our case, and certainly has been obscured due to the time resolution of our apparatus. The millisecond change agrees fairly well with identical changes observed in ***Chlorella*** cells (HILDRETH et al., 1966) and in spinach leaves (HILDRETH, 1968).

The results so far obtained justify the conclusion that electrophysiological studies on cell and organelle membranes of intact plant cells are useful means of studying fundamental aspects of the primary photosynthetic energy converting reactions of chloroplasts *in situ*, and of the bioenergetics of transport processes across cellular and organelle membranes.

D. Acknowledgements

This research was partly supported by the Netherlands Foundation for Biophysics, financed by the Netherlands Organization for the Advancement of Pure Research (Z.W.O.).

Thanks are due to Dr. P.H. Homann who participated in part of the research, and to Mr. W.J.M. Tonk for skilful technical and experimental assistance.

E. References

BULYCHEV, A.A., ANDRIANOV, V.K., KURELLA, G.A., LITVIN, F.F., Nature **236**, 175–176 (1972)
HIGINBOTHAM, N., Ann. Rev. Plant Physiol. **24**, 25–46 (1973)
HILDRETH, W.W., AVRON, M., CHANCE, B., Plant Physiol. **41**, 983–991 (1966)
HILDRETH, W.W., Biochim. Biophys. Acta **153**, 197–202 (1968)
KISHIMOTO, U., Plant Cell Physiol. **7**, 429–439 (1966)
SAITO, K., SENDA, M., Plant Cell Physiol. **14**, 147–156 (1973)
SPANSWICK, R.M., Biochim. Biophys. Acta **288**, 73–89 (1972)
VREDENBERG, W.J., Biochim. Biophys. Acta **274**, 505–514 (1972)
VREDENBERG, W.J., in Ion Transport in Plants (ed. W.P. Anderson), pp. 153–169, Academic Press, London and New York 1973
VREDENBERG, W.J., TONK, W.J.M., Biochim. Biophys. Acta **298**, 354–368 (1973)
VREDENBERG, W.J., HOMANN, P.H., TONK, W.J.M., Biochim. Biophys. Acta **314**, 261–265 (1973)
WITT, H.T., Quart. Rev. Biophys. **4**, 365–477 (1971)

Current-Voltage Relationship of the Electrogenic Pump in *Acetabularia mediterranea*

D. Gradmann and W. Klemke

A. Introduction

The giant marine algal cells of ***Acetabularia*** exhibit under normal conditions a resting potential of about -170 mV, which is far more negative than the equilibrium potential of potassium, $E_K \approx$ -90 mV, the most negative diffusion potential of this cell. The discrepancy seems to arise from an electrogenic Cl^- influx, operating in parallel with the passive diffusional system (GRADMANN, 1970), because the membrane potential drops toward E_K when the Cl^- influx is inhibited by darkness, low temperature or metabolic agents (SADDLER, 1970a).

In order to investigate the role of H^+ in the electrical properties of the membrane, we measured the intracellular pH by squeezing cell sap onto pH paper; it was 2.5. At the normal external pH of 8, $E_H \approx -300$ mV and this could account for the high resting potential. However, varying the external pH from 4 to 9 does not affect V_m (GRADMANN, 1970). Thus the H^+ permeability seems to be low enough to be ignored from the electrical point of view.

During the past ten years, electrogenic pumps have been reported in various cells and tissues, but little is known about how they work. Our present aim is to study the current-voltage relationship (I–V) of the electrogenic pump of ***Acetabularia***. First, we determined the general electrical cable properties of ***Acetabularia***; and second, we compared the system with and without the pump operating, in order to investigate the properties of the pump itself.

B. Symbols Used

c_m	membrane capacity ($\mu F/cm^2$)
E	electromotive force (V)
$E_{Cl,H,K}$	equilibrium potential of Cl^-, H^+, and K^+
I	input current (amp)
i_m	membrane current (amp/cm^2)
r_i	resistivity of cell interior (Ωcm)
r_m	membrane resistance (Ωcm^2)
V_m	transmembrane potential difference
V	change of V_m

Subscripts

o	refers to voltage response at zero distance from the current electrode
x	refers to voltage response at distance x from the current electrode
s	refers to a voltage step in voltage clamp experiments
p	refers to a voltage pulse in voltage clamp experiments
P	refers to the electrogenic pump
P'	refers to the entire Cl^- transport system

Superscripts
o refers to the small time constant, τ^{o}
∞ refers to the large time constant, τ^{∞}

C. Materials and Methods

Single cylindrical cells with a length of 2–4 cm and a diameter of about 0.5 mm were used. The outer medium was artificial seawater (BENTRUP, 1968). Conventional microelectrode techniques were used. The large size of the cells permitted the use of electrodes with a large tip diameter and a low resistance ($\approx$ 100 kΩ) to inject adequate current at the midpoint of the cells.

For resistance measurements we applied small square-wave current pulses, I_o, of either sign by a function generator ($\pm$ 10 V) with a high resistance (10^7 Ω) in series. The changes, V, of the transmembrane voltage V_m, at distances 0 (= V_o) and x (= V_x) from the current electrode, were registered by two intracellular electrodes and difference amplifiers. Fast events were recorded by a storage oscilloscope, slow ones additionally by a pen chart recorder.

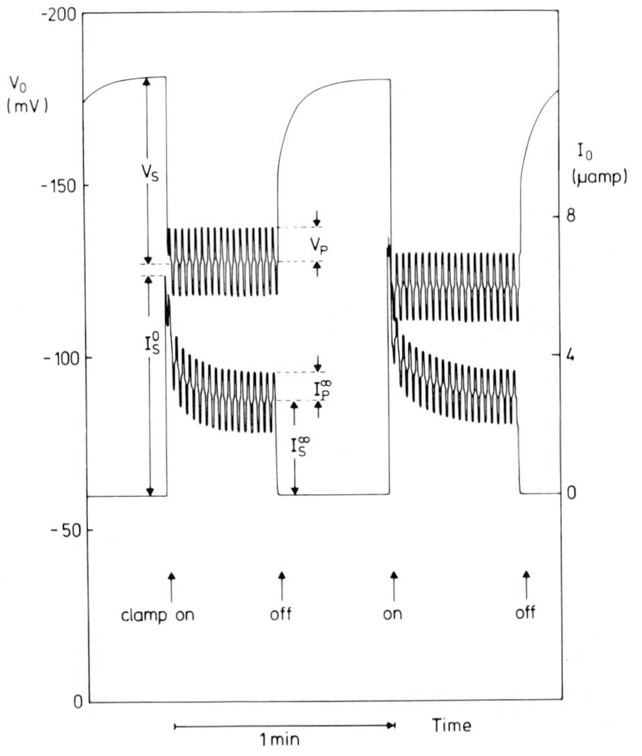

Fig. 1 Redrawing of an original voltage clamp recording. Upper trace: voltage V_o with and without voltage clamp; V_s, voltage step; V_p, voltage pulse. Lower trace: clamp current I_o; I_s^o, early current response (trace not resolved, real value determined by oscilloscope); I_s^∞, steady state level of clamp current; I_p^∞, steady state value of the early current response for V_p.

In voltage clamp experiments we used the same arrangement without the 10^7 Ω resistor and without the second (at x) voltage electrode. With our voltage clamp circuit (GRADMANN, 1970) the membrane potential V_o was varied by steps, V_s, to a new level upon which small rectangular voltage pulses, V_p, of

± 10 mV amplitude and ca. 0.5 s duration, were superimposed (Fig. 1). When the voltage clamp was first applied, no pulses were superimposed so as to record the early clamp current I_s^o (see Fig. 1) by the oscilloscope. The resulting total currents I_s to maintain V_s, and I_p to produce V_p, were registered on the pen recorder. After each step, the voltage clamp was turned off ($I_s = 0$) and the membrane potential returned slowly to its 'resting' value (see Fig. 1).

D. Results and Discussion

I. Time Constants and Resistances

As reported earlier (GRADMANN, 1970; SADDLER, 1971), the electrical system of ***Acetabularia*** has at least two clearly distinct time constants: a small one, τ_o, of the order of 1 ms, the time constant for charging the membrane capacity, c_m^o, and one or more large ones, τ^∞, of the order of 10 s, reflecting approach to a steady state potential. Accordingly, we mark our data with the superscripts 'o' and '∞'.

With short (ca. 10 ms) and small current pulses I^o, we obtained exponential voltage responses V_o at the distances 0 and x (1 to 3 mm) from the current electrode. Using linear cable theory (HOGG et al., 1969), analysis of the results from 7 cells under illumination (0.1 W/cm^2) and in darkness yields the resistance of the plasmalemma r_m^o, the resistivity of the cell interior r_i^o and the membrane capacity c_m^o; their values are given in Table 1 together with the measured resting potentials.

Table 1

Early and steady state cable parameters of ***Acetabularia mediterranea*** cells, obtained from square-pulse analysis. Results ± S.E. from 7 cells. 22°C, 0.1 W/cm^2 white light or dark. r_m, membrane resistance (kΩcm^2); r_i, resistivity of the cell interior (Ωcm); c_m, membrane capacity (μF/cm^2).

	Light	Dark
Resting potential (mV):	-180 ± 2	-167 ± 2
Early r_m, r_m^o	0.16 ± 0.03	0.31 ± 0.04
Steady state r_m, r_m^∞	1.6 ± 0.4	3.3 ± 0.9
Early r_i, r_i^o	39 ± 3	38 ± 3
Steady state r_i, r_i^∞	38 ± 8	38 ± 9
Early c_m, c_m^o	5.2 ± 0.9	4.8 ± 0.6
Steady state c_m, c_m^∞	3500 ± 600	3200 ± 500

Based on its ionic content (SADDLER, 1970b), the vacuole has a resistivity of 25 Ωcm. Because our value of r_i^o of 38.5 Ωcm was somewhat higher we believe that the electrode tips are located in the cytoplasm and that the tonoplast has an appreciable resistance.

Much longer small current pulses, I^∞, raise the voltage V further from V^o to V^∞, and exponentially with a time constant of the order of 10 s. The resulting resistance, r_m^∞, in the light is about half of its dark value, as with r_m^o (Table 1). However, r_m^∞ is an order of magnitude greater than r_m^o. The quasi-capacitance c_m^∞ is very large (ca. 3000 $\mu F/cm^2$) and light-independent. r_i^∞ does not differ from r_i^o.

If one constructs a I—V diagram (not shown) from these experiments, the linearly extrapolated curves (straight lines) from the light and dark experiments intersect at about −190 mV; this holds for both early, i_m^o, and late, i_m^∞, membrane currents. We interpret this by saying that in the membrane we have in parallel two voltage sources with their internal resistances: the passive diffusion channel(s) and the electrogenic pump. Upon illumination, the internal resistance of the pump decreases and the membrane potential therefore rises towards the voltage of the electrogenic pump (-190 mV).

II. Voltage Clamp Analysis

In order to carry the I—V analysis of the membrane into non-linear ranges, we did some voltage clamp experiments. Although a point clamp only was possible, the results could be correctly analyzed, using linear cable theory or COLE's theorem (COLE, 1961).

In Fig. 1 we present a section of an original voltage clamp record, so as to clarify our symbols. The upper trace is the voltage V_o, running free or clamped as denoted below. The voltage pulses, V_p, are chosen to be much shorter than τ^∞, and much longer than τ^o. The lower trace shows the course of the clamp current, I_s, to maintain the voltage step V_s, and I_p to produce the voltage pulses V_p.

When the electrogenic pump is blocked by low temperature (5°C), the ***Acetabularia*** membrane behaves almost like a pure diffusion system (GRADMANN and BENTRUP, 1970; GRADMANN, 1970; SADDLER, 1970a). Therefore, a comparison of the I—V curves at 5°C and 20°C, when the pump is operating, should give some information about the I—V characteristics of the pump itself. The following data are taken from an experiment on a particular cell at 5°C and 20°C in darkness.

1. Early Step Current I_s^o

Fig. 2A shows the early input I—V curves. They are linear at both temperatures. The slope I_s^o/V_s^o agrees well with the early input conductance I^o/V^o obtained from the square pulse experiments discussed above in section D.I. From the linear input function I_s^o (V_s^o) we obtain the function of the membrane current i_m^o (V_m^o) from linear cable theory. The resulting membrane resistance values of $r_m^o = V_m^o/i_m^o$ are indicated on the curves in Fig. 2A. The linear I—V curves of the membrane current i_m^o (V_m) at 20°C and 5°C intersect at -189 mV again (not shown), the input I_s^o (V_s^o) curves of Fig. 2 at higher voltages.

2. Early Pulse Current I_p^o

We usually omitted measuring and analyzing this parameter, since measurement of $I^o{}_p$ prevents a proper reading of I_s^o. When I_p^o was measured, the resulting early pulse conductance I_p^o/V_p^o turned out to be the same as the early step conductance I_s^o/V_s^o, as expected.

3. Steady State Step Current I_s^∞

The steady state input I—V curves for 5°C and 20°C are shown in Fig. 2B. At 5°C, I_s^∞ (V_o) resembles a diode characteristic with a low linear resistance for depolarising and a high linear resistance for hyperpolarising currents. The break is near E_K, and the resistance at this point, r_m^∞, equals the r_m^o in Fig. 2A. Thus the second time constant has disappeared in the cold. The diode characteristic probably mainly arises from the potassium concentrations on either side of the membrane and the polarity of the membrane current (GRADMANN and BENTRUP, 1970).

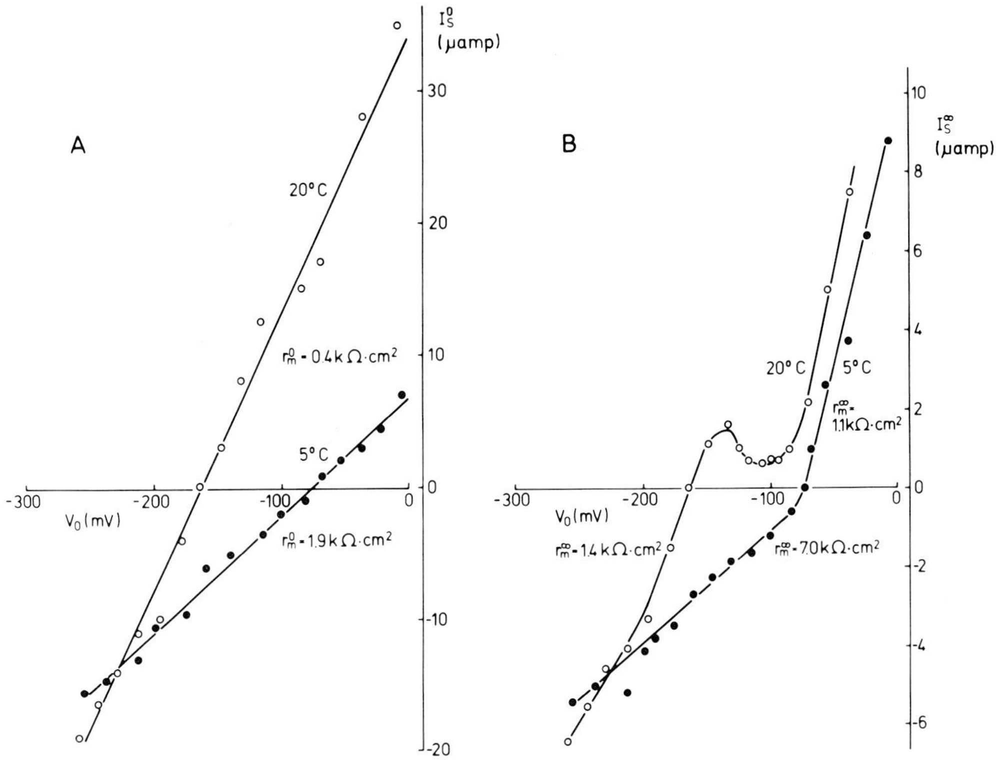

Fig. 2 A. early; B. steady state voltage clamp input I—V relationship. Recording from one cell in darkness at 20°C (○—○), and 5°C (●—●).

At 20°C, when the pump is operating, additional current passes through the pump channels. The most striking feature of the steady state I—V curve at 20°C is its N-shape under depolarisation; it finally merges with the high conductance potassium branch. This is certainly a rather complex phenomenon, since voltage- and space-dependent events in different pathways seem to be involved. At the moment this question will not be pursued further.

For hyperpolarisation at 20°C, the input characteristics are fairly linear up to -200 mV (Fig. 2B) and then change to a slightly smaller steady state conductance. We can again use linear cable theory to obtain the membrane currents i_m^∞ (V_m) in this range. These steady state, membrane, I—V curves at 20°C and 5°C intersect at -190 mV (again not shown), about the same intersection point (-189 mV) of the early step currents, mentioned above. We consider this value to be the electromotive force of the active system, because at this voltage all of the applied membrane current passes through the passive system and none through the pump (when there is no difference between the electromotive force of the pump and the transmembrane voltage). Since, however, the passive Cl^- channel (E_{Cl} = -6 mV) opens at the same time as the active electrogenic Cl^- channel (GRADMANN et al., 1973), E_p, = -190 mV represents not only the electrogenic Cl^- pump but the entire temperature- and light-stimulated Cl^- system, hence the electromotive force of the pump itself will be higher.

4. Steady State Pulse Current I_p^∞

Fig. 3 shows the data obtained by superimposing current pulses on the steady state clamp current during voltage clamp steps (Fig. 1). For the passive system at 5°C, we find again a high conductance for depolarising and a low conductance for hyperpolarising currents. This latter conductance (ca. 20 μmho at about -150 mV), I_p^{∞}/V_p, is the same as the steady state step conductance, I_s^{∞}/V_s, (Fig. 2B) in this range. In other words, the second time constant disappears at low temperatures with the activity of the electrogenic pump.

At 20°C, the steady state pulse conductance, I_p^{∞}/V_p, is always positive, and displays a maximum at a clamped voltage of -170 mV and a minimum at -100 mV.

III. Current-Voltage Relationship of the Electrogenic Pump

The difference between the steady state pulse conductances at 20°C and 5°C (Fig. 3) is again considered to be due to the I–V characteristic of the electrogenic pump. We derived this function (the electrogenic pump I–V characteristic) by first subtracting the smoothed curve at 5°C (Fig. 3) from the smoothed curve at 20°C, and then integrating the difference graphically from -260 mV to the functional value, V_o (cf. Eq. (1)).

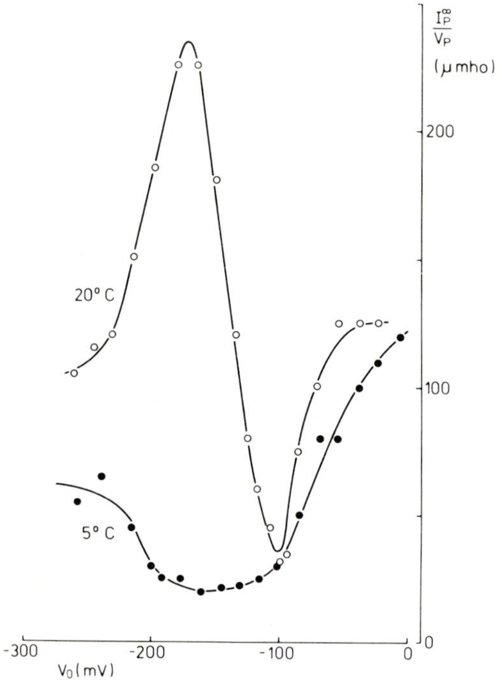

Fig. 3 Early input conductance I_p^{∞}/V_p in steady state at different clamped voltages V_o. Same cell as in Fig. 2, 20°C: (o—o), 5°C: (●—●), darkness.

$$I'' = \int_{-260}^{V_o} [\, I_p^{\infty}/V_p\,(20°C) - I_p^{\infty}/P_p\,(5°C)]\, dV_o \qquad (1)$$

We translated the resultant function to the origin, i.e. the electromotive force of the electrogenic pump, -190 mV, by subtracting 6 μamp, giving

$$I'(V_o) = I''(V_o) - 6, \text{ in } \mu\text{amp} \qquad (2)$$

This input I—V curve is plotted in the inset of Fig. 4.

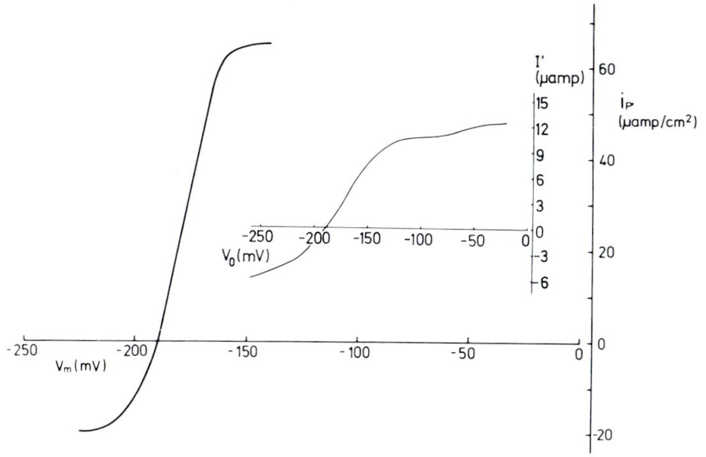

Fig. 4 Current-voltage relationship of the electrogenic pump per unit area of the membrane (cf. Eq. (3) in the text). Inset: input current-voltage relationship of the pump (cf. Eq. (2) in the text).

Calculating the membrane resistance from the input I—V relationship by linear cable theory, ignoring the finite length of our particular cell, leads to errors of less than 10 % for r_m if the input conductance is > 100 μmho. This condition is met for $I'(V_o)$ between -220 and -140 mV (inset of Fig. 4), and we can take the cell length as infinite in this range; COLE's theorem thus becomes applicable here. The pump current, i_P, per unit area of the membrane, is then given by

$$i_{P'}\ (V_m) = I'\ (dI'/dV_o)\ r_i/8\ \pi^2 a^3 \qquad (3)$$

I' can be obtained immediately from the results of Fig. 4 (inset), and dI'/dV_o from the slope at each point. r_i is 38 Ωcm and the radius, a, of the cell is 0.2 mm. The I—V curve of the electrogenic pump, $i_{P'}$ (V_m), thus obtained is shown in Fig. 4. For wider voltage ranges, we may assume according to I' (V_o) of the inset to Fig. 4, that $i_{p'}$ (V_m) will stay saturated, though exact analysis has not been worked out for these ranges.

Our calculation of $i_{P'}$ (V_m) is not quite correct. One should first calculate the membrane currents at 20°C and 5°C and then subtract them. However there is no way of calculating membrane currents from input currents for finite non-linear cables. A certain error was thus unavoidable. But this does not affect the quality of our final result.

A model of an electrogenic pump has been worked out by FINKELSTEIN (1964). It is a carrier model which exhibits basically the same I—V relationship as found in our case, i.e. limiting current density for large hyper- and de-polarisations. In our results E_P is more negative than the point of maximum conductance, whereas these entities coincide in FINKELSTEIN's model. The passive Cl^- channel might cause this discrepancy.

A different question is: why does the I_s^∞ (V_o) curve (Fig. 2B) show only a slightly increased resistance for large hyperpolarisation and a negative resistance range in the depolarisation region. Furthermore, the large time constant requires an explanation. It has been suggested that it arises from the coupling of the pump with the energy metabolism of the cell (GRAD-

MANN, 1970). We are therefore now investigating the role of ATP during the electrical events, as SLAYMAN et al. (1973) did in ***Neurospora***. In order to determine the energy consumption of the pump (our estimate is up to 30 $\mu W/cm^2$), the relationship between active and passive Cl^- pathways must be investigated. This work is in progress, since it is possible to measure electrical parameters and Cl^- efflux simultaneously on individual cells (GRADMANN et al., 1973).

E. Acknowledgements

This work was supported by the Deutsche Forschungsgemeinschaft.

F. References

BENTRUP, F.W., Z. Pflanzenphysiol. **59**, 309–339 (1968)
COLE, K.S., J. Gen. Physiol. **44**, 1055–1057 (1961)
FINKELSTEIN, A., Biophys. J. **4**, 421–440 (1964)
GRADMANN, D., Planta **93**, 323–353 (1970)
GRADMANN, D., BENTRUP, F.W., Naturwissenschaften **57**, 46–47 (1970)
GRADMANN, D., WAGNER, G., GLÄSEL, R.M., Biochim. Biophys. Acta **323**, 151–155 (1973)
HOGG, J., WILLIAMS, E.J., JOHNSTON, R.J., J. Theor. Biol. **24**, 317–334 (1969)
SADDLER, H.D.W., J. Gen. Physiol. **55**, 802–821 (1970a)
SADDLER, H.D.W., J. Exp. Bot. **21**, 345–359 (1970b)
SADDLER, H.D.W., J. Membrane Biol. **5**, 250–260 (1971)
SLAYMAN, C.L., LONG, W.S., LU, C.Y.-H., J. Membrane Biol. **14**, 305–338 (1973)

Preliminary Results of an Approach to the Quantitative Description of the Action of Light on the Membrane Potential of *Nitella*

U.P. Hansen

A. Introduction

There are several reasons for our interest in a quantitative description of the time course of the effect of light on the membrane potential of plant cells. In the first place, there is the problem of finding a set of appropriate parameters describing the time course; we can then study how these parameters are modified by various agents. Of course, maximum depolarisation or hyperpolarisation could be used for this purpose. But in complex systems such values are correlated with biophysical and biochemical events in a very complicated manner and thus, in most cases, provide no information. Secondly, plant physiologists study resistance, ion fluxes, photosynthetic activities etc. as well as the membrane potential. For comparing these measurements, it would be worthwile to have parameters 'labelling' the reactions involved, in order to find out whether they are involved in mediating different light effects. Thirdly, one could expect a quantitative description to lead to an identification of the biochemical reactions involved by providing the time constants of the reactions and their interrelations.

In this paper I have tried to give such a quantitative description, based on linear chemical kinetics.

B. Materials and Methods

The experimental set-up, the cultivating of ***Nitella mucronata*** and the sine-wave modulation of the light intensity have been described previously (HANSEN et al., 1973). However, instead of the xenon, high pressure, illuminator, a luminescence diode, MV4H of Monsanto, was used to provide red light (λ = 680 nm, bandwidth = 40 nm) modulated from 0.5 to 1.5 W/m^2.

Frequency responses were measured by the same procedure as that used by many workers in the field of sensory organs (ARDEN, 1969); it has been described in previous papers (HANSEN, 1971; HANSEN and GRADMANN, 1971; HANSEN et al., 1973). The light intensity was modulated by sine-waves with frequencies ranging from 8 cycles/min to 1 cycle/4 h. The amplitudes and phase-shifts of the changes in membrane potential induced by the sinusoidally modulated light intensity were recorded by a chart recorder and a tapepuncher. The tapes were fed into the X8-computer of the Kieler Rechenzentrum and the Fourier coefficients were calculated. The fundamental wave was used for the frequency response and the higher harmonics served as a proof of linearity. Only in the region of the notch (see discussion and Fig. 1), were they allowed to exceed 10 to 15 % of the fundamental wave.

Before Fourier-analysis, the drift in the membrane potential was eliminated by subtracting a polynomial of the third degree obtained by regression analysis. The calculation of the poles and zeros of the transfer functions (frequency responses) was carried out by means of the ALGOL-programme of STROBEL (1968), modified for dialogue with the PDP—10—computer.

C. Results

The circles and crosses in Fig. 1 show the frequency responses of the amplitude (A) and phase-sift (φ) obtained by the procedure described above. The smooth lines represent a computer approximation to the experimental data. The curve fitting procedure was based on the following function H (ALGOL programme of STROBEL, 1968):

$$H = H_0 \frac{(n_1^{-1} jf - 1)(n_2^{-1} jf - 1)(n_3^{-1} jf - 1)}{(p_1^{-1} jf - 1)(p_2^{-1} jf - 1)(p_3^{-1} jf - 1)(p_4^{-1} jf - 1)(p_5^{-1} jf - 1)(p_6^{-1} jf - 1)} \quad (1)$$

with f being the frequency, $j = \sqrt{-1}$, H_0 the 'd.c. amplification'.

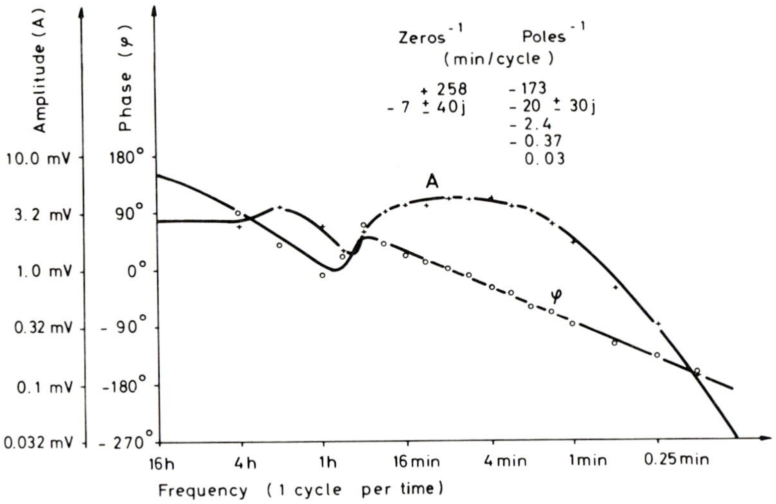

Fig. 1 Curve fitting applied to the frequency response of the light effect on the membrane potential of a single cell of *Nitella*. The inverse poles and the inverse zeros of the frequency response (smooth line) fitting the experimental data (crosses and circles for amplitude and phase, respectively) are given in the upper right-hand corner (in min/cycle, inverse of frequency).

The n_i's are called the 'zeros' (because the function H becomes zero, if $jf = n_i$), and the p_i's are called 'poles' (because H approaches infinity, if $jf = p_i$). The values of H_0, n_i, and p_i were adjusted to the experimental data by something like a least squares statistical method.

14 frequency responses obtained from 10 algae were fitted by Eq. (1). The mean values of the poles and zeros are shown in Table 1. In order to give a better appreciation of the time scale, the data in Table 1 as well as in Fig. 1 are given in inverse poles ($1/p_i$) and inverse zeros ($1/n_i$).

Table 1

Parameters of light frequency responses of the *Nitella* membranes. Mean values of the poles and zeros obtained from 14 frequency responses measured in red light (1 W/m^2), at a temperature of 17°C. The zeros and poles are listed by the numbers given in column 1. The values in columns 2 and 3 have to be multiplied or divided by the values displayed in columns 3 or 4 in order to get the range of the errors of the single measurement or of the mean value. The real components of the zeros 2 and 3 exhibit two signs, because both signs were found. Note that the parameters expressed in units of inverse of frequency (duration of period).

No.	(Zeros)$^{-1}$ (min/cycle)	(Poles)$^{-1}$ (min/cycle)	Time const. (s)	Multiplicative error variance		variance of mean	
1	145			2.3		1.24	
2 and 3	(+.−)7±51j			2.6	1.3	1.29	1.07
1		−68	650	2.5		1.27	
2 and 3		−20±60j	[19±57j]	2.5	1.6	1.27	1.13
4		−2.4	23	1.5		1.11	
5		−0.26	2.5	1.5		1.11	
6		−0.02	0.2	2.5		1.27	

D. Discussion

Contemplating the curve-fitting procedure of Eq. (1), several questions may arise.
1) Why is such a function as Eq. (1) used?
2) Do the parameters n_i and p_i result from the system under investigation or are they artefacts arising from the curve fitting procedure?
3) What is the biological meaning of the parameters n_i and p_i?

I. Transfer Functions and Chemical Kinetics

We start from the idea that the action of light on membrane potential is mediated by chemical reactions. It will be shown that Eq. (1) is closely related to linear chemical kinetics and this also provides an answer to question 3.

Chemical kinetics are based on rate laws which are normally non-linear (CASTELLAN, 1963). For instance, the time course of C in the reaction $A + B \overset{k_1}{\rightarrow} C$ and $C + D \overset{k_2}{\rightarrow} E$ is given by the following differential equation

$$\frac{dC}{dt} = k_1\, A\, B - k_2\, C\, D \qquad (2)$$

For the sake of simplicity, the brackets indicating concentrations are omitted.

Many equations like Eq. (2) are needed to describe a complete chemical system. This leads to very complicated mathematical problems which may be solved, if the system is known. The so-called 'inverse problem', however, that is finding the set of rate equations from the experimental results, is a job for a mathematical genius. However, for normal people, there is an approach to such problems; it is called linear relaxation theory in chemistry (EIGEN and DeMAEYER, 1963; WEI and PRATER, 1962; CAPELLOS and BIELSKI, 1972) and linear network theory in electronics (BODE, 1964; WOLFF, 1971).

The linear approach is achieved by making the variations in A so small that B and D may be assumed constant. In our experiments linearity was achieved by a small modulation depth of the light intensity and checked by the shape of the responses in membrane potential. (Normally the photoelectric effect in *Nitella* is non-linear (HANSEN, 1971; HANSEN and GRADMANN, 1971; HANSEN et al., 1973)). In this case Eq. (2) becomes a pseudo-first-order reaction

$$\frac{dC}{dt} = K_1 A - K_2 C \qquad \text{with } K_1 = k_1 B \text{ and } K_2 = k_2 D \tag{3}$$

As is known from electrical networks, frequency responses can be used instead of linear differential equations. (See 'operational method' in CAPELLOS and BIELSKI (1972)). Now four rules will be presented which show that all linearized chemical networks can be described by frequency responses as in Eq. (1).

Rule (1): The frequency response due to a (pseudo) first-order reaction like Eq. (3) is that of the well-known electrical low-pass filter

$$C(jf) = \frac{K_1/p_o}{(jf/p_o - 1)} A(if) \tag{4}$$

with $j = \sqrt{-1}$ and f the frequency (CAPELLOS and BIELSKI, 1972). In contrast to Eq. (3), C(jf) now is a frequency response, as is indicated by the argument jf. In Eq. (3), C was a function of the time, C(t). K_1 is the kinetic constant also seen in Eq. (3), and p_o, the 'pole' of the frequency responses, is related to the decay constant K_2 and the time constant τ, which is found from the square-wave response of a first-order reaction and given by the following equation:

$$p_o = \frac{K_2}{2\pi} = \frac{1}{2\pi\tau} \tag{5}$$

Rule (2): If there are two reactions in cascade, as in the scheme

$$A \overset{K_1}{\rightarrow} C \overset{K_C}{\rightarrow} E \overset{K_E}{\rightarrow} \qquad K_2 = K_C$$

then the single frequency responses of the two steps

$$C = \frac{K_1/p_C}{(jf/p_C - 1)} A \text{ and } E = \frac{K_2/p_E}{(jf/p_E - 1)} C \tag{4a, b}$$

are multiplied in order to describe the action of A on E. That is

$$E = \frac{\text{const.}}{(jf/p_C - 1)\,(jf/p_E - 1)}\,A \tag{6}$$

The poles p_C and p_E are related to K_C and K_E in the same way as shown for p_o and K_2 in Eq. (5). The 'const.' comprises all constants of the numerator which are of minor interest.

Rule (3): If there are two first-order reactions in parallel, then the sum S = C + E is measured and the frequency responses are added. As is well-known from elementary algebra the addition results in the creation of a zero, n:

$$S = E + C = \frac{\text{const.}\,(jf/n - 1)}{(jf/p_C - 1)\,(jf/p_E - 1)}\,A \tag{7}$$

with 'const' comprising the constant factors and

$$n = \frac{K_1 p_E + K_2 p_C}{K_1 + K_2} \tag{8}$$

Zeros may also arise in a chain of reversible reactions (CAPELLOS and BIELSKI, 1972).

From Eqs. (6) and (7) it can be seen that the poles remain unchanged whether the reactions are in series or in parallel and thus the poles provide direct information about the time constants of the single steps, from Eq. (5). However, this no longer holds, if a feed-back loop is involved in the system.

Rule (4): A feed-back loop may result in complex poles. (A complex pole means that it is no longer given by a real number, but by a complex one, comprising a real and an imaginary component.) A well-known formula for feed-back loops is (BODE, 1964)

$$V = \frac{M}{1 + B\,M} \tag{9}$$

where V is the frequency response of the whole system and M and B are frequency responses such as those in Eqs. (4), (6) or (7). In order to get a product of brackets, like those in Eq. (6) or (7), the sum in the denominator of Eq. (9) implies that the roots of a polynominal have to be calculated. Feed-back loops may create zeros, too. The four rules given above demonstrate that the frequency responses of linear chemical networks are always of the form shown in Eq. (1). That is why Eq. (1) was used to approximate the frequency responses obtained from the membrane potential in ***Nitella***.

II. The Significance of the Data Obtained by Curve Fitting

The second question, concerning the number of brackets and thus the number of poles and zeros in Eq. (1) used to fit the experimental data, will now be discussed. In the first place, at least three zeros are needed to describe the measured frequency responses. A pair of complex zeros is indicated by the notch in the frequency response of the amplitude (No. 2 and No. 3 in Table 1, see also Fig. 1). At low frequencies the phase-shift approaches 180° without an adequate change in amplitude. From network theory (BODE, 1964) it is known that this indicates the existence of a single real zero (No. 1 in Table 1). It can be shown by elementary mathematics that the difference between the number of poles and zeros must be 3, since the slope of the amplitude at high frequencies is f^{-3}. Thus at least 3 zeros and 6 poles are needed to describe the frequency response.

In order to find out whether more constants might be necessary to fit the experimental data, the computer was forced to use more than nine constants. Whereas the old poles and zeros shown in Fig. 1 kept their positions, the new ones arose at different frequencies in different cells, suggesting that within the range of the accuracy of our results they were superfluous. Often the real parts of these additional poles were positive; this is physical nonsense, since they indicate exploding systems.

If our results are compared with other curve fitting problems nine constants seem to be a lot. However, it should be kept in mind that this is a real benefit of the treatment in the frequency domain. There is one kind of frequency response that is capable of providing an enormous amount of data within a range of only one octave: spectroscopy, using visible light. Our approach is called spectroscopy of chemical relaxation (ALBERTY and HAMMES, 1960; CASTELLAN, 1963). Since relaxation spectra normally comprise real poles, instead of complex ones as in the case of spectral lines, the resolution is worse; however, the range of frequencies applied is greater, in our case more than three decades. The location of the poles of the 'photoelectric' effect in *Nitella* is very favourable, since the real poles, numbers 4, 5, and 6, are separated by about one decade in the frequency scale (see Table 1). At low frequencies, there are more constants to be determined. However, four of them are complex, and thus easily detected. Only the 'slow' real zero and the 'slow' real pole (No. 1, both) are determined with less accuracy.

III. The Biological Meaning of the Parameters n_i and p_i

The four rules discussed above show that the poles and the zeros of the frequency response are closely related to the kinetics of the biochemical reactions mediating the light effect on membrane potential and they are thus a better means of describing the photoelectric effect than maximum depolarization or hyperpolarization. This has been proved for the influences of temperature and light intensity on the light effect, and will be published elsewhere.

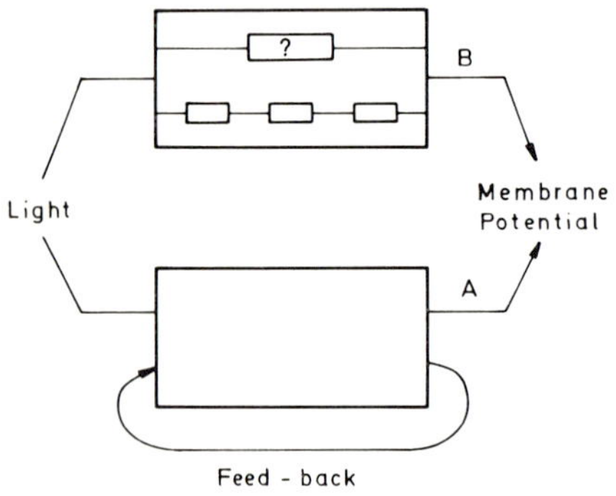

Fig. 2 A model based on the frequency response of the action of light on membrane potential in *Nitella* (for explanation see text).

Fig. 2 shows a model based on the frequency response data of Table 1. Rule (4) (and spontaneous oscillations of a period of about one hour) indicates that, because we have complex poles (No. 2 and No. 3), there must be a feed-back loop. This feed-back loop, located in pathway A, must be in parallel with pathway B, in order to create the complex zeros No. 2 and No. 3 (rule No. 3). The three boxes in series in pathway B represent the reactions associated with the poles 4, 5, and 6 (rule No. 2). There is a question mark in the fourth box because it is still unknown whether pole No. 1 which has been put in pathway B for mathematical reasons is really a third pathway or a feed-back loop controlling B.

The parallel pathways may be different ion transport systems, e.g. a hyperpolarizing pump B slowing down with light (HANSEN and KEUNECKE, 1974) and another pump A speeding up with light (VREDENBERG and TONK, 1973). It may also be the same pump, slowed down via the biochemical pathway B and accelerated via A.

This parallel pathway model is supported by several experimental findings. For instance the fundamental waves of the two pathways cancel each other in the region of the notch; highly non-linear distorted signals occur and positive as well as negative signs of the real components of the zeros are observed (Table 1). Equal apparent activation energies suggest an ultimate relationship of the zeros and poles in this model (unpublished results).

If we want to speculate about the nature of the three reactions in series in pathway B, at higher light intensity and temperature (unpublished results) poles 4 and 5 provide time constants similar to those of ATP– turnover (SLAYMAN et al., 1973) and transport reactions in the membrane (SLAYMAN and SLAYMAN, 1974).

E. Acknowledgements

This work was founded by the Deutsche Forschungsgemeinschaft. Thanks are due to Prof. Dr. W. Kroebel and Prof. Dr. K. Vanselow for their support and to Doz. Dr. R. Isermann and Dipl.Ing. P. Necker in Stuttgart for their help in getting the program for the calculation of the transfer-function. The computations were done in the Kieler Rechenzentrum.

F. References

ALBERTY, R.A., HAMMES, G.G., Z. Elektrochemie **64**, 124–128 (1960)
ARDEN, G.B., Progr. Biophys. Mol. Biol. **19**, 371–421 (1969)
BODE, H.W., Network Analyses and Feed-Back Amplifier Design, D. van Nostrand Comp. Inc., New York 1964
CAPELLOS, C., BIELSKI, B.H.J., Kinetic Systems, Wiley, New York 1972
CASTELLAN, G.W., Z. Elektrochemie **67**, 898–908 (1963)
EIGEN, M., DeMAEYER, L., in Technique of Organic Chemistry. Vol. 8/2 (ed. A. Weissberger), Interscience Publishers, New York 1963
HANSEN, U.P., Biophysik **7**, 223–227 (1971)
HANSEN, U.P., GRADMANN, D., Plant Cell Physiol. **12**, 335–348 (1971)
HANSEN, U.P., WARNCKE, J., KEUNECKE, P., Biophysik **9**, 197–207 (1973)
HANSEN, U.P., KEUNECKE, P., Atomkernenergie **23**, 5–8 (1974)
SLAYMAN, C.L., LONG, W.S., LU, C.Y.-H., J. Membrane Biol. **14**, 305–338 (1973)
SLAYMAN, C.L., SLAYMAN, C.W., Proc. Nat. Acad. Sci., in press (1974)
STROBEL, H., Systemanalyse mit determinierten Testsignalen. VEB Verlag Technik, Berlin 1968
VREDENBERG, W.J., TONK, W.J., Biochim. Biophys. Acta **298**, 354–368 (1973)
WEI, J., PRATER, C.D., Advances in Catalysis **13**, 203–392 (1962)
WOLF, H., Lineare Systeme und Netzwerke. Springer-Verlag, Berlin—Heidelberg—New York 1971

Dielectric Breakdown of Cell Membranes

U. Zimmermann, G. Pilwat, and F. Riemann

A. Introduction

A Coulter Counter is a measuring device for counting and sizing non-conducting particles suspended in a conducting medium. This instrument is based on the principle that the electrical resistance in a small orifice, across which a voltage is applied, will change when a suspension of non-conducting particles is sucked through it. The current or voltage pulses arising are, to a first approximation, proportional to the size of the particles (size = shape factor X volume); this means that after linear electronic amplification and subsequent pulse height analysis the size distribution of each population under investigation can be obtained. Using conventional Coulter Counters, however, the measured size distribution is generally strongly distorted (skewed) due to inhomogeneities in the electric field strength in and near of the orifice; the pulse height becomes dependent on the path and orientation of each particle in the orifice (GROVER et al., 1969a, b, 1972; THOM, 1972). As described previously (ZIMMERMANN et al., 1973) the distortion arising from the inhomogeneous field can be eliminated by hydrodynamic focusing of the suspension flow. Then the particles travel on the same pathway, parallelly oriented, along the central axis through the orifice, and a true measurement of the real size distribution of the population can be obtained.

When measuring the size distributions of bacteria and red blood cells with this improved detector system we found an apparent shift of the distributions towards smaller sizes when the electric field strength in the orifice was above a critical value (ZIMMERMANN et al., 1973). This effect was not observed using Latex particles or bacteria fixed with glutardialdehyde. We proposed that this apparent shift in the size distributions indicate an increase of the conductance of the cell membrane; when the current flows partly through the cell interior, the cell volume will be underestimated. Several observations confirmed that this increase in membrane conductance is not caused by a 'punch-through effect' (COSTER, 1965, 1969) but by a dielectric breakdown. The most significant evidence for dielectric breakdown is to be seen in the fact that red blood cells almost totally lost their haemoglobin after passing through the electric field (ZIMMERMANN et al., 1974a, b). The 'ghost-cells' showed resealing properties, since they were able to trap radioactively labelled albumin which had been added immediately after dielectric breakdown, whereas two hours later only a very small incorporation could be observed.

It seemed interesting to extend these investigations to algal cells because, with these cells too, one should in principle be able to incorporate non-permeating substances by dielectric breakdown and subsequent resealing. Furthermore, we wished to examine how far the critical external electric field strength, at which dielectric breakdown could be observed, is influenced by specific changes in the membrane properties. These experiments were started with erythrocytes in a plate condenser which formed part of a high-voltage discharge circuit, in order to simulate phenomena occuring in the orifice of the Coulter Counter in a homogeneous electric field (ZIMMERMANN et al., 1974a); the latter is more accessible to theoretical treatment. In such investigations, however, part of the information obtained using a Coulter Counter is lost, especially that concerning the behaviour of individual volumes in the electric field. To begin with red blood cells were chosen because the release of the haemoglobin can be used as a highly sensitive indicator of dielectric breakdown.

B. Materials and Methods

Ochromonas malhamensis cells were grown as described elsewhere (KAUSS, 1967; HUTNER et al., 1953). For measurements, *Ochromonas* cells were washed twice in a solution containing 0.5 % NaCl + 0.1 % glucose by centrifugation with about 1000 X g for 6 min.

Bovine blood was stored under sterile conditions in ACD buffer at 4°C for not more than 2 days. The red blood cells were prepared in the usual way (BODEMANN and PASSOW, 1972). In the experiment, the cells were suspended in isotonic NaCl-solution (0.9 %).

The electronic size measurements were carried out with an improved Coulter Counter (AEG-Telefunken, Germany) as previously described (ZIMMERMANN et al., 1973). For red blood cells and *Ochromonas* an orifice 40 µm in diameter and 40 µm in length was used. The flow velocity through the orifice was about 2 m/s. Electronic data are given in the legends of the figures. The details of the pulse method are given in section C. III.

C. Results and Discussion

I. The Dependence of the Size Distributions of *Ochromonas* on the External Electric Field Strength

In Fig. 1 typical size distributions[1]) of *Ochromonas malhamensis* at different detector currents (i.e. at different electric field strength in the orifice) are plotted. The size distri-

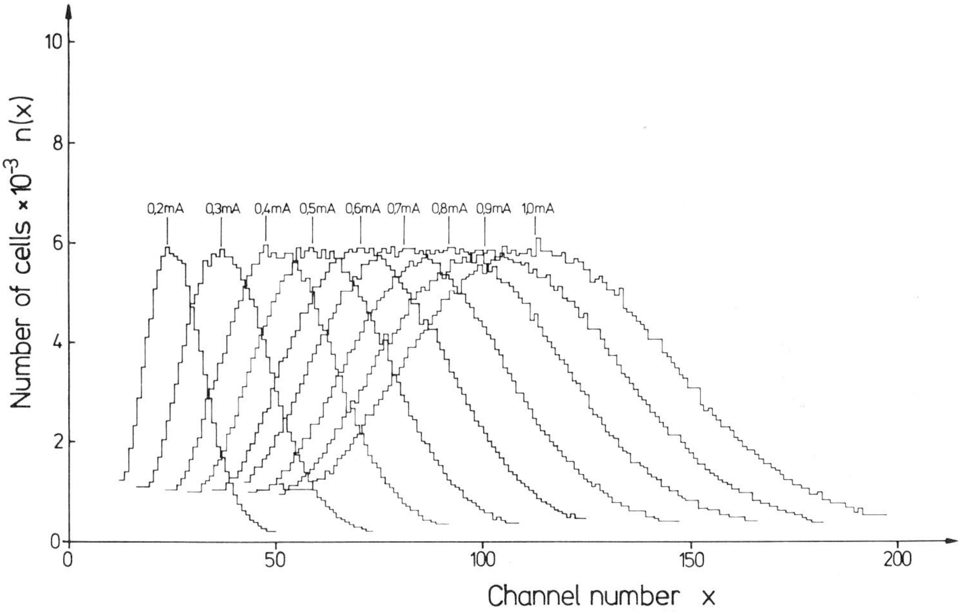

Fig. 1 Size (pulse-height) distribution of ***O. malhamensis*** as a function of the detector current (0.2 to 1.0 mA). Orifice: 40 µm in diameter and 40 µm in length; electrolyte: 0.5 % NaCl + 0.1 % glucose. Electronic amplification: 10.

---

[1])The pulse height is proportional to the size of the particles, the factor of proportionality depending on the detector current.

butions are normally distributed. On plotting different pulse heights (channel numbers) of the ***Ochromonas*** size distributions against detector voltage, two straight lines with a characteristic sharp bend are observed (Fig. 2). The correlation coefficients are 0.998 for all curves below and above the critical voltage at which the change in slope occurs. The critical voltage depends on the cell size, i.e. with increasing size the critical voltage decreases, the result of which is an apparent shift of the size distributions towards smaller sizes. In contrast, for Latex particles a linear correlation between pulse height and detector voltage is found, throughout the test range. The correlation coefficient is 0.998. From similar findings with bacteria and red blood cells we conclude that the discontinuities observed for ***Ochromonas*** cells in the electric field of the Coulter Counter orifice are due to dielectric breakdown of the cell membranes. A possible objection to this interpretation must be considered, however. As shown by ALLEN and MASON (1962) theoretically, and by GROVER et al. (1969b, 1972) experimentally, the shape of a cell can be altered in an electric field. Because the Coulter Counter measures sizes, the observed shift in the size distribution towards smaller values could be explained by a decrease of the shape factor, particularly as the shape factor can be volume- and age-dependent. The occurrence of a dielectric breakdown would on the contrary lead to an apparent decrease in the volume because the current lines partially penetrate the cells. Despite the fact that a change in shape due to electric fields was observed at much lower field strengths by GROVER et al. (1972),

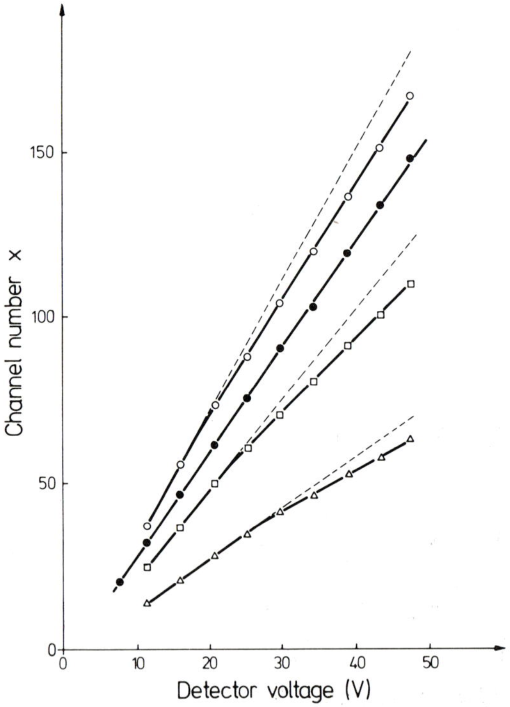

Fig. 2 Dielectric breakdown of the membrane of ***O. malhamensis***. Pulse heights (channel numbers) of the size distributions of ***O. malhamensis*** (presented in Fig. 1) are plotted as a function of the detector voltage. Squares: pulse heights corresponding to the modal volume of 178.3 μm^3. Circles: pulse heights corresponding to a smaller volume of 100.2 μm^3. Triangles: pulse heights corresponding to a larger volume of 267.7 μm^3. Filled circles: pulse heights corresponding to the modal volume of 3.34 μm^3 of Latex particles. Orifice and electrolyte as in Fig. 1; electronic amplification for ***O. malhamensis***: 10, and for Latex particles: 448.

this interpretation can be excluded in our experiments for the following reason. Red blood cells are known to be deformed into prolate ellipsoids by hydrodynamical focusing; their shape factor is 1.08 (THOM, 1972; ZIMMERMANN et al., 1974a). This value is near to that for an infinite cylinder (1.0) representing the lowest value for any shape factor. The shape factor for ***Ochromonas*** has not been determined yet, but since ***Ochromonas*** cells, like erythrocytes, have no cell walls, one can assume that these cells become prolate ellipsoids too. The modal value of the ***Ochromonas*** size distribution decreases by about 30 %, however, which would mean that the shape factor would become about 0.7, which is impossible.

II. Calculation of the 'Generated Membrane Potential'

The potential generated across the membrane (V_M) which is responsible for the observed dielectric breakdown can be calculated by applying Laplace theory. An analytical solution of the Laplace equation is not possible since the boundary conditions cannot be completely specified. For a rough estimation of the generated membrane potential we must introduce two simplifications. We replace the bell-shaped potential gradient along the axis of the orifice by a 'linear potential gradient', E_o, assuming that 97 % of the potential difference between the two electrodes drops linearly along the length of the orifice.

We further postulate that in the centre of the orifice there exists a small three-dimensional region in which the electric field is nearly homogeneous. In this region a solution of the Laplace equation is possible in spherical (r, ϑ, φ) and ellipsoidal coordinates (ξ_1, ξ_2, ξ_3) (cf. ZIMMERMANN et al., 1974a). First we consider a spherical particle, with an inner and outer radius a and b, respectively, and a membrane thickness d (d = b–a).

Let σ_1, σ_2, and σ_3 be the conductivities and Φ_1, Φ_2, and Φ_3 denote the potentials, where the suffixes 1, 2, and 3 refer to the interior of the cell, the membrane and the external solution, respectively.

For this arrangement the following boundary conditions hold:

$$\Phi \rightarrow - E_o\, r \cos\vartheta, \text{ for } r \rightarrow \infty; \ \Phi \text{ must be finite, for } r = 0;$$

$$\Phi_1(a) = \Phi_2(a) \text{ and } \frac{\sigma_1}{a} \cdot \frac{d\Phi_1}{dr} = \frac{\sigma_2}{a} \cdot \frac{d\Phi_2}{dr}, \text{ for } r = a; \tag{1}$$

$$\Phi_2(b) = \Phi_3(b) \text{ and } \frac{\sigma_2}{b} \cdot \frac{d\Phi_2}{dr} = \frac{\sigma_3}{b} \cdot \frac{d\Phi_3}{dr}, \text{ for } r = b.$$

The problem is now well defined and yields the following expression for the generated potential difference V_M across the membrane, superimposed upon the intrinsic membrane potential:

$$V_M = \frac{\frac{3}{2+\beta} E_0 \left[a - b + \frac{1-\alpha}{2+\alpha} a^3 \left(\frac{1}{a^2} - \frac{1}{b^2}\right)\right] \cos\vartheta}{1 + 2\left(\frac{1-\beta}{2+\beta}\right)\left(\frac{1-\alpha}{2+\alpha}\right)\left(\frac{a}{b}\right)^3} \tag{2}$$

where $\alpha = \frac{\sigma_1}{\sigma_2}$ and $\beta = \frac{\sigma_2}{\sigma_3}$.

Since the conductivities within the cell interior and the external solution are nearly equal and much larger than the conductivity of the membrane, Eq. (2) reduces to

$$V_M = 1.5\, b\, E_o \cos\vartheta \tag{3}$$

In ellipsoidal coordinates an analogous treatment with one semi-axis parallel to the external electric field yields the following equation (BERNHARDT and PAULY, 1973):

$$V_M = f_i\, a_i\, E_o \tag{4}$$

where f_i is the shape-factor expressed by

$$f_i = 2\left[2 - a_1 a_2 a_3 \int_0^{\infty} \frac{ds}{(s+a_i^2)\sqrt{\sum_{n=1}^{3} (s+a_n^2)}}\right]^{-1} \tag{5}$$

and where a_n, n= 1, 2, 3, denote the semi-axes, a_i the semi-axis parallel to the external electric field.

With the data of Fig. 2 for the linear gradient E_o, V_M can be calculated. For the modal volume of 178 μm^3 of the *Ochromonas* size distribution the critical value of the generated membrane potential is about 2.8 V using a spherical approximation (b = 3.5 μm) and equal about 3.7 V using an ellipsoidal approximation (major semi-axis 6.4 μm, minor semi-axis 2.6 μm). If we assume that the critical generated membrane potential depends only upon the specific properties of the cell or the cell membrane, and that these properties are not a function of the volume, we can calculate from Eq. (3) for different volumes the critical external field strength for dielectric breakdown for each volume. In Fig. 3 the theoretically and experimentally determined dependence of the external electrical field strength is plotted against the volume of *Ochromonas* and bovine red blood cells. The experimental data for the *Ochromonas* plot were taken from Fig. 2 and the data for the bovine red blood cells plot from results described elsewhere (ZIMMERMANN et al., 1974a). As can be seen the theoretically calculated curves with V_M = 2.8 V for *O. malhamensis* and V_M = 1.6 V for

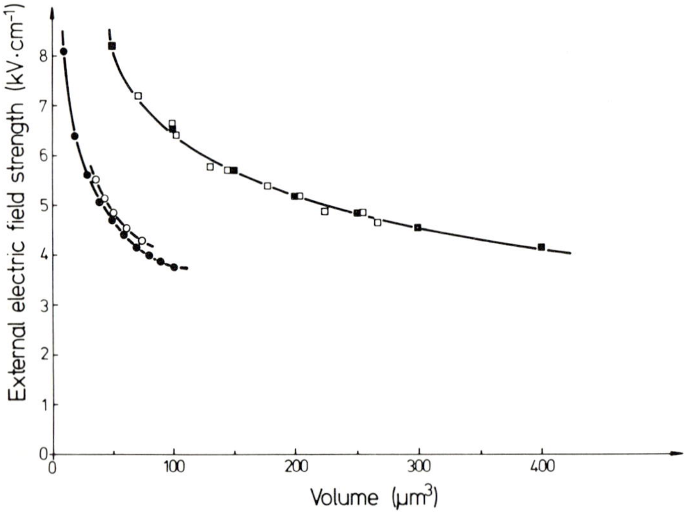

Fig. 3 The external electric field strength producing dielectric breakdown as a function of the particle volume. Filled squares: external field strength, calculated by Eq. (3) with V_M = 2.8 V for *O. malhamensis*, versus the volume. Squares: external field strength determined from the experimentally-determined critical detector voltages of *O. malhamensis* distributions. Data partly were taken from Fig. 2. Filled circles: external field strength, calculated by Eq. (3) with V_M = 1.6 V for bovine red blood cells, versus the volume. Circles: external field strength determined from the experimentally-determined critical detector voltages of bovine red blood cell distributions.

bovine red blood cells fit well the experimentally determined ones. If we assume that the generated potential difference of about 2.8 V drops across the bimolecular lipid leaflet the electric field strength in the membrane is in the order of 6–7 • 10^6 V/cm. This value is comparable with the breakdown voltages for bilayers of most oils which are of the order of 20 • 10^6 V/cm.

The agreement is in fact slightly better, since this generated membrane potential is superimposed on the intrinsic membrane potential during the very short time of passage through the orifice. This intrinsic membrane potential is of the order of 10^6 V/cm as was shown by COSTER (1973) in his analysis of the membrane as a double fixed charge system. Contributions from a polarisation effect due to the separation of the cations and anions in the ion cloud near the charged surface of the cells as discussed by NEUMANN and ROSENHECK (1973) can be excluded, since according to the SCHWARZ equation (SCHWARZ, 1962), the relaxation time for such an ion separation due to the external field would be of the order of 2.5 ms whereas the particles cross the orifice in about 20 μs.

III. Dielectric Breakdown in Homogeneous Electric Fields

For investigations of the dielectric breakdown in homogeneous electric fields, the suspension was placed in a temperature-controlled plexiglass chamber in which, at a distance of 1 cm, two flat platinum electrodes were mounted ($A = 3 \times 3$ cm^2). This chamber formed part of a high-voltage discharge circuit. The high voltage storage capacitor was discharged via a spark gap and in this way a potential was applied to the electrodes of the chamber. The voltage was varied between 2 kV and 6 kV, the decay constant was 27 μs, and the resistance between the electrodes when the chamber was filled with the suspension was 10 Ω. After dielectric breakdown the released haemoglobin was measured in the supernatant, the suspension having been centrifuged at 10 000 X g for 15 min.

In Fig. 4 the amount of haemoglobin released per 100 V/cm by dielectric breakdown is plotted against the external electric field strength. The maximum of the curve should correspond to the dielectric breakdown of the mean (or modal) volume of the size

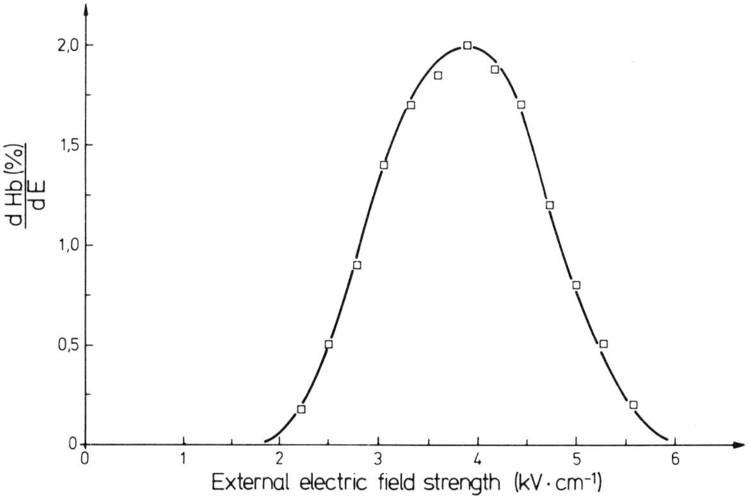

Fig. 4 Dielectric breakdown of bovine red blood cells in homogeneous electric fields. The release of % haemoglobin per 100 V/cm is plotted versus the external electric field strength.

distribution. It can be seen that the external field strength (3.9 kV/cm) is closely equal to that calculated for the critical field strength producing dielectric breakdown of the modal volumes in the orifice (4.2 kV/cm, ZIMMERMANN et al. (1974a)). When calculating the generated membrane potential one must take into account that shape and orientation in the two experiments are considerably different. In the plexiglass chamber the usually oblate red blood cells are randomly oriented relative to the direction of the electrical field, whereas in the Coulter Counter the prolate deformed red blood cells pass through the orifice with their major semi-axis oriented parallel to the axis of the orifice. Using Eqs. (3) and (4) it is found, however, that the difference for cells, parallel or perpendiculary oriented to the direction of the electrical field, is only a very small one (ZIMMERMANN et al., 1974a). The mean calculated value of 1.5 V for the generated membrane potential is nearly the same as that calculated for the dielectric breakdown membrane voltage in the orifice of the Coulter Counter.

IV. Effects of Membrane Stabilizing Substances on the Critical Voltage

It is well-known that low concentrations of tranquilizers, anaesthetics, anti-inflammatory compounds and other substances stabilize erythrocyte membranes against hypotonic haemolysis, but higher concentrations of these substances cause lysis. Therefore, the dose-response curves are in general biphasic. In Fig. 5 a typical haemolysis curve for bovine red blood cells subjected to hypotonic solutions is plotted against increasing concentrations of the local anaesthetic benzyl alcohol. The experimental conditions were chosen to be similar to those of METCALFE (1970), taking a 90 % release of haemoglobin in the absence of benzyl alcohol as the reference point. Benzyl alcohol at concentrations up to 60 mM protects the red blood cells against osmotic haemolysis in agreement with the finding of METCALFE. When haemolysis is caused by application of a homogeneous electric field a

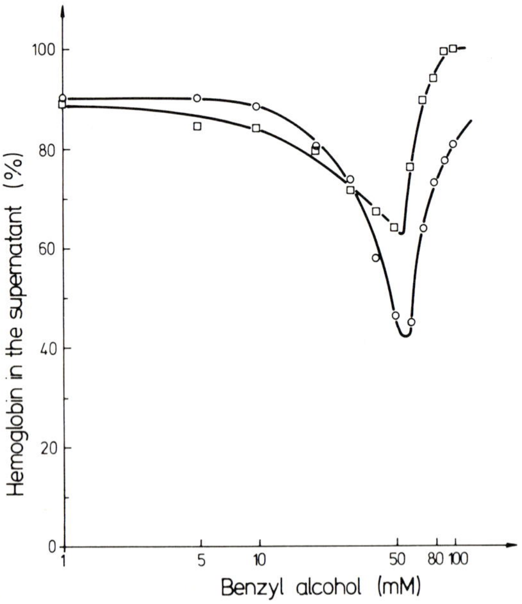

Fig. 5 Effect of benzyl alcohol on the membrane of bovine red blood cells. The amount of haemoglobin in the supernatant as a measure of haemolysis is plotted versus the concentration of benzyl alcohol. Circles: osmotic haemolysis. Squares: haemolysis by dielectric breakdown.

similar biphasic curve is obtained (Fig. 5). This preliminary result demonstrates that the critical voltage producing dielectric breakdown depends on the membrane properties, since NMR and ESR investigations have revealed that benzyl alcohol influences the interaction between proteins and lipids in the membrane (METCALFE, 1970). However, shape changes due to the drug certainly occur in both experiments (DEUTICKE and GERLACH, 1967; SEEMAN, 1970) and must be taken into account in extending such experiments to the Coulter Counter.

D. Acknowledgements

This work was supported by a grant from the Deutsche Forschungsgemeinschaft, Bonn–Bad Godesberg, Sonderforschungsbereich 160. The authors should like to thank H.J. Buers and H. Koch for expert technical assistance. We are very grateful to Prof. Kauss, University Trier/Kaiserslautern, for supplying us with the ***Ochromonas*** culture.

E. References

ALLEN, R.S., MASON, S.G., Proc. Roy. Soc. Ser. A. Math. Phys. Sci. **267**, 45–76 (1962)
BERNHARDT, J., PAULY, H., Biophysik **10**, 89–98 (1973)
BODEMANN, H., PASSOW, H., J. Membrane Biol. **8**, 1–26 (1972)
COSTER, H.G.L., Biophys. J. **5**, 669–686 (1965)
COSTER, H.G.L., Aust. J. Biol. Sci. **22**, 365–374 (1969)
COSTER, H.G.L., Biophys. J. **13**, 1119–1123 (1973)
DEUTICKE, B., GERLACH, E., Klinische Wochenschrift **19**, 977–983 (1967)
GROVER, N.B., NAAMAN, J., BEN–SASSON, S., DOLJANSKI, E., Biophys. J. **9**, 1398–1414 (1969a)
GROVER, N.B., NAAMAN, J., BEN-SASSON, S., DOLJANSKI, F., NADAV, F., Biophys. J. **9**, 1415–1425 (1969b)
GROVER, N.B., NAAMAN, J., BEN–SASSON, S., DOLJANSKI, F., Biophys. J. **12**, 1099–1117 (1972)
HUTNER, S.H., PROVASOLI, L., FILFUS, J., Ann. N.Y. Acad. Sci. **56**, 852–862 (1953)
KAUSS, H., Z. Pflanzenphysiol. **56**, 453–465 (1967)
METCALFE, J.C., in Permeability and Function of Biological Membranes (eds. L. Bolis, A. Katchalsky, R.D. Keynes, W.R. Loewenstein, B.A. Pethica), pp. 222–234, North-Holland Publishing Company, Amsterdam and London 1970
NEUMANN, E., ROSENHECK, K., J. Membrane Biol. **10**, 279–290 (1972)
SCHWARZ, G., J. Phys. Chem. **66**, 2636–2642 (1962)
SEEMAN, P., in Permeability and Function of Biological Membranes (eds. L. Bolis, A. Katachalsky, R.D. Keynes, W.R. Loewenstein, B.A. Pethica), pp. 40–56, North-Holland Publishing Company, Amsterdam and London 1970
THOM, R., AEG-Telefunken publication No. N1/EP/V (1972)
ZIMMERMANN, U., PILWAT, G., RIEMANN, F., Biophys. J., in press (1974a)
ZIMMERMANN, U., PILWAT, G., RIEMANN, F., Z. Naturforsch. **29c**, 304–305 (1974b)
ZIMMERMANN, U., SCHULZ, J., PILWAT, G., Biophys. J. **13**, 1005–1013 (1973)

The Effect of pH on the Low Frequency Capacitance of the Membranes of *Chara corallina*

H.G.L. Coster and J.R. Smith

A. Introduction

The electrical and ion transport properties of the plasma membrane of living cells have been shown to be sensitive to the pH of their external environment (CLARK and STRICKHOLM, 1971; COSTER, 1969; LANNOYE et al., 1970; STRICKHOLM et al., 1969). Such effects could arise through pH induced conformational changes in membrane proteins and/or changes in the degree of ionization of -COOH and -NH_2 groups incorporated in acidic and basic amino acids in these proteins.

As a first approximation the plasma membrane has traditionally been treated as if it behaved, electrically, simply as a capacitance, C, and a conductance, G, in parallel.

It has been long established that at high frequencies (e.g. $>$ 1 kHz for squid axons) the membrane impedance disperses with frequency, although the capacitance element of the impedance has a phase angle which is substantially frequency independent (COLE, 1968; SCHWANN, 1957). At very low frequencies ($<$200 Hz), however, additional dispersive effects are observed whereby the membrane capacitance increases with decreasing frequency (PALTI and ADELMAN, 1969; COLE, 1970).

At very low frequencies (1—200 Hz) the reactive impedance of the capacitance is very large in comparison with the total membrane impedance and so the associated phase angle is very small. Hence accurate measurements of membrane capacitance are inherently difficult at such low frequencies.

For accurate measurement of just the amplitude of the a.c. impedance, a four terminal method employing intracellular electrodes is essential. Some such experiments have been reported in the literature (CONTI, 1970; PALTI and ADELMAN, 1969). However, to explore the behaviour of the membrane capacitance in the very low frequency region very accurate measurements of both relative phase and amplitude of the injected sinusoidal current and membrane potential difference have to be made. In the present experiments this was achieved by the use of a high precision measuring technique outlined below.

A voltage controlled oscillator generated a pulse train whose frequency was controlled by a small on-line computer (DEC PDP 11) via a digital-to-analogue converter and a digital interface. These pulses were used to scan a read-only memory (which had been pre-programmed with a sine function) which, in conjunction with a digital-to-analogue converter, produced the desired sinusoidal current. The pulse train frequency could also be separately divided under computer control, thus producing a variable number of clocking pulses during each sinewave cycle. These tightly phase-locked clocking pulses triggered a multi-channel analogue-to-digital converter that sampled both the current and membrane potential in turn. The frequency of the sinewave injected into the cell was monitored by a frequency counter whose digital output could be read by the computer via the digital interface. At each of a programmed series of frequencies, a programmed number of cycles of current were injected into the cell and the results accumulated in the computer memory. In this fashion a good signal-to-noise ratio could be obtained with only a small perturbation of the membrane potential. In the final analysis, the method of least squares was employed to fit sinusoidal waveforms to the stored data points for current and potential. The relative phase and amplitude between these sinusoidal current and potential waveforms were then calculated. At below 200 Hz the system was capable of resolving phase angles of 0.01° and amplitude ratios of 0.1 %. A schematic diagram of the experimental set up is shown in Fig. 1.

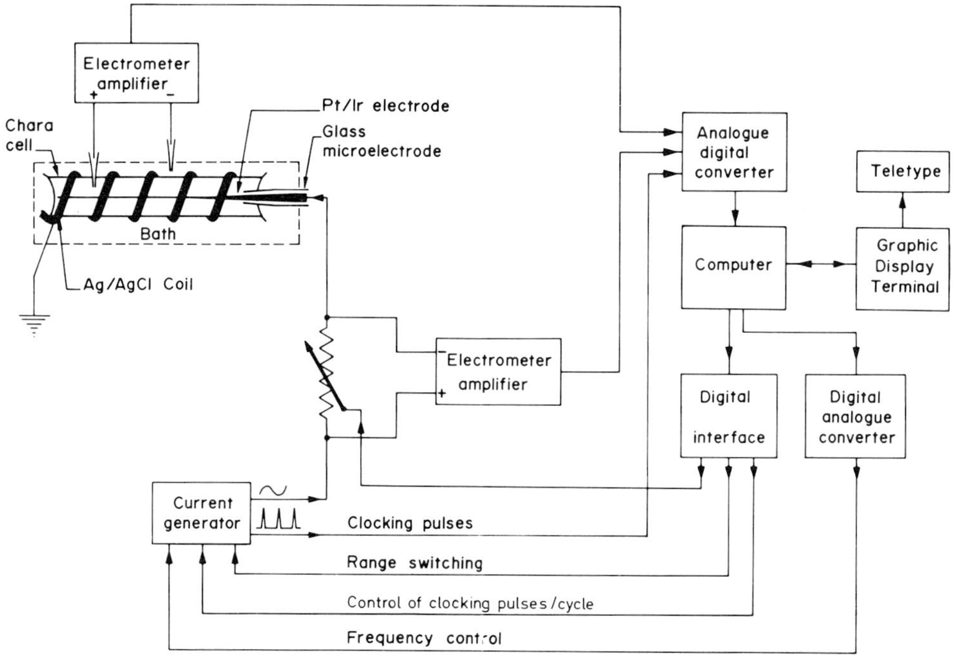

Fig. 1 A diagram of the experimental set-up. Membrane potentials were measured with 2N KCl filled intra and extra cellular microelectrodes. Longitudinal Pt/Ir wire electrodes, manipulated through a glass micropipette inserted into the cells, and the external Ag/AgCl coil were used for current electrodes. The cells were typically 2–5 mm long and were bathed in a flowing A.P.W. The computer controlled measuring technique is described in the text.

B. Results

I. Low Frequency Variation of Capacitance and Conductance

The typical variation of membrane capacitance, C, and conductance, G, with frequency for a cell of ***Chara corallina*** in artificial pond water[1]) A.P.W. at pH 9 is shown in Fig. 2. It is evident that at high frequencies (>100 Hz) the membrane capacitance is largely independent of frequency and has a value of 0.7 $\mu F/cm^2$, a figure very similar to that frequently quoted, though usually not acknowledged as a limiting high frequency value, for a variety of cell membranes (COLE, 1968).

---

[1])A.P.W.: 0.1 mN KCl, 2.0 mN NaCl, 0.2 mN $CaCl_2$. The cells were cultured in a medium (FORSBERG, 1965) which has pH ~ 8.5.

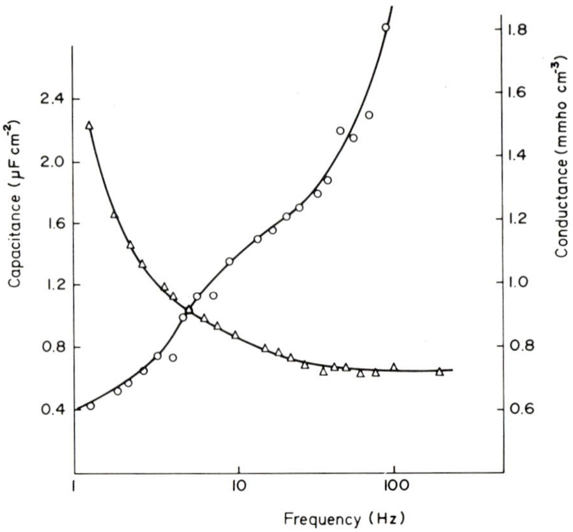

Fig. 2 The membrane capacitance C(△) and conductance G(o) as a function of frequency for a cell of *Chara corallina* in A.P.W. at pH 9.

The measured membrane capacitance increases dramatically with decreasing frequency, reaching a value of 2.4 μF/cm^2 (or sometimes more) at 1 Hz. This is in contrast to the behaviour of the membrane conductance, which increases with increasing frequency at low frequencies and continues to increase at high frequencies. It is immediately apparent that the membrane does not behave as an impedance element with a simple dielectric capacitance. Further, the time constant (τ = C/G) of the membrane is a function of frequency as shown in Fig. 3. This implies that estimates of the membrane capacitance, based on measuring the rise time of the membrane potential following a step change in the injected current, will not yield a very meaningful quantity.

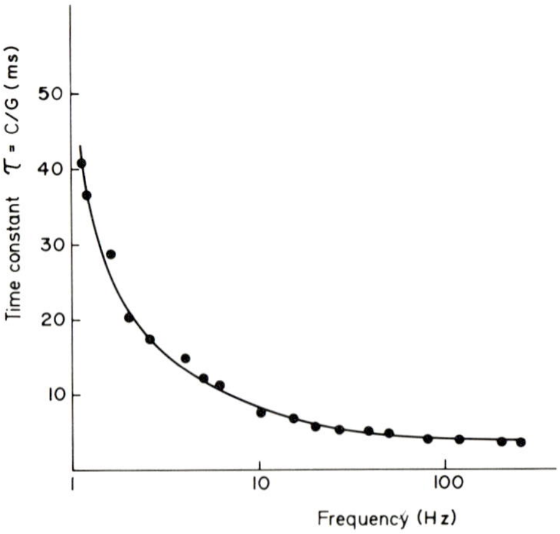

Fig. 3 The membrane time constant τ = C/G at pH 9 as a function of frequency.

II. The Effect of pH on the Membrane Capacitance

The membrane capacitance as a function of frequency for a cell of ***Chara corallina*** bathed in A.P.W. of pH 5.5 and pH 9 is shown in Fig. 4A. For frequencies below 4 Hz the membrane capacitance at pH 9 is much larger than at pH 5.5. At pH 5.5 the capacitance dispersed from a limiting low frequency value of 1.2 $\mu F/cm^2$ to a limiting high frequency value of 0.7 $\mu F/cm^2$ over the frequency range 5–50 Hz. At pH 9 the membrane capacitance was still increasing with decreasing frequency at 1 Hz (i.e. dispersion occurred over a wider frequency range).

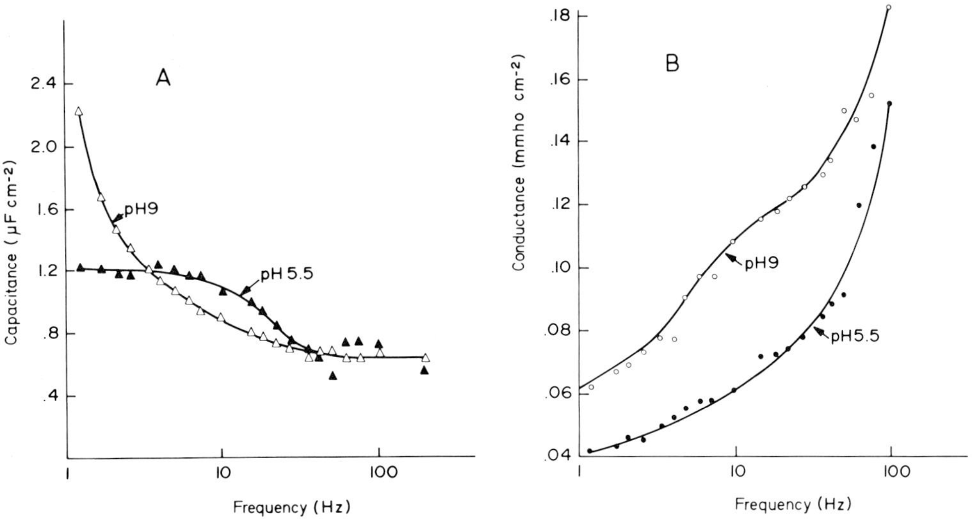

Fig. 4 The membrane capacitance (A) and conductance (B) as a function of frequency for a cell in A.P.W. at pH 9 and pH 5.5.

The limiting high frequency (>100 Hz) capacitance (0.7 $\mu F/cm^2$) was independent of pH. The membrane conductance, Fig. 4B, was found to be pH dependent over the entire frequency range investigated.

C. Theory and Discussion

It would appear from the frequency independence of the membrane capacitance above 100 Hz in the present results that any structure and hence component electrical elements present in the membrane cannot be detected in the impedance measurements at these frequencies. At high frequencies the reactive impedance of any capacitative elements of such impedance elements of the substructure will be small in comparison with their conductances. It is therefore not unexpected that the overall system then behaves as if it contains a single, frequency independent, capacitance. For a membrane of overall thickness 9 nm and dielectric constant of 10, this high frequency capacitance would be 0.98 $\mu F/cm^2$, which is close to the observed value.

At low frequencies (<100 Hz), the observed strong dispersion in membrane capacitance and conductance (Figs. 2 and 4) show that the membrane at low frequencies cannot be treated as a simple homogeneous system having an impedance with a single time constant. The strong increase in membrane capacitance with decreasing frequency reported here has been previously noted in other materials. For example, PALTI and ADELMAN (1969) reported that for squid axons the membrane capacitance, while virtually constant above 200 Hz, increased steeply below 200 Hz.

At sufficiently low frequencies the reactive impedances of some of the substructural layers within the membrane such as, for example, the three layers discernible under electron microscopy, will become comparable to their conductances. If the component layers of the membrane have different time constants, their presence will become evident in the increase in overall measured membrane capacitance with decreasing frequency. At these low frequencies, polarization effects in the diffusion of ions within the membrane can give rise to a further contribution to the capacitance which decreases rapidly with increasing frequency (COLE, 1965; OFFNER, 1969; COSTER, 1973a). The diffusion polarizations also give rise to a dispersion in the conductance.

The marked effect of pH on the low frequency membrane capacitance in the experimental results presented indicates that the electrical parameters of the substructural layers are pH sensitive. The absence of pH effects at high frequencies indicates that the overall dimensions and dielectric constant of the membrane are pH independent, or vary in exact direct proportion to each other.

The decrease in low frequency capacitance with decreasing pH points, in particular, to an increase with decreasing pH in the dimensions of the substructural layer of lowest conductivity, without affecting the total membrane thickness. The alternative possibility that a decrease in the dielectric constant of one of the substructural layers was responsible is unlikely, since this would have shown up in the high frequency capacitance.

The observed dispersion of membrane conductance and capacitance, and particularly their dependence upon pH, is very similar to the theoretical dispersion characteristics of a membrane containing two fixed charge regions, of opposite sign, in contact (COSTER, 1973a). When such a double fixed charge membrane (DFCM) is permeated by a dilute ionic solution, a transition region (depletion layer) is generated, at the junction of the two oppositely charged lattices, in which the mobile ion concentrations are very low (MAURO, 1962; COSTER, 1965). This region has a low conductivity. Outside this depletion layer, the

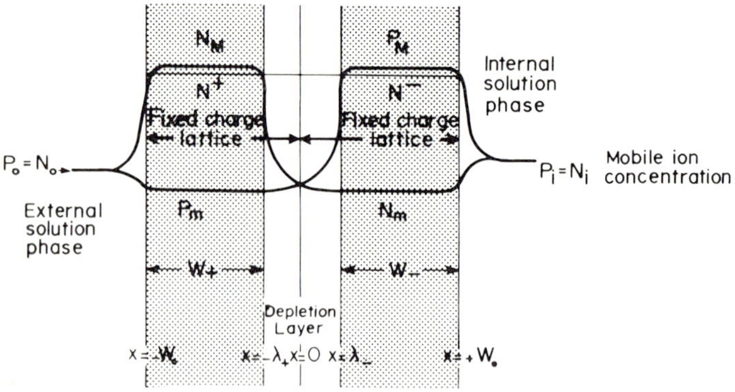

Fig. 5 Qualitative profiles of the mobile ion concentration for a DFCM. Three regions are generated in the membrane; the central depletion layer, which extends from $x = -\lambda_+$ to $x = \lambda_-$ (shown unshaded) in which the mobile ion concentrations are very low, the other two (shown shaded) are the regions outside the depletion layer where the mobile, counter, ion concentrations (either cations P_M or anions N_M) are high.

counter ions have high concentrations in each fixed charge lattice and hence these regions will have a high conductivity (see Fig. 5). Effectively, therefore, there are three distinct regions generated in the membrane, each associated with a capacitance. Using physiologically reasonable parameters for such a membrane (COSTER, 1973a), all three capacitances would be approximately equal, while the conductance of the depletion layer would be only 1/500 of the two outer regions.

One of the two contributions to the depletion layer capacitance is the junction capacitance associated with the changes in space charge due to modulations in the depletion layer width under the influence of an applied bias potential (MAURO, 1962). This capacitance, C_j, is given by:

$$C_j = \frac{\epsilon_r \epsilon_o}{\lambda_+ + \lambda_-},$$

where λ_+ and λ_- are the depletion layer widths in the positive and negative fixed charge regions respectively; ϵ_r is the relative permittivity and ϵ_o is the permittivity of free space.

The second contribution to the depletion layer capacitance arises from polarization effects in the diffusion of coions in the fixed charge lattices which leads to an out of phase component in the a.c. current. Each ion species contributes separately to this frequency dependent diffusion capacitance $C(\omega)$ (see Fig. 6). While this diffusion capacitance is negligible at high frequencies, it can exceed the dielectric capacitance of this region at sufficiently low frequencies.

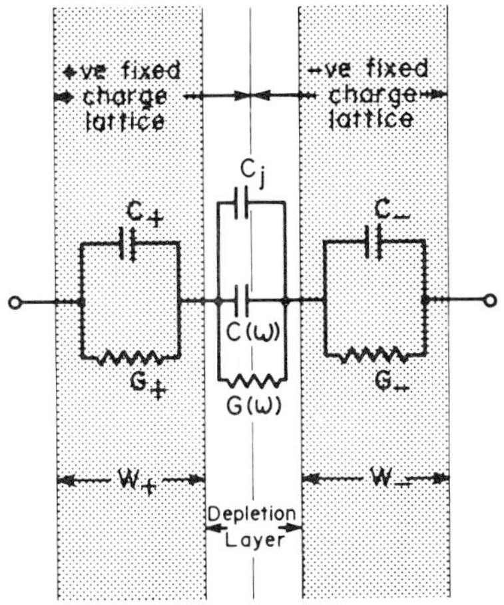

Fig. 6 An equivalent circuit representation of the DFCM. In the depletion layer C_j is the MAURO junction capacitance. The frequency dependent capacitance $C(\omega)$ and conductance $G(\omega)$ arise from diffusion polarization effects (COSTER, 1973a). Outside the depletion layer $C_\pm$ and $G_\pm$ are determined by the widths $W_\pm$, dielectric constant, diffusion constant and counterion concentrations.

The overall equivalent circuit for the DFCM membrane is shown in Fig. 6 and the dispersion characteristics for such a membrane are shown in Fig. 7. The capacitance disperses from a high value at very low frequencies, dominated by the diffusion capacitance and to a lesser degree the depletion layer dielectric capacitance, to a value of $\sim 0.9\ \mu F/cm^2$ at high frequencies which corresponds to the dielectric capacitance of the whole membrane.

Although the high frequency capacitance is independent of the widths, $W_{\pm}$, of the regions in the fixed charge lattices outside the depletion layer (see Fig. 6) the low frequency capacitance is strongly dependent on the widths of these regions.

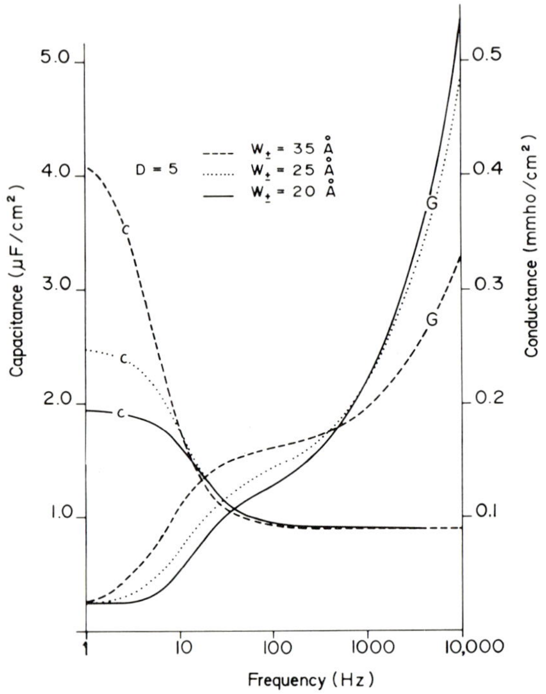

Fig. 7 The theoretically derived capacitance and conductance as a function of frequency for a DFCM for different values of the thickness of the depletion layer widths $\lambda_{\pm}$ and hence $W_{\pm}$ (= $W_o - \lambda_{\pm}$). Note the effect of W_+ on the membrane capacitance below 10 Hz. At high frequencies (>100 Hz) the capacitance is independent of $W_{\pm}$ and is simply equal to the overall dielectric capacitance of the membrane. In the calculations of the profiles the following parameters were used. $N^{\pm} = 0.1N, \epsilon_r = 10$; diffusion constant for cations and anions, $D_P = D_N = 5 \cdot 10^{-13}$ cm^2 s^{-1}, membrane thickness $2W_o$ = 10 nm, total membrane, d.c., resistance 40 kΩ cm^2.

It is evident that the theoretical dispersion curves for the DFCM shown in Fig. 7 are very similar to those observed experimentally. The effect of pH on the experimental dispersion curves for ***Chara*** have a parallel in the change in the theoretical dispersion in the DFCM with $W_{\pm}$. The depletion layer widths, λ_+ and λ_-, and hence the widths W_+ and W_-, are dependent on the fixed charge densities, N^+ and N^-. In biological membranes such fixed charges could arise from the ionisation of $-NH_2$ and -COOH groups of basic and acidic amino acids in the membrane proteins, and the fixed charge concentration would thus be pH dependent. A dependence of $W_{\pm}$ on pH has been previously suggested, on the basis of this model, in connection with another feature of the electrical characteristic: the punch-through effect (COSTER, 1969).

D. Conclusions

1) The results presented of experiments with cells of ***Chara corallina*** have shown that:
a) both the membrane capacitance and conductance undergo a strong dispersion at low frequencies (1–100 Hz);
b) above 100 Hz the membrane capacitance is constant and independent of pH;
c) the dispersion in capacitance below 100 Hz is much smaller at pH 5.5 than at pH 9;
d) at pH 9 the membrane time constant, $\tau = C/G$, decreases by a factor of 10 over the frequency range 1–100 Hz.

2) The results can be readily interpreted if the substructural layer of lowest conductivity in the membrane has a greater thickness at pH 5.5 than at pH 9, the overall dimensions of the membrane being independent of pH.

3) The dispersion characteristics and the effect of pH on these find a parallel in those derived theoretically for a double fixed charge membrane model in which the fixed charges arise from the ionization of $-NH_2$ and -COOH groups on membrane proteins.

E. Acknowledgements

The authors wish to thank Professor E.P. George for his reading of the manuscript and the Australian Research Grants Committee for support for the conduct of this research.

F. References

CLARK, H.R., STRICKHOLM, A., Nature **231**, 470–471 (1971)
COLE, K.S., Physiol. Rev. **45**, 340–379 (1965)
COLE, K.S., Membranes, Ions, and Impulses. University of California Press, Berkeley, 1968
COLE, K.S., in Physical Principles of Biological Membranes (eds. F. Snell, J. Wolken, G.J. Iverson, J. Lam), Gordon and Breach, Science Pub. Inc., New York, 1970
CONTI, F., Biophysik **6**, 257–270 (1970)
COSTER, H.G.L., Biophys. J. **5**, 669–686 (1965)
COSTER, H.G.L., Aust. J. Biol. Sci. **22**, 365–374 (1969)
COSTER, H.G.L., Biophys. J. **13**, 118–132 (1973a)
COSTER, H.G.L., Biophys. J. **13**, 133–142 (1973b)
FORSBERG, C., Physiol. Plant. **18**, 275–290 (1965)
LANNOYE, R.J., TARR, S.E., DAINTY, J., J. Exp. Bot. **21**, 543–551 (1970)
MAURO, A., Biophys. J. **2**, 179–199 (1962)
OFFNER, F.F., Bull. Math. Biophys. **31**, 359–381 (1969)
PALTI, Y., ADELMAN, W.T., J. Membrane Biol. **1**, 431–458 (1969)
SCHWANN, H.P., in Advances in Biological and Medical Physics (eds. C.A. Tobias, J.H. Lawrence), Vol. 5, Academic Press, London and New York 1957
STRICKHOLM, A., WALLIN, B.G., SCHRAGER, P., Biophys. J. **9**, 873–883 (1966)

Round Table Discussion 3

Chairmen: C. Gillet and F.W. Bentrup

The first part of the discussion was concerned with membrane current-voltage (I—V) relationships without and with, and of, the electrogenic pump. VREDENBERG stressed that when studying the membrane I—V relationships under different energy conditions we must try and separate the effects of light (the energy source) on the membrane itself, e.g. on the passive permeabilities, from the effects on the electrogenic pump. He did not think, for example, that it was easy to compare the dark and light I—V curves obtained by BENTRUP with ***Riccia***. He also pointed out that changes in ion concentrations may strongly alter the passive permeabilities. HANSEN thought that one could approach VREDENBERG's point (1) by changing the membrane resistance by the application of light and (2) by measuring the change in resistance when the cell (in the dark) is voltage-clamped at the same potential as that produced by light; in his opinion such resistance comparisons are helpful in trying to separate the effects of light on the membrane from the effects on the pump.

GRADMANN then discussed whether the electrogenic pump had a conductance of its own. He thought that it must. On theoretical grounds, if it did not, it would be an ideal current source with an infinite resistance and produce an extremely high voltage; and we cannot see why the voltage should be greater than, say, −500 mV which is the free energy of ATP hydrolysis. He pointed out that when the electrogenic pump in ***Neurospora*** is blocked the membrane resistance becomes very high. He also commented on the fact that the I—V curves, measured with and without inhibition of the pump, did not intersect at high voltages in ***Riccia*** and ***Neurospora***; maybe it was due to secondary processes arising during inhibition. On the question of a pump conductance, SLAYMAN mentioned two theoretical treatments by Finkelstein and Rapaport, one mechanistic and the other thermodynamic, which both showed that the pump ought to be represented by a voltage and a resistance in series. The voltage and resistance must be variable and both theories show that at very high voltages the currents should saturate, which is a simple consequence of a carrier model. He pointed out that GRADMANN's data on ***Acetabularia*** were consistent with this model, but that there were some unclarified, peculiar, features of his data on ***Neurospora*** and BENTRUP's on ***Riccia***. When WALKER asked at what voltage does the pump become energetically impossible, SLAYMAN replied that he could not really say from his own data, but it was generally accepted that there is only about 500 mV available from ATP hydrolysis.

BENTRUP and SLAYMAN asked for some comments on the, non-recoverable, very steep I—V curves in ***Neurospora***, beyond a potential of about −350 mV. COSTER did not think that this could be a 'punch-through' effect, which would be recoverable. COSTER also pointed out that at large hyperpolarizations there is a very high electric field strength in the membrane; the average field strength (E) is about 10^5 V cm^{-1} and, if one goes along with his three layer model, E can be as high as 10^6 V cm^{-1} in certain regions. In such a field water dissociates with the production of large quantities of H^+, which have to be pumped out. On depolarization such H^+ production would decrease and one would reach a point where production was so low that there would be no need to pump them out. BENTRUP, refering to COSTER's double fixed charge model, thought that the pH in the membrane must be quite critical. I—V characteristics obtained in the light, during proton pumping, must be associated with a decreasing membrane pH perhaps explaining BENTRUP's own data. At this point KEDEM reminded the audience of the rather nice behaviour of animal epithelial electrogenic pumps, which fit the theories mentioned by SLAYMAN; the pumps have a conductance which can, for instance, be affected by hormones; I—V intersects; and so on.

JAFFE then spoke of GRADMANN's linear I—V characteristics obtained within 5 ms of applying the current, which showed that the 'instantaneous' conductance is independent of voltage over a wide rage. JAFFE thought that such a result disagreed with the generally accepted theory, e.g. of Goldman and others, that the conductance of the passive channels is strongly voltage dependent. When COSTER said that with such a rapid application of the potential (5 ms) there is no time for the field to redistribute itself and thus GRADMANN's results are not surprising, JAFFE replied that this is not so, for old calculations done on nerve show that the time for redistribution of the ions in a channel is of the order of ns, not ms, i.e. there is a factor of 10^6 to account for. JAFFE and COSTER debated this point for some time, for the calculation is critically dependent on the number of passive channels per unit area of membrane. The figure (~ ns) quoted by JAFFE, is actually for the number of 'excitable' channels in nerve; even granted this, however, the factor of 10^6 needed to bring the time of ionic adjustment of ~ ms means, according to JAFFE, that practically the whole area of the membrane has to be covered with passive channels. COSTER did not see why not, because of the large amount of protein in the membrane. After VREDENBERG had described an experiment which showed that some time, 10 s or more, is needed for conformational changes, or reorganizations, of membrane channels, JAFFE (after agreeing that such architectural rearrangements needed a long time) came back to his initial point about the independence of the instantaneous conductance of voltage. He thought that somehow the electric field across the membrane could not be across the passive channels. SLAYMAN commented, in effect, that such a result could arise if the sum of diffusion potentials across a passive channel was zero. JAFFE could not accept such a restriction, and at this point the discussion moved to other topics.

SMITH put forward what he called two simple and naive propositions. First, if a current is passed through a membrane and hence through the pump one should be able to reverse the appropiate chemical reaction, say ATP hydrolysis: i.e. make ATP, which should be demonstrable. Secondly, if the current is carried by protons then quite large pH changes, both inside and outside the cell, should be observed. With respect to the first point, SLAYMAN said that both ***Neurospora*** and ***Acetabularia*** were unsuitable for demonstrating such ATP synthesis because the small space constants mitigated against applying a uniform voltage clamp; in ***Nitella*** such an experiment might be possible.As for looking for an external pH change, COSTER pointed out that one would probably need to have an initially high external pH in order to detect any changes due to current, and of course such a high external pH would cause a passive efflux of H^+. WALKER said that he and Thain, some years ago, had tried to detect a change in external pH consequent on passing a current through a ***Nitella*** (or ***Chara***) membrane, but had failed and the failure still puzzled him. GRADMANN mentioned that experiments had been started in Tübingen to investigate any correlations between pH changes and pump activity in ***Acetabularia***; it was too early to be definite about the results.

KYLIN next spoke at some length about his work, with ***Scenedesmus***, on the differential effects of DCMU on photophosphorylation. DCMU operates at different concentrations on the three sites of phosphorylation. Although he was making his comments principally with reference to the paper of VREDENBERG, he also thought that the different behaviour of the phosphorylation sites with respect to DCMU might resolve such questions as the opposite effects of DCMU on ion transport found by MacRobbie in ***Nitella*** and Jeschke on ***Elodea***. However JUNGE warned that KYLIN's remarks may not be relevant to VREDENBERG's work, because the important thing is not the absolute DCMU concentration but rather the DCMU to chlorophyll ratio.

WALKER brought up the question of the applicability of COSTER's double fixed charge membrane model to other ideas of membrane structure. He referred to the generally-accepted model of Singer in which lumps of protein float in a liquid lipid bilayer, and he wondered if COSTER had considered the relevance of his model to such a mosaic type of structure. COSTER replied that he himself favoured the Singer model. He commented that in many membranes there is a lot more protein than lipid, so that the protein 'blobs' would be fairly closely packed; if this is so, the double fixed charge model is quite appropriate, for it is particularly applicable to situations where there is a lot of protein units going right through the membrane. After SLAYMAN had suggested that you need both a series and a parallel formulation in such a membrane, COSTER said that this was not so as far as the pH dependence of the capacitance went. Only the protein units would generate three layers, one depletion layer and two non-depletion layers, which were strongly pH dependent.

BENTRUP then raised the question of the mechanism of dielectric breakdown. ZIMMERMANN pointed out that in his system there was hyperpolarization on one side and depolarization on the other side of the membrane. He thought that dielectric breakdown only occurred at one point in the membrane and he did not understand at present how the current goes through the cell. He asked for comments on the general question of the mechanism of dielectric breakdown in the cell membrane. COSTER commented that the field strengths in cell membranes are quite phenomenal and wondered why cell membranes preserved their integrity at all. It is not surprising that if field strengths are increased by an order of magnitude that things tear apart. KRISCHER asked ZIMMERMANN about the effect of surface charge on his erythrocyte membranes on the dielectric breakdown. ZIMMERMANN replied that this is a difficult matter to decide by experiment; they had found some dependence of the current necessary to produce breakdown on the red cell potential and some differences for bacteria depending on whether taken from the logarithmic or stationary phase of growth; but it is difficult to decide whether these effects are due to the membrane structure, or membrane potentials, or surface potentials. The session ended with an inconclusive debate between DAINTY, COSTER, and JUNGE on whether the field strength was the only parameter determining dielectric breakdown. Both DAINTY and JUNGE thought that the absolute potential difference must play an important role, in analogy with the avalanche theory of breakdown in macroscopic objects. COSTER was more concerned with the effect of the field strength on the electric dipole moments of the individual molecules.

Session 4

Solute Transport in Algae and Cellsuspension Cultures

Chairman: N. Higinbotham

Time Course of Chloride Fluxes in *Hydrodictyon africanum* during Alternating Light and Darkness

J.A. Raven

A. Introduction

One of the many differences between different green tissues and different ions with respect to light-stimulated active ion transport by whole cells concerns the time course of transport during transitions between light and darkness. While in some cases rapid transitions between light and dark steady states ($t_{1/2}$ less than a minute) are found (e.g. FINDENEGG et al., 1971), in many other instances much longer adjustment times, with complex time courses, have been reported (e.g. RAINS, 1967; LILLEY and HOPE, 1971).

Two major kinds of hypothesis can be advanced to explain the energetics of this slow decline to the dark steady-state active flux after illumination. One view is that there is a 'carry-over' of some photoproduced labile energy substrate (i.e. not a conventional respiratory substrate) into the dark period (e.g. SIMONIS and GIMMLER, 1969). This could be photoproduced ATP (or its precursor, the 'high energy state' of chloroplasts), or reductant, or, in the case of secondary active transport coupled to H^+ (OH^-) downhill fluxes at the plasmalemma, energy stored as a light-enhanced H^+ free energy gradient across this membrane (cf. SMITH, 1970, 1972).

The other possibility is that the energy for the stimulated active transport in the dark after illumination is powered by respiration, possibly because there is more respiratory substrate available for production of the high-energy compounds required for active transport. Whichever of these two explanations is correct for any particular case, an important point is whether the rate of transport is determined primarily by energy supply (i.e. additional energy supply from photosynthetic 'carry-over' or substrate-enhanced respiration) or by some other regulatory process. This could be a 'light activation' of the transport system by some photo-product; by activation is meant here an increased rate of transport in the presence of constant steady-state concentrations of energy (and transported) substrates (RAVEN, 1974a, b).

Previous work (RAVEN, unpublished) on ***Hydrodictyon africanum*** has shown that the active $^{36}Cl^-$ influx (RAVEN, 1967a, b; 1969a, b) shows a slow decrease from the light steady state flux upon darkening. The experiments described in this paper were designed to investigate the mechanism of this phenomenon in relation to the hypotheses mentioned above.

B. Materials and Methods

Hydrodictyon africanum was grown as previously described (RAVEN, 1967a). Short-term (15 or 30 minutes) influxes of $^{36}Cl^-$ (16°C, 155 erg m^{-2} s^{-1} visible irradiance or in darkness) were measured as described by RAVEN (1967a, b). In some experiments, particularly in the dark, a non-zero intercept of $^{36}Cl^-$ content of the cells at zero time was found. This was not a free space component, since it was not removed by the standard washing procedure (RAVEN, 1967a). The experiments reported here were carried out on cells in which this rapid, non-reversible uptake was less than 10 % of the total uptake over the experimental period.

Instead of boiling tubes, glass tubes of 25 ml capacity were used for the pretreatments and uptakes described in Figs. 1–3. These tubes were sealed with 'Suba-seals'; at the start of influx periods $^{36}Cl^-$ was injected via a syringe so that dark-pretreated cells were not exposed to the light (cf. RAVEN, 1969a). In experiments in which the pH or O_2 content of solutions were changed, the pretreatment solution was removed from the sealed tube with a syringe, and the new solution of different pH or gas content was added by syringe. The amount of the old solution remaining in the tube after three repetitions was inadequate to significantly change the pH or gas composition of the new solution. In parallel control experiments, the same procedure was followed. In experiments in which the gas phase was changed the tubes were almost full of solution (23 ml per tube); in other experiments 10 ml per tube was used.

Efflux experiments were carried out in an apparatus in which a single cell was held in a 1 ml capacity glass tube between two taps. The lower tap voided into a collecting tube; the upper tap was connected to a reservoir of solution. When the solution was to be changed, the lower tap then the upper tap was opened; when 2 ml of solution had run into the collecting tube the taps were shut again in the same order. The reservoirs containing the experimental solutions were equilibrated with air or N_2 as appropriate.

Individual influx and efflux experiments were carried out on parallel cultures of cells from the same net; cells grown in unlabelled Cl^- were used for the influx experiments, while cells cultured in $^{36}Cl^-$ were used for the efflux.

Solutions used were artificial pond water (1 mM NaCl, 0.1 mM KCl, 0.1 mM $CaCl_2$), with 5 mM MES adjusted to pH 6 with NaOH or 5 mM HEPES adjusted to pH 10 with NaOH.

C. Results

Fig. 1 shows the time course of $^{36}Cl^-$ influx and efflux in *Hydrodictyon africanum* during a dark-light-dark sequence. Both influx and efflux are steady during the initial dark period. Upon illumination the influx rises to the steady light level in 20 minutes; the (smaller) increase in efflux is rather slower. Upon darkening, both influx and efflux decline to the dark steady state with half-times of about 20 minutes. These cells were approximately in flux equilibrium in the dark; in the light, and for at least 15 minutes in the dark (see also Fig. 2) there is a net Cl^- influx.

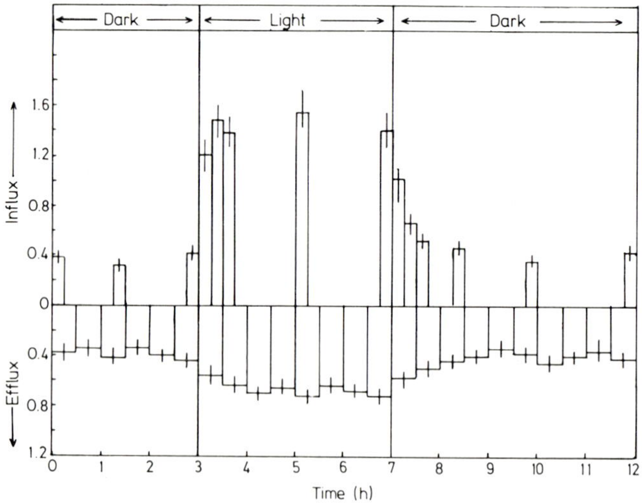

Fig. 1 Time course of $^{36}Cl^-$ influx and efflux in dark-light-dark treatment. All cells had 10 hours dark incubation before the start of the experiment (t = 0).

Fig. 2 shows the effect of anaerobiosis on the time course of $^{36}Cl^-$ influx and efflux in the post-illumination period and in the dark steady state. The absolute inhibition by removal of O_2 at the start of the post-illumination period is as great as that caused by anaerobiosis in the dark steady state, and greater than that caused by O_2 removal in the light. However, there is still a net Cl^- influx in the post-illumination period even under N_2. It is unlikely that this is entirely due to the slowness of the inhibition due to removal of O_2 in view of the rapid decline in the Cl^- influx after O_2 removal in the dark steady state. Essentially similar results were obtained when addition of 1 mM KCN, rather than removal of O_2, was used as the respiratory inhibitor (cf. RAVEN, 1969b).

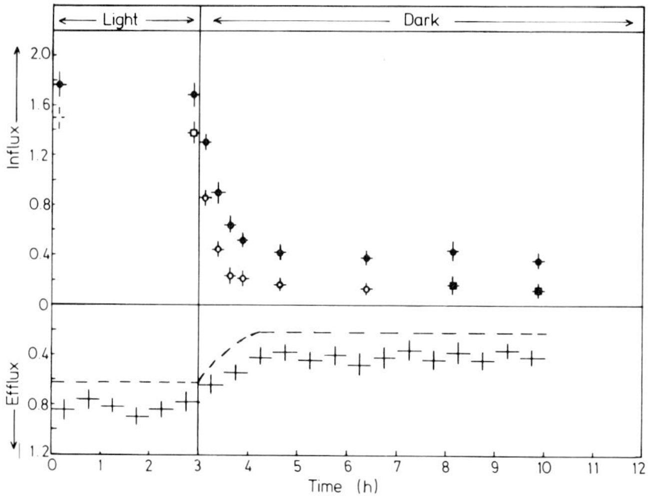

Fig. 2 Time course of $^{36}Cl^-$ influx and efflux in a light-dark transition as influenced by anaerobiosis. ✦ = influx of $^{36}Cl^-$ in air throughout; ✦ = influx of $^{36}Cl^-$, anaerobiosis from t = 0. ✧ = influx of $^{36}Cl^-$, anaerobiosis from t = 3 h. ✦ = influx of $^{36}Cl^-$, anaerobiosis from t = 8 h. + = $^{36}Cl^-$ efflux in air throughout. Dotted line lower limit of S.E. for anaerobic $^{36}Cl^-$ efflux, with anaerobiosis started at t = 0, 3, 8 h. The anaerobic efflux values were too similar to the aerobic ones for separate presentation in detail.

In terms of the third hypothesis discussed in the introduction, the post-illumination Cl^- net active influx could be due to the slow dark decay of a light-enhanced pH gradient across the plasmalemma; this pH gradient, according to SMITH (1970, 1972) powers the Cl^- influx in *Chara corallina*.

Fig. 3 shows that a pretreatment at high pH enhances dark $^{36}Cl^-$ influx upon return to a lower pH (cf. SMITH, 1972). It also shows that the enhanced influx is indeed a net influx, and that the enhanced flux decays with a $t_{1/2}$ of 15—30 minutes. The magnitude of the enhanced Cl^- influx produced by high pH pretreatment is smaller than in ***Chara***; the maximum enhanced efflux in the dark is always less than the steady state light influx (2.34 ± 0.27 pmoles cm^{-2} s^{-1} at pH 6 in the experiment shown in Fig. 3).

Fig. 3 also shows that the dark Cl^- influx enhanced by high pH pretreatment is partly inhibited by anaerobiosis. Thus in terms of both time course and sensitivity to anaerobiosis (and KCN, unpublished experiments), the enhanced Cl^- influx in the dark after high pH pretreatment resembles the enhanced dark Cl^- influx after light pretreatment, although it is of a smaller magnitude.

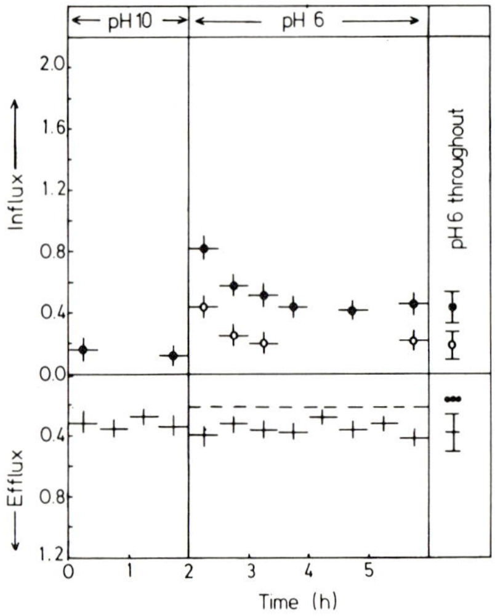

Fig. 3 Time course of $^{36}Cl^-$ influx and efflux during transition from pH 10 to pH 6 in the dark. ✦ = influx of $^{36}Cl^-$, air. ✧ = influx of $^{36}Cl^-$, anaerobic from t = 2 h. + = $^{36}Cl^-$ efflux, air.
--- = lower limit of S.E. for anaerobic $^{36}Cl^-$ efflux; the anaerobic $^{36}Cl^-$ efflux value were too similar to the aerobic ones for separate presentation in detail.
✦ = Mean $^{36}Cl^-$ influx and limits of S.E. for cells maintained at pH 6 in air throughout. ✧ mean $^{36}Cl^-$ influx and limits of S.E. for cell maintained at pH 6 from t = 0, with anaerobiosis from t = 2 h. + = mean $^{36}Cl^-$ efflux and limits of S.E. for cells maintained at pH 6 in air throughout. ... = lower limits of S.E. for efflux for anaerobic $^{36}Cl^-$ efflux for cells maintained at pH 6 from t = 0, with anaerobiosis from t = 2 h.

Influx experiments with the reverse pH change (pH 6 to pH 10) show that the $^{36}Cl^-$ influx decreases rapidly to the new value, with no significant 'undershoot' complementing the 'overshoot' found with pH decrease (RAVEN, unpublished). In some experiments an 'overshoot' in the efflux of $^{36}Cl^-$ is seen upon raising the pH from 6 to 10; in these cases there might be a temporary net efflux of Cl^- at the higher pH, possibly linked to OH^- influx. However, it is also possible that this temporary increase in passive Cl^- efflux is caused by the transient hyperpolarisation found when the pH of the medium bathing giant freshwater algal cells is increased (references in RAVEN and SMITH, 1974). Attempts to measure the time course of the light-dark transient at high pH, i.e. where the free energy gradient for OH^- efflux is low, or even negative, suggest that the transition may be more abrupt than at low pH, although the small light stimulation found at high pH makes accurate determinations of the time course difficult.

D. Discussion

The long half-time of decay of active Cl^- influx in the dark after illumination (Figs. 1 and 2) contrasts with the much more rapid decay of light-enhanced levels of ATP (SANTARIUS and HEBER, 1965), NADPH (HEBER and SANTARIUS, 1966) or 'high-energy intermediate' of photophosphorylation (HEBER, 1969; SIMONIS and GIMMLER, 1969). Thus the carry-over of these intermediates from the illumination period, or their maintenance at higher levels in the post-illumination period by enhanced respiration due to photo-produced respiratory carbon substrate, decays too quickly to account for the effect observed in *Hydrodictyon africanum*.

The remaining alternatives are either that the energy comes from a slow-decaying light-enhanced H^+ free energy gradient across the plasmalemma, or that there is some slow-decaying photo-activation of the transport mechanism, which increases energy supply from respiration by classical respiratory control (BEEVERS, 1970). The activation could itself be the H^+ gradient at the plasmalemma. The experiment shown in Fig. 2 shows that the light-enhanced active Cl^- influx in the dark is inhibited by anaerobiosis or the presence of cyanide. While this is consistent with energy supply from respiration, it is also consistent with the involvement of an inside-alkaline H^+ gradient. Anaerobiosis or cyanide induce fermentation in algae, and organic acids such as lactate are major fermentation products (e.g. BEGUM and SYRETT, 1970). This would lead to acidification of the cytoplasm (cf. DAVIES, 1973), and the transmembrane pH gradient would be lowered. The smaller inhibition of Cl^- influx by anaerobiosis or cyanide in continuous illumination in *H. africanum* (RAVEN, 1967b, 1969b) could be attributed to a smaller build-up of organic acids in illuminated green cells (HIRT et al., 1971), and to the availability of more energy for pH regulation in the cytoplasm in the light.

The involvement of a pH gradient across the plasmalemma in Cl^- influx in *H. africanum* is suggested by the results shown in Fig. 3, which shows that the phenomenon reported by SMITH (1972) of enhanced Cl^- influx at low pH in *Chara corallina* following a high pH pretreatment also occurs in *H. africanum*. In *H. africanum* it has been additionally shown that this stimulation is of a net Cl^- influx, and that its time course and sensitivity to respiratory inhibition resemble that of the post-illumination decay of Cl^- influx.

It must be remembered, however, that the evidence presented here by no means amounts to the complete test of SMITH's hypothesis which has been lucidly outlined by MacROBBIE (1974). In addition to these points, it must be remembered that the relative insensitivity of Cl^- influx in *H. africanum* to compounds which increase the H^+ permeability of biological membranes is not easy to reconcile with the involvement of a pH gradient at the plasmalemma in Cl^- transport. Another difficulty arises from the ability of *H. africanum* to actively transport Cl^- across the plasmalemma from solutions of high pH (RAVEN, 1974a). Under these conditions the inside-low free energy gradient of H^+ cannot be large, and it appears that a neutral Cl^-/OH^- antiport cannot account for active Cl^- influx under these conditions. Influx of one Cl^- for two OH^- lost from the cell would be energetically more favourable, but would imply a net inward transport of positive charge during Cl^- active transport. Evidence on the electrogenic or electroneutral nature of Cl^- influx in freshwater algal coenocytes is conflicting (PICKARD, 1973; WILLIAMS et al., 1972).

Despite these doubts as to the energetic adequacy of the H^+ gradient to power active Cl^- influx, especially at high external pH values, the regulation of cytoplasmic pH is likely to be an important factor in regulating salt accumulation (RAVEN and SMITH, 1974), and H^+ may be involved in the 'photo-activation' of transport mechanisms in green cells.

E. Acknowledgements

I should like to thank Dr. S.M. Glidewell for her critical reading of the manuscript, and Mr. S. Sinclair for technical assistance.

F. References

BEEVERS, H., in Prediction and Measurement of Photosynthetic Productivity (ed. I. Malek), pp. 209–214, Centre for Agricultural Publication and Documentation, Wageningen 1970

BEGUM, F., SYRETT, J.R., Arch. Mikrobiol. **72**, 344–352 (1970)

DAVIES, D.D., Symp. Soc. Exp. Biol. **27**, 513–529 (1973)
FINDENEGG, G.R., PASCHINGER, M., BRODA, E., Planta **99**, 163–173 (1971)
HEBER, U., Biochim. Biophys. Acta **180**, 302–319 (1969)
HEBER, U., SANTARIUS, K.A., Biochim. Biophys. Acta **109**, 390–408 (1966)
HIRT, G., TANNER, W., KANDLER, O., Plant Physiol. Lancaster **47**, 841–843 (1971)
LILLEY, R.M., HOPE, A.B., Biochim. Biophys. Acta **226** 161–171 (1971)
MacROBBIE, E.A.C., in Algal Physiology and Biochemistry (ed. W.D.P. Steward), pp. 678–713, Blackwell Scientific Publications, Oxford 1974
PICKARD, W.F., Can. J. Bot. **51**, 715–724 (1973)
RAINS, D.W., Science **156**, 1382–1383 (1967)
RAVEN, J.A., J. Gen. Physiol. **50**, 1607–1625 (1967a)
RAVEN, J.A., J. Gen. Physiol. **50**, 1627–1640 (1967b)
RAVEN, J.A., New Phytol. **68**, 45–62 (1969a)
RAVEN, J.A., New Phytol. **68**, 1089–1113 (1969b)
RAVEN, J.A., New Phytol. **73**, in press (1974a)
RAVEN, J.A., J. Exp. Bot. **25**, in press (1974b)
RAVEN, J.A., SMITH, F.A., Can. J. Bot. **52**, in press (1974)
SANTARIUS, K.A., HEBER, U., Biochim. Biophys. Acta **102**, 39–54 (1965)
SIMONIS, W., GIMMLER, H., in Progress in Photosynthesis Research (ed. H. Metzner), Volume III, pp. 1155–1161, International Biological Union, Tübingen 1969
SMITH, F.A., New Phytol. **69**, 902–917 (1970)
SMITH, F.A., New Phytol. **71**, 595–601 (1972)
WILLIAMS, E.J., MUNRO, C., FENSOM, D.S., Can. J. Bot. **50**, 2255–2263 (1972)

Chloride Transport to the Charophyte Vacuole

N.A. Walker

A. Introduction

It is the purpose of this paper to review the current evidence on Cl^- transport in charophyte cells, with two particular questions in mind: a) what Cl^- compartments need to be postulated to account for present results, and b) whether vesicular transport at either membrane needs to be postulated. New work will not be presented in full, but reference will be made as necessary to work in progress or in preparation for publication.

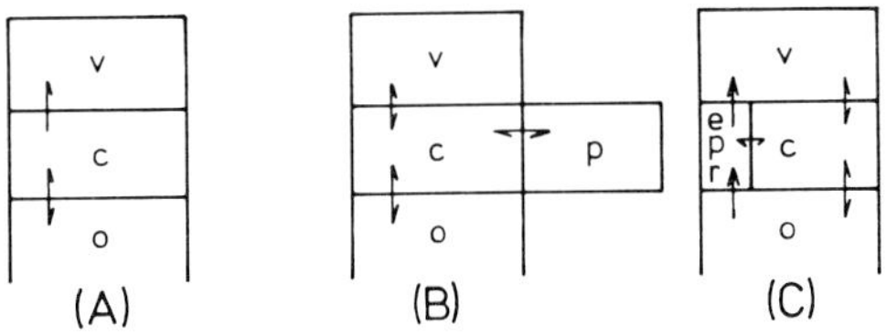

Fig. 1 Compartment models for Cl^- in charophyte cells. Labels: c cytoplasm, epr endoplasmic reticulum, o outside, p plastids, v vacuole. Half arrows represent molecular fluxes through membranes; full arrows represent mass transport in vesicles. A Original 3-series-compartment model. B 4-compartment model. C One of a number of possible vesicle models.

The starting point for this consideration will be the situation already well reviewed by MacROBBIE (1971a, 1973). The kinetics of Cl^- influx into charophyte cells, which used to be consistent with a 3-series-compartment model (Fig. 1A), was shown to require a 4-series-compartment model (Fig. 1B) and later was shown to require a vesicle transport model (Fig. 1C) (MacROBBIE, 1969, 1970, 1971b). The identification of the kinetic compartments of Fig. 1 with cellular compartments makes it relevant to attempt to determine their contents directly, as a check on the kinetic models. MacROBBIE's most cogent argument for rejecting the 4-compartment model was that the measured content of cellular compartments did not agree with the kinetic requirements.

B. Symbols Used

ϕ_{ab}	flux a to b (moles cm^{-2} s^{-1})
Q_a	quantity of chemical species in a (moles cm^{-2} of cell surface)
S_a	specific activity in a (cps $mole^{-1}$)
t	time (s)
t_a	half-time for change of specific activity in a (s)
Y_a	quantity of radioactivity in a (cps cm^{-2})

C. Results and Discussion

I. Chloride Content of Cytoplasm

The distinguishable major phases in the cytoplasm are the plastids and the remainder of the cytoplasm, most of which exhibits cyclosis. In practice it is easiest to try to separate the flowing cytoplasm from the non-flowing cytoplasm, of which the plastids constitute most of the volume and probably contain nearly all of the Cl^-. Since no permeability barrier has ever been observed between flowing and non-flowing cytoplasm one can roughly identify the flowing cytoplasm with the continuous water phase of the cytoplasm as a whole. The flowing cytoplasm can be brought to one end of the cell by gentle centrifugation, and removed for the measurement of Cl^- content with electrodes or by titration. Table 1 lists results by this method. It seems worthwhile to notice that KISHIMOTO and TAZAWA (1965) halved their estimate of the Cl^- concentration by exchanging the natural vacuolar sap for a solution containing no Cl^-. In an analogous way they reduced their estimate of the Na^+ concentration by first perfusing the vacuole with a solution containing no Na^+. Although the Cl^- and the Na^+ figures do not lead to the same quantitative conclusions, their work makes it seem likely that there is a considerable contamination of the centrifuged cytoplasm with vacuolar sap. In the absence of other work along these lines, their estimate of 35 mM may be taken as the best.

Table 1

Chloride concentration (or content) in flowing cytoplasm

Species	Conc. (mM)	Method	References
Nitella flexilis	35	Centrifugation: NO_3^- sap	KISHIMOTO and TAZAWA (1965)
Nitella flexilis	70	Centrifugation: Cl^- sap	KISHIMOTO and TAZAWA (1965)
Nitella translucens	65	Centrifugation: Cl^- sap	SPANSWICK and WILLIAMS (1964)
Nitella translucens	87	Centrifugation: Cl^- sap	HOPE et al. (1966)
Chara australis	10	Inserted AgCl electrode	COSTER (1966)
Nitella flexilis	20	Inserted AgCl electrode	LEFEBVRE and GILLET (1971)

A method which does not require physical separation of the phases is the insertion of a Cl^- electrode into the live cell. With this method COSTER (1966) got a value of about 10 mM, while LEFEBVRE and GILLET (1971) got values of 17–25 mM. Both refer presumably to the water phase of the cytoplasm, and so are to be compared with the 35 mM of KISHIMOTO and TAZAWA (1965). The two methods have never been compared on the same batch of cells. Table 1 suggests that, with no great confidence, one might take the Cl^- concentration of the cytoplasm to be around 30 mM.

The Cl^- content of the non-flowing cytoplasm is much less uncertain, since it is not sensitive to contamination from other phases of higher concentration; its value is of the order of 0.1 μmoles cm^{-2}.

II. Kinetic Compartments

1. Rapid Influx to Vacuole

It was shown by MacROBBIE (1969) that soon after a charophyte cell is exposed to radioactive halide solution a substantial fraction of the radioactivity in the cell is to be found in the vacuolar sap. This is not predicted by the 3-series-compartment model (Fig. 1A) in which the tracer is assumed to mix into the whole pool of cytoplasmic Cl^- (0.1 μmoles cm^{-2}). A sample experiment is that shown in Fig. 2. The curves are calculated from the model of Fig. 1B, with the following values of the parameters: Q_c, 3 nmoles cm^{-2}; Q_p, 0.1 μmoles cm^{-2}; Q_v, 3 μmoles cm^{-2}; ϕ_{oc}, 2.3 pmoles cm^{-2} s^{-1}; ϕ_{cp}, 10 pmoles cm^{-2} s^{-1}; ϕ_{cv}, 30 pmoles cm^{-2} s^{-1}.

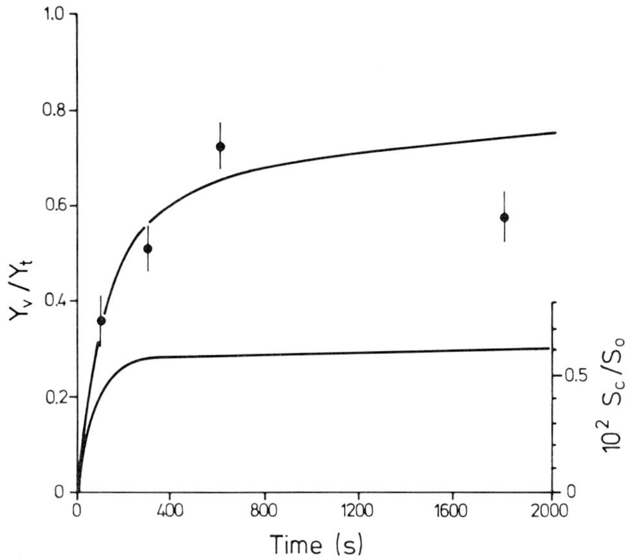

Fig. 2 Fraction of radioactivity in the vacuole (Y_v/Y_t) plotted against time (s), with points representing mean and S.E. of values for 10 cells of ***Chara corallina*** exposed to a solution of $^{36}Cl^-$. Upper curve: calculated time course of (Y_v/Y_t) from the model of Fig. 1B. Lower curve: similarly calculated time course of cytoplasmic specific activity (S_c/S_o). Parameter values used: Q_c = 3 nmoles cm^{-2}, ϕ_{oc} = 2.3 pmoles cm^{-2} s^{-1}; Q_p = 0.1 μmoles cm^{-2}, ϕ_{cp} = 10 pmoles cm^{-2} s^{-1}; Q_v = 3 μmoles cm^{-2}, ϕ_{cv} = 20 pmoles cm^{-2} s^{-1}.

Each point represents the mean and S.E. for 10 cells. The parameters Q_v and ($Q_p + Q_c$) are considered to be determined by the chemical methods already discussed; in fitting such an experiment as that of Fig. 2 the important parameters are $Q_c/(\phi_{co} + \phi_{cp} + \phi_{cv})$ and ϕ_{cv}/ϕ_{cp}. The value attributed to Q_c needs to be rather small in order to have a small value of the half-time t_c (since $t_c = Q_c/(\phi_{co} + \phi_{cp} + \phi_{cv}) \cdot 0.69$ s); but only if ϕ_{cv} is determined at the same time on similar cells does this experiment give a definite value for Q_c. The value chosen for the curves of Fig. 2 was 3 nmoles cm^{-2}, but a value ten times higher is not ruled out; it would require ϕ_{cv} to be 300 pmoles cm^{-2} s^{-1}, which does not seem impossible.

The value of t_c adopted depends on the early points on the graph – that at 2 min on Fig. 2 is the most important.

2. Quantization of the Fraction in the Vacuole

In a number of papers, MacROBBIE (1970, 1971, 1971b) interpreted data on the fraction of radioactivity in the vacuole as demonstrating that the individual values had a quantized rather than a unimodal distribution. Within a batch of similar cells, the values were said to have a multimodal distribution, with the modes at a, $2a$, $3a$, ... where a is the magnitude of the quantum, chosen for each batch from the data itself. This interpretation was rejected by FINDLAY et al. (1971) on the grounds that the analysis was not rigorous and that in experiments with large numbers of cells no quantization could be discerned. MacROBBIE has re-presented the original data (1973) but still without a convincing statistical treatment; WALKER (1973) mentioned rough tests on the original data which showed no sign of quantization. This question is relevant here because the quantized distribution could hardly be explained on the basis of the 4-compartment model (Fig. 1B) and might suggest a vesicle model like Fig. 1C (MacROBBIE, 1970).

FINDLAY and WALKER (in preparation) have been developing a statistical test which can be applied to this problem: it is a modification of one due to BROADBENT (1956). He derived a statistic (here called B) which measures the tendency of observations to cluster around integral multiples of a quantum a. B is defined by

$$B = \sqrt{n}(0.33 - 4s^2/na^2)$$

where a is the chosen quantum, n is the number of observations x_j in the batch, and s^2 is the sum given by

$$s^2 = \sum_j^n (x_j - n_j \cdot a)^2$$

where n_j is that integer which minimizes, for each x_j, the value of $|x_j - n_j|$.

The quantity s^2/a^2 seems reasonably to measure the tendency to cluster around modes at $n_j \cdot a$, and the statistic B seems to be roughly independent of n (BROADBENT, 1956). Given data x_j which may be quantized, but around an unknown quantum a, there are two problems: 1) to find a, (it seems reasonable to choose that value for each set of data which gives B its maximum value B_m) and 2) to compare these values of B_m with those predicted on the basis of some 'neutral' null hypothesis (which one must choose). BROADBENT provided a search procedure for finding the best value of a, and this was made the basis of a computer programme in Fortran. Before the search can begin, the limits between which a may lie must be decided on ***a priori*** grounds. Between these bounds, as a increases, there are many local maxima in B. We have no reason ***a priori*** to prefer any particular one, and must simply choose that which corresponds to the largest B, i.e. to B_m.

The null distribution of B_m can be found experimentally for a large number of sets of random numbers. Differing from BROADBENT, FINDLAY and WALKER have found a null distribution of B_m, not of all local maxima of B. The random numbers used had a Gaussian distribution, truncated at zero, with mean 0.2 and standard deviation 0.25. This is a good fit to the lumped distribution of data values in all of MacROBBIE's experiments. A computer programme generated pseudo-random numbers with this distribution, and for each set a search was made for B_m. The null distribution of B_m which resulted is compared in Fig. 3 with the distribution from experimental values of the fraction of halide in the vacuole of ***Nitella***, ***Chara***, and ***Tolypella***. No significant difference appears.

3. The Effect of Action Potentials

The values of fractions in the vacuole obtained by MacROBBIE do appear to be irregularly distributed, although their distribution is not quantized. That this 'odd' distribution may be in part due to the occurrence of action potentials was suggested by MacROBBIE (1973). In experiments on the partitioning of Cl^- between cytoplasm and vacuole, the cell is cut open at the end of the influx period. A tonoplast action potential (a.p.) always accompanies the plasmalemma a.p., and may well transfer Cl^- from cytoplasm to vacuole when the cell is cut, as MacROBBIE suggested (1969). Although the tonoplast action potential is in the direction to be expected if its rising phase is due to an increase in the permeability of the tonoplast to Cl^-, no information is available from which to estimate the likely increase. A plasmalemma a.p. is associated with a net outward movement of about 1 nmole cm^{-2}; and if this were the net inward transfer across the tonoplast from a phase containing say 5 nmoles cm^{-2} Cl^-, the fraction in the vacuole would be significantly affected. Recordings made while cells were cut open (WALKER, unpublished) showed that a plasmalemma a.p. happened on each occasion.

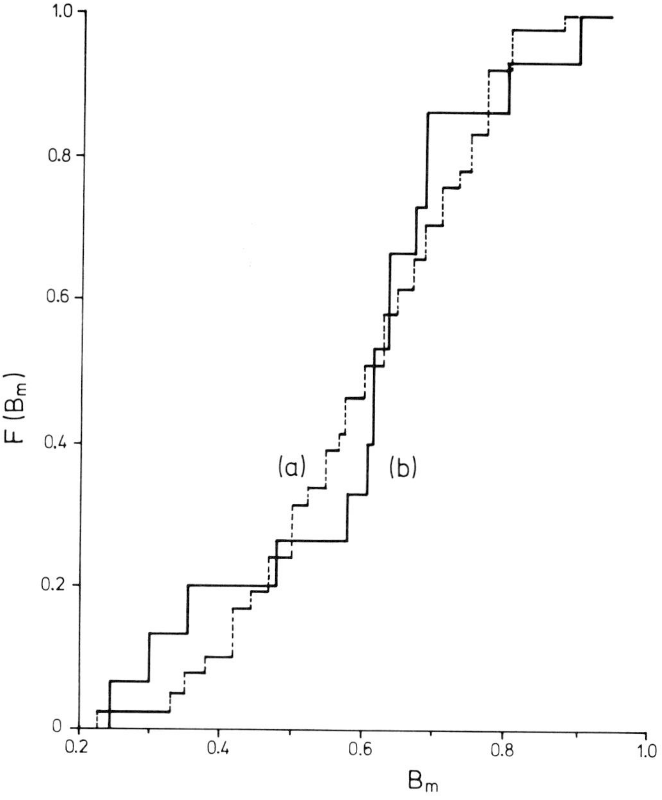

Fig. 3 Cumulative distributions $F(B_m)$ of the maximum values B_m of BROADBENT's statistic B, obtained from (a) sets of random numbers, and from (b) sets of experimental values of (Y_v/Y_t). Curve (a): 41 values of B_m obtained from 41 sets of 20 pseudo-random numbers having a truncated Gaussian distribution with mean 0.2 and S.D. 0.25, truncated at zero. Curve (b): 15 values of B_m obtained from sets of experimental values of (Y_v/Y_t). Most of the experimental sets are those of MacROBBIE (personal communication) and refer to experiments already published. In each case the quantum a was chosen to give the maximum value to B, so that some experiments had different values of a from those chosen by MacROBBIE.

The occurrence of the a.p. was found to be inhibited in about half the cases by an overnight soak in an A.P.W. containing Br^- in place of Cl^- and Mg^{2+} in place of Ca^{2+}. Batches of cells so treated were given a 10 min exposure to radioactive Br^-, and cut open for collection of a sap sample while supported by recording electrodes. Generally, it appeared that cells from which no a.p. was recorded had smaller fractions of radioactivity in the vacuole.

The conclusion was that there is a transfer of radioactivity when the cell is cut which increases the measured fraction in the vacuole. The points most affected will be the early ones when the true vacuole content is least; these are the points used to determine t_c.

The transfer of any significant amount of radioactivity to the vacuole in these experiments is most readily explained by the model of Fig. 1B, which predicts a high specific activity in compartment c at short times. A vesicle model along the lines of that of Fig. 1C would need an *ad hoc* assumption to explain the result, since at short times the specific activity in phase c should be small.

If the results obtained from sap sampling experiments were not so irregular, it might have been possible to develop the sap sampling experiment into a method of distinguishing between the models; but this does not seem possible with current methods.

4. Intercellular Kinetics

BOSTROM and WALKER have measured the rate of transport of radioactive halide out of an internodal cell of ***Chara*** into the adjoining node and the next internodal cell (WALKER and BOSTROM, 1973; BOSTROM and WALKER, in preparation). They converted the rate of transport of radioactivity into a flux, using as specific activity the value they measured for the whole cytoplasm of the 'source' cell. The resulting 'flux' apparently decreases sharply with labelling time, reaching a steady value of about 2 pmoles cm^{-2} s^{-1} at 22 ks (6 h). This is taken by BOSTROM and WALKER to be the true flux; they interpret the much higher early values as representing a flux labelled by a specific activity which is much greater than the average for the cytoplasm as a whole. In other words the results are as expected from the model of Fig. 1B, if the flux to the node arises from compartment c, whose specific activity follows a time-course like the lower curve in Fig. 2; the average specific activity of c and p together rises much more slowly. Neither of the other two models can readily explain this result, although the vesicle model might do so with the assumption that the plasmodesmata are fed with radioactivity by the endoplasmic reticulum. Electron micrographs of charophyte cells do not support such an assumption however.

Table 2

Parameters of the 4-compartment model (Fig. 1B)

Parameter	Value from flux to vacuole (Fig. 2)	flux to node (BOSTROM and WALKER, unpublished)
Q_c (nmoles cm^{-2})	3	10
$[Cl^-]_c$ (mM)	6	20[a]
Q_p (nmoles cm^{-2})	100	120
ϕ_{cp} (pmoles cm^{-2} s^{-1})	10	5–20
$\phi_{cp} + \phi_{cv}$ (pmoles cm^{-2} s^{-1})	40	50
ϕ_{oc} (pmoles cm^{-2} s^{-1})	2.3	1.8

[a]Values of $[Cl^-]_c$ are derived from those for Q_c assuming the flowing cytoplasm is 5 μm thick.

It is possible to derive from the curve of influx against time a rough estimate for the half-time of compartment p, and hence of ϕ_{cp}. Following the suggestion (WALKER and BOSTROM, 1973) that the rate of transfer of radioactivity across the node may be limited by the rate of arrival of radioactivity in the cytoplasmic stream, BOSTROM has shown that the transfer to the node during the first 30 min of labelling is proportional to streaming velocity. Cytochalasin B was used to reduce the velocity and from the slope of the regression line the half-time of compartment c may be obtained. From the flux as measured in a 6 h experiment the content of phase c may also be estimated. Table 2 compares these provisional estimates (converted to contents and fluxes) with those from Fig. 2. The agreement is satisfactory, though not exact.

D. Conclusions

The experimental information available is far from conclusive, and it suffers from various kinds of error and uncertainty. At the moment it seems possible to conclude that nearly all the data are consistent with the simple 4-compartment model Fig. 1B, if we take the parameters to have values like those of Table 2. These figures for Q may be converted to figures for concentration by assuming the flowing cytoplasm to be about 5 μm thick. Then $[Cl^-]_c$ = 20 mM.

The additional assumptions needed to fit this model to the data are:
1) that the method of centrifuging normal cells leads to an over-estimate of $[Cl^-]_c$ (this explains the values of 70 mM in Table 1 as in error),
2) that the plasmodesmata are in contact with the water of the cytoplasm, not with the endoplasmic reticulum (this allows the calculations leading to the values in the right-hand column of Table 2), and
3) that the occurrence of an action potential when the cell is cut open transfers some radioactivity from the flowing cytoplasm to the vacuole (this reconciles the left-hand column of Table 2 with the right-hand column).
These are all more or less plausible, and each seems more likely than its alternative; but all should be subjected to further checks.

While the transport of salt to the vacuole in vesicles is an interesting idea, it is not required or even strongly indicated by the evidence to date.

E. References

BROADBENT, S.R., Biometrika **43**, 32–44 (1956)
COSTER, H.G.L., Aust. J. Biol. Sci. **19**, 545–554 (1966)
FINDLAY, G.P., HOPE, A.B., WALKER, N.A., Biochim. Biophys. Acta **233**, 155–162 (1971)
HOPE, A.B., SIMPSON, A., WALKER, N.A., Aust. J. Biol. Sci. **19**, 355–362 (1966)
KISHIMOTO, U., TAZAWA, M., Plant Cell Physiol. **6**, 507–518 (1965)
LEFEBVRE, J., GILLET, C., Biochim. Biophys. Acta **249**, 556 (1971)
MacROBBIE, E.A.C., J. Exp. Bot. **20**, 236–256 (1969)
MacROBBIE, E.A.C., J. Exp. Bot. **21**, 335–344 (1970)
MacROBBIE, E.A.C., Phil. Trans. Roy. Soc. Lond. B. **262**, 333–342 (1971a)
MacROBBIE, E.A.C., J. Exp. Bot. **22**, 487–502 (1971b)
MacROBBIE, E.A.C., in Ion Transport in Plants (ed. W.P. Anderson), pp. 431–446, Academic Press, London and New York 1973
SPANSWICK, R.M., WILLIAMS, E.J., J. Exp. Bot. **15**, 190–201 (1964)
WALKER, N.A., in Ion Transport in Plants (ed. W.P. Anderson), pp. 459–461, Academic Press, London and New York 1973
WALKER, N.A., BOSTROM, T.E., in Ion Transport in Plants (ed. W.P. Anderson), pp. 447–458, Academic Press, London and New York 1973

NH_3 Efflux as a Means for Measuring H^+ Extrusion in *Nitella*

C.E. Barr, M.S. Koh, and Th.E. Ryan

A. Introduction

Widespread interest in H^+ extrusion as the principal electrogenic mechanism in plant cell membranes as proposed by KITASATO (1968) is very evident in the collection of papers presented at the recent Liverpool meeting (ANDERSON, 1973). KITASATO's experimental work was not directly concerned with H^+ extrusion itself; rather, this idea emerged as a necessary ***ad hoc*** measure to accomodate the problem of internal pH control in ***Nitella***. RAVEN and SMITH (1973) have subsequently argued, and convincingly so, that the primary role of the H^+ pump must be pH control rather than electrogenesis.

A third possible role for H^+ extrusion involves its participation in active K^+ uptake through a K^+/H^+ exchange mechanism. POOLE (1973) found evidence for this in red beet tissue and RYAN and BARR (1973) briefly reported similar findings for ***Nitella***. This contrasts with the concept of K^+ accumulation merely as a passive movement occurring in response to the generation of a membrane potential of the proper magnitude. Earlier SLAYMAN and SLAYMAN (1968) had described a comparable system in ***Neurospora*** in which K^+ uptake appeared to be coupled to the extrusion of both Na^+ and H^+. For such mechanisms to be electrogenic the exchange ratio must of course be different from unity.

Our present work was stimulated by a result inconsistent with the working hypothesis outlined above: no decrease in the external pH was observed during net K^+ influx (RYAN, 1973). This led us to test the experimental solution for the presence of some type of uncharged base, released during the putative exchange of K^+ for H^+. Buffering was detected in the neighborhood of pH 9.3, thus suggesting ammonia. Reported here are the results to date on this research.

B. Materials and Methods

Plants of ***Nitella clavata*** were grown in open culture tanks containing a small amount of soil, some charcoal, ***Elodea*** plants, and snails. The culture solution was similar to FORSBERG's (1965), containing 4 mM Tris at pH 7.0. Due to the degradation of Tris the NH_4^+ present varied between 0.5 and 2.8 mM, averaging about 1.0 mM. The ***Nitella*** plants did not grow at NH_4^+ concentrations above 2 mM.

Ammonia was determined in a total volume of 10 ml by the Nessler method (TARAS, 1958); 15 minutes after the reagents were added the absorbance was measured at 450 nm. To prevent turbidity 0.1 ml of 3.3 % (w/w) gum arabic was added to the reaction mixture before the Nessler's reagent was added. The gum arabic was purified by first preparing a 10 % solution, filtering, and allowing the filtrate to dry on glass.

Titrations of 50 to 100 isolated internodal cells were carried out in a 20 ml volume of K solution using a 5 ml microburette. Rapid mixing of the solution was effected by means of a medicine dropper.

The K solution contained 1.0 mM KCl, and 0.1 mM each of NaCl, $CaCl_2$ and $MgCl_2$. The isolated internodal cells were maintained in K solution under 350 lux cool-white illumination at 22°C.

C. Results and Discussion

I. NH_4^+ Accumulation during Growth

We have found that ***Nitella*** cells accumulate about 60 mM NH_4^+ in the vacuole during culturing. The source of the NH_4^+ is Tris, which is initially present in the FORSBERG-type (1965) medium at 4.0 mM. It is entirely possible that the degradation of Tris is bacterial.

In a culture having a pH of 7.1 and containing 1.0 mM NH_4^+, the vacuolar $[NH_4^+]$ of internodal cells was 62 ± 2 mM. If a vacuolar pH of 5.3 (as measured on other cells under different conditions) is assumed, the NH_4^+ accumulation can be attributed to the equilibration of NH_3 across the cell membrane and the trapping of the ammonia as NH_4^+. This would, of course, necessitate the internal production of an equivalent amount of H^+ to convert NH_3 to NH_4^+. The highest concentration of NH_4^+ accumulated was 135 mM.

II. NH_3 Loss by *Nitella* Cells at pH 5.7

When internodal cells are placed in K solution (a non-growth medium containing 1.0 mM K^+) NH_3 or NH_4^+ is lost exponentially, with a half-time of 2.8 days, as indicated in Fig. 1. From other experiments (RYAN, 1973) we know that the net K^+ influx approximately balances the ammonia loss while the vacuolar $[Cl^-]$ remains constant at 87 mM during the first 10 days following cell isolation.

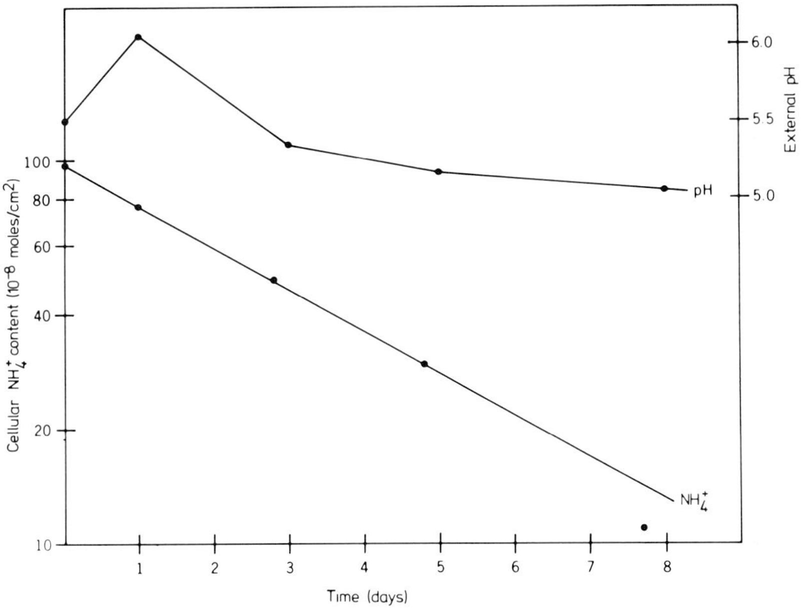

Fig. 1 The time course of the NH_4^+ content of internodal cells of ***Nitella clavata*** after transfer to K solution under 350 lux cool-white fluorescent illumination at 22°C. The NH_4^+ content is expressed on a surface area basis; the initial value of $97.3 \cdot 10^{-8}$ moles cm^{-2} corresponds to a mean vacuolar concentration of 61 mM. Ammonia analyses were carried out on aliquots of the external solution; the solution (100 ml) was completely replaced by fresh K solution at the intervals indicated by each point on the graph. A total of 85 internodal cells was used; these had a total surface area of 72 cm^2, a total volume of 1.36 ml, and a vacuolar volume of 1.15 ml. The initial pH of fresh K solution was 5.5.

Although the lack of a significant pH change might suggest a K^+/NH_4^+ exchange mechanism, there is some basis for proposing an alternative explanation. The exponential loss of NH_3 is quite consistent with the passive efflux of an uncharged species, i.e., unencumbered by the electrical restraints attending ionic movements. Moreover, the calculated NH_3 permeability coefficient of $4.4 \cdot 10^{-4}$ cm s^{-1} seems entirely reasonable (a vacuolar pH of 5.3 is assumed). On the other hand the calculated NH_4^+ permeability coefficient is $2.7 \cdot 10^{-5}$ cm s^{-1}, with the mean membrane potential being -152 mV; this value is 100 times greater than the P_K for *Nitella* (BARR, 1965). This would at least eliminate passive NH_4^+ efflux as the mechanism by which the cellular ammonia is lost.

Another argument favoring NH_3 as the exiting species is that the capacity of the cells to lower the external pH becomes greater after the internal ammonia has been depleted. The greatest acidification of the medium, to pH 4.7, was brought about by cells 3 to 4 weeks old.

From these results we conclude that H^+ extrusion simply keeps pace with the rate of NH_3 loss; this is as high as $2.4 \cdot 10^{-12}$ moles cm^{-2} s^{-1} during the first day, when the NH_3 efflux is greatest (Fig. 1). The slight pH rise occurring during this period suggests that these cells may be approaching their maximum capacity for H^+ extrusion under these conditions. There would, however, be no lack of H^+ within the cells since one H^+ is left behind as each NH_3 molecule exits.

An attempt was made to confirm the NH_3 permeability coefficient obtained above through a measurement of the net influx of NH_3 into 10-day old *Nitella* cells in the growth medium at pH 7.4. The value obtained was $7 \cdot 10^{-4}$ cm s^{-1}; although this is somewhat higher than the efflux-derived value, there are some uncertainties regarding the vacuolar pH and the effect the growth medium might have on the membrane permeability. The growth medium was chosen in order to provide a metabolic source of H^+ to neutralize the incoming NH_3.

III. NH_4^+ Loss by *Nitella* Cells at pH 4.7

In view of the above results we wished to reinvestigate previous work in which a large apparent H^+ uptake occurred when cells were bathed in K solution kept at pH 4.7 (RENT et al., 1972). That is, those results might possibly be accounted for by a large NH_3 efflux. However, Fig. 2 shows that during the first two hours of the acid treatment the amount of NH_3 released by the cells was much less than the amount of H^+ removed; the average rate of NH_3 loss was the same as that for pH 5.7.

After two hours of acid treatment, however, ammonia began to appear in the medium in large amounts. Even after the pH was returned to 5.7 by means of KOH addition, the heavy ammonia leakage continued for about two hours. It seemed likely that the low pH treatment was causing some membrane damage and that ammonia might be leaking out in ionic form, in association with Cl^-. This was confirmed by an hourly analysis of the solution for chloride. During the final 3 hours of the experiment the net Cl^- efflux exceeded the NH_4^+ efflux by about 30 %; during the first 2 hours the Cl^- loss was very small. It can be assumed that the extra Cl^- leaking out occurred mainly in association with K^+, as the vacuolar $[K^+]$ was about 30 mM, $[Na^+]$ was 16 mM, and $[NH_4^+]$ was 55 to 60 mM. The membrane potential measured on a number of cells some time ago under these conditions averaged about -90 mV.

Perhaps of significance is the fact that the heavy NH_4Cl leakage began only after the H^+ uptake capacity had been saturated, and this leakage continued even after restoration of the pH to 5.7. It suggests that the acid-caused damage may not be limited to the membrane but may involve internal protoplasmic components.

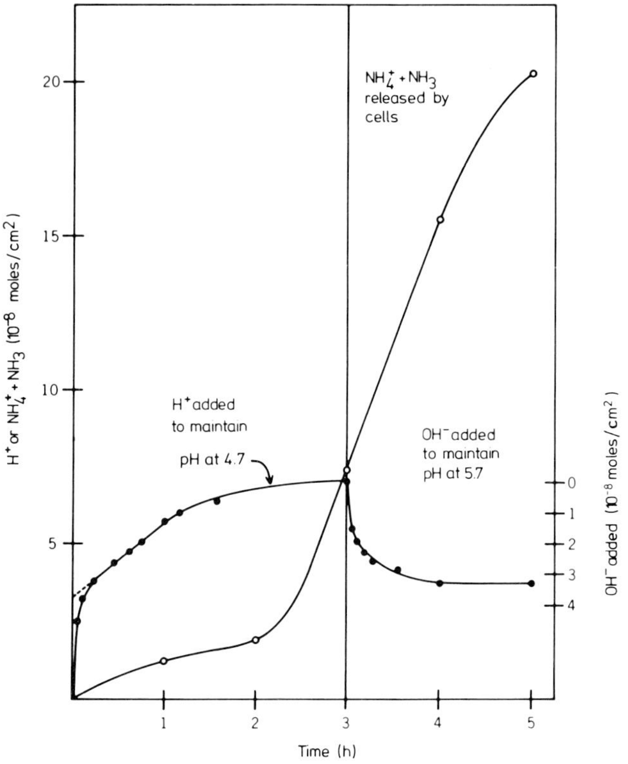

Fig. 2 The time course of the removal of external H^+ by internodal cells of ***Nitella clavata*** maintained at a pH of 4.7 in 20 ml of K solution. The pH was maintained through manual addition of 10 mM HCl at suitable intervals, with the entire 20 ml of solution being replaced by fresh K solution, pH 4.7, each hour. Also shown is the time course of the NH_4^+ loss by the cells (see text). At the end of the 3-hour acid treatment the pH was restored to 5.7 by addition of KOH and maintained at that pH. At the end of each hour the entire 20 ml was replaced by fresh K solution of pH 5.7. A total of 72 cells were used; these had a total surface area of 62 cm², a total volume of 1.15 ml; and an estimated vacuolar $[NH_4^+]$ of 62 mM.

The temporal characteristics of the H^+ uptake depicted in Fig. 2 are of considerable interest since there appear to be 2 distinct components: a fast component with a half-time of 3 minutes, and a linear component extending over a period of about an hour before saturation effects are evident. The fast fraction amounts to $3.3 \cdot 10^{-8}$ moles cm^{-2}, and all of this H^+ can be recovered afterwards by adding OH^-; this neutralization also has a half-time of 3 minutes. Presumably the buffering material is in the cell wall or on the external surface of the membrane, or both.

The linear uptake of H^+ occurs at a rate of $6.7 \cdot 10^{-12}$ moles cm^{-2} s^{-1}. The absence of any buffering material in the solution other than ammonia makes it appear rather certain that the H^+ is actually getting into the protoplast. The limitation of the total uptake to $4.1 \cdot 10^{-8}$ moles cm^{-2} perhaps is related to the exhaustion of the buffering capacity of the protoplasm. Assuming that one-tenth of the cell volume is protoplasm, the calculated buffering capacity is 22 mequiv per litre. There is some question, however, as to the validity of the assumption that H^+ comes to electrochemical equilibrium across the membrane as this would require a protoplasmic pH of 3.2.

The above results tend to confirm KITASATO's (1968) conclusion that the *Nitella* membrane is quite permeable to H^+. However, the calculated H^+ permeability coefficient obtained here, $1.0 \cdot 10^{-4}$ cm s^{-1}, is about one-tenth the value he arrived at on the basis of other types of experimental data.

IV. Electrical Effects of External NH_4^+

In very preliminary work the replacement of K^+ with NH_4^+ in the K solution at pH 5.7 caused a 45 mV depolarization in old cells containing negligible NH_4^+, but the effects on young cells were variable. It is possible that the depolarization is caused by an inhibition of the suggested K^+/H^+ pump rather than by the influx of NH_4^+ since addition of 1.0 mM NH_4Cl to the K solution had no effect.

D. Conclusions

Initially viewed as a nuisance, the abundance of NH_4^+ in the vacuole of *Nitella* under the conditions described here eventually came to be recognized as a useful tool for assessing the capability of the H^+ extrusion pump. If KITASATO (1968) and RAVEN and SMITH (1973) are correct in making the pump responsible for maintaining a constant internal pH, then the rapid loss of NH_3 from the cell would force the pump to eliminate H^+ at a comparable rate. Under this assumption the NH_3 efflux can be used as a rather precise measure of the H^+ extrusion rate. Under the present conditions, the highest rate observed is only $2.4 \cdot 10^{-12}$ moles cm^{-2} s^{-1}; the apparent limiting factors here are the NH_3 permeability and the internal concentration of NH_4^+. The NH_3 permeability coefficient of $4.4 \cdot 10^{-4}$ cm s^{-1} properly applies to the movement of NH_3 from the vacuole to the external medium and therefore is a measure of the ease of movement through the entire layer of protoplasm including the two delimiting membranes.

Implicit in the above is the hypothesis that the H^+ extrusion pump operates only to the extent required for the maintenance of a suitable internal pH. Moreover, the passive H^+ influx at pH 5.7 may be smaller than that calculated on the basis of results obtained at pH 4.7 (see below). If the P_H of 10^{-4} cm s^{-1} obtained at pH 4.7 is used to calculate the passive H^+ influx at pH 5.7, the value is $1.2 \cdot 10^{-12}$ moles cm^{-2} s^{-1}. We suggest, however, that the H^+ pump may be quite inactive at pH 5.7, i.e., for cells in which K^+ is the major cation.

It is possible at this time to conclusively rule out the alternative explanation that a K^+/NH_4^+ exchange mechanism is operating here. The evidence for acid production at the external surface of characean cells (SPEAR et al., 1969; SMITH, 1970) should be considered as well as the simple kinetics of the ammonia release. This possibility cannot be unequivocally rejected, but it seems less probable.

Another main conclusion of this study is that *Nitella* cells are quite permeable to H^+ at pH 4.7 (P_H of 10^{-4} cm sec^{-1}), but this does not necessarily mean that the H^+ permeability is constant over a wide range of pH values. After 2 hours at pH 4.7 *Nitella* cells become quite leaky, and this does suggest some damage to the protoplasm. However, the ability of the cells to survive the ordeal would indicate that the bulk pH of the protoplasm, although perhaps slightly reduced, cannot be anywhere near the equilibrium pH of 3.2. The question arises as to what happens to the $4.1 \cdot 10^{-8}$ moles cm^{-2} of H^+ which gets into the cells but cannot be recovered by neutralizing the medium back to pH 5.7. That the H^+ is not immediately extruded suggests the possibility of its removal to the vacuole, where the normal pH is probably 2 units lower than that of the protoplasm. In earlier work(RENT et al., 1972) the same conclusion was drawn on the basis of a slight vacuolar pH decrease;

the compatibility of this conclusion with the vesicular mechanism of uptake was noted. Such results suggest that a simple type of permeation may not be responsible for the H^+ uptake.

E. Acknowledgements

We gratefully acknowledge the assistance of Mrs. Terace Rohland and Mrs. Mari Ann Melnick in preparing the manuscript and the technical assistance of Mr. Daniel Holland. This research was supported by grant GB-18069 from the U.S. National Science Foundation.

F. References

ANDERSON, W.P. (ed), Ion Transport in Plants, Academic Press, London and New York 1973
BARR, C.E., J. Gen. Physiol. **49**, 181–197 (1965)
FORSBERG, C., Physiol. Plant. **18**, 275–290 (1965)
KITASATO, H., J. Gen. Physiol. **52**, 60–87 (1968)
POOLE, R.J., in Reference 1, pp. 129–134 (1973)
RAVEN, J.A., SMITH, F.A., in Reference 1, pp. 271–278 (1973)
RENT, R.K., JOHNSON, R.A., BARR, C.E., J. Membrane Biol. **7**, 231–244 (1972)
RYAN, T.E., The Relation of the Membrane Potential and Resistance of ***Nitella*** to Ion Fluxes, M.S. Thesis, State University of New York, College at Brockport 1973
RYAN, T.E., BARR, C.E., Plant Physiol. **51S**, 16 (1973)
SLAYMAN, C.L., SLAYMAN, C.W., J. Gen. Physiol. **52**, 424–443 (1968)
SMITH, F.A., New Phytol. **69**, 903–917 (1970)
SPEAR, D.G., BARR, J.K., BARR, C.E., J. Gen. Physiol. **54**, 397–414 (1969)
TARAS, M.J., in Colorimetric Determination of Nonmetals (ed. D.F. Boltz), pp. 75–160, Interscience, New York-London-Sydney 1958

Light-Dependent Ion Fluxes in *Mougeotia*: Control by Photosynthesis, not by Phytochrome

G. Wagner

A. Introduction

Compartmental analysis of ion fluxes has been successful in many studies on plant cells (HIGINBOTHAM, 1973; LÜTTGE, 1973); yet it is uncertain whether large cell organelles, such as chloroplasts or cytoplasmic vesicles, constitute autonomous ion flux compartments and hence are not represented by the current model of three compartments in series (free space, cytoplasm, vacuole). Such additional flux compartments could lead to wrong estimates of the intracellular ionic relations (MacROBBIE, 1971).

A study of ion fluxes in the green alga ***Mougeotia*** should help to clarify this situation, since in this plant cell the single chloroplast constitutes the major part of the cytoplasmic phase, actually 60—70 % of its volume.

At the moment, the primary action of the photomorphogenic pigment phytochrome is still unknown (BRIGGS and RICE, 1972). The hypothesis generally favoured assumes that phytochrome controls membrane parameters (WEISENSEEL and HAUPT, 1974), which seems reasonable for ***Mougeotia*** for two reasons. The work of HAUPT and coworkers (HAUPT and SCHÖNBOHM, 1970) shows that the phytochrome molecules, governing the orientation of the ***Mougeotia*** chloroplast towards the light, are located at or within the plasmalemma; and during the phototransformation $P_r \rightarrow P_{fr}$, and ***vice versa***, the orientation of the first singlet transition moment of the dichroic phytochrome molecule will turn by 90 degrees relative to the cell surface (ETZOLD, 1965; HAUPT, 1968). These findings suggested the present study.

B. Results and Discussion

I. Fluxes and Compartmentation of Cl^- and K^+

The Cl^- ionic relations of ***Mougeotia*** will be presented as the chief example. Experimental details and a more extended treatment are given by WAGNER (1973)[1]). The wash-out kinetics of $^{36}Cl^-$ (Fig. 1) show the usual three different rate constants, k_0, k_1, and k_2, indicating the corresponding apparent amounts of radioactivity, Y_0, Y_1, and Y_2, respectively.

We must test whether additional rate constants exist in the system and whether k_0 can be neglected, as is usually done (CRAM, 1968). Thus we must first ask whether k_1 and k_2, taken as efflux rate-constants from a serial model for two compartments, suffice to predict quantitatively the influx kinetics, measured in flux equilibrium, (HOPE, 1971).

---

[1])For explanation of symbols see contributions of PITMAN and WALKER in this book.

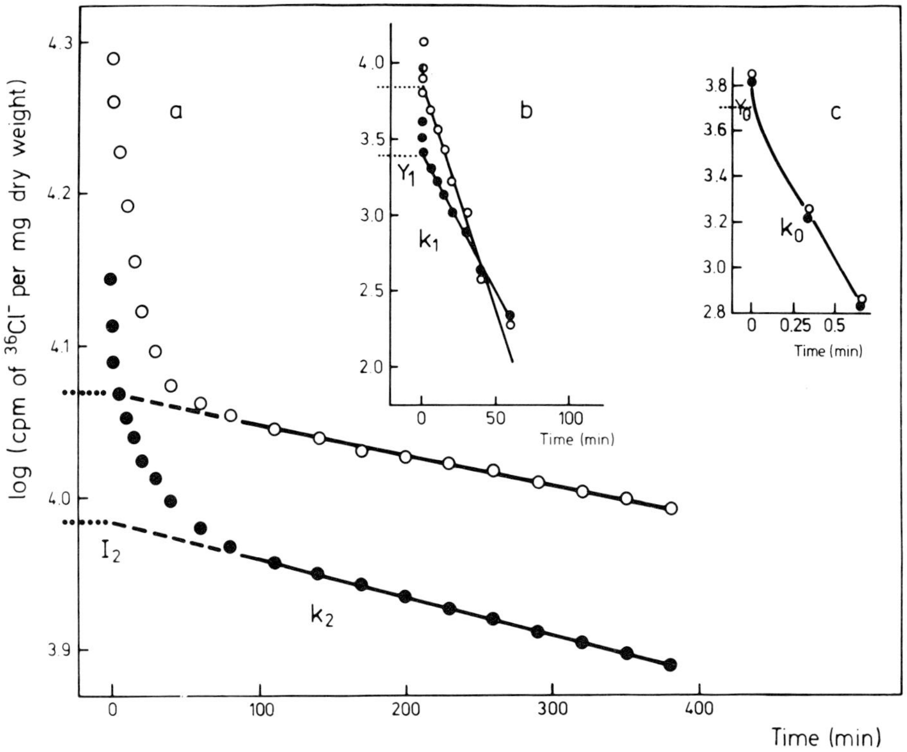

Fig. 1 Loss of $^{36}Cl^-$ from ***Mougeotia*** in white light of 4000 lux (o) or darkness (•) starting (t = 0) 330 min after incubation with $^{36}Cl^-$, plotted as the logarithm of the activity remaining in (a) the whole cell, (b) compartment 0 + 1, (c) compartment 0. Q_{total} was 34 mM both in light and dark.

Fig. 2 shows that the measured influx in the dark can be fitted satisfactorily to computer-calculated theoretical curves. The values used for this purpose were in s^{-1}: $\phi_{01}/Q_1 = 3.7 \cdot 10^{-4}$; $\phi_{12}/Q_1 = 1.66 \cdot 10^{-4}$; and $\phi_{21}/Q_2 = 6.25 \cdot 10^{-6}$. With these values the influx kinetics can be simulated for the whole cell between t = 0 and t = 270 min (Fig. 3). Similarly, the kinetics for the Cl^- influx in the light have been obtained using the following values in s^{-1}: $\phi_{01}/Q_1 = 8.6 \cdot 10^{-4}$; $\phi_{12}/Q_1 = 1 \cdot 10^{-4}$; $\phi_{21}/Q_2 = 4.4 \cdot 10^{-6}$.

Thus, even for the morphologically peculiar ***Mougeotia*** cell the conventional model of two intracellular compartments in series seems applicable, at least for the Cl^- fluxes. There is no need to assume an additional flux compartment or vesicular transport, as has been inferred for ***Tolypella*** and ***Nitella*** by LARKUM (1968) and MacROBBIE (1971), respectively. The measured values of k_1 and k_2 agree well with those of other plants (CRAM, 1968; PIERCE and HIGINBOTHAM, 1970; JESCHKE, 1971). Hence it seems reasonable to assign compartment 1 to the cytoplasm and compartment 2 to the vacuole. Thus the large chloroplast is probably part of the cytoplasmic compartment. In fact k_1 compares well with rate constants from isolated chloroplasts (WINOCUR et al., 1968; NOBEL, 1969). Consequently the measured fluxes ϕ_{01} etc. are to be identified with ϕ_{oc} etc.. ϕ_{oc} and ϕ_{co} show a 3-fold light-dependent increase. A similar increase was found for the fluxes of K^+ ($^{86}Rb^+$).

In Table 1 the ratios ϕ_{oc}/ϕ_{co} and ϕ_{cv}/ϕ_{vc} are compared with the corresponding ratios of the electrochemical activities (*a*) according to the Ussing-Teorell criterion (HOPE, 1971). The

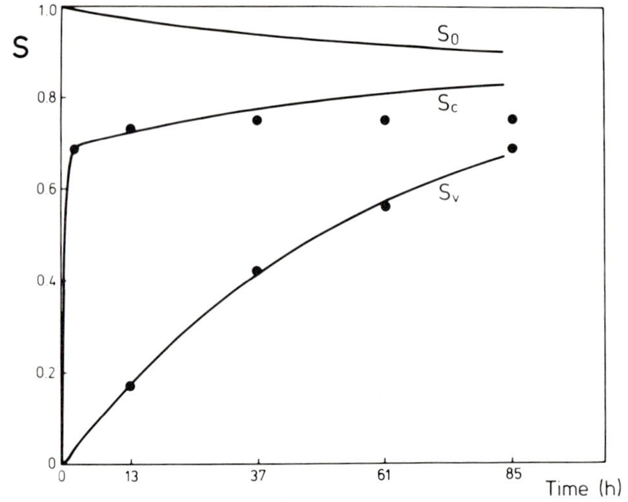

Fig. 2 Theoretical curves and experimental data (dots) for the influx of $^{36}Cl^-$ into ***Mougeotia*** cells, plotted as the increase of relative intracellular specific radioactivity, S_c and S_v, in the corresponding compartments 1 ('cytoplasm') and 2 ('vacuole'). S_o = relative specific activity of the medium. Data from two to four influx experiments with subsequent tracer wash-out (see Fig. 1) to get Y_c and Y_v. Q_{total} = 21 ± 3 mM.

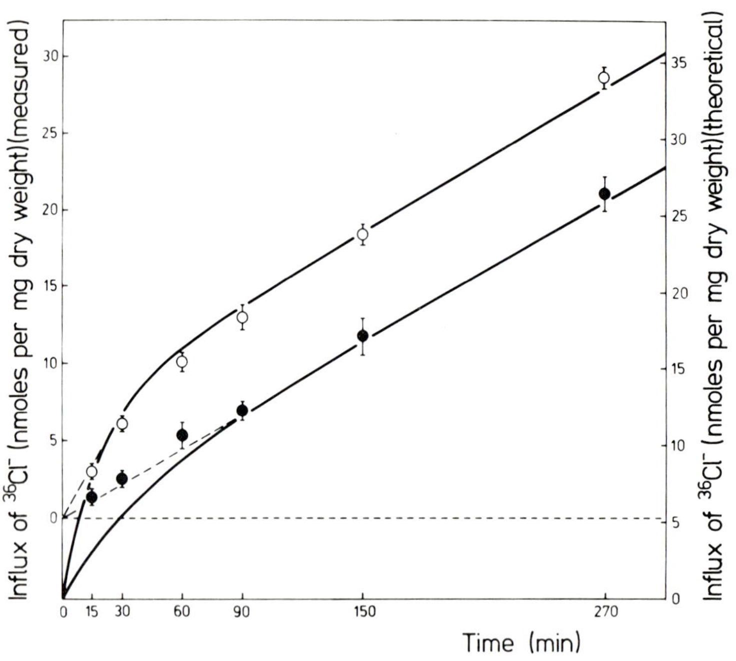

Fig. 3 Theoretical curves and experimental data (dots; ± S.E.) of the influx of $^{36}Cl^-$ into cells of ***Mougeotia*** in white light of 4000 lux (o) and in darkness (●) plotted as the total Cl^- uptake, i.e. $S_c \cdot Q_c + S_v \cdot Q_v$. ($Q_{total}$ = 26 mM (light), 27 mM (dark).) Pretreatment: 1.5 h in normal culture light (200 lux), then 30 min at 4000 lux or dark.

data show that both ions, but especially Cl^-, are actively pumped inward across the plasmalemma, whereas the deviation from electrochemical equilibrium is small and nearly equal for both ions at the tonoplast.

Table 1

Flux ratios of K^+ and Cl^- across plasmalemma and tonoplast and corresponding ratios of electrochemical activities (***a***) according to the Ussing-Teorell equation.

Ion	Plasmalemma ϕ_{oc}/ϕ_{co}	Plasmalemma a_j^o/a_j^c	Tonoplast ϕ_{cv}/ϕ_{vc}	Tonoplast a_j^c/a_j^v
K^+				
light	1.0	0.067	1.0	0.13
dark	1.0	0.067	1.0	0.13
Cl^-				
light	1.0	0.01	1.0	0.28
dark	1.0	0.014	1.0	0.24

II. Photosynthesis versus Phytochrome

Experiments with the inhibitor DCMU (dichlorophenyl-dimethylurea) suggest that the light-dependent uptake of Cl^- is driven by photosystem II, whereas the uptake of K^+ additionally depends on energy from photosystem I. These results agree with those on ***Nitella*** and ***Hydrodictyon*** (MacROBBIE, 1970; RAVEN, 1969a, b). Hence, the light-dependent active ion fluxes in ***Mougeotia*** are probably due to photosynthesis only, and not

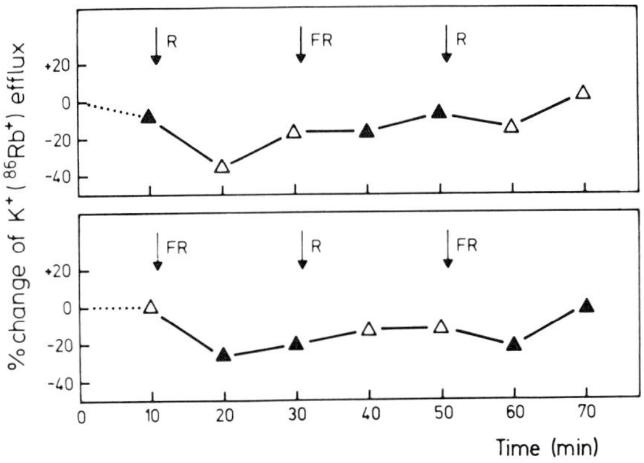

Fig. 4 K^+ ($^{86}Rb^+$) efflux during the cytoplasmic wash-out period under alternating light pulses of red and far-red, i.e. 300 s of 1.2 nE cm^{-2} s^{-1} of 653 and 720 nm, respectively. The wash-out was sampled twice after each red (△) and far-red (▲) pulse. Pretreatment: after 12 h in dark the cells were incubated for 5 h with $^{86}Rb^+$. All manipulations in green safelight.

to phytochrome (see also SWEET and HILLMAN, 1969; SATTER et al., 1970; YUNGHANS and JAFFE, 1972). However, are there passive ion fluxes and hence passive membrane properties, which depend on the state of phytochrome?

If phytochrome-dependent changes of membrane permeability exist, then these changes should be as stable as P_{fr} (SCHÄFER et al., 1973) and induce changes of the passive efflux ϕ_{co}. Consequently, k_c should be affected for $k_c = (\phi_{co} + \phi_{cv})/Q_c$. The experiments did not show such changes of k_c, and hence $k_c(P_{fr}) = k_c(P_r)$.

However, the phytochrome effect could be rather short-lived. To test this, pulses of red or far-red light were applied during efflux experiments. One typical result for K^+ is shown in Fig. 4. Again no significant changes of the efflux with the changing state of phytochrome could be detected. Also no effect of red or far-red upon the membrane potential has so far been found (BENTRUP; WEISENSEEL; personal communications).

C. Conclusion

Theoretical considerations have shown (WAGNER, 1973) for the mobile ions, K^+ and Cl^-, that appreciable phytochrome-induced intracellular ion gradients which could control chloroplast movement (BENTRUP, 1968), should be associated with transmembrane fluxes large enough to be detected by the techniques used. Perhaps rather immobile ions such as Ca^{2+} or Mg^{2+} are involved in phytochrome-controlled movements (RASMUSSEN, 1970). This is probable since chloroplast movement of ***Mougeotia*** is sensitive to cytochalasin B, suggesting an involvement of contractile filaments (WAGNER et al., 1972). Moreover, an intracellular pH-gradient is not excluded by this study.

D. Acknowledgements

This work was supported by the Deutsche Forschungsgemeinschaft.

E. References

BENTRUP, F.W., Ber. Dtsch. Bot. Ges. **81**, 311–314 (1968)
BRIGGS, W.R., RICE, H.V., Ann. Rev. Plant Physiol. **23**, 293–334 (1972)
CRAM, W.J., Biochim. Biophys. Acta **163**, 339–353 (1968)
ETZOLD, H., Planta **64**, 254–280 (1965)
HAUPT, W., Z. Pflanzenphysiol. **58**, 331–346 (1968)
HAUPT, W., SCHÖNBOHM, E., in Photobiology of Microorganisms (ed. P. Halldal), pp. 283–307, Wiley-Interscience, New York–London–Sydney–Toronto 1970
HIGINBOTHAM, N., Bot. Rev. **39**, 15–69 (1973)
HOPE, A.B., Ion Transport and Membranes. A Biophysical Outline, Butterworth, London 1971
JESCHKE, W.D., Proc. I. Europ. Biophys. Congr. (Baden/Vienna), **3**, 111–117 (1971)
LARKUM, A.W., Nature **218**, 447–449 (1968)
LÜTTGE, U., Stofftransport der Pflanzen. Springer-Verlag, Berlin–Heidelberg–New York 1973
MacROBBIE, E.A.C., Quart. Rev. Biophysics. **3**, 251–294 (1970)
MacROBBIE, E.A.C., Ann. Rev. Plant Physiol. **22**, 75–96 (1971)
NOBEL, P.S., Biochim. Biophys. Acta **172**, 134–143 (1969)
PIERCE, W.S., HIGINBOTHAM, N., Plant Physiol. **46**, 666–673 (1970)
RASMUSSEN, H., Science **170**, 404–412 (1970)
RAVEN, J.A., New Phytol. **68**, 45–62 (1969a)
RAVEN, J.A., New Phytol. **68**, 1089–1113 (1969b)

SATTER, R.L., MARINOFF, P., GALSTON, A.W., Am. J. Bot. **57**, 916–926 (1970)
SCHÄFER, E., SCHMITT, W., MOHR, H., Photochem. Photobiol. **18**, 331–334 (1973)
SWEET, H.C., HILLMAN, W.S., Physiol. Plant. **22**, 776–786 (1969)
WAGNER, G., Ionenflüsse und Phytochrom-abhängige Chloroplastenbewegung bei ***Mougeotia***. Ph.D. Thesis, University Tübingen, W.-Germany (1973)
WAGNER, G., HAUPT, W., LAUX, A., Science **176**, 808–809 (1972)
WEISENSEEL, M., HAUPT, W., in Membrane Transport in Plants (eds. U. Zimmermann, J. Dainty), Springer-Verlag, Berlin–Heidelberg–New York 1974
WINOCUR, B.A., MACEY, F.I., TOLBERG, A.B., Biochim. Biophys. Acta **150**, 32–40 (1968)
YUNGHANS, H., JAFFE, M.J., Plant Physiol. **49**, 1–7 (1972)

Carbonic Anhydrase and the Driving Force of Light-Dependent Uptake of Cl^- and HCO_3^- by *Scenedesmus*

G.R. Findenegg

A. Introduction

It is easy to investigate ion uptake by plant cells for Cl^- but difficult for HCO_3^-. When photosynthesis proceeds at a high pH in the medium, HCO_3^- is apparently used, but we cannot really decide whether HCO_3^- or CO_2 enters the cell. Even if ($^{14}CO_2$ + $H^{14}CO_3^-$) uptake and pH change are measured simultanuously, we cannot be sure that HCO_3^- is taken up because of the possibility of a HCO_3^-/OH^--exchange mechanism.

Until now, our knowledge of HCO_3^- uptake has come from measurements of the membrane potential (HOPE, 1965; DENNY and WEEKS, 1970) or from differences of photosynthetic characteristics at high and low pH (RAVEN, 1970). With ***Scenedesmus obliquus*** we have found another approach to studying the uptake of HCO_3^- by cells. When this green alga is grown in air plus 1,5 % CO_2 a drastic reduction of the activity of the enzyme carbonic anhydrase occurs, compared to air-adapted cells. The lack of this enzyme makes HCO_3^- unavailable for photosynthesis, as was seen from inhibition of photosynthetic O_2 production at high pH. This is not unexpected, because HCO_3^- must be converted to CO_2 prior to fixation by ribulosediphosphate-carboxylase (COOPER et al., 1969). It was surprising, however, that unlike air-adapted cells, CO_2-adapted cells did not take up Cl^- in the light.

It thus seems that carbonic anhydrase participates in light-induced uptake of HCO_3^- and Cl^-; and Cl^- uptake may be an alternative process to HCO_3^- uptake (FINDENEGG, 1974).

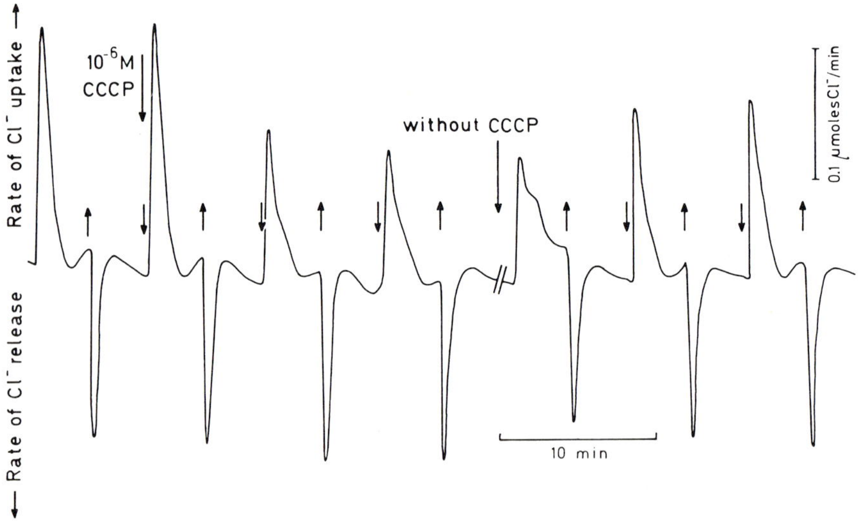

Fig. 1 Rate of Cl^- uptake and release by air-adapted ***Scenedesmus*** cells in alternating light-dark periods. 5 mM HEPES-NaOH, pH 7.1; 30°C; 0.1 mM KCl. Light on ↓, light off ↑.

Besides its dependence on a high carbonic anhydrase level, light-induced uptake of Cl^- is sensitive to inhibitors of energy metabolism. The time course of Cl^- uptake was measured with a Cl^--sensitive electrode in the flowing external medium as described by FINDENEGG (1974). When the light was switched on about 0.3 μmoles Cl^- were taken up by 1 g algae (fresh weight). The same amount was released after darkening. This uptake corresponded to an internal $[Cl^-]$ of 3 times the external $[Cl^-]$.

1 μM DCMU inhibited this light-induced Cl^- uptake completely, but 1 μM CCCP had more of a retarding than suppressing effect (Fig. 1). Cl^- release after the end of the light period was not affected by CCCP. This suggests that Cl^- efflux is passive.

In view of the above results it is of some interest to look for a specific action of carbonic anhydrase in the accumulation process.

B. Results and Discussion

I. Role of Carbonic Anhydrase in Cl^- Transport

Because of the above-mentioned difficulties of demonstrating HCO_3^- uptake some of our studies were carried out with $^{36}Cl^-$. If Cl^- uptake is an alternative event to HCO_3^- uptake the following results should also apply to the uptake of HCO_3^-.

Carbonic anhydrase activity can be measured in both homogenized and whole ***Scenedesmus*** cells. This suggests the enzyme is located at the cell surface. However, addition of an excess of bovine carbonic anhydrase to the medium did not restore the ability of carbonic anhydrase deficient cells to use HCO_3^- in photosynthesis, thus excluding the possibility of a free space location for the algal enzyme. Carbonic anhydrase in the free space would convert excess HCO_3^- to CO_2 which, because of its low equilibrium concentration, would only very slowly enter the cell (ENNS, 1967).

A comparison of $^{36}Cl^-$ uptake between air- and CO_2- adapted cells showed that a) cells with zero initial Cl^- and low carbonic anhydrase activity take up Cl^- more slowly than air-adapted cells with high enzyme activity, and b) if $^{36}Cl^-$ was added 10 min before or after addition of unlabelled Cl^-, isotopic equilibrium was reached within 5 min with air-adapted cells (high carbonic anhydrase) but not within 15 min with CO_2-adapted cells. The same results were found in the light and in the dark.

Accordingly, we believe that the carbonic anhydrase is probably located at the plasmalemma and its effect is to increase the uptake of Cl^-. Diamox, an inhibitor of carbonic anhydrase, also cuts down Cl^- uptake in the light and in the dark (FINDENEGG, 1974).

II. Counterion of Light-Dependent Cl^- Uptake

Carbonic anhydrase has neither a directional nor a light-dependent effect on Cl^- transport. We must thus ask what is the driving force for the light-induced Cl^- uptake. It could be a coupling with the active transport of another ion in the light. It is not certain what this other ion could be.

The extrusion of an organic anion to drive Cl^- uptake is very unlikely, for after $^{14}CO_2$ photosynthesis no release of labelled substances into the medium could be detected.

Moreover, K^+ and Na^+ are also unlikely to drive light-dependent Cl^- uptake. First, K^+ transport in alternating light-dark periods is the same whether K_2SO_4 or KCl are used; second, light-induced Cl^- uptake is unchanged if Na^+ instead of K^+ is the counterion; third, K^+ is released at the beginning of light periods, i.e. at the time of light-induced Cl^- uptake. Therefore, K^+ and/or Na^+ cannot compensate the Cl^- influx electrically.

The only other ion which could drive light-induced Cl^- uptake is OH^- (or H^+), and in fact at the time of light-induced Cl^- uptake there is a transient increase in external pH (Fig. 2). For air-adapted cells this apparent release of OH^- was greater in the presence than in the absence of Cl^- (Fig. 3). So it seems likely that light-induced Cl^- uptake is balanced by OH^- efflux (or H^+ uptake).

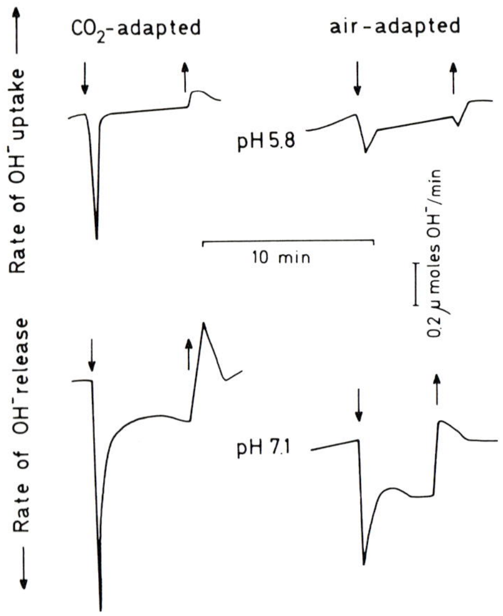

Fig. 2 pH-transients in the medium of ***Scenedesmus*** cells in alternating light-dark periods. pH 5.8 : 5 mM MES-NaOH; pH 7.1 : 5 mM HEPES-NaOH. 25°C; 0.1 mM KCl.

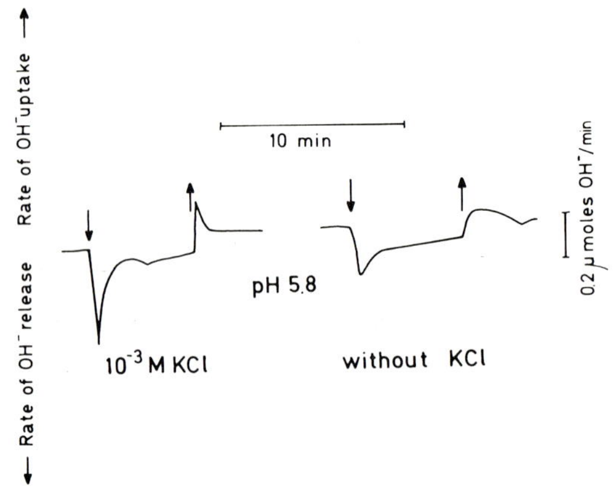

Fig. 3 pH-transients in the medium of air-adapted ***Scenedesmus*** cells in alternating light-dark periods. 5 mM MES-NaOH; 25°C.

Fig. 2 shows the pH-transients of air- and CO_2-adapted algae at two different pH values. Both air-adapted cells with high, and CO_2-adapted cells with low, carbonic anhydrase level show a transient alkalinisation of the medium at the beginning of light. However, the effect is more pronounced with CO_2-adapted algae which take up neither HCO_3^- nor Cl^- at the time of transient alkalinisation. As these algae are restricted to CO_2 uptake, at high pH the remaining HCO_3^- will dissociate to give OH^-.

Dissociation of HCO_3^- in the medium cannot be the whole reason for the alkalinisation. In the experiment with CO_2-adapted algae shown in Fig. 2, photosynthetic O_2 evolution was only about 30 % higher at pH 5.8 (78 % CO_2/22 % HCO_3^-) than at pH 7.1 (24 % CO_2/76 % HCO_3^-). Assuming a photosynthetic CO_2/O_2 quotient of 1 it can easily be calculated that there must be an efflux of OH^- independent of the CO_2 uptake of the algae.

It is consistent with this idea that a small transient alkalinisation of the medium on illumination, with air-adapted cells (Fig. 4) or with CO_2-adapted algae, is also found when Cl^- and HCO_3^- are absent. Although the nature of the counterion of this remaining OH^- release is unknown we must conclude that on illumination OH^- ions are extruded from the interior of *Scenedesmus*.

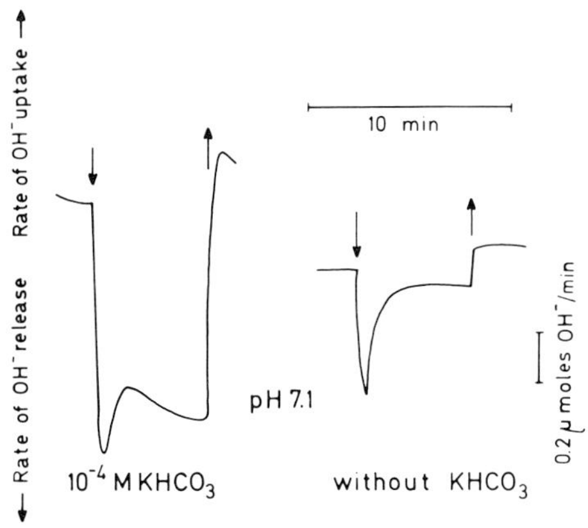

Fig. 4 pH-transients in the medium of air-adapted *Scenedesmus* cells in alternating light-dark-periods. 5 mM HEPES-NaOH; 25°C; solution 'without $KHCO_3$' was bubbled with CO_2-free air.

C. Conclusion

From the data here presented we propose a model for light-induced uptake of HCO_3^- and Cl^- of *Scenedesmus.* Carbonic anhydrase may act as a permease for these ions in the plasmalemma. The action of carbonic anhydrase is neither light-dependent nor directional. When light is switched on a pH gradient is established, as was suggested by SMITH and LUCAS (1972) for Characean cells. This may be an active step. In air-adapted cells this pH gradient causes a passive uptake of HCO_3^- or Cl^-. In CO_2-adapted algae this cannot happen because of the lack of carbonic anhydrase, necessary as permease.

D. References

COOPER, T.G., FILMER, D., WISHNIK, M., LANE, D.M., J. Biol. Chem. **244**, 1082–1083 (1969)
DENNY, P., WEEKS, D.C., Ann. Bot. **34**, 483–496 (1970)
ENNS, T., Science **155**, 44–47 (1967)
FINDENEGG, G.R., Planta, in press (1974)
HOPE, A.B., Aust. J. Biol. Sci. **18**, 789–801 (1965)
RAVEN, J.A., Biol. Rev. **45**, 167–221 (1970)
SMITH, F.A., LUCAS, W.J., in Ion Transport in Plants (ed. W.P. Anderson), pp. 223–231, Academic Press, London and New York 1973

Photoinduced Changes in Electrical Potentials and H^+ Activities of the Chloroplast, Cytoplasm, and Vacuole of *Phaeoceros laevis*

R.F. Davis

A. Introduction

It is becoming increasingly clear that in green plant cells H^+ ions play an important role both in photoinduced changes of the membrane potential and in photophosphorylation. Transients of the membrane potential after light-dark (L-D) and dark-light (D-L) changes are thought to be dependent on electron flow and ion movements (possibly H^+) across the chloroplast thylakoid membranes (LÜTTGE and PALLAGHY, 1969). KITASATO (1968) and SPANSWICK (1972) have considered electrogenic H^+ pumps in ***Nitella***. There is evidence with broken chloroplasts (HIND and JAGENDORF, 1965) for involvement of H^+ in photophosphorylation in accordance with the chemi-osmotic hypothesis of MITCHELL (1966). However, knowledge of the extent of light-induced changes in H^+ concentration in intact chloroplasts and the possible effects of these changes on the membrane potential is meagre. Nor is there any direct experimental evidence linking changes in cytoplasmic or vacuolar pH to photoinduced changes in the membrane potential. Therefore, the purpose of the present study was to make measurements with microelectrodes of the pH and electrical potential of chloroplasts ***in situ*** and, as well, of the cytoplasm and vacuole.

B. Materials and Methods

Greenhouse-grown gametophytes of ***Phaeoceros laevis***, a hornwort, were used for these studies. Each cell of this unusual plant has a single, large chloroplast that is about 35μm in diameter and which is suitable for microelectrode insertion. In addition, the cytoplasm is frequently distinguishable from the vacuole thereby permitting separation of plasmalemma and tonoplast electrical potentials.

The nutrient solution used throughout contained the following concentrations of salts in mM: 0.1 KCl; 0.1 $Ca(NO_3)_2$; 0.025 $MgSO_4$; Na_2HPO_4 and NaH_2PO_4 in appropriate amounts to make Na^+ 0.10 mM and buffer the solution to pH 5.7. Nutrient solutions with other pH's were of the same composition but had added HCl or NaOH to adjust the pH to the desired value; the adjustment of pH in this manner did not have any significant effect on the membrane potential.

Cell membrane electrical potentials were measured using glass microelectrodes as previously described (DAVIS, 1972). For measurement of chloroplast membrane potentials a microelectrode was inserted into the chloroplast with a reference microelectrode positioned in the cytoplasm, i.e. the electrical potential of the chloroplast was measured with respect to the cytoplasm.

The measurement of intracellular pH was made with antimony covered glass or stainless steel microcapillaries (models 801 and 814, respectively; Transidyne General Corp., Inc., Ann Arbor, Michigan, USA). Stable readings were obtained by positioning the pH microelectrode and a reference microelectrode (of conventional construction) as close as possible to each other and thrusting them simultaneously into a cell. The electrical potential difference between the two electrodes was measured with a differential amplifier. Calibration of the pH microelectrodes was made with buffered tissue macerate.

C. Results

I. Cell Membrane and Chloroplast Electrical Potentials

The upper tracing in Fig. 1 is a typical time course for the membrane potential (vacuole) of a gametophyte cell. In the dark the potential is -210 mV. When the microscope lamp is switched on the potential depolarizes, after a transient, to about -185 mV. With some cells in the dark the potential slowly returned to the level in the light after 30 to 60 min; in other cells a steady hyperpolarized potential was maintained for several hours. The rapid depolarizing transient with D-L change shown in Fig. 1 was not observed in all cells. The results from several similar experiments are collected in Table 1. With the microelectrode in the cytoplasm the potential is similar to that with it in the vacuole.

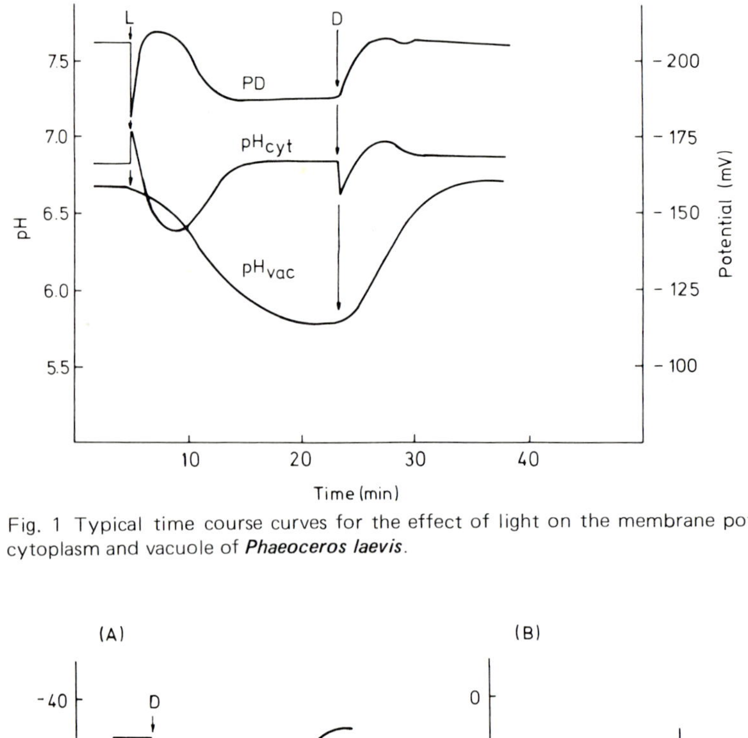

Fig. 1 Typical time course curves for the effect of light on the membrane potential and the pH of the cytoplasm and vacuole of ***Phaeoceros laevis***.

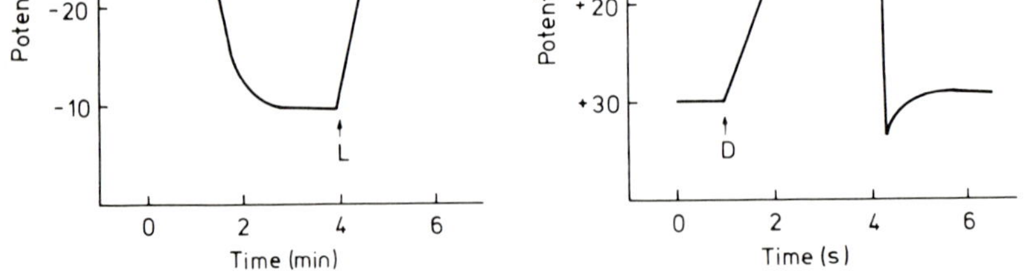

Fig. 2 Typical time course curves for the effects of light on the chloroplast electrical potential of ***Phaeoceros laevis***; (A) the slow response; (B) the fast response.

Fig. 2 depicts the two types of responses observed for chloroplast electrical potentials with L-D and D-L transitions. In one response, the potential in the light is -35 mV and with darkness depolarizes in about 1.5 min to -10 mV (Fig. 2A). With the other type of response the chloroplast potential is +30 mV in the light and shifts in the dark in about 2 s to +10 mV. The rise time for the fast response in the light is very fast and beyond the resolution of the instruments used. Very similar results to those shown in Fig. 2 have been found with chloroplasts in leaf mesophyll cells of ***Peperomia metallica*** by BULYCHEV et al. (1972). Table 1 presents the averages for several experiments with chloroplasts. A typical experiment with step increases in the intensity of the microscope illuminator gave potentials of +18, +34, and +42 mV, respectively. Slow photoinduced responses were not much affected by change in light intensity. Recent experiments in this laboratory with very fine-tipped microelectrodes (resistance 20 to 30 MΩ) have given fast response potentials in the light of

Table 1

The effects of light and dark on the membrane potential and pH of the chloroplast, cytoplasm, and vacuole of ***Phaeoceros laevis***. All values represent the maximum steady levels attained after changes in external pH or illumination. Chloroplast potentials were measured ***in situ*** and are relative to the cytoplasm. Cells were treated in solutions of a given pH for at least 30 min before making potential or pH measurements. Numbers in parentheses represent the number of measurements.

	External	Light – Potential or pH		Dark – Potential or pH	
	pH	Average	Range	Average	Range
		Potential (mV)			
Chloroplast					
slow response	5.7	-34.3(11)	-20 to -50	-15.7(11)	-9 to -23
fast response	5.7	+28.4(17)	+20 to +65	+ 9.1(14)	+3 to +12
Cytoplasm	5.7	-193(12)	-180 to -202	-212(8)	-190 to -220
Vacuole					
	4.7	-160(7)	-140 to -165	-165(7)	-155 to -178
	5.7	-186(26)	-178 to -196	-203(26)	-183 to -219
	6.7	-206(6)	-185 to -209	-234(6)	-215 to -238
		pH			
Chloroplast	5.7	4.6(12)	4.5 to 5.0	6.2(8)	6.0 to 6.3
Cytoplasm					
	4.7	6.6(5)	6.5 to 6.7	–	–
	5.7	6.7(6)	6.6 to 6.9	6.8(5)	6.6 to 6.8
	6.7	6.8(4)	6.6 to 6.8	–	–
Vacuole					
	4.7	5.7(7)	5.6 to 5.7	–	–
	5.7	5.8(14)	5.7 to 6.0	6.6(14)	6.4 to 6.7
	6.7	5.8(6)	5.6 to 5.9	–	–

up to +65 mV. The slow response was found infrequently with the fine-tipped microelectrodes which may suggest that the slow response results when the chloroplast is damaged by the microelectrode.

II. pH of the Chloroplast, Cytoplasm, and Vacuole

Table 1 gives pH values for the chloroplast, cytoplasm, and vacuole. For chloroplast pH measurement the two microelectrodes were positioned about 5 μm across the outer chloroplast membrane, presumably into the outer thylakoids. Recordings of rapid changes in pH with D-L or L-D transitions were not possible due to chloroplast swelling and shrinking, respectively. Table 1 shows that light reduces the chloroplast pH by about 1.6 units.

Typical photoinduced pH changes in the cytoplasm and vacuole are shown in the two lower traces in Fig. 1. (Each curve in Fig. 1 is from a different cell.) The vacuolar pH decreases under the influence of light due to H^+ influx from the cytoplasm, with the reverse occuring in the dark. Similar changes have been shown by ANDRIANOV et al. (1971) with ***Nitella***. In contrast to the vacuole, Fig. 1 shows an initial rapid increase in the cytoplasmic pH in the light followed by a slower decrease and then a settling-down to the former dark level. When the light is switched off the cytoplasmic pH becomes more acid again as a consequence of H^+ loss from the chloroplasts but then comes back to the steady level after several minutes (Fig. 1).

D. Discussion

In the present study the membrane potential of ***Phaeoceros*** was found to depolarize with increasing external H^+, suggesting high permeability to H^+ and a contribution of H^+ diffusion to the membrane potential. However, the conditions for a contribution from H^+ diffusion to the membrane potential in either the light or in darkness cannot be met because the cytoplasmic (or vacuolar) pH is close to neutrality which makes the potential too positive. This may be seen from the familar Goldman equation which has been modified to include terms for the passive diffusion of H^+ (KITASATO, 1968):

$$E = \frac{RT}{F} \ln \frac{P_K[K^+]_o + P_{Na}[Na^+]_o + P_H[H^+]_o + P_{Cl}[Cl^-]_i}{P_K[K^+]_i + P_{Na}[Na^+]_i + P_H[H^+]_i + P_{Cl}[Cl^-]_o} \tag{1}$$

where o and i designate the external and internal concentrations of the various ions, respectively; P is the permeability coefficient for each of the ions as indicated by the appropriate subscript; and the other symbols have their usual meaning.

Electrogenic pumps have been considered in attempts to explain the membrane potential of ***Nitella*** (KITASATO, 1968; SPANSWICK, 1972) and of higher plant cells (HIGINBOTHAM et al., 1970). With ***Phaeoceros*** an electrogenic pump also seems reasonable since the membrane potential is significantly more negative than the equilibrium potential for K^+ (-165 mV) or for any other of the major ions (DAVIS, 1973). But the present data give no indication as to the ion(s) pumped in this manner.

Contrary to the hyperpolarizations of membrane potential in the dark reported here, it has been reported that darkness induces depolarizations in ***Nitella*** (SPANSWICK, 1972), ***Atriplex*** leaves (LÜTTGE and PALLAGHY, 1969), and barley leaves (LÜTTGE, 1973). However, similar responses to those described here have also been demonstrated in two moss species (SINCLAIR, 1968; LÜTTGE and PALLAGHY, 1969), suggesting the possibility that bryophytes may differ from other plants in the effects of light on the membrane potential.

However, there is some evidence that the nature of photoinduced responses with algae is dependent on the external solution pH (THROM, 1970).

The cytoplasm of ***Phaeoceros*** appears to have a mechanism for maintaining essentially constant pH, as is shown by the lack of fluctuation of the cytoplasmic pH in response to pH change in the external solution (Table 1). This may be explained by H^+ and OH^- loss (possibly by pumps) to the external medium (RAVEN and SMITH, 1973). In the case of D-L change the initial potential transients are likely accounted for by light-induced pumping of H^+ into the chloroplast and vacuole (Table 1 and Fig. 1). Following this, the cytoplasmic pH levels off due to H^+ influx from the external solution (HOPE et al., 1972) which is counterbalanced by the continued loss of H^+ to the chloroplast and vacuole.

The similarity of transients in membrane potential and pH with D-L changes suggests that the potential transients are diffusion potentials resulting from photoinduced changes in pH of the cytoplasm. Eq. 1 would predict this when the only term changing is H^+_i. However, a similar case cannot be made for the L-D change, i.e. a transient depolarization would result from the initial decrease in cytoplasmic H^+ concentration. Thus, to properly describe the potential transients, data are needed on changes in the concentration of the other major ions and on the membrane resistance.

The exact location of the microelectrode tips in chloroplasts in this study was unfortunately not known. This information will be needed before the chloroplast pH and electrical potential data can be used quantitatively as a test of the MITCHELL hypothesis. At a minimum, however, the magnitude and sign of the chloroplast potential (fast response) and the sensitivity of both the potential and pH to illumination lend support to the MITCHELL hypothesis.

E. Acknowledgements

The author is grateful to Professor N. Higinbotham who suggested using an anthocerote for these studies. Support for this work was through a Rutgers University Research Council grant to the author.

F. References

ANDRIANOV, V.K., BULYCHEV, A.A., KURELLA, G.A., LITVIN, F.F., Biofizika **16**, 1031–1036 (1971)
BULYCHEV, A.A., ANDRIANOV, V.K., KURELLA, G.A., LITVIN, F.F., Nature **236**, 175–177 (1972)
DAVIS, R.F., Plant Physiol. **49**, 451–452 (1972)
DAVIS, R.F., Plant Physiol. Suppl. (Abstr.) **51**, 15 (1973)
HIGINBOTHAM, N., GRAVES, J.S., DAVIS, R.F., J. Membrane Biol. **3**, 210–222 (1970)
HIND, G., JAGENDORF, A.T., J. Biol. Chem. **240**, 3195–3201 (1965)
HOPE, A.B., LÜTTGE, U., BALL, E., Z. Pflanzenphysiol. **68**, 73–81 (1972)
KITASATO, H., J. Gen. Physiol. **52**, 60–87 (1968)
LÜTTGE, U., PALLAGHY, C.K., Z. Pflanzenphysiol. **61**, 58–67 (1969)
LÜTTGE, U., in Ion Transport in Plants (ed. W.P. Anderson), pp. 205–221, Academic Press, London and New York 1973
MITCHELL, P., Biol. Rev. **41**, 445–502 (1966)
RAVEN, J.A., SMITH, F.A., in Ion Transport in Plants (ed. W.P. Anderson), pp. 271–277, Academic Press, London and New York 1973
SINCLAIR, J., J. Exp. Bot. **19**, 254–263 (1968)
SPANSWICK, R.M., Biochim. Biophys. Acta **288**, 73–89 (1972)
THROM, G., Z. Pflanzenphysiol. **63**, 163–180 (1970)

Active Hexose Transport in *Chlorella vulgaris*

W. Tanner, D. Haass, M. Decker, E. Loos, B. Komor, and E. Komor

A. Introduction

When photosynthetic phosphorylation was studied *in vivo* by following light-dependent glucose assimilation of ***Chlorella vulgaris***, it was observed (TANNER and KANDLER, 1967; TANNER, 1969) that glucose uptake in this organism is an inducible process. Fig. 1 shows that the induction is prevented, when actidion ($1.5 \cdot 10^{-5}$M) is added together with the inducer (here glucose); the addition of the poison to already-induced cells remains without effect on the uptake.

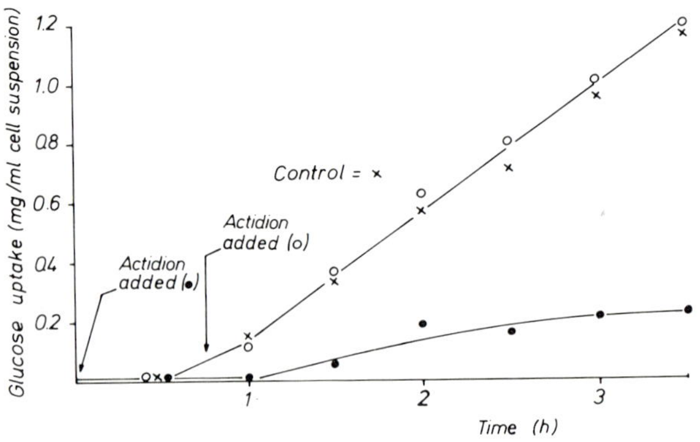

Fig. 1 Effect of actidion (15 μM) on the induction of glucose uptake. Actidion is added either together with glucose at time zero, •, or 45 min after the addition of glucose, ○; x = control without addition of actidion.

A number of glucose analogues like 3-O-methylglucose, 6- and 1-deoxyglucose act as inducers and as substrates for uptake as well (TANNER, 1969; KOMOR and TANNER, 1971), although these compounds are not metabolized at all. Some sugar analogues are accumulated more than one thousand fold as free sugar. Thus ***Chlorella vulgaris*** possesses an inducible system for active hexose uptake.

Besides the actidion experiment, results from double labelling experiments (Fig. 2) have been interpreted as evidence for the formation of transport protein(s) during induction. Fig. 2 shows the ratios of ^{14}C- to ^{3}H-radioactivity in membrane components after separation on disc gels. During the induction with 6-deoxyglucose ***Chlorella*** cells were supplied with ^{14}C-phenylalanine and were mixed after induction with cells supplied with ^{3}H-phenylalanine for the same length of time, but without the inducer. Then membranes were prepared from the mixed culture. As can be seen, at least in one region (slice 26–29) ^{14}C-phenylalanine has been incorporated preferentially in the presence of the inducing sugar. In a control experiment two samples incubated with ^{14}C- and ^{3}H-phenylalanine, respectively, were treated identically, except that neither sample had been induced.

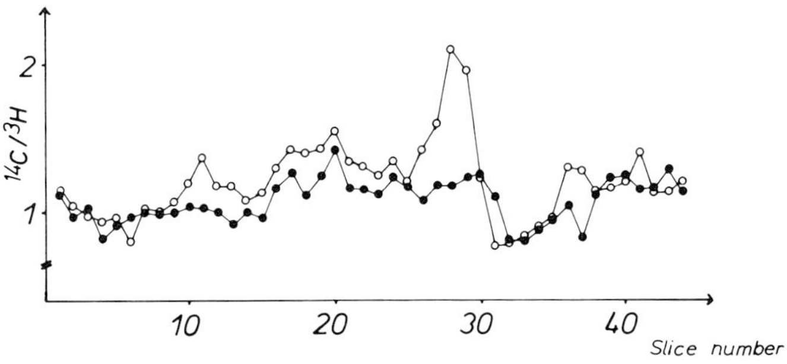

Fig. 2 Membrane components after incorporation of ^{3}H-phenylalanine and ^{14}C-phenylalanine and subsequent separation by disc gel electrophoresis. The cells were broken in a French press in 0.14 M NaCl. The membrane fraction obtained between 8000 and 48 000 X g was treated according to HINMANN and PHILLIPS (1970). ○ = induced cells; ● = non-induced cells.

B. Results and Discussion

I. Characteristics of the Active Hexose Uptake System

1. Specificity and Non-phosphorylating Uptake
The uptake of hexoses by induced cells is stereospecific (TANNER et al., 1970) and phosphorylation of the sugar is neither required for translocation nor for accumulation (KOMOR and TANNER, 1971).

2. Steady State Uptake
An accumulation plateau can either be caused by a steady state equality of influx and efflux or by a lack of any fluxes. ***Chlorella*** possesses a typical steady state system (KOMOR and TANNER, 1971); both influx and efflux need the carrier and the steady state influx is 2 to 3 times greater than net influx, thus showing a positive transmembrane effect (Table 1).

3. A Strictly Coupled Transport System
A number of active transport systems, especially in eucaryotic cells, have been described, which are strictly coupled (KOMOR et al., 1972; KOTYK, 1973). No translocation of substrate even 'down' the concentration gradient is observed with such systems in the absence of energy. These cells – and ***Chlorella*** belongs to them – do not seem to possess a system of facilitated diffusion either in parallel or in series with the system responsible for accumulation of the substrate in question. In the presence of a poison of energy metabolism, e.g. an uncoupler, loaded ***Chlorella*** cells (Fig. 3) do not lose the accumulated sugar as ***E. coli*** cells for example their β-galactoside (KEPES, 1971), nor does ***Chlorella*** show the overshoot phenomenon typical of the existence of a facilitated diffusion component (WINKLER and WILSON, 1966).

Recently it has been possible, however, to transform the strictly coupled sugar uptake system of ***Chlorella*** into one of facilitated diffusion by the addition of the polyene antibiotic nystatin (KOMOR et al., 1974). When nystatin, which binds to sterols and only affects membranes containing sterols (LAMPEN, 1966), is added to ***Chlorella*** cells preloaded with ^{3}H-6-deoxyglucose, they rapidly lose the sugar (Fig. 3). This efflux, which can now be stimulated also by FCCP, proceeds still via the carrier, since nystatin treated cells show the classical overshoot phenomenon (Fig. 4). When the sterols of ***Chlorella*** have reacted with nystatin, the cells obviously are able to carry out facilitated diffusion. Thus it seems possible that sterols are responsible in part for the phenomenon of strict coupling. However, other explanations are possible.

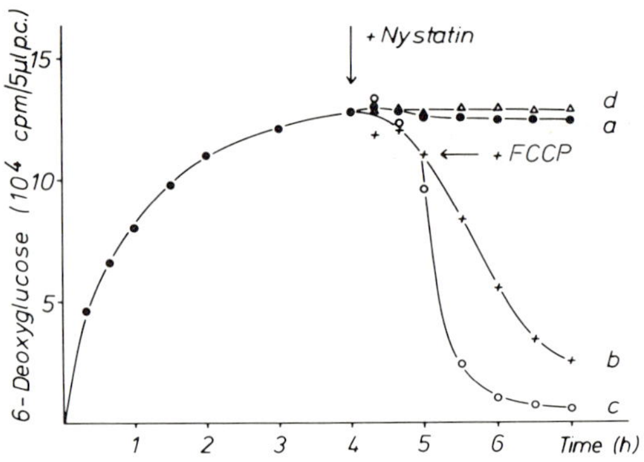

Fig. 3 Efflux of 6-deoxyglucose after addition of nystatin and of FCCP. To cells preloaded with ^{3}H-deoxyglucose, nystatin (6 µg/ml) has been added to samples b and c, and FCCP ($5 \cdot 10^{-5}$ M) to sample d. To sample c FCCP ($5 \cdot 10^{-5}$ M) was added 1 h after the addition of nystatin. Sample a is the control.

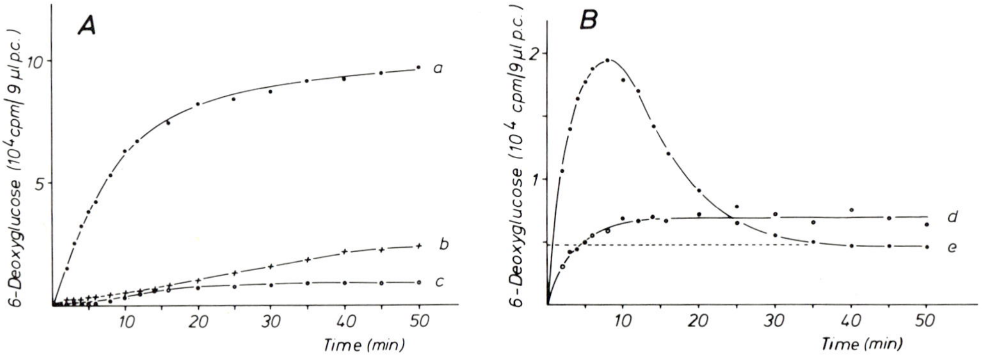

Fig. 4 Overshoot experiment. The algae (36 µl packed cells/ml) were preloaded in sodium phosphate buffer containing 10^{-1} M 6-deoxyglucose and 12 µg nystatin/ml for 2–3 h. Then the cells were centrifuged and washed once with ice-cold buffer to remove external 6-deoxyglucose. During this washing procedure more than 75 % of the internal 6-deoxyglucose remains in the cells. The cells were resuspended in buffer containing ^{3}H-6-deoxyglucose ($1 \cdot 10^{-5}$ M; 160 mCi/mmole) to give 4.5 µl packed cells/ml.
A: influx into cells not treated with nystatin; (a) cells not preloaded (preloaded ones give similar uptake kinetics, see KOMOR and TANNER, 1971); (b) with FCCP $5 \cdot 10^{-5}$ M present, cells preloaded; (c) with FCCP $5 \cdot 10^{-5}$ M present, cells not preloaded.
B: Influx into cells treated with nystatin (12 µg/ml) before and after dilution; (d) with FCCP $5 \cdot 10^{-5}$ M, cells not preloaded; (e) with FCCP $5 \cdot 10^{-5}$ M, cells preloaded. The broken line indicates concentration equilibrium.

II. Energetics of Hexose Uptake and Accumulation

This topic will be dealt with in much more detail in the subsequent paper. Here only the following aspects of it will be briefly mentioned: 1. energy sources; 2. stoichiometries; 3. evidence for efflux-energy; 4. what contributes to accumulation?

1. Energy Sources

The fastest rate of uptake (≈ 300 µmoles/ml packed cells per h corresponding to 17 pmoles cm^{-2} s^{-1}) is observed under aerobic conditions in the dark. Under anaerobic conditions a low rate of transport is observed, which can be increased by a factor of 7 to 10 by saturating light intensities (1500 erg cm^{-2} s^{-1}, 712 nm; in the presence of DCMU, $2 \cdot 10^{-6}$M). The rate under the latter condition can reach half the aerobic rate (KOMOR et al., 1973b). The energy for uptake, therefore, can be derived from respiration, fermentation and photophosphorylation.

2. Stoichiometries

A good correlation between the uptake of sugar analogues and an increased oxygen uptake due to the addition of the sugar has been observed (DECKER and TANNER, 1972). In the case of 6-deoxyglucose, 5.1 molecules of sugar are taken up for each extra O_2 consumed, which corresponds well with one ATP or an equivalent required for each sugar. With glucose a value of 2.5 ATP per glucose has been found, however, it is known that in this case 1.5 ATP are required for the assimilation of glucose to starch and sucrose.

The amount of additional O_2 consumed remains constant during sugar uptake (Table 1). Since the influx during steady state is increased at least by a factor 2, the cell seems to become more efficient with time. This is not the case, however, as will be shown in the next paragraph.

Table 1

Stoichiometry of 6-deoxyglucose uptake and increased respiration during inital uptake and steady-state uptake.

Net uptake was measured in a total volume of 7.5 ml 0.025 M sodium phosphate buffer (pH 6.5) which contained 180 µl ***Chlorella*** cells (packed cells) and 2.5 µCi ^{3}H-6-deoxyglucose (0.01 M). Steady state influx was measured in a parallel sample which initially contained only 0.01 M nonradioactive 6-deoxyglucose. 2.5 µCi ^{3}H-6-deoxyglucose were added to this sample after the 'net uptake sample' was in the steady state.

	Initial uptake	Steady-state uptake
6-Deoxyglucose uptake (µmoles/ml p.c. per min)	5.9[1])	12.5[1])
Respiration increase (µmoles O_2/ml p.c. per min)	1.12	1.12
Sugar taken up per O_2	5.3	11.3
'ATP' required per sugar transported	1.13	0.53

3. Evidence for Efflux Energy

In contrast to net uptake steady state uptake in the dark is inhibited only to about 50 % when O_2 is replaced by N_2 (Fig. 5). This seemed to indicate that about half of the steady state flux proceeds by a non-energy-requiring exchange transport. Surprisingly this flux can be completely prevented by the addition of uncoupling agents (Fig. 5), which shows that it also is energy dependent. However, where can energy come from under these conditions, i.e. in the dark plus N_2? The generation of some form of energy by efflux of sugar has been postulated, therefore. This energy, however, can only be used for transport, since with ^{14}C-glucose as counter-transported sugar under N_2 in the dark (the cells being prefilled with 6-deoxyglucose) considerably more free glucose is found in the cells, but further metabolism of glucose is not at all increased (Fig. 6).

[1]) Corresponds to 20 and 42 pmoles cm^{-2} s^{-1}, respectively.

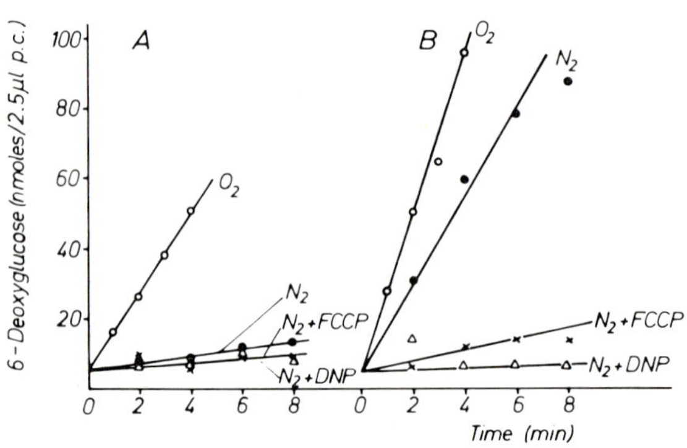

Fig. 5 Net (A) and steady-state (B) uptake of 6-deoxyglucose under aerobic and anaerobic conditions in the dark; where indicated, DNP ($4 \cdot 10^{-4}$ M) and FCCP ($5 \cdot 10^{-5}$ M) have been added. For further details see KOMOR et al. (1973b).

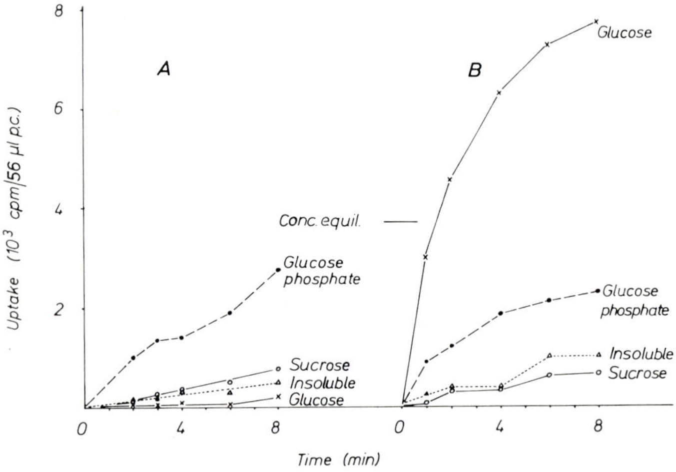

Fig. 6 Uptake and metabolism of ^{14}C-D-glucose by ***C. vulgaris*** (A), and by ***C. vulgaris*** preloaded with 0.1 M 6-deoxyglucose (B) under anaerobic conditions in the dark. 300 µl packed cells of induced algae were incubated anaerobically in 12 ml 0.025 M sodium phosphate buffer pH 6.5. The uptake experiment was started by addition of 0.8 µCi ^{14}C-glucose at a final concentration of $5 \cdot 10^{-3}$ M. Samples of 2 ml were withdrawn, filtered and the cells were extracted in 80 % ethanol. The radioactive compounds were separated by paper chromatography and the amounts determined. The algae used for experiment (A) did not have detectable amounts of free hexoses inside, whereas the algae for experiment (B) had been incubated aerobically for several hours in $1 \cdot 10^{-2}$ M non-radioactive 6-deoxyglucose and were then centrifuged to remove external 6-deoxyglucose.

4. What Contributes to Accumulation?

In the simplest of all models accumulation of non-electrolytes is explained by a difference in K_s values of the carrier at the outer ($K_{s,e}$) and inner ($K_{s,i}$) sides of the cell (WINKLER and WILSON, 1966). ***Chlorella*** clearly shows such a difference (KOMOR et al., 1973a); a value of 70 fold has been determined (Table 2). The actual accumulation, however, is very much higher at low substrate concentrations (1500 fold for 10^{-5} M 6-deoxyglucose). When the relative constants for the various translocation reactions were estimated (Table 2; for details

Table 2

Coefficients involved in the accumulation of 6-deoxyglucose. For details see KOMOR et al., (1973a).

	Experimentally determined values	Estimated or calculated values
K_m for net influx (mM)	0.21	
K_m for net efflux (mM)	21	
$K_{s,e}$ (mM)		0.60
$K_{s,i}$ (mM)		43
Translocation constants (t^{-1})		
k_{-2}		$\approx k_2$
k_1		$\approx 0.02\ k_2$
k_{-1}		$\approx 0.5\ k_2$
Accumulation factor f		
at 100 μM outside	500	
at 10 μM outside	1520	
at 1 μM outside	1430	

see KOMOR et al., 1973a), a difference between the rate constant for the flux of empty carrier to the outside and that for the flux of empty carrier to the inside of at least a factor 25 was obtained. The accumulation f of a nonelectrolyte according to REGEN and MORGAN (1964) is given by

$$f = \frac{\alpha_1 \cdot k_2 \cdot \beta_2 \cdot k_{-1}}{\beta_1 \cdot k_{-2} \cdot \alpha_2 \cdot k_1},$$ where $\frac{\beta_1}{\alpha_1}$ corresponds

to the true dissociation constant $K_{s,e}$ for the carrier substrate complex at the outside, and $\frac{\beta_2}{\alpha_2}$ to the true dissociation constant $K_{s,i}$ for the carrier substrate complex inside the cell. k_1, k_{-1}, k_2, and k_{-2} are the translocation constants as used in Fig. 7. Since the accumulation f can be rewritten as

$$f = \frac{K_{s,i}}{K_{s,e}} \cdot \frac{k_2 \cdot k_{-1}}{k_{-2} \cdot k_1},$$

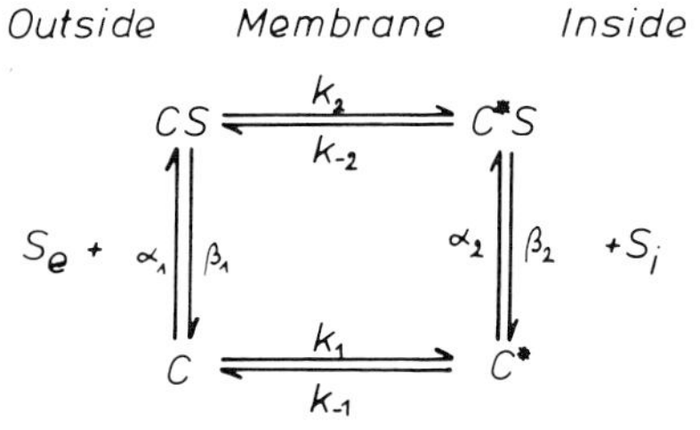

Fig. 7 General scheme for sugar transport (see text and Table 2).

the accumulation of 6-deoxyglucose would be expected to amount to f = 70 · 25 = 1750, assuming that k_2 and k_{-2} are rather similar. This value agrees well with the value experimentally found. The metabolic energy is assumed, therefore, to be required mainly to yield the difference in the K_S-values and the difference in the translocation constants for empty carrier fluxes (KOMOR et al., 1973a).

C. References

DECKER, M., TANNER, W., Biochim. Biophys. Acta **266**, 661–669 (1972)
HINMANN, N.D., PHILLIPS, A.H., Science **170**, 1222–1223 (1970)
KEPES, A., J. Membrane Biol. **4**, 87–112 (1971)
KOMOR, E., HAASS, D., KOMOR, E., TANNER, W., Eur. J. Biochem. **39**, 193–200 (1973a)
KOMOR, E., HAASS, D., TANNER, W., Biochim. Biophys. Acta **266**, 649–660 (1972)
KOMOR, E., LOOS, E., TANNER, W., J. Membrane Biol. **12**, 89–99 (1973b)
KOMOR, E., TANNER, W., Biochim. Biophys. Acta **241**, 170–179 (1971)
KOMOR, E., KOMOR, E., TANNER, W., J. Membrane Biol., in press (1974)
KOTYK, A., Biochim. Biophys. Acta **300**, 183–210 (1973)
LAMPEN, J.O., Symp. Soc. Gen. Microbiol. **16**, 111–130 (1966)
REGEN, D.M., MORGAN, H.E., Biochim. Biophys. Acta **79**, 151–166 (1964)
TANNER, W., Biochem. Biophys. Res. Commun. **36**, 278–283 (1969)
TANNER, W., GRÜNES, R., KANDLER, O., Z. Pflanzenphysiol. **62**, 376–385 (1970)
TANNER, W., KANDLER, O., Z. Pflanzenphysiol. **58**, 24–32 (1967)
WINKLER, H.H., WILSON, T.H., J. Biol. Chem. **241**, 2200–2211 (1966)

Proton Movement Associated with Hexose Transport in *Chlorella vulgaris*

E. Komor and W. Tanner

A. Introduction

The alga *Chlorella vulgaris* is able to accumulate non-metabolizable hexose analogues more than a thousandfold (KOMOR et al., 1973a), whereby the uptake work is achieved by stoichiometric use of metabolic energy (DECKER and TANNER, 1972; KOMOR et al., 1973b). The nature of the energy directly responsible for the work of sugar uptake is unknown, as in most cases where sugars are accumulated chemically unchanged within the cell. MITCHELL has suggested that a proton motive force (i.e. pH-gradient plus membrane potential) could drive active nonelectrolyte transport. For ***E. coli*** there are experiments which at least sustain this point of view (WEST and MITCHELL, 1972).

In this paper we have investigated sugar and the concomitant uptake of H^+ by *Chlorella* and, by studying the pH dependence of the uptake system, have estimated the kinetic parameters involved.

B. Results and Discussion

I. The Cotransport of Protons with Hexoses

When 6-deoxyglucose is added to a suspension of induced *Chlorella* cells, an immediate pH-shift to more alkaline values occurs in the suspending medium. This pH-shift cannot be observed with non-induced cells. Furthermore only sugars which are transported give the pH-shift (like glucose, 6-deoxyglucose, 1-deoxyglucose); sugars not transported, such as α-methylglucoside, do not produce any alkalinization (Fig. 1). The sugar-induced pH-shift not only occurs with aerobic cells, but also with anaerobic ones when they are illuminated, whereas in darkness in nitrogen only a very low rate of H^+ uptake is seen (Fig. 2). This completely parallels sugar uptake. Uncouplers also inhibit sugar uptake as well as the pH-shift.

The rate of H^+ uptake is dependent on the concentration of sugar added. A saturation behaviour is observed with a K_m-value identical with the K_m-value for sugar uptake. This is shown in Table 1 for three sugars with largely differing affinities for the transport system.

Thus the uptake of sugars for all parameters tested agrees very well with sugar-induced net uptake of H^+ under various experimental conditions.

II. The Stoichiometry of Proton Cotransport and Sugar Uptake

When the initial rate of H^+ uptake after the addition of sugar is compared with the initial uptake rate of the sugar, a fixed stoichiometry of one H^+ per sugar molecule is observed. This is true for glucose as well as for 6-deoxyglucose and it also holds for different energy

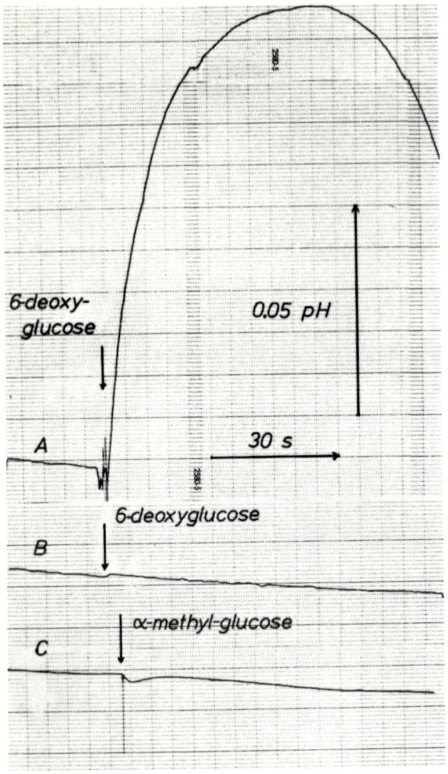

Fig. 1 The change in pH of the external medium due to the addition of sugar to a suspension of ***Chlorella*** cells under aerobic conditions in the dark. A and C: induced algae. B: non-induced algae. Experimental conditions as previously described (KOMOR, 1973).

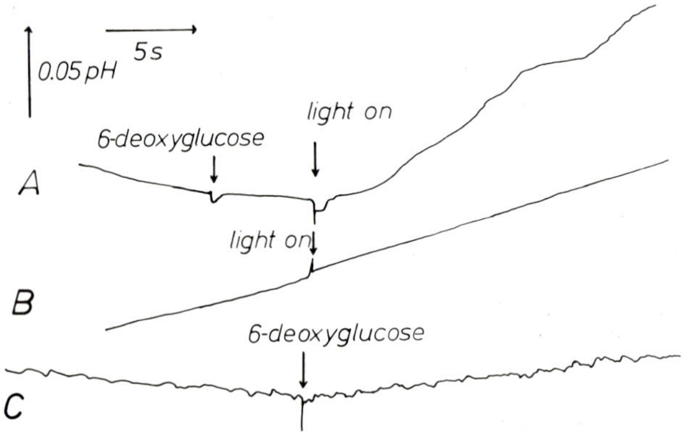

Fig. 2 pH-Tracings of suspensions of algae under various energetic conditions. 6-deoxyglucose was added (arrow) to a final concentration of 4 mM. A: anaerobiosis, dark; after addition of sugar light was switched on. B: anaerobiosis, dark; light was switched on without addition of sugar. C: anaerobiosis, sample kept in the dark all the time.

Table 1

K_m-values for sugar uptake and sugar-induced pH-shift. Experimental procedure as described previously (KOMOR et al., 1973a).

Sugar	K_m for sugar uptake (μM)	K_m for pH-shift (μM)
Glucose	10 – 20	8
6-Deoxyglucose	200 – 300	300
3-O-Methylglucose	1000 – 2000	1900

sources, i.e. respiratory, photosynthetic or fermentation energy supply (Table 2). As has been pointed out in the preceding paper, an increased respiration is observed when hexoses or their analogues are added to ***Chlorella***. In the case of 6-deoxyglucose each sugar molecule taken up induces the additional consumption of 0.2 molecules of O_2 (corresponding to 1.2 ATP per sugar) and for glucose the extra O_2 uptake is 0.41 per sugar (corresponding to 2.5 ATP). From the metabolic fate of glucose within the cells (DECKER and TANNER, 1972) it is known that 1.5 ATP per glucose are required for assimilation reactions. Thus for 6-deoxyglucose as well as for glucose the same amount of energy, i.e. close to 1 ATP or an equivalent, is required for the actual transport work. The H^+ stoichiometry reported above indicates, therefore, that H^+ uptake is related to sugar transport only through the membrane and not to subsequent metabolic reactions.

Table 2

Stoichiometry of H^+/sugar uptake. Experimental conditions as described previously (KOMOR, 1973).

Sugar	Conditions	H^+/sugar
Glucose	O_2, dark	0.98
	N_2, light	0.80
	N_2, dark	0.87
6-Deoxyglucose	O_2, dark	1.06
	N_2, light	1.00
	N_2, dark	1.00

High concentrations of K^+ and Na^+, when added immediately before the sugar, can strongly reduce sugar uptake and H^+ uptake (Table 3). But the stoichiometry of H^+ taken up per sugar molecule remains the same. The possibility is excluded, therefore, that these alkali ions can substitute for H^+ to achieve transport (otherwise a lower stoichiometry would be expected). The reduction in uptake might have been caused by a change in the membrane potential.

Table 3

Amount of H^+ taken up per 6-deoxyglucose in the presence of 0.2 M KCl or NaCl. The experimental conditions were as described previously (KOMOR, 1973). The salts were added 2 min or 5 min before the addition of sugar.

	Uptake of 6-deoxyglucose (μmoles/h per ml packed cells)	H^+/sugar
Control	220[1]	1.09
+ K^+ (2 min before sugar)	48	1.23
+ K^+ (5 min before sugar)	92	0.98
+ Na^+ (2 min before sugar)	68	1.01
+ Na^+ (5 min before sugar)	80	1.11

[1]) Corresponds to 12.5 pmoles cm^{-2} s^{-1}

III. The Affinity of Protonated and Deprotonated Carrier to Sugar

The net uptake of H^+ together with sugar can be explained by protonation of the carrier molecule at the outside of the membrane, followed by binding of the sugar and, after movement to the inner side, by deprotonation and sugar release into the cell interior. This would imply that the different properties of the outside carrier as compared to the inside carrier (KOMOR et al., 1973a) might be due to protonation and deprotonation, respectively, of the carrier. We therefore tried to simulate inside conditions at the outer side of the cell by increasing the pH-value in the medium so as to deprotonate the carrier.

When the affinity of the carrier for 6-deoxyglucose is measured at acid pH a single uptake system with a K_m of 0.3 mM is observed, but at more alkaline pH-values a mixed K_m is seen, until at a very alkaline pH an uptake system with a K_m of about 30 mM (Fig. 3) is solely present. The latter indeed resembles the affinity of the inside carrier of 'normal' cells. In fact a high-affinity system, which is the only one present at acid pH, can be gradually converted to a low-affinity system by alkalinization (KOMOR and TANNER, 1974b). With increasing pH-values the high-affinity system is decreased and in parallel the low-affinity system increases. At neutral pH this phenomenon quite strikingly resembles the biphasic uptake pattern obtained for ions in plants (EPSTEIN, 1972). Here the two phases originate from two populations of the same carrier molecule, the protonated and the unprotonated forms. If for example the high-affinity system, solely present at acid pH, is taken as one hundred percent and the low-affinity system, solely present at very alkaline pH, is also taken as one hundred percent, the two uptake activities always add up to one hundred percent at intermediate pH-values (KOMOR and TANNER, 1974b).

IV. The Affinity of the Carrier to Protons

When the amount of the high-affinity uptake system (expressed as V_{max}) is plotted against H^+ concentration, a K_m-value for H^+ of about 0.14 μM, i.e. pH 6.85, is observed. This means that at pH 6.85 about 50 % of the carrier molecule is protonated (Fig. 4). Since half of the

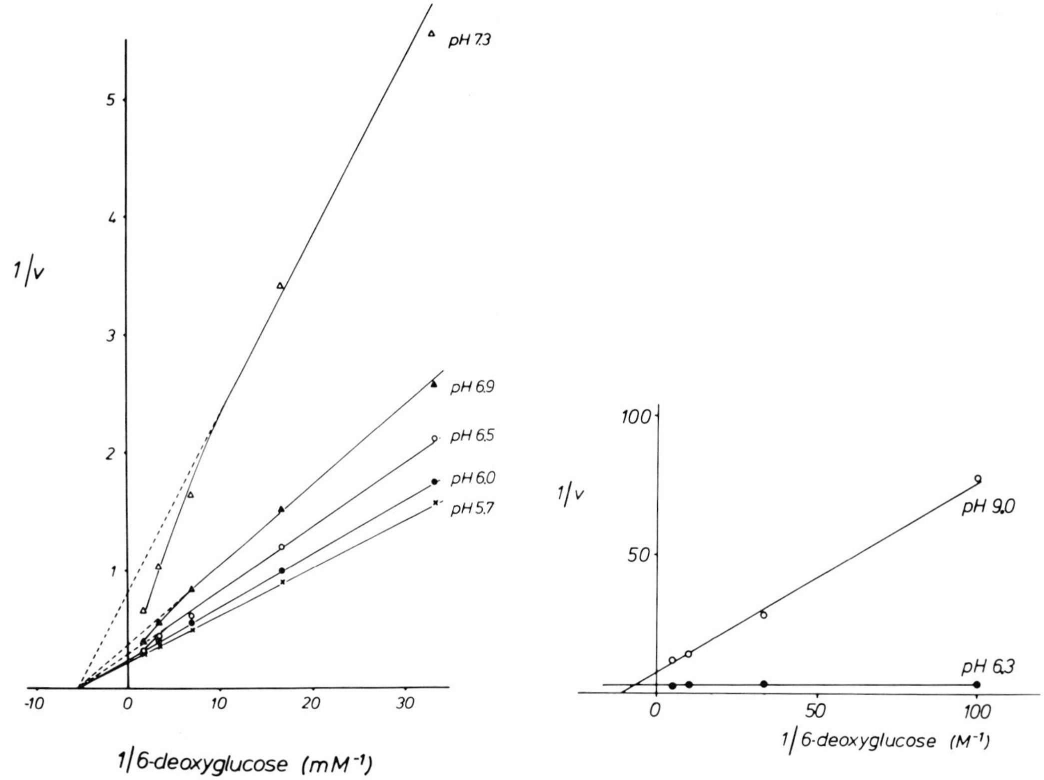

Fig. 3 Lineweaver-Burk plots of the rate of 6-deoxyglucose uptake as a function of 6-deoxyglucose concentration. Left picture: at low sugar concentrations, v is expressed as μmoles/min per ml packed cells. Right picture: at high sugar concentrations, v is expressed as mmoles/h per ml packed cells. Experimental procedure as described previously (KOMOR et al., 1973a).

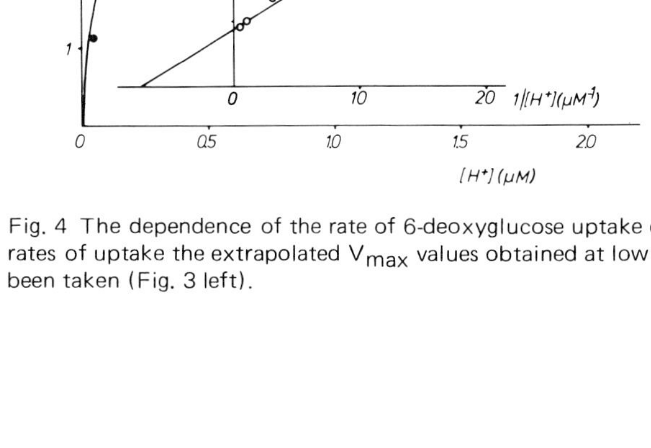

Fig. 4 The dependence of the rate of 6-deoxyglucose uptake on the H^+ concentration in the medium. For rates of uptake the extrapolated V_{max} values obtained at low sugar concentrations (0.01 to 0.2 mM) have been taken (Fig. 3 left).

low-affinity system is also observable at this pH-value, the latter uptake activity most likely corresponds to the activity of the unprotonated carrier. The affinity constant of the uptake system to H^+ also explains the pH-curve of sugar uptake, which shows a steep decline above pH 6.5 when measurements are carried out at low sugar concentrations (Fig. 5).

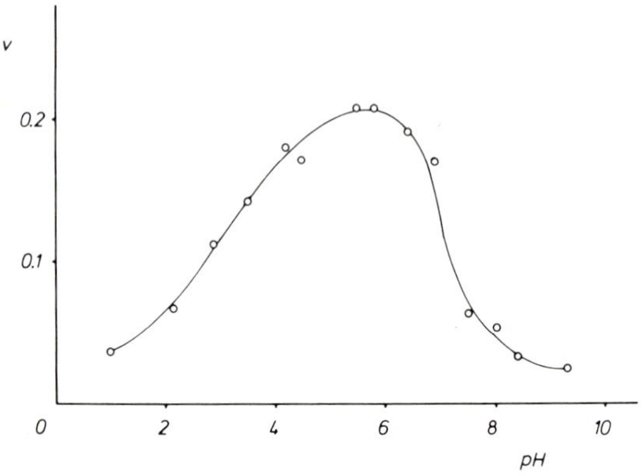

Fig. 5 pH-profile of 6-deoxyglucose uptake at a sugar concentration of 1 mM. (v is expressed as mmoles/h per ml packed cells; 0.1 corresponds to 5.6 pmoles cm^{-2} s^{-1}.)

The binding of a H^+ not only alters the affinity to sugar, it also alters the translocation velocity of free carrier through the membrane. Thus an accelerated net efflux of accumulated sugar occurs at alkaline pH. This also has consequences for sugar uptake and especially for accumulation properties.

V. The Transport of Sugar: a Model

The transport of sugar by ***Chlorella*** is depicted in a model (Fig. 6) in which the carrier molecule (C) outside is rapidly protonated (C^H) due to an acid pH-value in the medium, whereas inside the carrier is rapidly deprotonated because of an alkaline pH-value. The apparent $K_{s,e}$ at the outside for sugar had been previously found to be 0.3 mM, whereas inside $K_{s,i}$ had been shown to be 30 mM (KOMOR et al., 1973a). This difference in K_s-values can now be explained by the protonation of the carrier on the outside and its deprotonation at the inside.

This implies that a H^+ gradient through the membrane exists which causes the change of a protonated to a deprotonated carrier. Since a stoichiometric amount of H^+ is transported inwardly together with sugar, a H^+ pump has to exist which counteracts this sugar-dependent H^+ influx.

When in the outside medium an alkaline pH is maintained and the H^+ gradient is thus abolished, the carrier is deprotonated and the affinity to sugar is the same on both sides of the membrane. No sugar accumulation occurs under this condition (KOMOR and TANNER, 1974b). The model of Fig. 6 is very over-simplified (for example no reverse reactions have been introduced). It must be pointed out that the translocation rate of C^H to the inner side is very low, in contrast to the translocation rate of C. The asymmetry of free carrier movement is most likely also caused, therefore, by H^+ binding outside. The reasons why the movement of C^H without substrate does not occur, or why uncouplers inhibit the influx of C^HS, (protonated carrier-substrate-complex) are unknown. It can only be speculated that here the membrane potential or the 'energy state' of the membrane are playing an important role.

In summary a H^+ gradient can be regarded as the primary energy source responsible for active sugar transport. We do not know how H^+ gradients are built up at the cytoplasmic membrane, but so far no evidence for the involvement of ATP has been found (KOMOR and TANNER, 1974a).

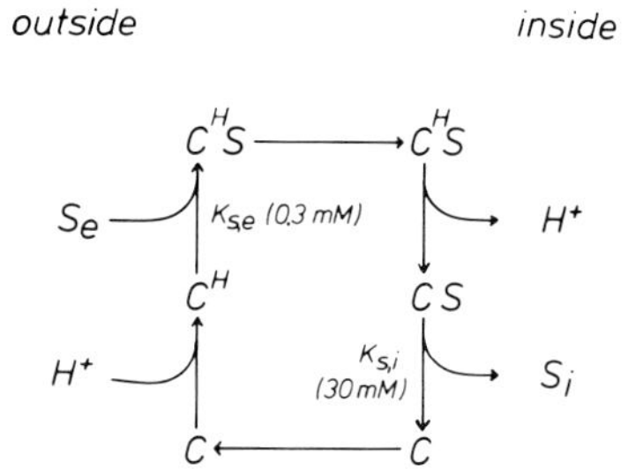

Fig. 6 Model for active sugar uptake by ***Chlorella*** (reverse reactions are disregarded). C = carrier; C^H = protonated carrier; C^HS = protonated carrier-substrate-complex.

C. References

DECKER, M., TANNER, W., Biochim. Biophys. Acta **266**, 661–669 (1972)

EPSTEIN, E., Mineral Nutrition of Plants. Principles and Perspectives. John Wiley and Sons, New York–London–Sydney–Toronto, 1972

KOMOR, E., FEBS Letters **38**, 16–18 (1973)

KOMOR, E., HAASS, D., KOMOR, B., TANNER, W., Eur. J. Biochem. **39**, 193–200 (1973a)

KOMOR, E., LOOS, E., TANNER, W., J. Membrane Biol. **12**, 89–99 (1973b)

KOMOR, E., TANNER, W., Z. Pflanzenphysiol. **71**, 115–128 (1974a)

KOMOR, E., TANNER, W., J. Gen. Physiol., in press (1974b)

WEST, I.C., MITCHELL, P., Bioenergetics **3**, 445–462 (1972)

Carrier Turnover and Phosphate Uptake in *Chlorella pyrenoidosa*

B. Jeanjean and G. Ducet

A. Introduction

In a previous publication (JEANJEAN, 1973) it was shown that ***Chlorella*** cells growing in a culture medium depleted of phosphate exhibit an increased capacity for phosphate absorption. This increase can be inhibited by impairing protein synthesis in a number of ways. The addition of phosphate to starved ***Chlorella*** rapidly results in a reduction in the absorption rate. These results were interpreted as a repressing effect of phosphate on phosphate carrier synthesis.

The data reported in this paper support the idea that, in cells grown on a normal medium, a continuous synthesis of phosphate carrier occurs which is balanced by a continuous destruction of the transport carrier.

B. Materials and Methods

The material and the techniques were the same as previously described (JEANJEAN, 1973). Cycloheximide (CH) was used at $4 \cdot 10^{-4}$ M. Base analogues tested were: 8-bromoguanidine, 10^{-4} M, and 6-methylpurine (MP), $5 \cdot 10^{-4}$ M.

C. Results

Fig. 1 shows the rate of phosphate absorption after various times of phosphate starvation. Curve 1 shows the increase resulting from starvation, curve 2 shows that CH completely inhibits this increase, whilst curve 3 shows that when CH is added after the maximum rate is reached, the absorption rate is slowly inhibited (the half-time of the decrease is more than 8 h) and curve 4 shows the rapid decrease in the absorption rate after phosphate is added to the medium. The half-time for the decrease in rate is less than 2 h. After phosphate addition, the absorption rate decreases to nearly the same value as the initial rate (Fig. 2, curve 2). When CH is added with phosphate (Fig. 3A, curve 2) the absorption rate also decreases rapidly. This last result demonstrates that the addition of phosphate to starved ***Chlorella*** does not induce the synthesis of an enzymic protein capable of destroying the phosphate carrier.

The results shown in Fig. 2 confirm this finding. Curve 2 shows the decrease in the absorption rate after phosphate addition. After 6 h (curve 3) or 8 h (curve 5) the cells were harvested and again resuspended in a depleted medium. The rate of absorption increases but more slowly than that observed at the beginning of the starvation period. Curve 4 (same experimental conditions as curve 3) shows that CH again inhibits this increase. These results clearly indicated that the decrease after phosphate addition is not the result of an inactivation of the phosphate carrier; if this were true, CH would not have inhibited the reactivation implied by curves 3 or 5.

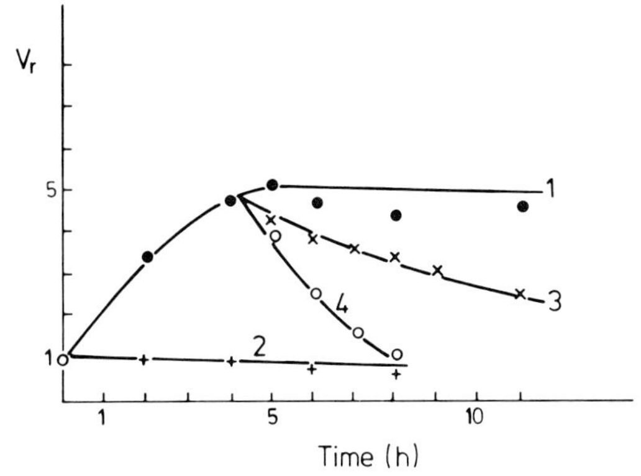

Fig. 1 V_r = Relative phosphate absorption rate (non-starved ***Chlorella*** cells taken as 1) as a function of time (h). 1: time course of increase in phosphate absorption during phosphate starvation. 2: CH ($4 \cdot 10^{-4}$ M) addition at the beginning of phosphate starvation. 3: CH ($4 \cdot 10^{-4}$ M) addition after 4 h of phosphate starvation. 4: phosphate ($5 \cdot 10^{-4}$ M) addition after 4 h of phosphate starvation.

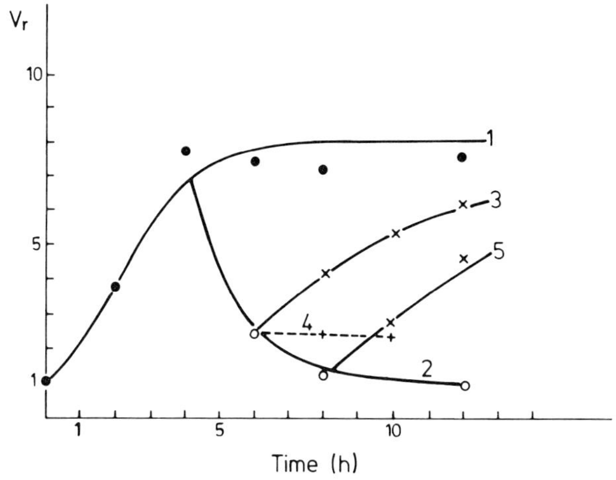

Fig. 2 V_r = Relative phosphate absorption rate (non-starved ***Chlorella*** cells taken as 1) as a function of time (h). 1: time course of increase in phosphate absorption during phosphate starvation. 2: phosphate ($5 \cdot 10^{-4}$ M) addition after 4 h of phosphate starvation. 3: after 2 h of incubation in phosphate ($5 \cdot 10^{-4}$ M), ***Chlorella*** cells were resuspended in a medium without phosphate. 4: after 2 h of incubation in phosphate ($5 \cdot 10^{-4}$ M), ***Chlorella*** cells were resuspended in a medium without phosphate, but with CH ($4 \cdot 10^{-4}$ M). 5: after 4 h of incubation in phosphate ($5 \cdot 10^{-4}$ M), ***Chlorella*** cells were resuspended in a medium without phosphate.

Another possibility is that the phosphate carrier is used up during transport; if this were true, the addition of CH to ***Chlorella*** grown on a normal medium (containing phosphate) would result in a decrease in the absorption rate (because CH would inhibit the carrier synthesis). Such an effect was observed (Fig. 3A). Curve 4 (Fig. 3A) shows that CH does not significantly change the absorption rate of starved ***Chlorella*** while curve 5 shows that CH decreases the absorption rate of ***Chlorella*** growing on phosphate.

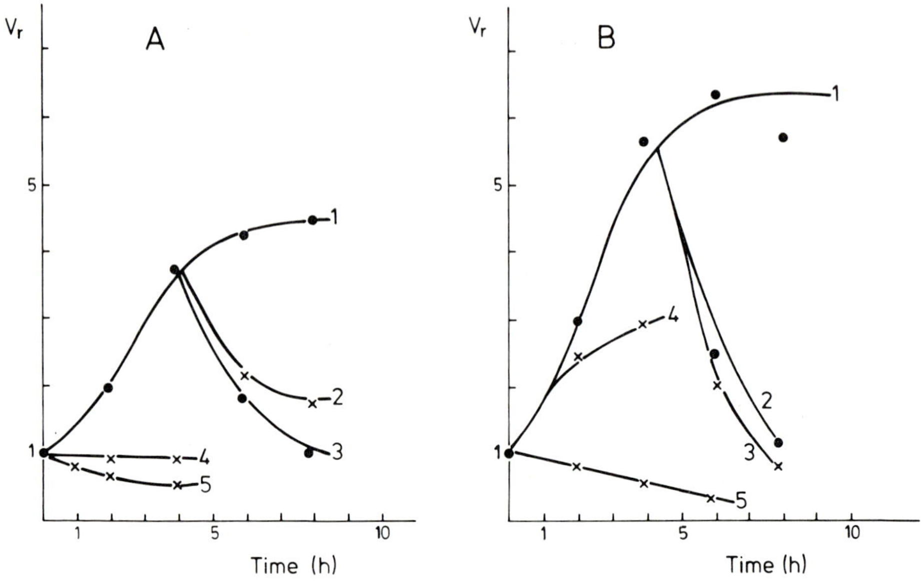

Fig. 3 V_r = Relative phosphate absorption rate (non-starved *Chlorella* cells taken as 1) as a function of time (h).
A. 1: time course of increase in phosphate absorption during phosphate starvation. 2: phosphate ($5 \cdot 10^{-4}$ M) and CH ($4 \cdot 10^{-4}$ M) addition after 4 h of phosphate starvation. 3: phosphate ($5 \cdot 10^{-4}$ M) addition after 4 h of phosphate starvation. 4: CH ($4 \cdot 10^{-4}$ M) addition at the beginning of phosphate starvation. 5: phosphate ($5 \cdot 10^{-4}$ M) and CH ($4 \cdot 10^{-4}$ M) addition to non-starved *Chlorella* cells.
B. 1: time course of increase in phosphate absorption during phosphate starvation. 2: phosphate ($5 \cdot 10^{-4}$ M) addition after 4 h of phosphate starvation. 3: phosphate ($5 \cdot 10^{-4}$ M) and MP ($5 \cdot 10^{-4}$ M) addition after 4 h of phosphate starvation. 4: MP ($5 \cdot 10^{-4}$ M) addition at the beginning of phosphate starvation. 5: phosphate ($5 \cdot 10^{-4}$ M) and MP ($5 \cdot 10^{-4}$ M) addition to non-starved *Chlorella* cells.

Fig. 3B shows that the base analogue 6-methyl-purine (MP) inhibits (although not immediately) the rate increase on a depleted medium (curve 4) and also inhibits the absorption rate on normal medium (curve 5). When it is present with phosphate, the decrease in the absorption rate (curve 3) is somewhat more marked than that observed with phosphate alone (curve 2). Only 6-methyl-purine ($5 \cdot 10^{-4}$ M) was an active inhibitor. 8-Bromoguanidine at the concentration used (10^{-4} M) had no effect.

D. Discussion

It seems that our results can best be interpreted by assuming that a phosphate carrier is continuously synthesized in *Chlorella* cells (in non-synchronized cultures) and that phosphate transport results in 'carrier being used up'. If no transport occurs (phosphate starvation) the capacity for phosphate transport increases. If there is no carrier synthesis, this capacity for phosphate transport is either unimpaired (phosphate starvation) or impaired (phosphate present) for the carrier is used up.

Unpublished results from this laboratory have shown that phosphate starvation does not greatly alter the content of intracellular inorganic phosphorus (P_i); after 4 h starvation, P_i levels decreased by about 30–35 %. This more or less agrees with the results of AITCHISON and BUTT (1973) who observed a decrease of about 20 % in P_i in ***Chlorella vulgaris*** cells after 20 h of phosphate starvation. Furthermore after phosphate is added to the medium, the internal P_i decreases very slightly. Such small changes in P_i concentration imply that P_i cannot be directly involved as a repressor of protein synthesis. We have also observed, in agreement with MIYACHI et al. (1964) and AITCHISON and BUTT (1973), the pool of polyphosphate disappears during phosphate starvation and is quickly replenished when phosphate is again added to the medium; but these changes do not affect the growth rate at least for the first 12 h of phosphate starvation (GAUDIN, 1972). Recent work of COCUCCI and MARRE (1973) shows that CH induces an increase in nucleotide triphosphate in aerobic yeast. If the same increase occurs in ***Chlorella*** then it may have some influence on transport but this is not immediately apparent from our results; after adding phosphate to starved cells with or without CH, the transport rate decreases at about the same speed for the first 2 h. Experiments in our laboratory have shown that CH ($4 \cdot 10^{-4}$ M) has no effect on the phosphate absorption rate for short absorbing times (1 to 10 min). ***Chlorella*** cells grown in the presence of CH ($4 \cdot 10^{-4}$ M) incorporate sulphate in proteins at 12 to 15 % of the rate of control. Thus even at this high concentration, CH does not completely inhibit protein synthesis. This may be due to a low permeability of ***Chlorella*** cells to antibiotics; 8-bromoguanidine has no effect on the increase of the phosphate absorption rate induced by phosphate starvation and MP acts more slowly than CH on this increase.

However, some effects of CH on transport are to be expected from our results. In Fig. 3A, curve 2 indicates that the effects of CH decrease with time, as compared with curve 3. This may be related to the observations of MacDONALD and ELLIS (1969), ELLIS and Mac DONALD (1970) and COCUCCI and MARRE (1973) who found that CH can act on the energy metabolism. If this is so, then the energy pool is reduced, less phosphate transport occurs and phosphate carrier is not so rapidly destroyed. The same effect seems to be observed for ***Chlorella*** cells in a normal medium; curve 5 in Figs. 3A, B shows that the decrease of absorption rate is more prolonged with MP than with CH. The slower effect of MP, than of CH, on the increase in the absorption (curve 4, Fig. 3B), may result either from a permeability effect on the absorption of MP or from the presence of a relatively long-lived messenger for the carrier synthesis.

The plateau observed in the absorption rate (Figs. 1, 2, 3) can be interpreted as a cessation of carrier synthesis. This may be the result of impaired biosynthesis due to prolonged phosphate starvation; however the results of GAUDIN et al. (1973) on the effect of osmotic shocks on phosphate transport rule out this hypothesis. It may be better to assume that the cell membrane cannot accomodate all the newly synthesized carriers; i.e. there is a limited number of sites to attach a carrier to the cell surface.

Recently, GAZZOLA et al. (1973) have observed that the half-time of the alanine carrier in chick embryo cells was shorter when the cells were growing on this amino acid than when they were starved; this is comparable with our results.

E. References

AITCHISON, P.H., BUTT, V.S., J. Exp. Bot. **24**, 497–510 (1973)

COCUCCI, M.C., MARRE, E., Plant Science Letters **1**, 293–301 (1973)

ELLIS, R.J., MacDONALD, I.R., Plant. Physiol. **46**, 227–232 (1970)

GAUDIN, C., These de Specialite, Universite d'Aix-Marseille, 1972

GAUDIN, C., JEANJEAN, R., BLASCO, F., C.R. Acad. Sci. Serie D, **277**, 301–304 (1973)

GAZZOLA, R.F., GAZZOLA, G., RONCHI, P., SAIBENE, V., GUIDOTTI, G.G., Biochim. Biophys. Acta **291**, 291–301 (1973)

JEANJEAN, R., FEBS Letters **32**, 149–151 (1973)

MacDONALD, I.R., ELLIS, R.J., Nature **222**, 791–792 (1969)

MIYACHI, S., KANAI, R., MIHARA, S., MIYACHI, S., AOKI, S., Biochim. Biophys. Acta **93**, 625–634 (1964)

Phosphate Uptake and Photophosphorylation in the Blue-Green Alga *Anacystis nidulans*

W. Simonis, T. Bornefeld, J. Lee-Kaden, and K. Majumdar

A. Introduction

Many efforts have been made to clarify the relations between phosphate uptake and photophosphorylation using higher plants and green algae (SIMONIS and URBACH, 1973). Blue-green algae are structurally simpler than cells of unicellular green algae and do not possess a chloroplast membrane system. Moreover in these algae photosystems I and II can be separated more clearly than in green algae. Therefore experiments on light-dependent phosphate uptake with reference to energy supply were carried out with *Anacystis*.

B. Materials and Methods

Anacystis nidulans (Richt.) Dr. strain 1402–1 (Algal Culture Collection Göttingen) was cultured in the medium of KRATZ and MYERS (1955), as modified by ESSL (1969), at 38°C and at an intensity of 4500 lux, 14 h daily. For phosphate incubations algae were grown in light for 8 h in a medium lacking phosphate. When uptake was measured, phosphate was added to the algae together with the radioactivity. For experimental details see recent publications (ULLRICH-EBERIUS and SIMONIS, 1970; BORNEFELD and SIMONIS, 1974).

C. Results and Discussion

The uptake of phosphate by the phosphate-depleted *Anacystis* cells after addition of 10 μM $KH_2{}^{32}PO_4$ is light-dependent and proceeds in a nearly constant manner for 20 min (Fig. 1). A steep increase in the first 5 min is followed by a somewhat lower slope beginning between 5 and 10 min. Uptake in the dark tends from the outset to saturation.

The phosphate uptake for 5 min from various phosphate concentrations added to the medium results in a sequence of isotherms differing in light and dark (Fig. 2). These isotherms show different saturation states in light and in dark in the range 1 to 100 μM; we call this System I with three Subsystems I_1–I_3. In contrast to System I, System II uptake (higher concentration ranges) eventually shows a higher uptake in the dark than in the light. The Subsystems I_1, I_2, I_3 of System I can be expressed in terms of Michaelis-Menten kinetics, demonstrated by the solid lines in Fig. 2. These multiphasic absorption isotherms in combination with the pH dependence (ESSL, 1969; ULLRICH-EBERIUS, 1973) indicate a carrier-mediated mechanism in *Anacystis*. Therefore the working hypothesis of a single multiphasic absorption mechanism at the plasmalemma, as postulated by NISSEN (1973) and LINASK and LATIES (1973), seems applicable. System II in the dark can be represented by a straight line while the values in the light approach a hyperbolic curve.

The light-dependence of System I suggests that this uptake is an energy-dependent process. It was roughly calculated (E = -40 mV, cf. BARBER, 1968; DEWAR and BARBER, 1973;

$[PO_4{}^{3-}]$ = 70 μM; cell volume = 38.4 μm^3) from the Nernst equation that the phosphate uptake would be an up-hill transport at least up to 500 μM. This agrees with the absorption experiments in the region of System I.

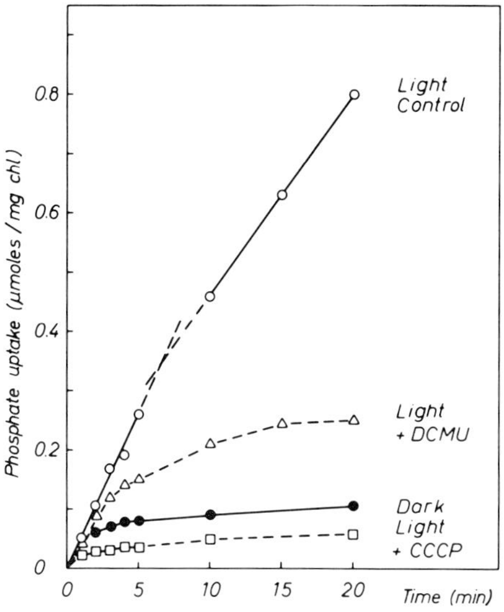

Fig. 1

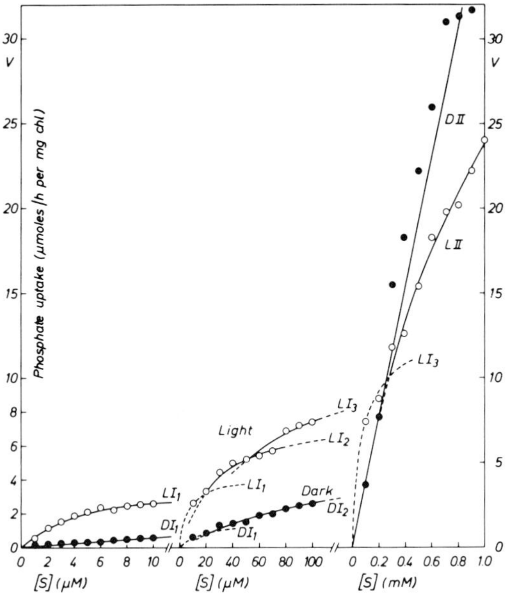

Fig. 2

Fig. 1 Time-course of phosphate uptake by ***Anacystis nidulans***. Experimental conditions: 7000 lux of white light, or dark, and air + 1.5 % CO_2. Reaction mixture: 2.0 ml algal suspension (25 μg chlorophyll); 0.5 ml Tris -H_2SO_4-buffer (pH 8.0, 0.5 M); 1.5 ml distilled water or inhibitor solution; 0.5 ml KH_2PO_4 (10^{-4} M) to give a final phosphate concentration of 10^{-5} M; 0.5 ml ^{32}P. Inhibitors, $5 \cdot 10^{-6}$ M DCMU and $1 \cdot 10^{-5}$ M CCCP, were added to the medium 5 min prior to the simultaneous addition of radioactivity and inactive phosphate. Open symbols: light; full symbols: dark.

Fig. 2 Uptake of phosphate from various concentrations added to the medium. Experimental conditions and reaction mixtures, except phosphate concentration, as in Fig. 1.

The application of inhibitors confirms this assumption. Addition of CCCP or DCMU strongly inhibit this process (Fig. 1). CCCP inhibition suggests that products of non-cyclic electron transport and also generation of ATP are involved in the light-dependent uptake process. Since DCMU considerably inhibits $^{14}CO_2$ fixation and phosphate uptake, leaving the ATP pool unchanged up to high DCMU concentrations (Fig. 3A), the ATP pool ***per se*** cannot be a limiting factor for phosphate uptake. CCCP (Fig. 1B) inhibits the $^{14}CO_2$ fixation more severely than the ATP level, although the ATP pool is much reduced. The position of the curve of phosphate uptake between the two others (Fig. 3) further supports the assumption made above.

In System II the dark uptake eventually exceeds that in the light (Fig. 2). Therefore it is of interest to examine the effects of CCCP and DCMU on System II (Fig. 4). Contrary to all expectations both inhibitors promote phosphate absorption in the light; however, in the dark the uptake is somewhat reduced by CCCP. These results indicate a passive component predominates in this range. It should be mentioned that incorporation in the organic soluble phosphate fraction remains blocked from the outset.

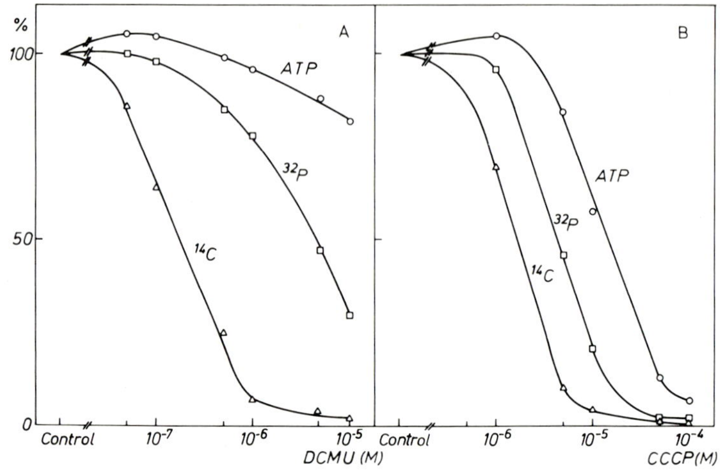

Fig. 3 ATP level and ^{32}P uptake in air + 1.5 % CO_2 and ^{14}C fixation (10 μmoles $NaHCO_3$/ml reaction mixture) vs. DCMU (A) and CCCP (B) concentration in white light of 20 000 lux and at pH 8.0 as % of the controls. ^{32}P and ^{14}C experiments were carried out for 5 min with 7 min pre-illumination. The ATP level was measured after 12 min pre-illumination. Medium phosphate concentrations: in A) $5 \cdot 10^{-5}$ M for ATP and ^{32}P measurements; $7.5 \cdot 10^{-5}$ M for ^{14}C measurements; in B): $7.5 \cdot 10^{-5}$ M for ATP and ^{14}C measurements, and 10^{-5} M for ^{32}P measurements.

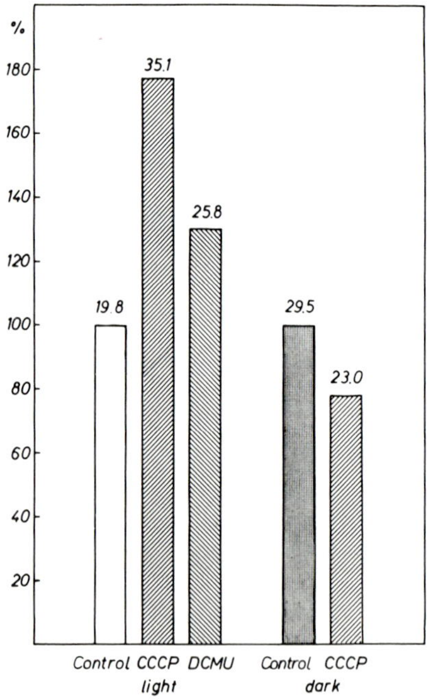

Fig. 4 Effects of $5 \cdot 10^{-6}$ M DCMU and $5 \cdot 10^{-5}$ M CCCP on the uptake from 700 μM phosphate as % of the controls. Experimental conditions and reaction mixture (except phosphate concentration) as in Fig. 1. The values at the top on the columns are μmoles phosphate taken up per mg chl per h.

The pronounced enhancement of phosphate uptake in the light in the upper region of System II in the presence of inhibitors could have arisen in several ways. DCMU and CCCP may depolarize a negative membrane potential, hyperpolarised by light as proved in several plants (JESCHKE, 1970; BENTRUP et al., 1973), and would therefore enhance a passive anion uptake. (But a depolarisation of the membrane potential in the light has also been reported (THROM, 1970)). Efflux changes may occur in the presence of inhibitors and regulation mechanisms could be involved. PELROY et al. (1972) have shown that in the presence of ribulosediphosphate the oxidation of glucose through the pentose phosphate pathway in blue-green algae is controlled by inhibition of glucose-6-phosphate dehydrogenase, inhibiting glucose oxidation in the light. This effect is reversed by the presence of DCMU. Therefore the respiratory chain can get into action and the phosphate influx be enhanced.

To clarify further the energy dependence of phosphate uptake in the light, the activity of non-cyclic electron transport was studied by measuring O_2 evolution and phosphate uptake at different wavelengths (Table 1). The highest values of O_2 evolution are obtained at 630 nm, the maximum of phycocyanin absorption. In the absorption range of photosystem I (717 nm) O_2 production drops to very low values, indicating the small relative quantum efficiency of non-cyclic electron transport. In contrast phosphate uptake (630 nm = 100 %) remains at a high level at 717 nm and is clearly higher than in the dark. Therefore light contributes energy to the uptake under conditions of both non-cyclic and cyclic electron transport. The participation of cyclic electron transport can best be seen in the effect of DCMU on phosphate incorporation at 717 nm in contrast to 630 nm (Fig. 5). Even in air + CO_2, DCMU inhibition is only 20 % at a concentration of 10^{-5} M.

Table 1

Relative quantum efficiency, measured by O_2 production (630 nm = 100 %) relative to phosphate uptake at different wavelengths in air + 3.0 % CO_2.

	630 nm	683 nm	717 nm	dark
O_2 production	100	17.3	0	–
Phosphate uptake	100	74.3	69.3	35.0

The use of desaspidin as an uncoupler, especially of photosystem I, confirms the contribution of cyclic electron transport to phosphate uptake (Fig. 6). As the figure shows this observation is valid in the presence of air and in N_2 at 717 nm; the inhibitory effect at 630 nm in air is distinctly lower. In the dark no inhibition in N_2 is observed.

These experiments indicate an energy requirement for phosphate uptake at least in the range of the multiphasic System I. The uptake is enhanced by light and suppressed by inhibitors of photosynthesis and uncouplers of photosynthetic and oxidative phosphorylations. The results in far-red light, in combination with inhibitors, prove that ATP generated by cyclic photophosphorylation contributes to the light-enhanced uptake of phosphate by ***Anacystis*** under nitrogen and in the presence of oxygen, in contrast to anion uptake in higher plants (JESCHKE, 1972). The ATP level itself, due to its relative insensitivity to DCMU and CCCP, cannot be the limiting factor for phosphate uptake. However, pH gradients, products of non-cyclic electron transport, or compounds of the glycolytic pathway in far-red light or in dark may be involved.

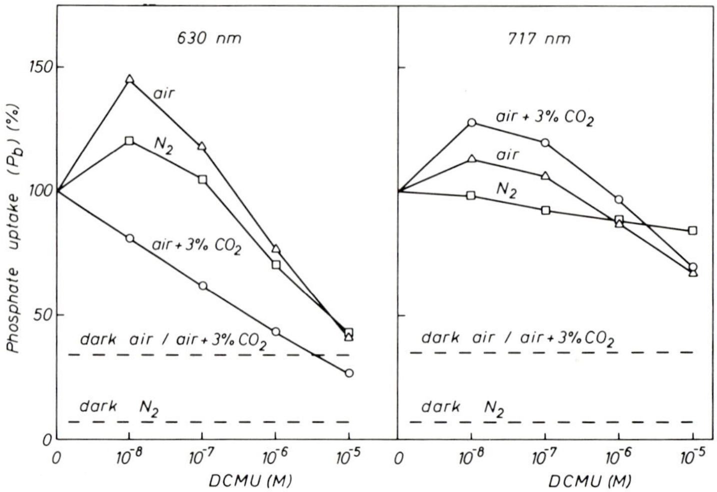

Fig. 5 Inhibition of phosphate uptake by DCMU as % of the controls in air + 3 % CO_2, air, and N_2. Dotted lines represent phosphate uptake in the dark in the respective gas supply. Left side: light of 630 nm at an intensity of $1.5 \cdot 10^4$ erg cm^{-2} s^{-1}. Right side: light of 717 nm at the same intensity. Phosphate concentration was 10^{-4} M. Experimental conditions: K-phthalate-buffer pH 7.0, 0.01 M; uptake time 10 min; P_b = total bound phosphate.

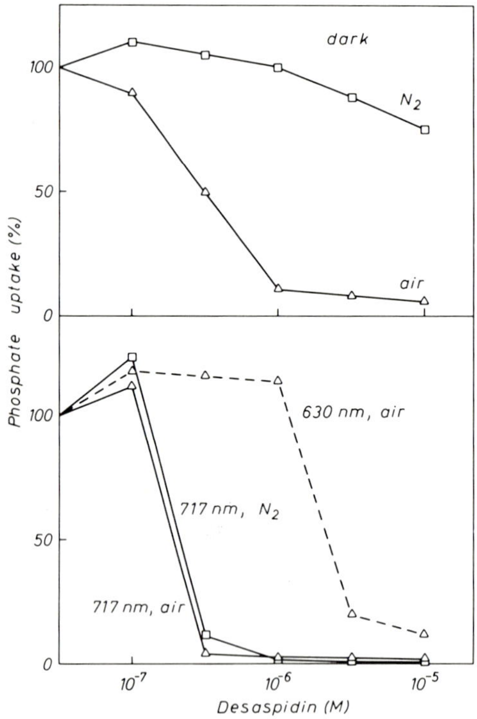

Fig. 6 Inhibition of phosphate uptake by desaspidin as % of the controls in air and in N_2. Upper part of diagram: in darkness; lower part of diagram: in light of 630 nm and 717 nm. Other conditions as in Fig. 5.

D. Acknowledgements

The study was supported by Deutsche Forschungsgemeinschaft. Thanks are due to Priv.Doz. Dr. W.D. Jeschke and Dr. M. Grünsfelder for constructive discussions. The technical assistance of Miss G. Borgwedel and Miss E. Wagner is acknowledged. K. Majumdar thanks the Friedrich-Ebert-Foundation for a grant.

E. References

BARBER, D.A., Biochim. Biophys. Acta **150**, 618–625 (1968)

BENTRUP, F.W., GRATZ, H.J., UNBEHAUEN, H., in Ion Transport in Plants (ed. W.P. Anderson), pp. 171–187, Academic Press, London and New York 1973

BORNEFELD, T., SIMONIS, W., Planta **115**, 309–318 (1974)

DEWAR, M.A., BARBER, J., Planta **113**, 143–155 (1973)

ESSL, A., Vergleichende Untersuchungen über die Wirkungsspektren der Photophosphorylierungsprozesse und der photosynthetischen Sauerstoffproduktion bei der Blaualge ***Anacystic nidulans***. Dissertation, Würzburg (1969)

JESCHKE, W.D., Z. Pflanzenphysiol. **62**, 158–172 (1970)

JESCHKE, W.D., Planta **103**, 164–180 (1972)

KRATZ, W.A., MYERS, J., Am. J. Bot. **42**, 282–287 (1955)

LINASK, J., LATIES, G.G., Plant Physiol. **51**, 289–294 (1973)

NISSEN, P., Physiol. Plant. **28**, 304–316 (1973)

PELROY, R.A., RIPPKA, R., STANIER, R.Y., Arch. Mikrobiol. **87**, 162–180 (1972)

SIMONIS, W., URBACH, W., Ann. Rev. Plant Physiol. **24**, 89–114 (1973)

THROM, G., Z. Pflanzenphysiol. **63**, 162–180 (1970)

ULLRICH-EBERIUS, C.I., SIMONIS, W., Planta **92**, 358–373 (1970)

ULLRICH-EBERIUS, C.I., Planta **109**, 161–176 (1973)

Transcellular Currents and Ion Fluxes through Developing Fucoid Eggs

L.F. Jaffe, K.R. Robinson, and R. Nuccitelli

A. Introduction

Our broadest interest is in the role of ionic movements in development particularly in the central questions variously referred to as localization, patterning and differentiation. As a model or prototype of this broad problem we are studying the role of ion movements in the development of the fucoid egg, particularly that of ***Pelvetia fastigiata***. This common seaweed egg has two major advantages. First, it is free of the enveloping tissues, accessory cells, and impermeable shells which so commonly surround eggs. Second, it begins with a minimum of preformed pattern, practically lacking the primary polarity found in most plant (and all animal) eggs. However, it soon exhibits the essential features of localization. Within a day after fertilization it 'germinates', i.e. initiates growth at one pole, visibly polarizes and divides into two quite different cells: a rhizoid, or attachment cell at the growth pole and a thallus cell at its antipode.

Before discussing the involvement of localized membrane changes in this process, we provide some information on over-all changes. The plasma membrane of the unfertilized fucoid egg, like that of other such idling cells, exhibits a relatively low ion selectivity and a low potential (ROBINSON and JAFFE, 1973; WEISENSEEL and JAFFE, 1972), and maintains no osmotic pressure difference (ALLEN et al., 1972). After fertilization it returns (in the course of some hours) to a state typical of active, somatic cells. Thus the 8 h old ***Pelvetia*** egg, at a stage just before germination, has a membrane potential of about -70 mV which can be largely accounted for as a K^+ diffusion potential (WEISENSEEL and JAFFE, 1972). This latter is shown by the thirty-fold ratio of internal to external K^+ (ALLEN et al., 1972), the 44 mV change in membrane potential per ten-fold change in external K^+ (WEISENSEEL and JAFFE, 1972), and the relatively high K^+ permeability inferred from tracer fluxes (ROBINSON and JAFFE, 1973) (Table 1). At this stage, too, the ***Pelvetia*** egg has developed an excess osmotic pressure of 210 mosM or about 5 atm. This large pressure is quantitatively accounted for by the uptake of Cl^- (and K^+) after fertilization (ALLEN et al., 1972) and is maintained with a minimum of energy expenditure by virtue of some very effective barrier to the outward leakage of Cl^- (ROBINSON and JAFFE, 1973).

B. Results and Discussion

I. Localized Changes: Nature of Transcellular Currents

1. Electrical Evidence

Transcellular electrical currents were first measured in developing fucoid eggs by a series method (JAFFE, 1966, 1968). This involved lining up hundreds of eggs in a long loose-fitting capillary, inducing parallel development with light from one end and then measuring the voltage between the ends of the tube. Such experiments indicated that a current of at least 60 pamp begins to flow through the normally developing ***Pelvetia*** or ***Fucus*** egg at or somewhat before the time that it begins to elongate, and continues to flow as long as it continues to elongate. Considered as a flow of positive ions, current enters the egg at its growth or rhizoid pole and leaves elsewhere. Within the egg it has a density of the order of 1 μamp/cm².

Recently we have developed a new, ultrasensitive, vibrating probe system with which we can measure currents external to single developing fucoid eggs (JAFFE and NUCCITELLI, 1974). The probe has a 25 μm diameter, platinum black electrode at its tip which is vibrated laterally at about 200 Hz between two extracellular points about 30 μm apart. This vibration converts any voltage difference between these points into a sinusoidal output whose amplitude is measured with a lock-in amplifier. Using this new tool we have already confirmed the presence of a relatively steady current loop which enters the egg's growth pole with a surface density of about 1 μamp/cm² and leaves elsewhere.

In addition to this steady component, the endogenous current proves to contain a pulse component. Fig. 1A illustrates a group of such pulses recorded by a vibrating probe positioned to detect currents moving normal to the growing tip and along the embryo's axis. Throughout their approximately 100 s duration these pulses indicate current entering the growing point. Fig. 1B, on the other hand, shows a group of pulses recorded by a probe vibrating normal to a region of the embryo which is far from the growing tip. At all such points, the pulses are likewise of about 100 s duration but indicate current leaving the embryo. As is illustrated, pulses tend to occur in groups. These groups are separated by quiet periods lasting of the order of an hour, so that the overall average pulse frequency is lower than would be suggested by Fig. 1. This average frequency rises from about one to about five pulses per hour during the first day after germination.

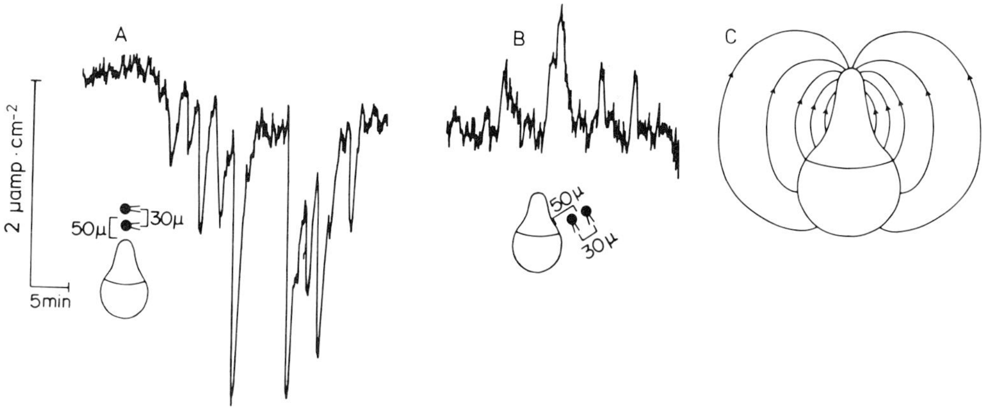

Fig. 1 Current pulses through growing, one day old, ***Pelvetia*** eggs. A) Inward pulses recorded in front of the growing point. The indicated current densities are at the point of measurement; at the cell's surface they are estimated to be 8 times greater. B) Outward pulses recorded at the base of the rhizoid cell. C) Map of the current pattern during a pulse.

From a study of hundreds of such pulses we conclude that they are non-propagating since they never display the biphasic form of a propagating impulse. Rather, both the pulse and steady components of the current simply enter the growth region and leave elsewhere as is shown in Fig. 1C. We estimate that the pulse component rises from about 5 % of the total transcellular current to about 30 % of this total during the first day following germination.

One remarkable feature of these pulses is their lack of a refractory period. The embryos frequently generate current shapes interpretable as pairs (or larger groups) of pulses which overlap to any degree, from slightly to almost completely. Such overlapping groups seem almost inexplicable unless they are produced by repetition of the same process in different parts of the membrane of the growing tip.

Another remarkable feature of these endogenous current pulses is their similarity to those recently reported to traverse regenerating (and initially depolarized) stalk segments of ***Acetabularia*** (NOVAK and BENTRUP, 1972). Like the fucoid pulses, those through ***Acetabularia*** always enter the growth pole, last for 100 s, rise faster than they fall, and show a similar spontaneous frequency, starting at about 1 per hour and increasing after about a day to 3–5 per hour.

2. Tracer Evidence

As will be discussed below, the size of the cytoplasmic field generated by the endogenous current should depend strongly on which ions make up this current. To find this out, we have begun to employ a method that makes use of a very thin (25 μm) nickel screen which contains many, almost perfectly round, holes (Fig. 2A). These are available from Perforated Products, Brookline, Massachusetts, and can be obtained with holes that are slightly smaller than a ***Pelvetia*** egg. We can fill every hole in such a screen with an egg and then, using unilateral light, induce every egg to grow towards one side or the other (Fig. 2B). Since the developing eggs secrete a glue, they stick very tightly in the holes. This glue also prevents medium from flowing through the egg-filled screen.

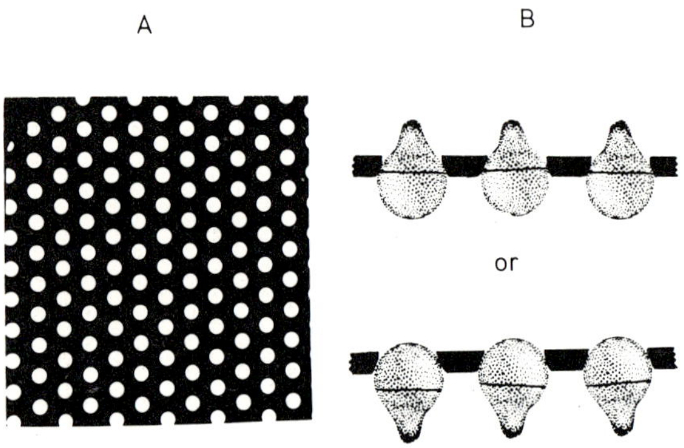

Fig. 2 Screen used to measure flux asymmetries in ***Pelvetia*** eggs. Holes are 75 μm in diameter. Left: face view photograph. Right: sectional views of screens bearing day-old ***Pelvetia*** eggs. (These are based upon photographs of both screen sections and of eggs.)

By putting two such screens in appropriate chambers, radioisotopes of the various ions in sea water can be applied to the rhizoid or thallus ends of the eggs and the amount of ion entering these two ends compared. On the other hand, the amounts which leave them may be compared by completely preloading the eggs in a screen with tracer and then measuring the amount escaping into 'cold' media on both sides of the screen.

The most important result so far obtained with this screen method is shown in Fig. 3. It shows the asymmetries in both Ca^{2+} influx and Ca^{2+} efflux measured on ***Pelvetia*** eggs while being polarized by unilateral light. At 6 h after fertilization, when the eggs are first stuck tightly enough in the screens to do the experiment, Ca^{2+} is found to enter the dark (and future rhizoid) pole five times faster than the lighted (and future thallus) pole. The efflux asymmetry is reversed; Ca^{2+} leaves the future thallus pole about four times faster than the future rhizoid end.

Clearly, there is a Ca^{2+} current traversing the photopolarizing egg. It is in the same direction as the net electrical current but is only a relatively small component of it. At 6 h it is calculated to have an average value of 2 pamp and by 10 h after fertilization, when the very first eggs have begun to germinate, it has fallen by half to about 1 pamp. These values may be compared with the net current estimated from the series method: from 6 to 10 h it rises from about 4 to 30 pamp (JAFFE, 1968).

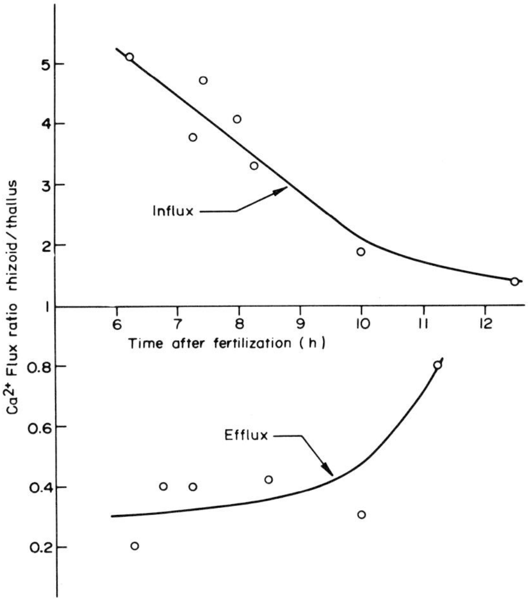

Fig. 3 Ca^{2+} flux asymmetries in ***Pelvetia*** eggs during polarization by 1000 foot-candles of unilateral white light. Each influx point is the ratio of the amount of $^{45}Ca^{2+}$ that entered the future rhizoid pole to the amount that entered the future thallus pole during a 30 min exposure; each efflux point indicates the same ratio for $^{45}Ca^{2+}$ leaving the two poles of fully labelled eggs. Eggs were under 1000 foot-candles of unilateral light from 2 h on, but no measurements were possible before 6 h because the eggs had not yet stuck firmly enough to the screen. Half the eggs had germinated by about 11 h.

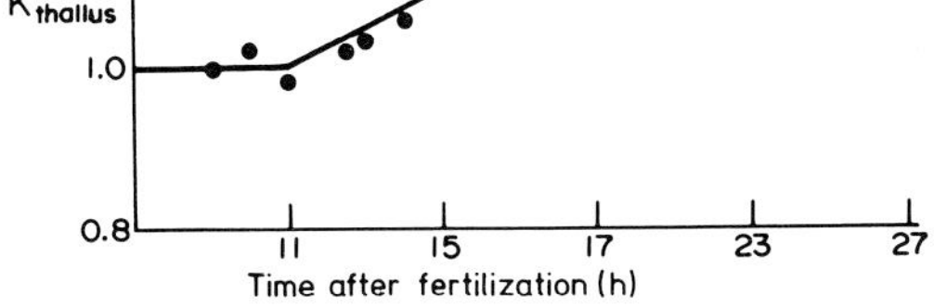

Fig. 4 The ratio of K^+ uptake at the rhizoid end to K^+ uptake at the thallus end of ***Pelvetia*** eggs growing under even illumination (points). The solid line indicates the comparable ratio of surface areas. Illumination was by 100 foot-candles of white light (ROBINSON, 1973).

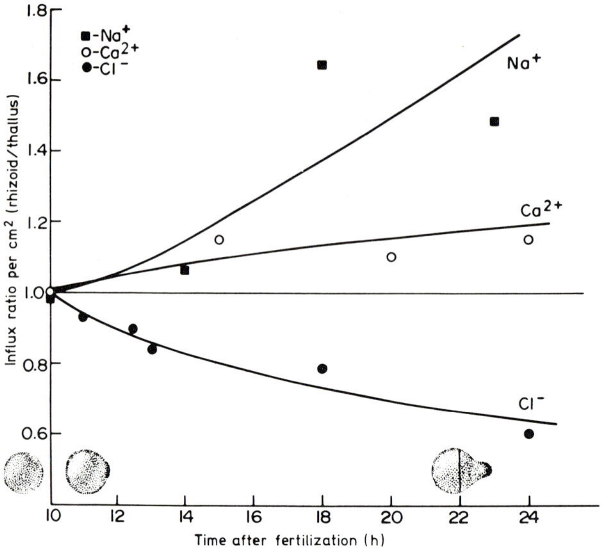

Fig. 5 Ion uptake asymmetry in ***Pelvetia*** eggs developing under even illumination by 100 foot-candles of white light. Each data point represents a ratio of influxes during a 30 min exposure to $^{42}K^+$ as well as to $^{22}Na^+$, or $^{45}Ca^{2+}$ or $^{36}Cl^-$. Each of the latter ratios was divided by the $^{42}K^+$ influx ratio to correct for the growth in area on the rhizoid side. The pictures symbolize the stage of development and indicate that half the eggs have germinated by about 11 h while half have reached the two cell stage by about 22 h.

By now, we have also measured influx ratios (but not efflux ratios) for Ca^{2+}, as well as for K^+, Na^+, and for Cl^- on older, already polarized eggs germinating under diffuse light. No detectable asymmetry in K^+ influx per unit area develops (Fig. 4). However, as Fig. 5 shows, Ca^{2+} again come to enter the rhizoid pole faster (though no more than 20 % faster), Na^+ likewise come to enter this end faster (the influx ratio rising to about 1.6 by 18 to 24 h), while Cl^- come to enter this pole slower (the influx ratio falling to about 0.6 by 24 h). These influx asymmetries all indicate specific ion currents in the same direction as the net electrical

Table 1

Concentrations, pump directions, and plasma membrane permeabilities of the major ions in 8 h old ***Pelvetia*** eggs.

	K^+	Cl^-	Na^+	Mg^{2+}	Ca^{2+}	References
Egg concentration (mM)	320	310	. 30	25	4	ALLEN et al. (1972)
External concentration (mM)	10	565	483	55	10	ALLEN et al. (1972)
Pump direction	In?	In	Out	Out	Out	WEISENSEEL and JAFFE (1972)
Permeability (nm/s)	5.0	—[a]	0.06	—	0.02	ROBINSON and JAFFE (1973)

[a]Efflux too low to measure

current (although they cannot be quantitatively converted into current values until efflux data are obtained). Moreover, if one compares the available flux asymmetry results with the pump directions listed in Table 1 a tentative rule emerges: ions leak faster at the growing or rhizoid end (thus Ca^{2+} and Na^{+} leak in faster there), but are pumped faster at the opposite or thallus end (thus Ca^{2+} and Cl^{-} are respectively pumped out and in faster there).

II. Developmental Role of Transcellular Currents

1. Electrophoretic Theory

Whether steady or pulsed, the transcellular current is likely to produce cytoplasmic fields strong enough to profoundly affect the fucoid egg. This may be particularly true for the current's Ca^{2+} component since (judging from its behavior in nerve and in muscle cells) Ca^{2+} should be strongly bound to fixed negative charges in the egg's cytoplasm. Thus only about $3 \cdot 10^{-7}$ M Ca^{2+} is free in resting nerve and muscle, while the mobility of $^{45}Ca^{2+}$ is immeasurably low in such cells (EBASHI and ENDO, 1968; KUSHMERICK and PODOLSKY, 1969; BAKER, 1972). Such Ca^{2+} binding is also suggested by recent microprobe evidence of a zone of concentrated Ca^{2+} binding sites just within the plasmalemma of the squid giant axon (PRUSCH et al., 1973).

The maintenance of a Ca^{2+} current through a cytoplasmic zone which binds Ca^{2+} strongly will require the maintenance of a correspondingly high gradient of fixed Ca^{2+} across this zone. To preserve electroneutrality, each bound Ca^{2+} must displace one or two mobile counter ions, mainly K^{+}; so an equal countergradient of K^{+} should result. Such a countergradient in turn would produce a diffusion field with its positive end towards the pole of Ca^{2+} entry. Finally, this field would tend to concentrate the cell's larger components (which should have a negative surface charge) towards this end. It can be calculated that such self-electrophoretic effects could be much larger than those produced by a flux of unbound ions such as K^{+} (JAFFE et al., 1974). So it now seems well worthwhile looking for the transcytoplasmic gradient of Ca^{2+}, of K^{+} and of mobile polyanions predicted by this theory.

2. Effects of Imposed Currents

Another test of the developmental role of the endogenous current is to observe the effect of driving a current through the eggs by imposing some external gradient on them. The results of such experiments with fucoid eggs are shown in Table 2. In all cases the growth or rhizoid pole is induced to form at the end into which cations are driven. Thus the polarizing currents

Table 2

Polarization of fucoid eggs by imposed gradients

Gradient	Resultant growth end	Difference per egg to half polarize	Resultant current	References
Electrical	Positive	25 mV	2 μamp/cm^2	LUND (1923)
K^{+}	High K^{+}	50 %	1 μamp/cm^2	BENTRUP et al. (1967)
H^{+}	High H^{+}	33 %	?	BENTRUP et al. (1967)
Dinitrophenol	High DNP	$2 \cdot 10^{-5}$	?	WHITAKER and BERG (1944)

are in the same direction as the endogenous one. This is obvious for the electrical, K^+, and H^+ ion gradients; for dinitrophenol (DNP) we interpret the results this way since DNP is believed to affect its biological actions by shuttling H^+ ions across membranes (FINKELSTEIN, 1970). The size of the resultant currents when they are driven by electrical or by K^+ gradients can be estimated from tracerderived conductance data (ROBINSON and JAFFE, 1973), and are seen to be comparable in size to the endogenous current. So the limited data now available is consistent with the idea that the endogenous current acts back to further polarize the cell.

However, one ambiguity should be discussed. An imposed electrical field may localize growth by virtue of the field thus placed across the egg's plasma membrane (or even its wall) rather than its cytoplasm. This consideration may explain observations that fucoid eggs can actually show a strong counter-response, growing towards the (previous) negative pole if the field is applied early enough. (This effect peaks at only three hours after fertilization; BENTRUP, 1968; unpublished data.) However, since this is long before the endogenous current begins (JAFFE, 1968), it may not be so surprising that an imposed field may then act by another mechanism.

3. Potassium Bursting

Finally, we will describe one last phenomenon that may be a lead towards understanding the role of the endogenous current. We observe that when recently germinated ***Pelvetia*** eggs are shifted to an artificial sea water containing 10 to 30 times the normal K^+ concentration (and a correspondingly reduced Na^+ concentration) they tend to eventually burst at their growing tips. The time course of one such study is shown in Fig. 6. Tracer studies show no detectable change in the internal K^+ concentration in such high-potassium media; so bursting is unlikely to result from such a cytoplasmic change. Furthermore, we find that up to 97 % of the external Na^+ may be replaced with Tris without any detectable effect on development. Hence, bursting cannot result from the concomitant reduction in Na^+. On the other hand, a ten fold increase in K^+ yields a 46 mV fall in membrane potential in recently germinated eggs (ROBINSON and JAFFE, 1973). So it seems quite likely that high K^+ bursts eggs by lowering the membrane potential; this depolarization in turn somehow weakening the wall at the growth tip. This in turn suggests that during development in natural sea water there may be some smaller variations in the membrane potential at the tip which are involved in the natural or physiological control of tip growth.

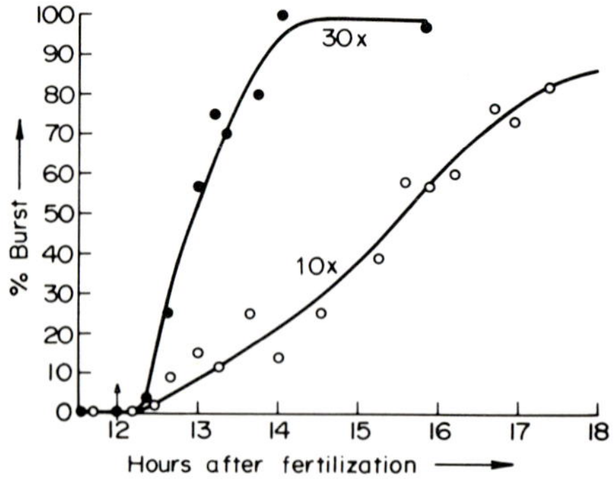

Fig. 6 Typical time course of bursting of recently germinated ***Pelvetia*** eggs by high K^+ sea water. At 12 h after fertilization (as indicated by the arrow) the natural sea water medium was replaced by an artificial sea water with either 10 or 30 times the natural K^+ concentration, thus 100 or 300 mM instead of 10 mM; T = 15°C.

C. Acknowledgements

This work was supported by a NSF grant to L.F.J., a N.I.H. postdoctoral fellowship to K.R.R. and a N.I.H. predoctoral training grant to R.N.

D. References

ALLEN, R.O., JACOBSEN, L., JAFFE, L.F., Dev. Biol. **27**, 538–545 (1972)
BAKER, P.F., Progr. Biophys. Mol.Biol. **24**, 177–223 (1972)
BENTRUP, F.W., Z. Pflanzenphysiol. **59**, 309–339 (1968)
BENTRUP, F.W., SANDAN, T., JAFFE, L.F., Protoplasma **64**, 254–266 (1967)
EBASHI, S., ENDO, M., Progr. Biophys. Mol. Biol. **18**, 123–183 (1968)
FINKELSTEIN, A., Biochim. Biophys. Acta **205**, 1–6 (1970)
JAFFE, L.F., Proc. Nat. Acad. Sci. **56**, 1102–1109 (1966)
JAFFE, L.F., Advan. Morphog. **7**, 295–328 (1968)
JAFFE, L.F., NUCCITELLI, R., J. Cell Biol., in press (1974)
JAFFE, L.F., ROBINSON, K.R., NUCCITELLI, R., Ann. N.Y. Acad. Sci., in press (1974)
KUSHMERICK, M.J., PODOLSKY, R.J., Science **166**, 1297–1298 (1969)
LUND, E.J., Bot. Gaz. **76**, 288–301 (1923)
NOVAK, B., BENTRUP, F.W., Planta **108**, 227–244 (1972)
PRUSCH, R., WALL, B.J., OSCHMAN, J.L., Biol. Bull. **145**, 451 (1973)
ROBINSON, K.R., Ion Movements and Membrane Asymmetries in the Developing Fucoid Egg. Ph.D. Thesis, Purdue University Lafayette, Indiana (1973)
ROBINSON, K.R., JAFFE, L.F., Dev. Biol. **35**, 349–361 (1973)
WEISENSEEL, M.H., JAFFE, L.F., Dev. Biol. **27**, 555–574 (1972)
WHITAKER, D.M., BERG, W.E., Biol. Bull. **86**, 125–129 (1944)

A Shock Effect on the Permeability to Sulphate of *Acer pseudoplatanus* Cell-Suspension Cultures

A. Thoiron, B. Thoiron, J. Le Guiel, J. Guern, and M. Thellier

A. Introduction

The cellular ionic exchanges of higher plant species are often studied using multicellular tissues. In such tissues most of the cells are not in direct contact with the known external medium, and important free space effects, diffusive processes, fixed charges, etc., can obscure the true cellular absorption processes. Hence we decided to use cell-suspension cultures as experimental material; they are in fact loose cellular aggregates in which it can be assumed that each cell exchanges freely with the external medium, with only the cell walls interfering as free spaces. Such cultures were originally obtained by LAMPORT (1964) from *Acer* cambial tissue culture and a few authors have already used them for transport studies (DOREE et al., 1972; HELLER et al., 1973).

We have tried to use them to study sulphate exchange in higher plant cells; but, in so doing, we encountered a 'shock effect' on the permeability to sulphate (somewhat comparable to the one which has been described for the same material by DOREE et al. (1972) for adenine and leucine exchange). It was thus first necessary to study and overcome this effect.

B. Materials and Methods

The cell suspensions were routinely grown at 25°C, on LESCURE's (1966) germ-free nutritive solution, 250 ml per flask, with a rotatory agitation of 60 r.p.m., and under white light (1500 lux with a photoperiod of 12/12 h). Every 12 days, 70 ml of cell suspension were transferred to 160 ml of new nutritive medium. Usually, 7 day-old cultures were used for the experiments (exceptionally 8 or 9 days old); under the above conditions, except when explicitly indicated otherwise, the pH of the suspension rises from 5.8 to 6.5, while its sulphate concentration drops from 1.5 mM to 1.3 mM, during the 7 days of growth.

For an experiment the suspension flask was opened and, when necessary, cell aggregates that were too large were eliminated by filtration through a 300 μm sieve and rinsed with 500 ml of a 2 % sucrose solution. The remaining cells were then separated by filtration on a 30 μm sieve, sometimes transferred to a 2 % sucrose intermediary incubation medium, and finally to the experimental medium (generally 2 % sucrose solution with a given concentration of $Na_2{}^{35}SO_4$, pH 5.6). The beginning of the filtration on the 30 μm sieve was taken as time zero.

C. Results and Discussion

I. Time Course of Sulphate Uptake

An apparently paradoxical result was obtained (Fig. 1): the accumulation of sulphate exhibited a concave-upwards shape with no tendency to plateau over a period of 3 h, and the efflux of absorbed radioactivity remained practically unmeasurable. To ensure that this behaviour corresponded to the same type of shock effect (due to the opening of the flasks, the filtration of the cells or the shaking of the flask), as that found by DOREE et al. (1972)

it was first necessary to measure the initial velocity of sulphate absorption. As a quasi-linear behaviour obtains during the first 5 min, the slope in this period is taken as the initial velocity.

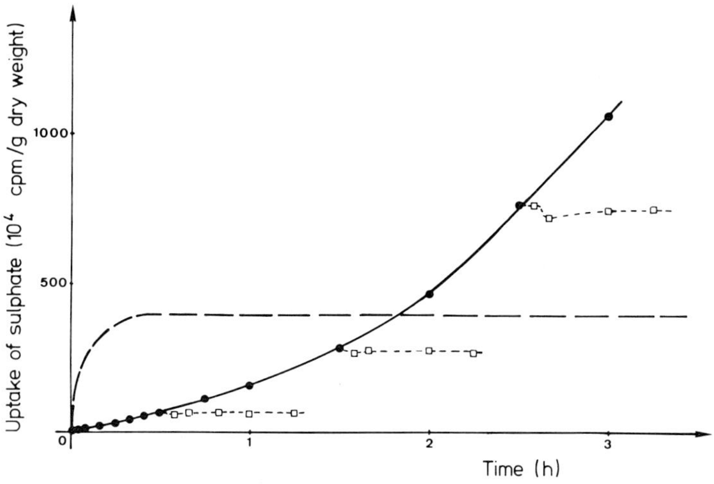

Fig. 1 Time course of accumulation of $^{35}SO_4^{2-}$. Full line, accumulation: experiment conducted at 22°C and $5 \cdot 10^{-5}$ M (specific radioactivity 4.54 mCi/mmole) with 9 day-old suspension cultures. Dotted lines, efflux of absorbed radioactivity obtained by transferring an aliquot of the cells to a similar, but unlabelled, sulphate sucrose medium. The dashed line is a reminder of the classical accumulation curve.

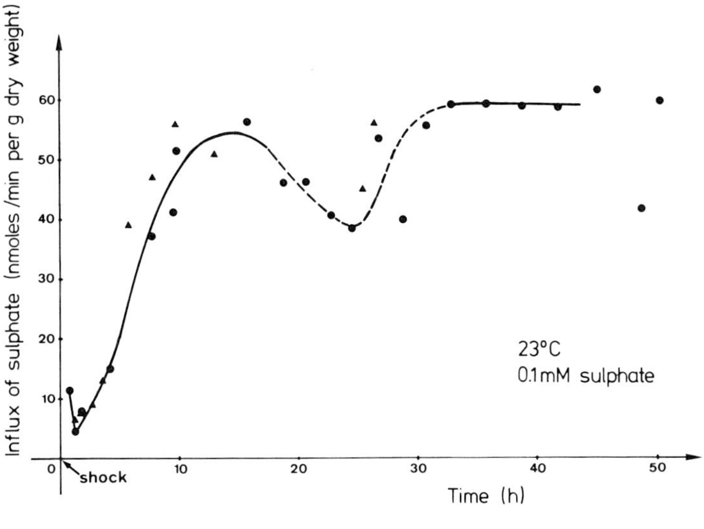

Fig. 2 Initial rates of sulphate uptake at different times following the filtration of the cells. Two experiments were made (represented by ▲ and ●); the cells were filtered at time zero, resuspended in the incubation medium (sucrose 2 %) and, at increasing times of incubation, transferred to the absorption medium (sucrose 2 %, $Na_2{}^{35}SO_4$ 0.1 mM) where the initial rate of uptake was measured during the first 5 min.

II. The 'Shock' under the Usual Experimental Conditions

Fig. 2 summarizes the results of two typical experiments. During the first hours following the opening of the flask and the filtration of the cells, the initial velocity of uptake is very low; then it slowly rises and, after perhaps an oscillatory behaviour between the 10th and the 25th h (though this needs confirmation), attains a constant value. It can be seen that this 'shock' effect produces changes in the uptake of sulphate of at least a factor 10. The time elapsing between the opening of the flask and the filtration of the cells has practically no influence on the initial velocity of uptake; it thus seems that the filtration is important for initiating the phenomenon, hence the choice of time zero at the moment of filtration. It was also observed that energetic shaking of the suspension, without any filtration, could promote the same kind of shock effect.

III. Determination of the Initial Velocity of Sulphate Uptake before the Shock

Experiments conducted under the usual conditions, as in Fig. 2, show the recovery of the cells after the shock, but they do not allow a measurement of the uptake of sulphate just before the shock (as a filtration is naturally necessary for the transfer from the culture medium to the sucrose sulphate absorption medium). We have therefore performed an experiment, under different conditions, that would allow this measure. The cells were first transferred onto a sulphate deficient medium (0.2 mM instead of 1.5 mM) so that, after the 7 days of culture, the sulphate concentration of the medium had dropped to $1.7 \cdot 10^{-2}$ mM. An aliquot of the suspension was taken on opening of the flask and then other aliquots, at increasing intervals of time after the opening, for measuring the initial velocities of absorption. The initial velocities were measured by adding 0.1 mM of $Na_2{}^{35}SO_4$ to the aliquots of the suspension themselves, without any filtration or transfer to new solutions, so as to avoid the superposition of the filtration shocks on to the shaking shock.

Fig. 3A gives the results obtained, which confirm and complete those of Fig. 2; there is, at the moment of the shock, a rapid drop in uptake, with a minimum after 1 h, and then a

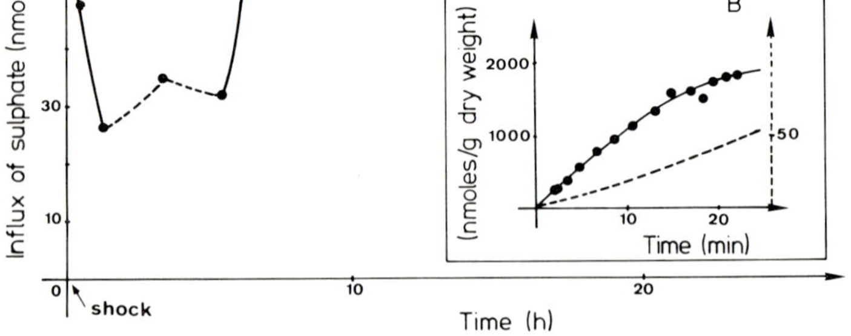

Fig. 3 Experiments conducted in the culture medium. A. Sulphate uptake measured at different times after opening the flask without filtration or shaking. B. Accumulation curve without shock (continuous line and points •), as compared to an accumulation curve after a shock (dotted line, redrawn from Fig. 1).

recovery to values almost equal to those before the shock. With this experimental procedure, it is possible to study the uptake of sulphate without any shock effect, and it is then seen (Fig. 3B) that the accumulation curve exhibits a classical shape, and no longer the paradoxical one obtained after the shock (Fig. 1). However cells which had recovered were liable to exhibit a second shock effect following a new filtration.

IV. Tentative Interpretation of the Shock-Effect

1. Hypothesis

According to DOREE et al. (1972), the shock effect arises from a CO_2 outflow from the cells, and is more pronounced the more rapid the outflow (tough shaking, air bubbling, filtration). There remain however different possible mechanisms to explain this effect of the CO_2 outflow on the shock and the further recovery (DOREE et al., 1972): e.g. direct flux coupling, and modification of the number of specific sites of absorption or of the structure of these sites. The first case would be a force-effect, and the other two would be effects on the catalytic devices for the functioning of the absorption process. Although no definitive answer can yet be given, some information can be gained from the following.

2. Flow-Force Relations

We may assume that, not too far from sulphate flux equilibrium between the cells and the external medium, there is a linear relation between the net flux ϕ of sulphate and the force X:

$$\phi = L\,X$$

And under quasi-stationary conditions the force X will depend linearly on the logarithm of the concentration of sulphate in the external medium:

$$X = A \log (\Pi[SO_4{}^{2-}])$$

in which A stands for 2.3 RT/nF and Π summarizes all the constant contributions to the force X (THELLIER, 1968, 1973). We therefore get

$$\phi = AL \log (\Pi[SO_4{}^{2-}])$$

in which Π is a parameter characteristic of forces only, while AL is a parameter characteristic only of the devices that catalyse sulphate absorption. Thus, according to which of these two parameters is affected by the shock, we can obtain information on the nature of the perturbation.

With the procedure we have used we could obtain sulphate influxes in initial stationary conditions. The efflux could not be determined very accurately, but remained smaller than 5 nmoles/min per g dry weight; we could thus calculate approximate values of the net fluxes from the measured influxes. Fig. 4 gives the results obtained for cells used either just after the filtration-shock or 25 h later, when recovery was complete. In both cases, a satisfying linearity obtains over a concentration range of 0.02 mM to 0.2 mM; and it can be seen that the log Π values, intercepts on abscissa, equal to 4.68 and 4.90 are very similar, but the AL values (slopes equal to 11.4 and 71.7) are very different. This means that the shock does not much affect the forces, but strongly affects the sites of sulphate absorption.

3. Temperature Effect

The rate of sulphate accumulation at low temperature is very low for the shocked as well as for the recovered cells; however, the rate of sulphate absorption is practically independent of the temperature for the shocked cells, whereas it is very temperature dependent after the cells have recovered. This perhaps means that the shock almost totally suppresses the specific sites catalysing absorption of sulphate, to leave only a passive process.

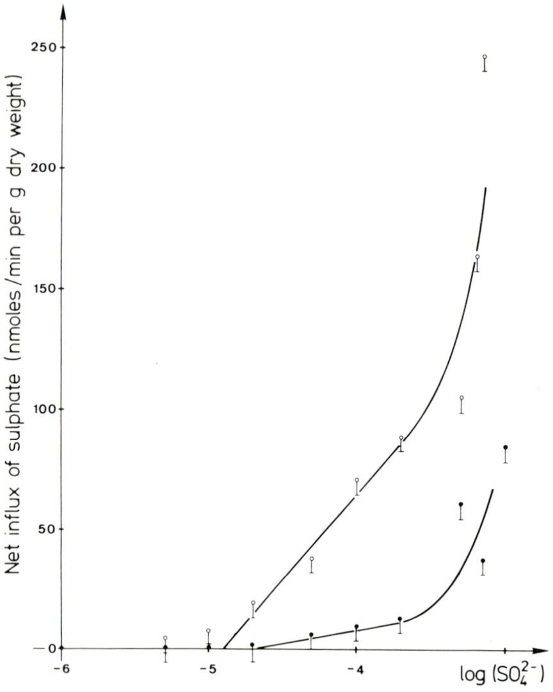

Fig. 4 Flow-force relations for cells just after the shock ● or after recovery ○. The vertical line, for each point, arises from the uncertainty in the sulphate efflux.

D. Acknowledgements

This work was supported by grants of the C.N.R.S. (L.A.203, ERA, and RCP 285) and of the D.G.R.S.T. (72 7 0300). We thank Prof. Bourriquet and Dr. Morvan for their help in preparing the cell suspension cultures.

E. References

DOREE, M., LEGUAY, J.J., TERRINE, C., Physiol. Veg. **10**, 115–131 (1972)

HELLER, R., GRIGNON, C., SCHEIDECKER, D., in Ion Transport in Plants (ed. W.P. Anderson), pp. 337–353, Academic Press, London and New York 1973

LAMPORT, D.T.A., Exp. Cell Res. **33**, 195–206 (1964)

LESCURE, A.M., Physiol. Veg. **4**, 365–378 (1966)

THELLIER, M., C.R. Acad. Sc. **266**, 826–829 (1968)

THELLIER, M., in Ion Transport in Plants (ed. W.P. Anderson), pp. 47–63, Academic Press, London and New York 1973

Importance of the Cell Wall in the Thermodynamic Equilibrium of Ions in Free Cells of *Acer pseudoplatanus*, L.

R. Heller, C. Grignon, and J.-P. Rona

A. Introduction

Cell suspensions provide an abundant, homogeneous, and relatively simple material well suited to the study of the movement of ions in the cells of higher plants. The possibility of obtaining protoplasts from them renders their use even more interesting. This article describes original findings on the potential differences across membranes measured in free cells, protoplasts and free vacuoles of ***Acer pseudoplatanus***. The results are related to the concentration and intracellular activity of potassium as evaluated by compartmental analysis. In the discussion that follows, the role played by the pecto-cellulosic wall in establishing potential differences across membranes and in the maintenance of ionic equilibria is pointed out.

B. Materials and Methods

Cell suspensions ***(Acer pseudoplatanus,*** L.*)* were maintained on a modified Heller's medium (GRIGNON, 1969) at 25°C in 1-litre flasks inclined at 45° and rotated at 60 r.p.m. The cells used for experiments, were taken during their exponential phase of growth.

Protoplasts were prepared by enzymatic hydrolysis after plasmolysis (RONA and GRIGNON, 1972). The yield of protoplasts amounted to 80 %. They were washed and resuspended in a molar solution of sucrose. Vacuoles could be isolated by slightly lowering the osmotic pressure or increasing agitation.

The cells were held for 24 h at 21°C in a solution of 50 mM sucrose, 0.5 mM $CaCl_2$, and 10 mM KCl labelled with $^{42}K^+$. A 20 ml sample of the suspension was then filtered and introduced into a device which could both be agitated and rapidly drained. The experiment started with the removal of the absorption medium ($^{42}K^+$) and its prompt substitution by 20 ml of an exchange solution identical to the absorption medium but containing non-radioactive K^+ instead. The medium was continually renewed for 24 h at intervals of time progressing gradually from 30 s at the beginning to 20 h at the end of the experiment. The radioactivity of each sample removed was measured by liquid scintillation and the values summed to plot the 'efflux' curve.

The measuring device consisted of two microelectrodes held in a de Fonbrune micromanipulator and connected to a measuring circuit (RONA, 1973). The cells were placed in a drop of medium on an agar disc (Difco agar, 8 g/l; KCl, 2.5 μM) on a glass slide under the microscope. The particular cell chosen was held in place with a glass microinstrument. One microelectrode was used as the reference electrode in the medium while the other was gradually inserted into the cell.

C. Results

I. Electrical Potential Values

1. In the Culture Medium

Cells

We obtained three successive values during the gradual penetration of the microelectrode into the cell, in agreement with MORVAN (1972) who has also studied the cells of ***Acer***: 1st potential (contact with the cell), ψ_a = −8 to −12 mV; 2nd potential (beginning of penetration), ψ_b = −14 to −18 mV; 3rd potential (deep penetration), ψ_c = −22 to −40 mV.

Protoplasts
Simple contact with the protoplast had no effect at all; the gradual penetration of the microelectrode, however, always gave rise to two successive values: 1st potential (14 determinations), ψ'_a = +10.4 ±0.9 mV; 2nd potential (18 determinations), ψ'_b = + 15.7 ±1.2 mV.

Isolated Vacuoles
No potential was produced on contact, but progressive penetration brought about the sudden appearance of a single value: +15.5 ±4.4 mV. Readings were taken on 6 vacuoles isolated from the same protoplasts used for the above determinations of ψ'.

2. In KCl Solutions
Determinations were made after 3 h of incubation in KCl solutions. For protoplasts the solution contained 0.5 M sucrose, 0.5 M mannitol and the enzyme solution containing 0.19 mM K^+. Increasing levels of KCl (Fig. 1) cause a depolarization in both cells and protoplasts (for all potentials).

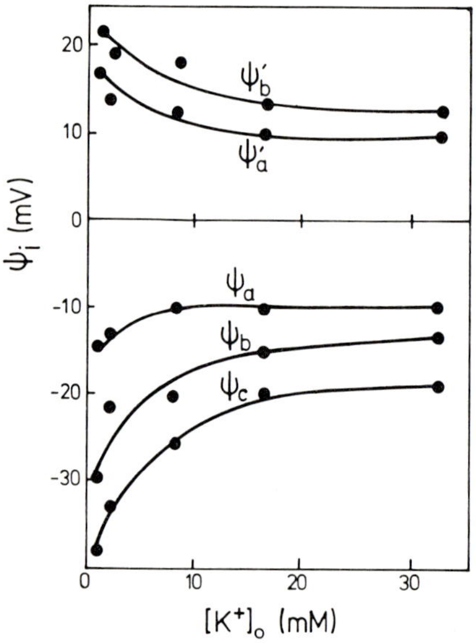

Fig. 1 Potentials measured in cells (ψ_a, ψ_b, ψ_c) and in protoplasts (ψ'_a, ψ'_b) after 3 h incubation in different KCl solutions.

II. Intracellular Distribution of Potassium

The efflux curves of $^{42}K^+$ into KCl solution have been analysed and show three components:

$$Y = Y_1 + Y_2\ (1 - e^{-\lambda_2 t}) + Y_3\ (1 - e^{-\lambda_3 t})$$

where λ_2 = 0.25 to 0.36 h^{-1}, λ_3 = 0.021 to 0.024 h^{-1}, and Y_1, Y_2 and Y_3 are respectively equal to 14 to 21 %, 48 to 69 %, and 10 to 37 % of the initial radioactivity in the cell.

We have interpreted these results, using PITMAN's (1963) compartmental analysis, in terms of three compartments in series and in a quasi-stationary state. The validity of the model has been verified by complementary experiments. A study of efflux into pure sucrose supported our adoption of a series disposition (wall, cytoplasm, and vacuole). A net efflux of K^+ during $^{42}K^+/K^+$ exchange is always present. It is greatly reduced by the addition of Ca^{2+} to the medium.

Table 1 gives the concentrations in the three compartments calculated from their contents (determined by compartmental analysis) and their volumes as evaluated from measurements of the mean cell diameter and the thickness of the wall and of the cytoplasm. The latter, as determined by electron microscopy, is about 1 μm, which is a minimal estimate because of shrinkage during fixation. The real value is probably between 1 and 2 μm. Table 1 also gives the corresponding values of the activities. Activity coefficients were estimated by the Debye-Hückel equation.

Table 1

K^+ concentrations, c, activities, a, (in mM), and electrical potentials ψ (in mV). The concentrations c_w and c_c are calculated from the mean thicknesses of the wall and cytoplasm of 1.5 μm; activities are based on minimal and maximal estimations of the volumes of the cell-wall and the cytoplasm, and on a mean estimation of the vacuolar volume. Suffixes w, c, v refer to wall, cytoplasm and vacuole, respectively. c = 10, a =9, and ψ = 0 in the medium.

Concentrations		Activities			Potentials
c_w	69 ± 3	a_w	min	29 ± 2	ψ_1 −18 ± 4
c_c	266 ± 36	a_w	max	54 ± 2	ψ_2 −41 ± 1
c_v	79 ± 13	a_c	min	87 ± 15	ψ_3 −31 ± 3
		a_c	max	170 ± 23	
		a_v		54 ± 9	

D. Discussion

I. Electrical Potentials

1) The gradual penetration of the microelectrode into a free cell leads to the detection of three distinct values of electrical potential, all of which are negative with respect to the medium. MORVAN (1972) considers them as the parietal, cytoplasmic, and vacuolar potentials respectively. We do not agree with this interpretation. The diameter of the microelectrode tip is of the same order of magnitude as the thickness of the cell wall or the cytoplasm (1 μm). The cell is not fixed rigidly and the cell wall is somewhat indented under the pressure of the microelectrode. It seems highly probable that, when the cell wall is pierced and springs back to its convex shape, the tip of the microelectrode penetrates the cytoplasm too rapidly for its potential to be measured. We therefore suggest that the three potential values recorded correspond to, first, the external surface, then the 'superficial zone', i.e. the bulk of the cell wall itself which is difficult to pierce, and finally the vacuole.

On the other hand the two values measured in protoplasts, where penetration is easy, may be attributed to the cytoplasm and the vacuole; this would bring our results into agreement with the well-established findings in most plants, where ψ_v is more positive than ψ_c.

2) The vacuolar potential, measured with respect to the medium, is negative in a cell and positive in a protoplast. It is indeed negative in a protoplast that is still surrounded by a cell wall, but changes sign rapidly when the shell is torn off by micromanipulation. When the microelectrode tip lies between the wall and a shrunken protoplast it indicates a negative potential which gradually disappears as the wall is ruptured. The potential difference between the medium and the vacuole therefore includes a considerable negative component of parietal origin. This component could be a diffusion potential arising from the presence of negative sites in the cell-wall which lower its permeability to anions. This is in keeping with the fact that the value is more negative in the cell wall than on its outer surface. It is, however, too early to speculate about the origin of positive potentials in protoplasts.

3) The dependence of the potentials in cells on external KCl concentration involve ψ_a and $(\psi_b - \psi_a)$ but practically never $(\psi_c - \psi_b)$. For protoplasts the dependence is essentially of ψ'_a, rarely $(\psi'_b - \psi'_a)$. It is, moreover, opposite for cells and protoplasts, which is a further argument against attributing ψ_b to the cytoplasm (Fig. 1). Hence we suggest the hypothetical scheme of Fig. 2 for the association of the potential components with the zones in the cell. Similar conclusions have been reached with cell suspensions and protoplasts of ***Rubus fructicosus*** L. and of ***Vitis vinifera*** L. crown-gall. In all three species the vacuolar potential is positive with respect to the cytoplasmic potential and the presence of the cell-wall makes both these values negative.

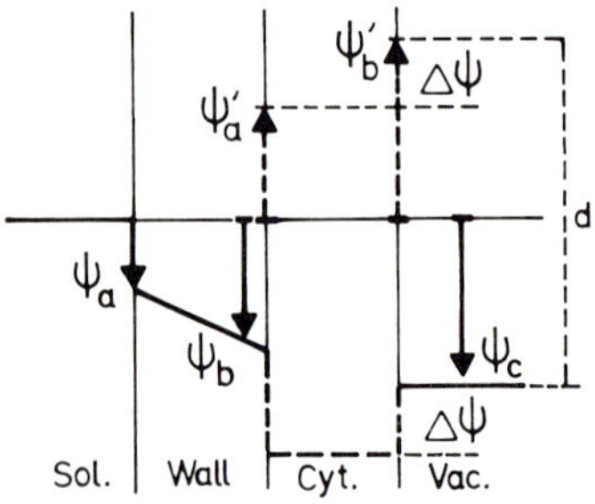

Fig. 2 Assumed profile of the electrical potential between the external medium and the vacuole. Solid lines represent values obtained by directly measuring ψ_a, ψ_b, and ψ_c in cells. Dashed lines represent those deduced from the difference $\Delta\psi = \psi'_a - \psi'_b$ in protoplasts. d illustrates the cell-wall effect.

4) The profile of ψ proposed in Fig. 2 has been recently directly observed using stiffer microelectrodes, whose movement was easier to control. We have obtained proof that the intermediate compartment (cytoplasm) has the lowest potential.

II. Estimations of Electrochemical Potentials

The calculation of the electrochemical potentials $(\bar{\mu})$ for K^+ in the various phases is somewhat difficult because of uncertainties in the thicknesses of the cell wall and the cytoplasm and in the activity coefficients, particularly in the cell wall and cytoplasm. ($\bar{\mu}$ is given by $RT\ln a + zF\psi$, where a is the ion activity and the other symbols have their usual meaning). From the data of Table 1, $\bar{\mu}$ has been calculated for the three cell phases — wall, cytoplasm and vacuole — taking the external medium as the zero of electrochemical potential. The values are relatively low: 0.26 to 0.60 for the wall; 0.40 to 0.75 for the cytoplasm; 0.26 for the vacuole, all in kcal $mole^{-1}$. The uncertainties arise largely because of the above-mentioned doubts about the volumes of the cell wall and the cytoplasm. The $\bar{\mu}$ values in the cytoplasm and wall are possibly even smaller than the above, for the activity coefficients in these phases have probably been overestimated. We thus conclude that the K^+ distribution approaches electrochemical equilibrium, showing perhaps a slight internal excess which could be the cause of the net flux observed. (Ussing-Teorell flux-ratio analysis is compatible with this conclusion and suggests that the plasmalemma transport of K^+ is passive.)

For the protoplasts, since the potentials are positive, electrochemical equilibrium no longer holds. $\bar{\mu}$ for the cytoplasm is 1.32 kcal $mole^{-1}$ and for the vacuole 1.25 kcal $mole^{-1}$.

In our opinion there is no doubt that the thermodynamic equilibrium of K^+ in the cell (cytoplasm and vacuole) is conditioned to a great extent by the presence of the pecto-cellulosic wall which, with its numerous anionic sites, creates an environment for the cytoplasm and vacuole which is different from the external solution (GRIGNON and LAMANT, 1973).

The ionic equilibrium that we have described may be peculiar to cells under our experimental conditions and cannot be indiscriminately extended to cells in culture. In general, we always observe considerable permeability modifications which increase the efflux of K^+ and the influx of Na^+. Inhibition of the mechanism of absorption of sulphate (THELLIER, personal communication) seems to occur as a result of removing cells from their culture vessel. It is therefore quite possible that higher values of $\bar{\mu}$ than those we have calculated exist under culture conditions, but this does not affect our conclusions on the importance of the wall, which is at least partly responsible for the accumulation of cations in the cytoplasm and the vacuole.

E. References

GRIGNON, C., Absorption du potassium par des suspensions cellulaires d'Erable *Acer pseudoplatanus,* L. These 3° cycle, Paris (1969)
GRIGNON, C., LAMANT, A., C.R. Acad. Sci. Paris (D), **276**, 1685–1688 (1973)
MORVAN, H., Mesure du potentiel intracellulaire des suspensions cellulaires d'Erable (*Acer pseudoplatanus*, L.). These 3° cycle, Lille (1972)
PITMAN, M.G., Aust. J. Biol. Sci. **16**, 647–668 (1963)
RONA, J.P., C.R. Acad. Sci. Paris (D), **277**, 185–188 (1973)
RONA, J.P., GRIGNON, C., C.R. Acad. Sci. Paris (D), **274**, 2976–2979 (1972)

Round Table Discussion 4

Chairman: L.F. Jaffe

The important technical question of the exact location of a microelectrode tip in a plant cell occupied the first part of this discussion. WALKER opened with his 'dogma': you can't push a sharp (fine) electrode through a membrane unless that membrane is pressed up against a wall. In other words, although one can get a coarse electrode through the cell wall and into the vacuole, through the tonoplast, by a kind of explosive 'pop', a fine electrode, once in the cytoplasm, cannot be pushed through the tonoplast, which simply wraps itself around the tip. DAVIS and HIGINBOTHAM both commented on this statement, particularly with respect to inserting microelectrodes into higher plant cells. Although one can't exactly see where the tip is, evidence from the existence of two relatively stable potentials appearing as the tip penetrates the cell, the second one being 8 to 12 mV more positive than the first, convinces HIGINBOTHAM, in particular, that he does get into the vacuole of his higher plant cells. LÜTTGE then talked about some experiments he and his colleagues had carried out on the moss, ***Mnium***. By centrifugation they can get the cytoplasm to one end of the cells and hence insert a microelectrode into this cytoplasm which gives a potential of about −140 mV. An electrode apparently inserted into the vacuole, now at the other end of the cells, gives the same potential. LÜTTGE thus believed, despite being challenged by JAFFE, that the PD across the tonoplast is close to zero, and that at least they had obtained a good answer for the cytoplasmic potential. VREDENBERG then described an experiment with ***Nitella***. He inserted two electrodes; the first was driven by a 'rapid pulse' into the vacuole and the second put into the cytoplasm. He found, in agreement with SPANSWICK, that the electrode he thought was in the vacuole was 15 mV more positive than the electrode he thought was in the cytoplasm. Now applying a rapid pulse to the cytoplasm electrode, thus putting it, hopefully, into the vacuole, led to it recording the same potential as the first, 'vacuolar', electrode.

VREDENBERG was therefore convinced that his electrodes were, in fact, where he thought them to be. GILLET said that in his experience with ***Nitella***, a glass microelectrode seems to go into the vacuole whereas a silver microelectrode remains covered with cytoplasm. HANSEN described some experiments using external electrodes with ***Nitella*** which could be used as a check on the position of an inserted microelectrode tip; these experiments were performed by KEUNECKE. Measuring the resistance of a cell using two external electrodes is equivalent to measuring the resistance between an electrode in the vacuole and an external electrode, because most of the current flow between two external electrodes is along the vacuole. Thus one can check, from a resistance measurement comparison, whether an inserted microelectrode tip is in the vacuole or not. Apparently, sometimes it is and sometimes it isn't.

MERTZ described a piezoelectric device, developed by HIGINBOTHAM and his colleagues at Washington State University, which ensured the precise placing, at high speed, of a microelectrode tip exactly where it is desired in a plant cell. He naturally felt, although SLAYMAN had reservations, that this device was a technical answer to the problem of inserting a microelectrode into the chosen phase of a plant cell.

There was some discussion between DAVIS, JAFFE, GRADMANN, and WALKER about setting up criteria for finding out where microelectrode tips were inside a plant cell. It seemed to be finally agreed that one would need potential and resistance measurements made with more than 2 electrodes before one could be satisfied about the location of one of these electrode tips.

The discussion next moved on to the subject of the inter-relations between membrane potentials, H^+ concentration gradients and transport rates. WALKER asked: what are the kinetics of a carrier system which is sensitive to both the electric potential and to the H^+ concentrations? SLAYMAN thought there was a treatment of this problem in a recent issue of the Journal of Mathematical Biophysics. He pointed out that the depolarization of ***Neurospora***, caused by 1 mM glucose or 3—O—methyl glucose, by about 150 mV is good evidence that the carrier is partly driven by a voltage gradient. TANNER pointed out that if the proton-sugar cotransport in ***Chlorella*** was driven by a pH gradient, only, then there would have to be at least 3 pH units difference between inside and outside the cell; he believes that the pH difference is only about 1 unit and thus there must be an electric potential component of the driving force. About 100 mV of PD, inside negative, would be needed and BARBER agreed, from his own direct measurements, that such a PD probably existed.

HÖFER introduced another organism, the obligatorily aerobic yeast ***Rhodotorula gracilis***, possessing an active sugar-transporting membrane system. He postulated, in a model of this system which shares some of the properties of ***Chlorella***, that the free carrier can only move through the membrane when it is 'energized'; this would mean that after the carrier has taken sugar across the membrane into the cell, it must be protonated again to travel back to the outside face of the membrane. KOMOR said that in the ***Chlorella*** system the mobility of the protonated unloaded carrier is low whereas the mobility of the protonated sugar-loaded carrier is very high. They tentatively suggest a sterol binds to the protonated carrier, preventing its movement; when sugar is bound to the protonated carrier a conformational change of the protein takes place resulting in a lowering of the binding to the sterol and hence the complex can move, driven by the membrane potential; nystatin can weaken the binding to the sterol, thus enabling the free protonated carrier to move and the observed 'overshoot' results. In answer to a question by HÖFER, KOMOR said that the unprotonated sugar-carrier complex can move if the pH is very high, but at normal pH's the proton-loading process is much faster than sugar loading and the movement of sugar without protons is not seen. SLAYMAN would interpret KOMOR's nystatin experiment quite differently; he suggested that nystatin simply short-circuits the membrane potential against which the loaded carrier has to work to come back out. He also said that in ***Neurospora*** the singly-loaded carrier, with sugar alone, can cross the membrane; the evidence is that there is less current flow across the membrane than would result from a one to one stoichiometry between glucose and H^+. TANNER replied that their data also show that the unprotonated carrier, with sugar, is able to influx quite well for, at alkaline pH's, the V_{max} for sugar entry is not very different from the V_{max} at acidic pH's. As for the nystatin-sterol hypothesis, it is very tentative.

JAFFE suggested that the discussion now move more to the connection between the energy sources, particularly photosynthesis, and ion movements. SMITH brought out the spatial problem, e.g. in ***Characean*** cells the energy is produced in the chloroplasts and the Cl^- pump is at the plasmalemma. The spatial connection is not easy to envisage, although it is more difficult to get reducing power out of the chloroplasts than ATP. And the pH gradient hypothesis, for the Cl^- pump, is no help for this depends on the active transport of H^+ which still needs ATP or reducing power. SMITH's remarks about the spatial problems were not really taken up by the discussants, but the question of the nature of the energy source was vigorously debated. RAVEN raised the question of the stoichiometry of ATP breakdown to H^+ transported; some people, for chloroplasts and mitochondria, say $2H^+$/ATP; Witt's group says maybe $3H^+$/ATP. On the other hand KOMOR and TANNER's data suggest that the ratio is only one and this would mean that they have a Mitchell's type 1 ATPase. SLAYMAN and RAVEN agreed that, since in ***Neurospora*** there is about 310 mV per ATP, there may, just, be a type 2 ATPase in this organism.

JESCHKE said that in ***Elodea*** both Cl^- and K^+ uptakes are similarly inhibited by CCCP, in the light, but one must be careful not to say that this proves that ATP is the energy source. He also has measured decay of both K^+ and Cl^- uptakes after switching ***Elodea*** from light to dark. The same $T_{1/2}$ of about 0.5 min was obtained and this is similar to SPANSWICK's decay time of the light-stimulated electric potential hyperpolarization; connections between these phenomena are thus indicated. LÜTTGE commented that the nature of the energy source is somewhat fugitive. He in fact has the feeling that most cells use whatever kind of energy source they happen to have for ion transport. JESCHKE then pointed out that it is quite frustrating that the levels of ATP pools do not change between light and dark and he asked whether a process can be powered by an energy source which does not change. RAVEN took this point up by saying that the carbon cycle people had long since given up trying to get 100 % increases in rates of processes from 5 % increases in the ATP or NADPH levels. We should follow them and begin looking for non-energetic signals from the thylakoids such as proton or Mg^{2+} levels which might activate enzymes. He did not think appealing to concepts of energy charge was of much help. JAFFE, much interested in the possibility of a non-energy signal, suggested that an effective signal might be a small change, in absolute terms, in the level of Ca^{2+}, an ion which is maintained at a very low, free, level normally. JAFFE, RAVEN, and SMITH argued whether or not the Ca^{2+} level was a better candidate than the H^+ level, the latter two preferring the H^+ level which is not buffered in any static sense. HODGES, however, said that oat root plasma membrane ATPase is very sensitive to Ca^{2+}. HANSEN supported RAVEN's idea of a signal; he quoted experiments by his group which showed that for the first 3 min or so after switching on light, active transport in ***Nitella*** is slowed down, presumably in response to a signal, and the ATP level therefore rises.

SLAYMAN stressed the necessity to know something about the compartmentation of ATP. Even if the cell ATP level stays constant there may be an increase in the chloroplast ATP level with a concomitant decrease in the cytoplasmic ATP level. However BORNEFELD said that in ***Anacystis***, where the thylakoids are lying in the cytoplasm, the ATP level is very stable. It remains the same at low and high temperatures, at different pH values, and whether or not photosynthesis is taking place. LEIGH commented that, from his experiments on maize roots, the turnover time (about 1 s) for ATP is so small that it would be difficult to determine changes in the ATP level; the cell can keep its ATP level constant because it can turn over ATP so rapidly. SIMONIS, supporting LEIGH's comments, said that in ***Anacystis*** the ATP level does change from a low, dark, level to high, light, level within 1 s. After it has reached its high value in the light, the ATP level stays constant. Also, inhibiting phosphate uptake results in a high ATP level and therefore the ATP level ***per se*** cannot be the cause of phosphate uptake. LÜTTGE then told of his experiments on the greening of etiolated leaves of barley; during this greening, for 4 h, the ATP level and the Cl^- flux increase together, with a good correlation between them.

After KAUSS had said that energy charge, not ATP level, is the parameter to look at, MARRE spoke about the relevance of the concept of energy charge to the problem of the energy source for H^+ extrusion in auxin-stimulated growing cells from etiolated plants. In this system, the ATP levels sometimes increase and sometimes decrease and the energy charge, usually, remains unchanged. In his opinion any energy charge above 0.8 is alright. He, as others, stressed that it is impossible to draw conclusions about whether ATP is necessary for active transport, from ATP levels. Finally, SLAYMAN pointed out that energy charge is not really relevant because it was developed by Atkinson with respect to regulatory mechanisms connected with maintaining nucleotide levels. The important parameter, in his opinion, for the question of how much energy is available from ATP for a transport process, is the phosphate potential, i.e. $[ATP]/[ADP]\,[P_i]$.

Session 5

Transport in Isolated Chloroplasts

Chairman: H. Ziegler

Proton Transport and its Relation to Energy Conservation in Isolated Chloroplasts

M. Avron

A. Introduction

Chemiosmotic coupling (MITCHELL, 1966; JAGENDORF, 1967) predicts that oxidation-reduction energy is converted via H^+ movements into energy stored in the form of a H^+ gradient and a membrane potential. These secondary forms can, in turn, serve as the driving force for ATP formation.

During the last few years we have concentrated our efforts in designing and utilizing techniques to enable us to measure accurately the size of these H^+ gradients and membrane potentials under a variety of conditions (ROTTENBERG et al., 1971, 1972; SCHULDINER et al., 1972a). It was found that under conditions optimal for ATP synthesis, H^+ concentration gradients as large as 10 000 ($\Delta pH = 4$) could be measured, but only insignificant membrane potentials were observed. This seemed in agreement with our earlier results (KARLISH et al., 1968, 1969) which indicated that agents which would be expected to completely abolish a membrane potential, had one existed, were ineffective in decreasing the efficiency of ATP production by isolated chloroplasts.

In this report I will describe some of our most recent work on the ability of chloroplasts to utilize externally imposed membrane potentials as the driving force for ATP formation, and on a more quantitative study of the relation of ATP formation to the size of ΔpH across the thylakoid membrane.

B. Methods

All methods have been described in full in the references indicated.

C. Results and Discussion

The elegant experiments of JAGENDORF and collaborators (JAGENDORF, 1967) have clearly illustrated that chloroplasts are capable of utilizing energy in the form of H^+ gradients as a driving force for ATP formation in the absence of any other independent energy source. Our initial attempts to utilize energy stored in the form of a membrane potential as a driving force for ATP formation have resulted in failure. However, when such an externally produced membrane potential was superimposed on a suboptimal H^+ gradient, chloroplasts clearly could use the membrane potential energy as a driving force for ATP formation (SCHULDINER et al., 1972b, 1973). As can be seen in Table 1, it was immaterial whether the suboptimal H^+ gradient was achieved by lowering the light intensity, or by changing the pH of the dark stage. In either case, the imposition of a membrane potential by including KCl+valinomycin in the dark stage, resulted in a marked increase in the observed yield of ATP. It should be noted (Table 1) that this increase in yield was not observed when the H^+ gradient was itself maximal. Similar observations were also reported by URIBE and LI

Table 1

Membrane potential as a driving force for ATP formation by post-illumination ATP synthesis in chloroplasts.
From SCHULDINER et al. (1972b). The reaction mixture of the light stage contained, in a final volume of 1.2 ml: sorbitol, 100 mM; choline-chloride, 66 mM; pyocyanine, 15 μM; phenylene-diamine-di HCl, 5 mM, at pH 6.5; MES-Tris, pH 6.5, 5 mM; and chloroplasts containing 160 μg chlorophyll. In the dark stage the volume was 1.5 ml (final volume 2.7 ml) and the final concentrations (in 2.7 ml): $Mg(H_2PO_4)_2$, 0.8 mM (containing $1.6 \cdot 10^7$ cpm ^{32}P); ADP, 1 mM; Tris-HEPES, 5 mM; choline chloride or KCl, 100 mM; valinomycin, where indicated, 3 μM. Final pH was adjusted by addition of Tris base. Chloroplasts were illuminated 2 min at the indicated intensity before injection into the dark reaction mixture. 15 s later the reaction was stopped by the addition of trichloroacetic acid to a final concentration of 3 %, and the contents assayed for the ATP formed.

pH			ATP formed	
Light stage	Dark stage	Light intensity (erg cm^{-2} s^{-1})	Choline chloride	KCl + valinomycin
6.5	8.5	$3.5 \cdot 10^5$	154	160
6.5	8.5	$9.0 \cdot 10^3$	9	29
6.5	7.5	$3.5 \cdot 10^5$	61	144
6.5	7.5	$9.0 \cdot 10^3$	3	26

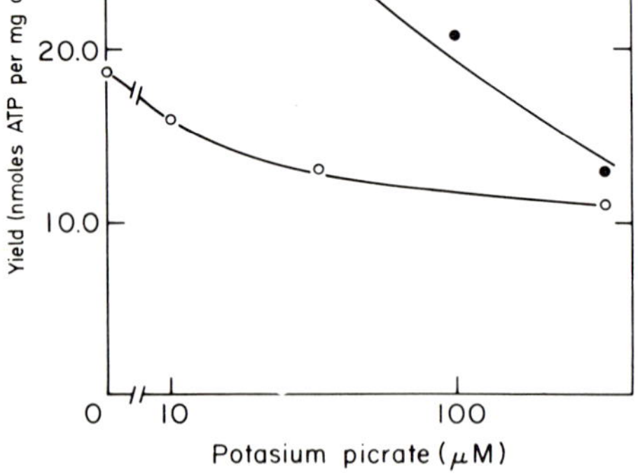

Fig. 1 Inhibition of the membrane potential driven ATP formation by picrate. From SCHULDINER et al. (1973). Details as described under Table 1, except for the addition of picrate to the dark stage, at the indicated final concentrations.

(1973). In order to convince oneself that these stimulations are indeed due to the imposition of a membrane potential, we have run two types of control experiments. In one the chloroplasts were preincubated with KCl, so that no significant K^+ movements would be expected during the light to dark transition, and so no membrane potential should develop. Under these conditions no increase in ATP formation was observed (SCHULDINER et al., 1972b). Alternatively a permeable anion, picrate, was introduced together with the K^+. Under these conditions one would expect the movement of potassium picrate, rather than K^+, during the light-dark transition and therefore no membrane potential should be formed. As can be seen in Fig. 1, the introduction of a sufficient concentration of picrate indeed fully inhibited the increase in ATP formation due to the K^+ + valinomycin (SCHULDINER et al., 1973).

Thus, we can conclude that chloroplasts have the capability to use energy stored either in the form of H^+ gradients or membrane potentials as a driving force for ATP formation.

We can now turn our attention to the relation between the magnitude of these potentials and the ability to synthesize ATP.

First, we may ask, does phosphorylation vary the magnitude of the measured ΔpH values? If the energy stored in the H^+ gradient indeed drives ATP formation, it would be expected that ATP formation would create an additional drain on the size of ΔpH and so decrease its magnitude. Fig. 2 indicates that this is clearly the case (PICK et al., 1973). When ADP is

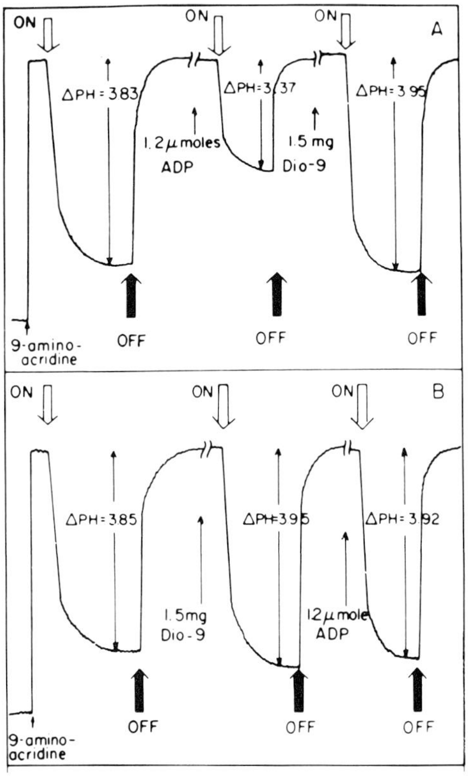

Fig. 2 Decrease in the magnitude of ΔpH by photophosphorylation. From PICK et al. (1973). The reaction mixture contained, in a total volume of 3.0 ml: Tricine, pH 8.6, 50 mM; KCl, 20 mM; $MgCl_2$, 5 mM; arsenate, 1 mM; pyocyanine, 5 μM; 9-aminoacridine, 1 μM; and chloroplasts containing 30 μg of chlorophyll. Where indicated, ADP, 0.4 mM; and Dio-9, 0.5 mg/ml, were added.

added to an otherwise complete photophosphorylating system, a marked decrease in the magnitude of ΔpH is observed. This decrease is reversed or prevented by the post- (Fig. 2A) or pre- (Fig. 2B) addition of the energy transfer inhibitor Dio-9.

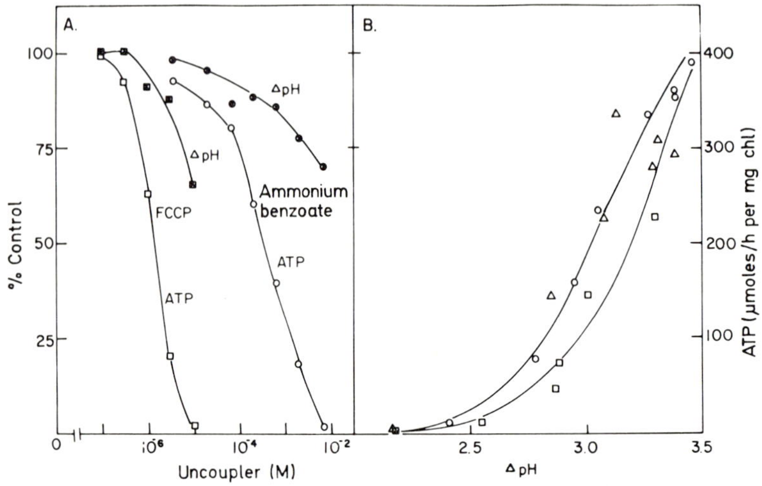

Fig. 3 Dependence of the rate of ATP formation on the magnitude of ΔpH varied by uncouplers.
The reaction mixture contained in a total volume of 3.0 ml: Tricine, pH 8.0, 60 mM; magnesium acetate, 2 mM; inorganic phosphate, 1 mM (containing 6 · 10^6 cpm ^{32}P); ADP, 1 mM; pyocyanine, 30 μM; 9-aminoacridine, 1 μM; chloroplasts containing 20 μg of chlorophyll, and the uncouplers at the concentrations indicated (A). ATP formation and ΔpH were determined. In B, the rate of phosphorylation is plotted as a function of ΔpH for ammonium benzoate (○), FCCP (□) and dianemycin (△).

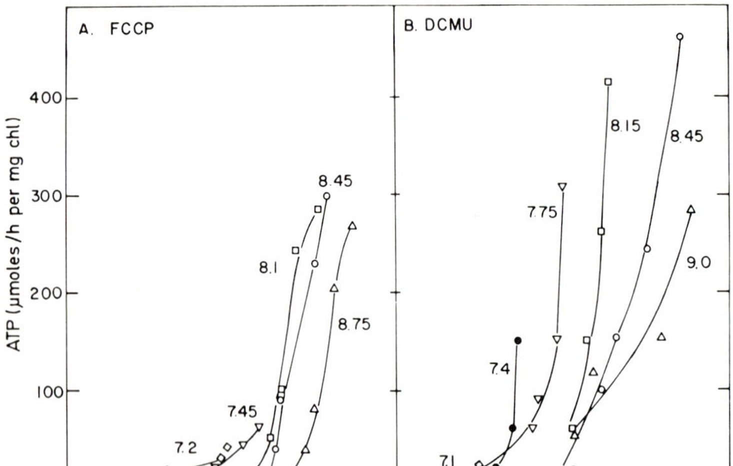

Fig. 4 Dependence of the rate of ATP formation on the magnitude of ΔpH at different initial pHs.
Details as described under Fig. 3. The initial pH is indicated by the numbers next to the curves. ΔpH and the rate of ATP formation were varied in A by varying the concentration of FCCP between 0.1–2 μM, and in B by varying the concentration of DCMU between 3–600 nM.

Second, we may ask, does the rate of ATP formation depend on the magnitude of ΔpH, and what is the form of that dependence? Fig. 3A illustrates the effect of two uncouplers on the rate of ATP formation and the magnitude of ΔpH. Fig. 3B shows this data plotted to show the rate of ATP formation as a function of ΔpH. It is clear that with all 3 uncouplers tested, two phases can be clearly distinguished: (a) in the range of ΔpH up to a threshold value of about 2.5, no ATP formation was observed, and (b) above this threshold value there was a sharp dependence of the rate of ATP formation on the magnitude of ΔpH. The data obtained by varying several other parameters (light intensity, electron transport inhibitors, pH of light or dark stage in post-illumination experiments and ΔpH in acid-base experiments) resulted in a similar dependence of ATP formation on ΔpH. Fig. 4 illustrates the effect of an electron transport inhibitor (DCMU) and an uncoupler (FCCP) in this kind of plot when measured at different initial pH's. Under all the tested pH's the phenomenon of a threshold followed by a sharp dependence of ATP formation on ΔpH is apparent. In addition these types of experiments permit one to analyse the dependence of ATP formation on medium pH at constant ΔpH. It is clear from Fig. 4 that at constant ΔpH phosphorylation is maximal at the lower pH range, around pH 7. Thus, one may conclude that the optimal pH of around 8 commonly observed for photophosphorylation is the result of two opposing effects: (a) the optimum of the phosphorylation reaction at constant ΔpH which is around pH 7 and (b) the maximal ΔpH which is observed around pH 9 (SCHULDINER et al., 1972a).

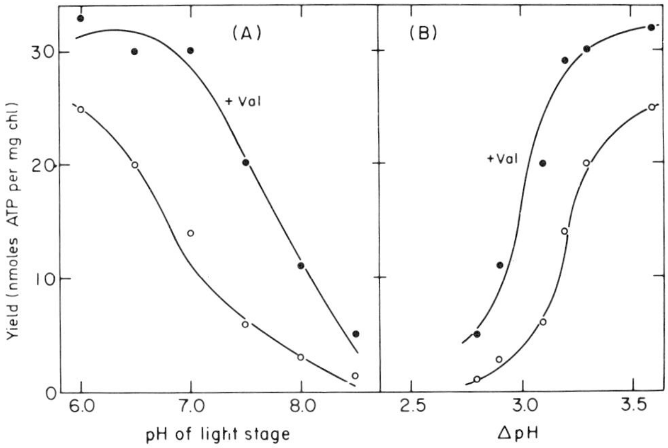

Fig. 5 The effect of a membrane potential (+val) on the ΔpH dependence of ATP formation. From SCHULDINER et al. (1973).
Details as described under Table 1, except for the pH of the light stage which was as indicated, and the light intensity which was 9000 erg cm^{-2} s^{-1}. ΔpH plotted equals the difference between the chloroplasts' internal pH in the light stage and the pH of the dark stage (8.5).

We can now return to the effect of a membrane potential on the yield of ATP formation, and ask how does the effect affect the ΔpH dependence of ATP formation. Fig. 5 shows such an analysis, where ΔpH was varied by varying the pH of the light-stage in a post-illumination experiment. In Fig. 5B the data are plotted to show the yield of ATP as a function of ΔpH (SCHULDINER et al., 1973). As can be seen, when a membrane potential was superimposed (+val.), the curve shifted towards lower ΔpH values resulting in a lower threshold value. Thus, we may conclude that the superposition of a membrane potential lowers the ΔpH value required for a given rate of ATP synthesis.

Finally, the question should be asked, are the values of ΔpH measured sufficient to drive ATP synthesis by the proposed chemiosmotic mechanism (MITCHELL, 1966)? In order to form a molecule of ATP by the transfer of 2 H^+, ΔpH must exceed 5.5, when no membrane potential exists, and if we accept the measured phosphate potential in chloroplasts of 17 kcal/mole (ROTTENBERG et al., 1972). Since the latter is a maximal value, and most of the reported experiments were performed under a lower phosphate potential, we designed an experiment to test whether the ΔpH dependence of ATP formation is sensitive to changes in phosphate potential. Fig. 6 illustrates the results of such an experiment, where ΔpH was controlled by varying the light-intensity, and the phosphate potential was changed by a factor of 400 by varying the ADP/ATP ratio. It is clear that the dependence of the rate of ATP formation on the magnitude of ΔpH was only marginally, if at all, dependent on the phosphate potential during the reaction. We must conclude, therefore, that the ΔpH values of 3–4 which permit measurable rates of ATP formation, are thermodynamically insufficient for a movement of 2 H^+ to provide sufficient energy for the synthesis of an ATP molecule. Nevertheless, it is amply clear that the energy stored in the form of H^+ gradients or membrane potentials can be used by chloroplasts to drive ATP formation.

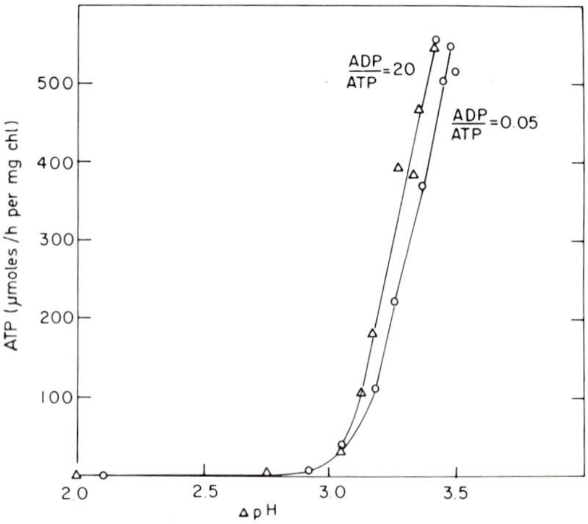

Fig. 6 Insensitivity of the phosphorylation dependence on ΔpH, to changes in the phosphate potential. The reaction mixture contained in a total volume of 3.0 ml: Tricine, pH 8.5, 30 mM; KCl, 30 mM; $MgCl_2$, 5 mM; inorganic phosphate, 1 mM (containing $1 \cdot 10^7$ cpm ^{32}P); $K_3Fe(CN)_6$, 0.2 mM; 9-aminoacridine, 1 μM; and chloroplasts containing 20 μg chlorophyll. The initial concentration of ADP was 0.2 mM and of ATP 4 mM or 0.01 mM. The rate of phosphorylation and the size of ΔpH were varied by changing the light intensity from 10^2 to $4 \cdot 10^5$ erg cm^{-2} s^{-1}.

The answer to this dilemma may lie either in introducing a variation in the chemiosmotic hypothesis which will permit the coupled transfer of more H^+ per electron per site, and so a higher H^+ to ATP ratio, or in suggesting that the H^+ gradients and membrane potentials are energy storage devices in equilibrium with another, yet undefined, high-energy intermediate state, which directly drives the synthesis of ATP.

D. References

JAGENDORF, A.T., Fed. Proc. **26**, 1361–1369 (1967)
KARLISH, S.J.D., AVRON, M., FEBS Letters **1**, 21–24 (1968)
KARLISH, S.J.D., SHAVIT, N., AVRON, M., Eur. J. Biochem. **9**, 291–298 (1969)

MITCHELL, P., Chemiosmotic Coupling in Oxidative and Photosynthetic Phosphorylation. Glyn Research Ltd., Bodwin 1966
PICK, U., ROTTENBERG, H., AVRON, M., FEBS Letters **32**, 91–94 (1973)
ROTTENBERG, H., GRUNWALD, T., AVRON, M., FEBS Letters **13**, 41–44 (1971)
ROTTENBERG, H., GRUNWALD, T., AVRON, M., Eur. J. Biochem. **25**, 54–63 (1972)
SCHULDINER, S., ROTTENBERG, H., AVRON, M., Eur. J. Biochem. **25**, 64–70 (1972a)
SCHULDINER, S., ROTTENBERG, H., AVRON, M., FEBS Letters **28**, 173–176 (1972b)
SCHULDINER, S., ROTTENBERG, H., AVRON, M., Eur. J. Biochem. **39**, 455–462 (1973)
URIBE, E.G., LI, B.C.Y., Bioenergetics **4**, 435–444 (1973)

Proton Transport and the High-Energy State in Isolated Chloroplasts

A.B. Hope and W.S. Chow

A. Introduction

As is well known, MITCHELL's theory (1966) for the mechanism of ADP phosphorylation postulates that a 'high-energy state', equivalent to an electrochemical potential gradient for protons (= proton-motive force, PMF), is an intermediate between electron transport and phosphorylation.

In chloroplasts the thylakoid membranes are the probable sites of the gradient, since H^+ are apparently transported into the intrathylakoid spaces during light-driven electron transport (NEUMANN and JAGENDORF, 1964). Estimates of the pH difference across the thylakoids are 2–3 (RUMBERG and SIGGEL, 1969; SCHULDINER et al., 1972) and the electric potential difference is reckoned to be 100–200 mV (SCHLIEPHAKE et al., 1968; BARBER, 1972).

There is a general correlation between the ΔpH part of the PMF and conditions suitable for photophosphorylation under a variety of conditions: ± uncouplers, electron transport inhibitors, and so on. So far, so good, but the kinetics and stoichiometry of electron transport, H^+ translocation and phosphorylation do not seem to be agreed upon. SCHWARTZ (1971) defends his finding (1968) that the radio of H^+ translocation rate to electron transport rate in the steady-state ($\phi_{H^+,ss}/\phi_{e^-}$) is 2 at pH_o = 7–8 and using ferricyanide or $NADP^+$ as electron acceptors.

Others have found ratios less than this, and that the ratio varies with pH_o. Recently TELFER and EVANS (1972) found $\phi_{H^+,ss}/\phi_{e^-} = 1$ using methyl viologen and 'correcting' in an elaborate way the lag in response of their glass pH electrode. The 'initial ratio', calculated from rates found when the light is first switched on may be higher than the steady-state ratio. At least that $\phi_{H^+,init.} > \phi_{e^-,ss}$ is agreed upon. There is disagreement about the change in electron transport rate in the first few seconds of the light period.

The nearest integer to this infamous ratio is taken as indicating the number of coupling sites or points of contact between components of the electron transport chain requiring H^+ and e^- for their reduction, and the external and internal solution phases (Fig. 1). MITCHELL's theory is felt to require an integer ratio but non-integer ratios may still be possible (see below). Occasionally the same ratio, in what seems to be a ***non sequitur***, is taken as indicating the number of coupling sites between the thylakoid ATPase and the electron transport chain components, per quantal photosynthetic unit.

The level of the '1 ms delayed fluorescence' (DF) is also held to be an indicator of the high-energy state (WRAIGHT and CROFTS, 1971; EVANS and CROFTS, 1973). Recently a new technique was devised for directly estimating the ΔpH component of the PMF, as opposed to relying on a measurement of ΔH^+_o. We have undertaken a fresh investigation of the high-energy state using several techniques to make concurrent measurements of $[H^+]_i$, DF, ΔH^+_o, phosphorylation rate and ϕ_{e^-}.

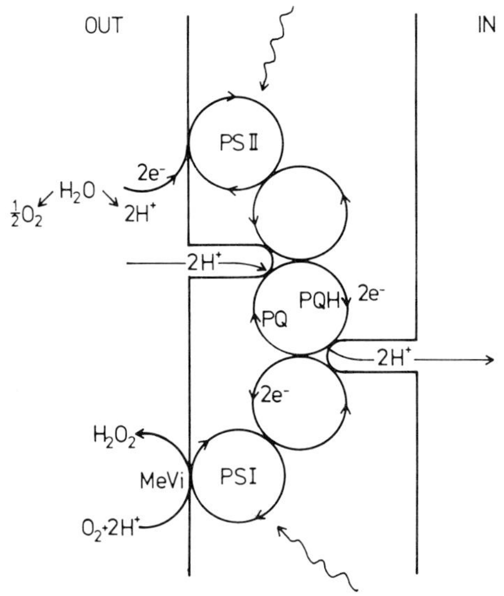

Fig. 1 Partial reactions of photosynthesis, using methyl viologen as electron acceptor, which may lead to a translocation of H^+ from outside to inside a thylakoid membrane during reduction and oxidation of a substance such as plastoquinone ($PQ + 2e^- + 2H^+ \rightleftharpoons PQH_2$). Partly after MITCHELL (1966)

B. Materials and Methods

Chloroplasts were isolated from leaves of spinach grown in a culture medium under fluorescent light supplemented with incandescent light. Just before use, a sample from the chloroplast suspension was diluted in distilled water to rupture the chloroplast envelopes. A reaction medium was then added to give the following final composition: 330 mM sorbitol, 3 mM $MgCl_2$, 0.5 mM K_2HPO_4, 10 mM KCl, 0.15 mM tricine, 1 mM NaN_3, and 25 μM methyl viologen. 1 mM ADP was added when needed. When $K_3Fe(CN)_6$ (1 mM) was used, methyl viologen and NaN_3 were omitted.

The glass electrode (Horiba combination electrode) used had a half-time of response to a step change in pH of less than 0.6 s, and estimates of H^+ translocation rates are either not in error due to the finite speed of response, or else corrections are possible (IZAWA and HIND, 1967); except that the rate of decay of ΔH^+_O in the dark in the presence of some higher concentrations of uncoupler is underestimated.

The light-induced acidification of the intrathylakoid spaces was followed independently of the glass electrode by measuring changes in the level of UV-induced fluorescence from certain amines, mainly N-(1-naphthyl)-ethylene diamine (NED), following actinic light-on or light-off (SCHULDINER et al., 1972). For estimation of $[H^+]_i$ from these observations, the osmotic volume of the chloroplasts per mg chlorophyll must be known. This was estimated using the procedure of ROTTENBERG et al. (1972) based upon the use of ^{14}C-sorbitol and HTO. The mean for control conditions of 6 μl (mg chl)$^{-1}$ is less than found for lettuce by these authors; no difference in osmotic volume between light and dark conditions was apparent. Certain assumptions in the use of the fluorescence quenching of NED, atebrin and 9-aminoacridine to calculate $[H^+]_i$ were considered by SCHULDINER et al. (1972) to be valid. We confirmed that NED in the concentrations used did not uncouple photophosphorylation or alter any other measured parameters. The speed of response of the amine probe for pH_i was measured by setting up a pH difference in a chloroplast suspension using actinic red light and then adding NED to the rapidly -stirred suspension. The half-time for the quenching of the fluorescence signal under these conditions was about 1 s, varying somewhat with pH_O and volume concentration of chlorophyll. The half-time for the relevant pH and chloroplast volume was used where necessary to correct the time-course of change of $[H^+]_i$.

The level of DF induced by sudden unshuttering of red actinic light was measured using a phosphoroscope, photomultiplier tube, phase-lock amplifier and 10 Hz pen recorder.

Measurements of O_2 evolution and changes in H^+ concentration were carried out in the vessel of a Rank oxygen electrode, volume 3.0 ml, those of NED fluorescence quenching in a quartz cuvette, volume 1.5ml, and delayed fluorescence in a quartz sample tube, volume 0.4 ml. Care was taken to use the same (saturating) light intensity, from 150 W quartz-iodine bulbs and RG 630 or 645 Schott low pass filters and water (IR) filters in each parallel experiment, and also to use fast, comparable, stirring rates. However, the delayed fluorescence cuvette was unstirred.

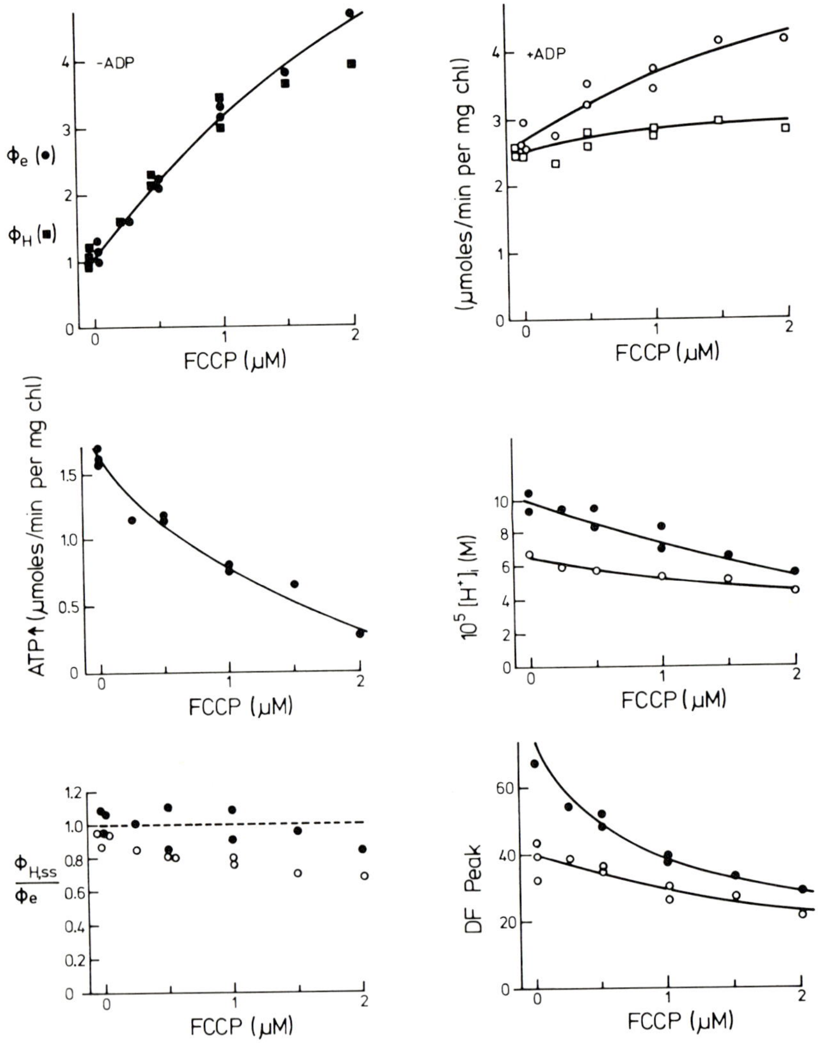

Fig. 2 Effect of FCCP on electron transport rate, ϕ_{e^-}, and H^+ translocation rate, ϕ_{H^+}, both in the steady state, in the absence of ADP (top left) and in its presence (top right); middle left – rates of photophosphorylation; middle right – calculated steady-state internal proton concentrations in the absence of ADP (•) and in the presence of ADP (○); bottom left – the ratio of electron transport rate to H^+ transport rate in the steady-state in the absence of ADP (•) and in its presence (○); bottom right – peak values for 1 ms delayed fluorescence, -ADP (•) and +ADP (○) in arbitrary units.

C. Results

Fig. 2 shows a series of results using the uncoupler FCCP. Other series have been obtained with DCMU, with varied pH_o, valinomycin and nigericin.

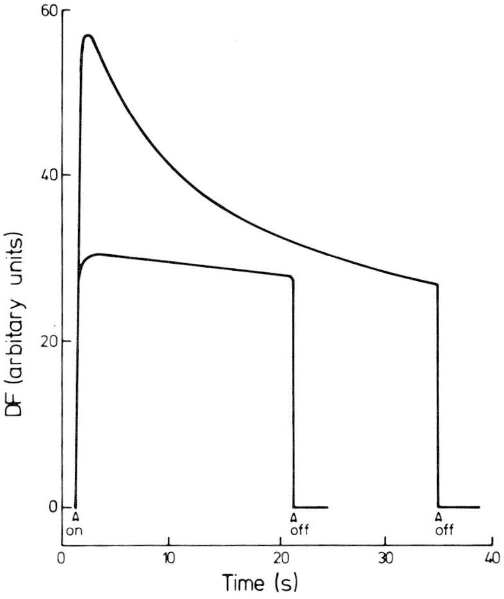

Fig. 3 The 1 ms delayed fluorescence signal -ADP (top curve) and +ADP (bottom curve) from chloroplasts of the same experiment as Fig. 2.

The general shape of the DF induction was that shown in Fig. 3, for -ADP and +ADP conditions. The effect of FCCP was increasingly to reduce the peaked signal for -ADP until it resembled that for +ADP, and both signals took increasing times to reach a peak (see Fig. 4).

Details of the onset and decay of ΔH^+_o and $[H^+]_i$ are shown in Fig. 5A and, using the same data, Fig. 5B shows the shape of plots of $\log_{10} \Delta H^+_o(t)$, $\log_{10} [\![\Delta H^+_o(\infty) - \Delta H^+_o(t)]\!]$, $\log_{10} [H^+]_i(t)$, $\log_{10} [\![[H^+]_i(\infty) - [H^+]_i(t)]\!]$, for decay and onset of the respective quantities.

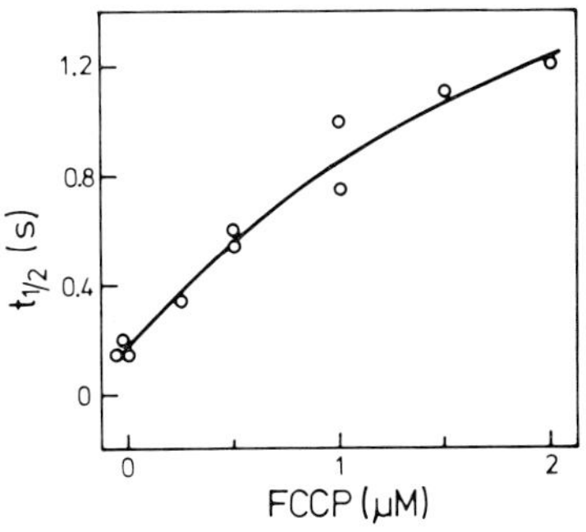

Fig. 4 The rate at which the 1 ms DF signal builds up, in media containing ADP, expressed as the half-time to the peak, as a function of concentration of FCCP. -ADP results were similar.

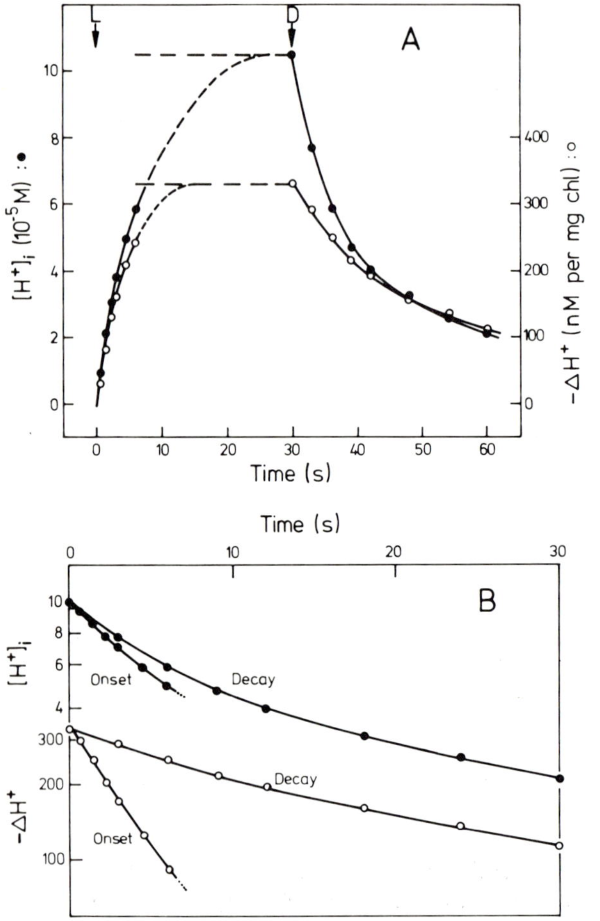

Fig. 5 A. The changes in internal H^+ concentration (•) and the amount of H^+ leaving the external solution (○), as functions of time after illumination commenced (L) or ceased (D).
B. The same data as A replotted on a log scale against time. The units and symbols as for A.

Values of $\phi_{H^+,ss}$ were calculated for control experiments, -ADP, by two methods: from the maximum slope of ΔH^+_o immediately after light off, and from log plots of the decay of ΔH^+_o, which enabled the initial slope to be estimated independently since $(dy/dt)_{t=0} = ky_\infty$ when $y = y_\infty \exp(-kt)$. The estimates for $\phi_{H^+,ss}$ (+ADP) are based on the assumption that the phosphorylation rate does not change instantaneously at the time the light is extinguished. $\phi_{H^+,ss}$ was taken as the sum of the rate of disappearance of H^+ due to phosphorylation just before dark, and the initial rate of decay in the dark (SCHWARTZ, 1968). The mean value of the ratio $\phi_{H^+,ss}/\phi_{e^-}$ for all controls, -ADP, was 1.00 ± 0.28 (S.E., n = 16), when the slope was used to estimate $\phi_{H^+,ss}$. In 13 experiments when $\phi_{H^+,ss}$ was estimated by the two methods mentioned, $\phi_{H^+,ss}/\phi_{e^-} = 0.98 \pm 0.16$ (visible slope) or 1.09 ± 0.16 (slope from decay rate constant). For control experiments with ADP, the ratio was 0.82 ± 0.08 (S.E., n = 13). For [DCMU] = 0.1 –0.25 μM, -ADP, $\phi_{H^+,ss}/\phi_{e^-} = 0.96_5 \pm 0.15$ (6) (visible slope) or 1.07 ± 0.17 (6) (slope from rate constant); for 3 experiments, +DCMU, +ADP, the mean ratio was 0.90.

D. Discussion

I. The Coupling between Light-Induced Electron Transport Rate and Proton Influx

In nearly all experiments, using methyl viologen or, occasionally, ferricyanide as electron acceptor, and under both non-phosphorylating and phosphorylating conditions and considering steady values attained 30–60 s after illumination, $\phi_{H^+,ss}/\phi_{e^-}$ was 1.0. The ratio fell below unity (a) when using concentrations of uncoupler that caused the decay of the H^+ signal to be so rapid that the glass electrode could not follow the H^+ efflux in the dark or (b) with $pH_o = 9$, using methyl viologen, when the ratio was often about 0.5. This is puzzling since values of about unity were obtained with ferricyanide at pH 6–9.

It is concluded that normally the reduction of a single compound such as plastoquinone is accompanied by the uptake of one H^+ per electron on the outside of the thylakoids and, upon oxidation, the H^+ is lost to the interior, as suggested by MITCHELL (1966) (Fig. 1). It is possible that under some conditions (high pH?) the components of the electron transport chain may move relative to one another such that there is a higher probability that a H^+ will be discharged back into the exterior, upon the oxidation of PQH_2. The value of 1 for $\phi_{H^+,ss}/\phi_{e^-}$ agrees with that found by TELFER and EVANS (1972) also using spinach chloroplasts and methyl viologen, but is not compatible with the results of SCHWARTZ (1968) ($\phi_{H^+,ss}/\phi_{e^-} = 2$). In his study, though a promptly-responding glass electrode was used, ϕ_{H^+} was probably overestimated because the rate of change of pH near the electrode, covered by a thin layer of chloroplasts, did not reflect the rate of change in the whole, unstirred volume.

II. Light-Induced Proton Uptake, and Acidification of the Osmotic Space

The half-time for the onset of internal acidification in control experiments was about 4 s, and that for ΔH^+_o, 2 s. The kinetics and extent of the change in H^+ concentration in the medium, and of the change within the osmotic space, estimated from quenching of NED fluorescence, suggest that under our conditions, in an osmotic space of about 6 μl (mg chl)$^{-1}$, there is an average maximum concentration of about 90 mequiv l^{-1} of indiffusible anions, with a pK of 3.5–4. These estimates were obtained by considering the osmotic space as a Donnan system, ignoring bivalent cations and applying equations given by BRIGGS et al. (1961).

In control experiments, when an uptake of about 0.3 μmoles H^+ (mg chl)$^{-1}$ causes $[H^+]_i$ to rise to about 10^{-4} M, most of the translocated H^+ is in the unionized form HA. The pH gradient established in the light was up to 4 units. This is somewhat higher than in the experiments of SCHULDINER et al. (1972). The indefiniteness in both these studies is probably ± 0.4 pH units; atebrin indicated a lower, and 9-aminoacridine, a higher $[H^+]_i$ than NED.

III. Photophosphorylation and the Light-Induced pH Difference

The steady-state ΔpH was somewhat lowered in conditions (+ADP) where phosphorylation was possible. When uncouplers had reduced the rate of phosphorylation to 20% of control rates, the pH difference between inside and outside was as much as 66 % control (FCCP), or 67 % (nigericin), and ΔpH not less than 3.3. This seems inconsistent with the expected action of these uncouplers. Uncoupling seems more to be related to inability of the ATPase to use protons rather than to a reduced proton motive force, unless an electric PD is the chief component in the PMF and is specifically much reduced by uncouplers. SCHULDINER et al. (1972) found photophosphorylation and ΔpH to be reduced in parallel when methylamine was the uncoupler, which we also observed. A model to relate the PMF to phosphorylation rate is needed.

IV. The Delayed Fluorescence Induction, and the High-Energy State

It was observed that although the size of the 1 ms delayed fluorescence signal was related in a general way to conditions favourable for a high rate of photophosphorylation, there did not seem to be a direct connection between any component of the DF (or its logarithm), and the steady-state ΔpH, or $[H^+]_i$ directly. Thus, the onset of the peak of DF occurred in control experiments with a half-time of 0.1–0.2 s (cf. 0.3 s, EVANS and CROFTS, 1973) whereas the light-induced increase in $[H^+]_i$ had a half-time of 4 s. The onset of ΔH^+_o is faster but $t_{1/2}$ is still about 2 s.

The peak of the DF signal was always reduced by uncouplers, ± added ADP, and the half-time to reach the peak increased from about 0.15 s (control) to 1 s (+ 2 μM FCCP). The half-time for the increase in $[H^+]_i$ increased only moderately between these conditions, from 4 s to 5–6 s. As DCMU reduced net electron transport, the DF (± ADP) was reduced in size but the shape remained relatively unchanged. A reliable division into 'fast' and 'slower' components of the DF in the first few seconds could not be made in our experiments (WRAIGHT and CROFTS, 1971). Thus most of the observations lack a model in terms of control by ϕ_{e^-}, ϕ_{H^+} or ΔpH. Further, the prominent decay of the DF (-ADP), between 1–5 s and 30–60 s (not emphasized by WRAIGHT and CROFTS (1971), and not found by ITOH et al. (1971) is unexplained, but may be related to light-induced changes in thylakoid permeability to H^+, or the decreasing rate of active H^+ influx during this period. The decay of DF (+ADP) is absent or small (Fig. 3) while ϕ_{H^+} still declines during the light period, though less than in non-phosphorylating conditions.

The most consistent finding is that the difference in the peak of the 1 ms DF signal between nonphosphorylating and phosphorylating conditions is well correlated with (but not obviously mathematically related to) rates of photophosphorylation. This was the case when photophosphorylation was varied using DCMU, uncouplers and phloridzin. The results with phloridzin show that the reduction in DF in the presence of ADP, compared with -ADP, is related to ongoing photophosphorylation since the signal (+ADP) increased and approached the peaked signal (-ADP) as phloridzin inhibited phosphorylation.

V. The Shape of Onset and Decay Curves for ΔH^+_o and $[H^+]_i$

The shape of these time-courses is expected to depend on (1) the permeability of the thylakoids to H^+, (2) volume of the osmotic space as a function of time, (3) geometry of the membranes in relation to intra- and extra-thylakoid spaces (4) the rate of active transport of H^+ and (5) buffering capacity of the osmotic and outside spaces as functions of pH. In general none of these curves (Fig. 5B) shows a negative exponential approach to the steady level in the light, expected if there is a constant active influx, constant H^+ permeability and constant volume (onset) or constant permeability and volume (decay). Increasing concentrations of FCCP straighten out the log plot for ΔH^+_o decay, which straightening does not depend on the finite speed of the glass electrode response. In control experiments, the decreasing slope of the log plot with time could be due to decreasing proton permeability following light out. Volume changes have not been assessed.

Further experiments in this study will attempt to assess the relationship between the P518 signal and the high-energy state, and to obtain more details about delayed fluorescence than given by the 1 ms samplings.

E. Acknowledgements

The support of the Australian Research Grants Committee and Flinders University is gratefully acknowledged. W.S. Chow is grateful for a Flinders Postgraduate Scholarship. Mr. D. Martin gave able assistance in some experiments.

F. References

BARBER, J., FEBS Letters **20**, 251–254 (1972)
BRIGGS, G.E., HOPE, A.B., ROBERTSON, R.N., 'Electrolytes and Plant Cells', Botanical Monographs Vol. 1 (ed. W.O. James), Blackwell Scientific Publications, Oxford 1961
EVANS, E.H., CROFTS, A.R., Biochim. Biophys. Acta **292**, 130–139 (1973)
ITOH, S., MURATA, N., TAKAMIYA, A., Biochim. Biophys. Acta **245**, 109–120 (1971)
IZAWA, S., HIND, G., Biochim. Biophys. Acta **143**, 377–390 (1967)
MITCHELL, P., Biol. Rev. **41**, 445–502 (1966)
NEUMANN, J.S., JAGENDORF, A.T., Arch. Biochem. Biophys. **107**, 109–119 (1964)
ROTTENBERG, H., GRUNWALD, T., AVRON, M., Eur. J. Biochem. **25**, 54–63 (1972)
RUMBERG, B., SIGGEL, U., Naturwissenschaften **56**, 130–132 (1969)
SCHLIEPHAKE, W., JUNGE, W., WITT, H.T., Z. Naturforsch. **23b**, 1571–1578 (1968)
SCHULDINER, S., ROTTENBERG, H., AVRON, M., Eur. J. Biochem. **25**, 64–70 (1972)
SCHWARTZ, M., Nature **219**, 915–919 (1968)
SCHWARTZ, M., Ann. Rev. Plant Physiol. **22**, 469–484 (1971)
TELFER, A., EVANS, M.C.W., Biochim. Biophys. Acta **256**, 625–637 (1972)
WRAIGHT, C.A., CROFTS, A.R., Eur. J. Biochem. **19**, 386–392 (1971)

Structural Aspects of the Electrochemical Generator in Photosynthesis of Green Plants

W. Junge, W. Ausländer, and A. Eckhof

A. Introduction

Evidence has now accumulated that stimulation of photosynthetic electron transport leads to the generation of a proton electrochemical potential difference across the inner membrane of chloroplasts. Light causes H^+ uptake from the outer aqueous phase (NEUMANN and JAGENDORF, 1964), H^+ release into the inner phase (HAGER, 1969; RUMBERG and SIGGEL, 1969; ROTTENBERG et al., 1971) and the generation of an electric potential across the functional membrane of photosynthesis (JUNGE and WITT, 1968). The structural properties of the electrochemical generator are the subject of this communication.

B. Results and Discussion

I. The Electric Generator

The generation of an electric potential difference across the inner membrane of chloroplasts on illumination was detected via special absorption changes peaking around 515 nm which were sensitive to alteration of the electric conductance of the inner membrane of chloroplasts (JUNGE and WITT, 1968). Two lines of evidence led to the conclusion that these absorption changes reflect the electrochroic response of chloroplast bulk pigments to a high electric field strength across the membrane:
Kinetic evidence came from the fact that the decay of these absorption changes after excitation with a light flash was accelerated by changes in the membrane's electric conductivity (by osmotic shock, detergents, ionophores). This acceleration depended on the concentration of permeating ions in a chloroplast suspension; it was specific for certain ions in the presence of ionophores but unspecific with osmotic shock or detergent treatment. It was therefore concluded that the above absorption changes were sensitive to the electric charge but not to the chemical nature of ions. Since acceleration of the decay was observed when only one molecule of gramicidin (an ionophore) was added to a chloroplast suspension equivalent to 10^5 chlorophyll molecules, it was concluded that these absorption changes reflect a light-induced electric field across a functional membrane, the size of which is characterized by an ensemble of at least 10^5 chlorophyll molecules (JUNGE and WITT, 1968).

Spectroscopic evidence for the electrochroic nature of these absorption changes was provided by a comparison of the light flash induced difference spectrum of those absorption changes, having the above kinetic properties, resulting from isolated chloroplasts (EMRICH et al., 1969) with the spectrum of the electrochroic response of extracted chloroplast bulk pigments *in vitro* (SCHMIDT et al., 1971). Reasonable agreement was obtained. Similar absorption changes observed on excitation of bacterium chromatophores were attributed to electrochroic effects, too, because of the similarity between their difference spectrum and the difference spectrum obtained by submitting chromatophores to a salt jump causing a diffusion potential (JACKSON and CROFTS, 1969).

The rise of the electrochroic absorption changes in isolated chloroplasts is extremely fast. The 90 %-rise time on laser excitation is less than 20 ns (WOLFF et al., 1969), suggesting that the electric potential generation involves one of the primary photochemical reactions. The contribution of the two light reactions to the electric potential was studied by excitating with single-turnover-flashes (saturating and short compared with the relaxation time of the light reactions). The result, illustrated in Fig. 1, shows a relative value of 1 for the electrochroic absorption change at 515 nm, if both light reactions were active, but a relative value of about 1/2 if only one of them was active. Thus it was concluded that both light reactions contribute about equal amounts to the electric potential (SCHLIEPHAKE et al., 1969).

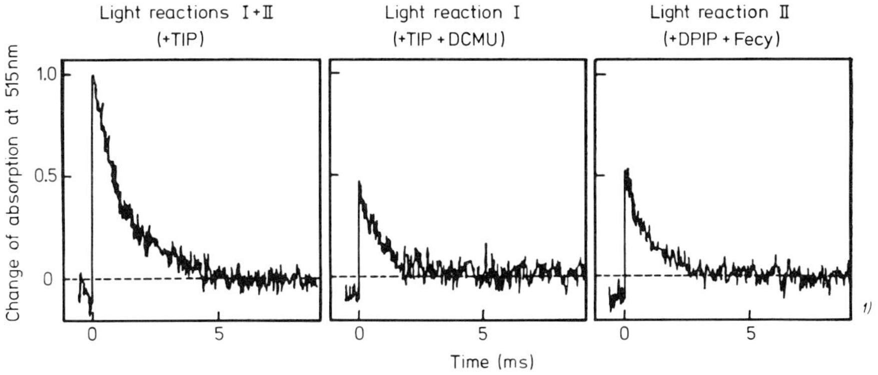

Fig. 1 Electrochroic absorption changes at 515 nm on excitation of isolated spinach chloroplasts with a short flash of light at t=0 (SCHLIEPHAKE et al., 1968).
Left: both light reactions active; middle: light reaction I active only; right: light reaction II active only. (TIP: thymolindophenol.)

II. Vectorial Electron Transport and Protolytic Reaction Sites

MITCHELL (1966) postulated that a proton electrochemical potential difference across the functional membrane of photosynthesis is generated by an alternating electron-hydrogen-transport as illustrated in Fig. 3. Light reaction II drives an electron transport from water (water-splitting system at inner side of the membrane) to the outer side of the membrane, thus causing H^+ release into the inner phase and electric potential generation. At the outer side an electron carrier (plastoquinone) becomes reduced and binds one H^+ from the outer aqueous phase. If this carrier can be oxidized at the inner side of the membrane it would act as a H^+ carrier across it. On reduction of a non-proton-binding electron acceptor at the inner side a H^+ would be released into the inner aqueous phase. Light reaction I finally drives the transport of one electron from the inner to the outer side across the membrane and so contributes to the electric potential generation as well as to H^+ binding in the outer phase as a consequence of the reduction of the terminal electron acceptor.

This model accounts for the above result that both light reactions contribute about equal amounts to the electric potential on excitation with a single-turnover-flash of light. The vectorial properties of this model are supported by the following results.
1) Electron carriers attributed to the reducing side of light reaction I have been located at the outer side of the membrane by immunological methods (BERZBORN et al., 1966; BERZBORN, 1968; REGITZ et al., 1970).

2) The rate limiting step of the electron transport between the two light reactions is sensitive to the pH of the internal phase (RUMBERG and SIGGEL, 1969).
3) Plastocyanin, an electron carrier between the light reactions seems to be loosely bound to the inner side of the membrane (HAUSKA et al., 1971).

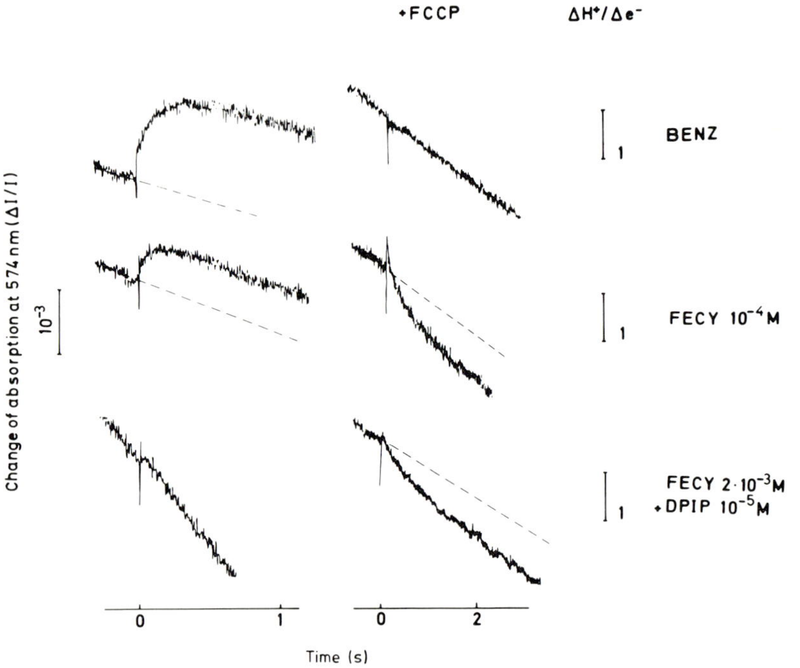

Fig. 2 pH-indicating absorption changes of cresol red at 574 nm on excitation of isolated spinach chloroplasts (broken type) with a short flash of light at t=0 (JUNGE and AUSLÄNDER, 1974).
Left: in the absence of the proton carrier FCCP; right: in the presence of the proton carrier FCCP. Different electron acceptors: BENZ: proton binding, after light reaction I; FECY (low): nonproton binding, after light reaction I; FECY + DPIP: non proton binding, after light reaction II.

The protolytic properties of the electron transport chain have been studied by excitation with single-turnover-flashes. The rather small pH-changes in the outer aqueous phase of the inner membrane system of chloroplasts were resolved by using pH-indicating dyes, the absorption changes of which, with certain precautions (JUNGE and AUSLÄNDER, 1974), are linear indicators of the number of H^+ bound or released by chloroplast inner membranes. Four different protolytic reaction sites were identified: two for H^+ binding from the outer phase (SCHLIEPHAKE et al., 1968; JUNGE and AUSLÄNDER, 1974) and two for H^+ release into the inner one (JUNGE and AUSLÄNDER, 1974). If oxygen, via benzyl viologen (BENZ), was used as terminal electron acceptor the H^+/e^- stoichiometry of each site was about one. The four protolytic reaction sites were attributed to certain redox reactions based on experiments one of which is illustrated in Fig. 2 (JUNGE and AUSLÄNDER, 1974). The traces in Fig. 2 and their interpretation (Fig. 3) will be discussed in some detail.

The traces on the left of Fig. 2 represent pH-indicating absorption changes observed on excitation of isolated chloroplasts with a short flash of light at t=0 in the presence of the dye cresol red. The extent of these absorption changes is proportional to the number of H^+

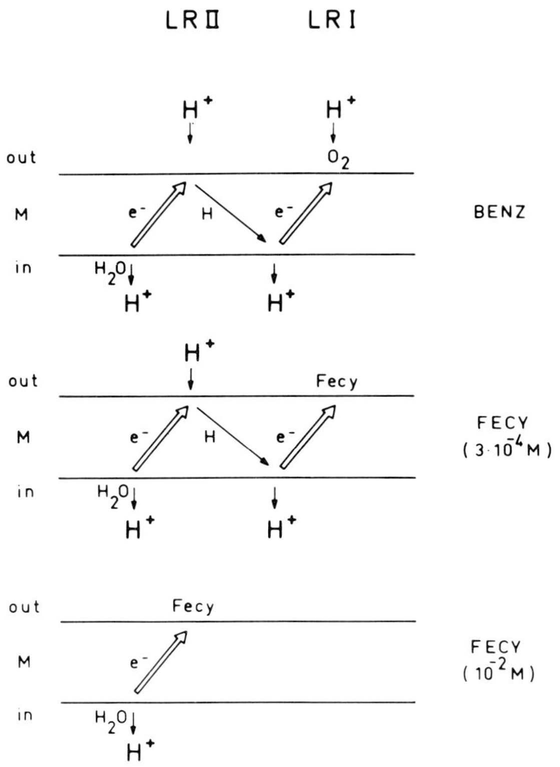

Fig. 3 Vectorial electron-hydrogen-transport scheme compatible with the experimental results depicted in Figs. 1 and 2 (JUNGE and AUSLÄNDER, 1974); see legend to Fig. 2.

bound (positive signal) in the outer aqueous phase of the inner chloroplast membrane. The bars on the right indicate the equivalent of one H^+ bound per electron transferred to the terminal acceptor. If FCCP is added to the chloroplast suspension the H^+ permeability of the functional membrane is increased and any pH-difference across the membrane relaxes within 1 s. Then the absorption changes of cresol red indicate the net H^+ production or H^+ consumption from both sides of the functional membrane (right-hand side of Fig. 2). The results depicted in Fig. 2 have been used to deduce the vectorial electron-hydrogen-transport-system illustrated in Fig. 3 (JUNGE and AUSLÄNDER, 1974). The argument being somewhat lengthy, we will only show here that the experimental results are compatible with the above model.

The upper two traces in Fig. 2 resulted if oxygen, via BENZ, was the terminal electron acceptor. The H^+ uptake from the outer phase (left) was two H^+/e^-, while the net H^+ production was zero (right), corresponding to the release of two H^+/e^- into the inner phase (see upper row in Fig. 3). The middle traces of Fig. 2 were obtained if oxygen, which binds one H^+ upon reduction, was substituted by the non-proton-binding electron acceptor ferricyanide at relatively low concentrations. Here only one H^+ was taken up from the outer phase, the net H^+ production was plus one and consequently the H^+ release into the inner phase was still two H^+/e^- (see Fig. 3, middle). If ferricyanide (FECY) in higher concentrations together with the lipophilic cofactor DPIP (2,6-dichlorophenolindophenol) was present, it accepted electrons efficiently from light reaction II thus competitively inhibiting light reaction I activity (see Fig. 3, below). Under these conditions no H^+ uptake from the outer phase was observed but a net production of one H^+/e^- from the inner phase. These results are easily understood in terms of the model illustrated in Fig. 3, if the four protolytic reaction sites are attributed to the following redox reactions: the oxidation of water at the inner side of the functional membrane (release of 1 H^+/e^-), the reduction of plastoquinone at the outer side (binding of 1 H^+/e^-), the oxidation of plastohydroquinone at the inner side (release of 1 H^+/e^-) and the reduction of the terminal electron acceptor at the outer side (binding of $\nu H^+/e^-$) (JUNGE and AUSLÄNDER, 1974).

III. The Structure of the Functional Membrane

The above evidence for vectorial electron transport across the functional membrane stimulated studies as to the structure of the membrane across which the vectorial reactions take place. The structural models for the inner membrane of chloroplasts are conflicting, as yet. While small angle X-ray scattering studies have suggested a quasi-crystalline mosaic model – a sandwich of proteins, pigments and lipids (KREUTZ, 1970) –, electron micro-

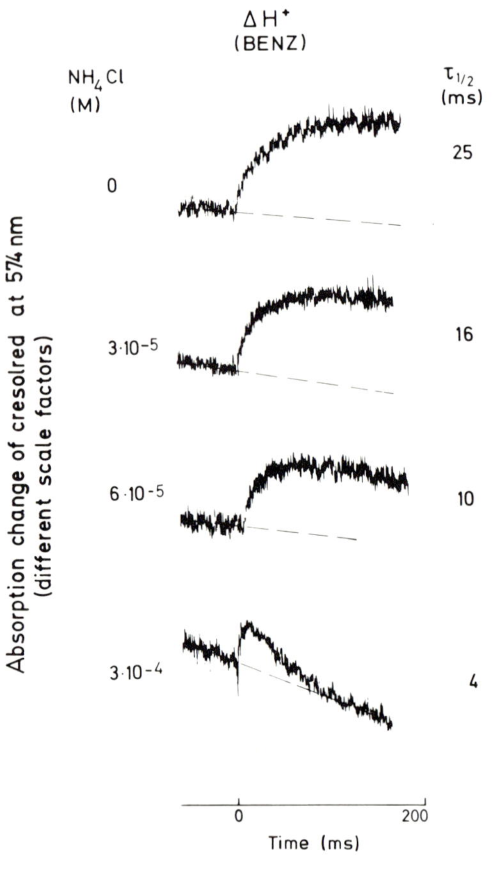

Fig. 4 Dependence of the pH-indicating absorption changes of cresol red at 574 nm on the concentration of the uncoupler NH_4Cl in chloroplasts which were fragmented by grinding with sand (AUSLÄNDER and JUNGE, 1974). An acceleration of the protolytic reaction between the reducing sites of both light reactions at the outer side of the membrane and the outer aqueous phase is obvious.

scopy suggested irregularly distributed protein 'bulbs' embedded at different depths in a lipid matrix (MÜHLETHALER, 1972).

We have chosen another approach to studying the structure of the membrane. We used the rate of H^+ uptake from the outer phase as an indication of the 'depth' of the respective redox reaction sites in the functional membrane. The two redox reactions which lead to H^+ binding in the outer aqueous phase, i.e. the reduction of plastoquinone and of the terminal acceptor, proceed in less than 1 ms. However, H^+ uptake as monitored via the absorption changes of cresol red is considerably slower (see upper left of Fig. 2). We found that mechanical (sand grinding) or chemical (digitonin treatment) disintegration of broken chloroplasts diminished the delay between redox reactions and H^+ uptake from the outer phase. The half-time of H^+ uptake shortened from 60 ms (broken chloroplasts) to 30 ms (ground chloroplasts) to 20 ms (digitonin) in the presence of BENZ (AUSLÄNDER and JUNGE, 1974). H^+ uptake was also accelerated in the presence of electron acceptors like DPIP, which can also act as uncouplers, or on addition of H^+ carriers like NH_4Cl and FCCP. The acceleration of H^+ uptake from the outer phase by NH_4Cl is illustrated in Fig. 4. The four traces show the pH-indicating absorption changes of cresol red on excitation of a suspension of ground chloroplasts with a short flash of light at t=0. The rise time of the pH-signal decreased from 25 ms (no NH_4Cl) to about 4 ms (NH_4Cl, 300 μM). At the highest concentration (lowest trace) not only H^+ binding from the outer phase, but the relaxation of the light induced pH-difference across the functional membrane was accelerated.

The observation that structural disintegration plus uncoupling agents, which can act as H^+ carriers across lipophilic barriers, reduce the delay between the redox reactions and the corresponding H^+ uptake from the outer phase, led us to postulate that these redox reaction sites are shielded from the outer phase by a diffusion barrier for H^+ (AUSLÄNDER and JUNGE, 1974). Moreover, we had to postulate the existence of a H^+ reservoir beneath this diffusion barrier providing rapid protonation of the reduced electron acceptors, while refilling by H^+ from the outer phase is considerably slower.

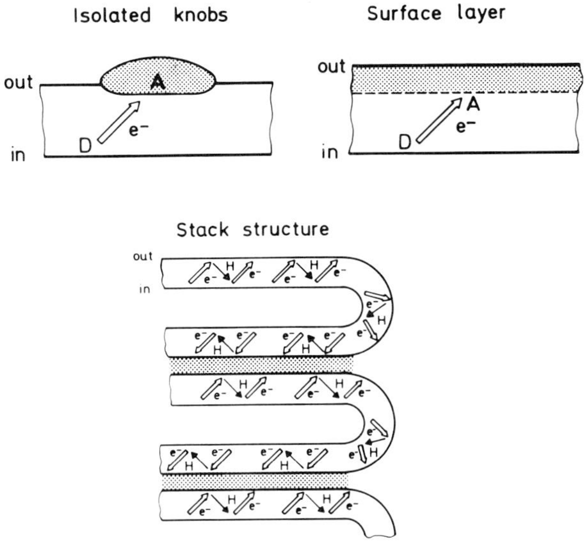

Fig. 5 Three possible structures of the diffusion barrier for H^+ which shields the redox reaction sites at the outer side of the functional membrane against the outer aqueous phase.
D: electron donor; A: electron acceptor for the primary charge translocation across the membrane; white: core layer; gray: shielding layer.

The structural and chemical nature of this diffusion barrier is still an open question. Three different structures, illustrated in Fig. 5, may be responsible for the observed delay between redox reactions and protolytic reactions. Isolated knobs covering the reducing sites for both light reactions, a continous layer covering the whole membrane, or lipophilic material filling the interthylakoid space in grana stacks are possible candidates for the H^+ diffusion barrier. No matter which of these structures is correct we must visualize the functional membrane as made up from a core layer across which light generates an electrochemical potential difference plus some shield, which delays the access of H^+ to the redox reaction sites on the outer side of the core layer. The core of the membrane and the shield are kinetically well distinguished; in broken chloroplasts (with BENZ) it takes several s for H^+ from the inner aqueous phase to leak into the outer one, while diffusion of H^+ across the shield takes only some ten ms.

IV. The Location and the Orientation of Chlorophyll a_1

It has been shown above that the vectorial electron transport crosses the core of the functional membrane but ends beneath the shielding diffusion barrier for H^+ at the outer side (see Fig. 5). Questions arise as to the location of chlorophyll a_1, which drives the vectorial electron transport of light reaction I at either side of the membrane, and to its orientation with respect to the membrane plane.

Chlorophyll a_1 is characterized by absorption changes at 700 nm (KOK, 1961), at 438 nm (RUMBERG and WITT, 1964) and by a minor component at 682 nm (DÖRING et al., 1968). It represents an aggregate of at least two chlorophyll a molecules interacting via excitons and charge transfer.

The location of chlorophyll a_1 can be inferred from kinetic arguments. The electric potential rises in less than 20 ns (WOLFF et al., 1969). Chlorophyll a_1 is oxidized in less than 20 ns (WITT and WOLFF, 1970), while its reduction takes some ms (WITT et al., 1963). Thus it is obvious that the oxidation of chlorophyll a_1 generates the electric potential. The polarity of the electric potential is positive inside (JUNGE and WITT, 1968). Thus we conclude that chlorophyll a_1 is located at the inner side of the membrane core.

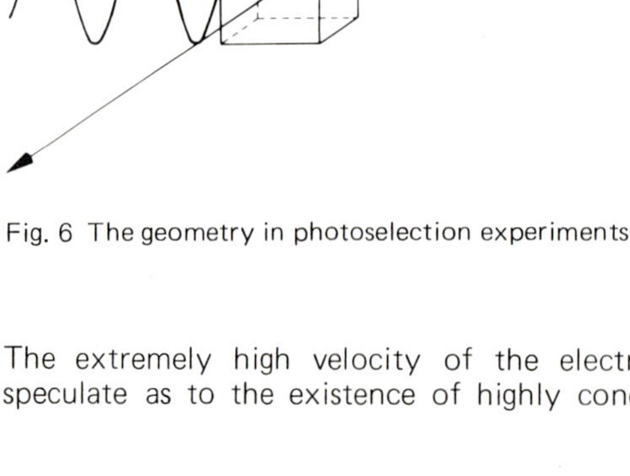

Fig. 6 The geometry in photoselection experiments.

The extremely high velocity of the electron transport across the membrane made us speculate as to the existence of highly conducting channels bridging the membrane core which separates chlorophyll a_1 from the electron acceptors at the outer side of the membrane. Possible candidates for this bridge are special proteins (TRIBUTSCH, 1972) or possibly carotenoids or chlorophyll-aggregates with overlapping π-electron systems of their porphyrin rings. To obtain information on the architecture of the interacting chlorophyll system in the membrane we have started linear dichroism studies on the orientation of chlorophyll a_1 by photoselection (JUNGE and ECKHOF, 1973, 1974). These studies were carried out with isotropic suspensions of broken chloroplasts. The suspension was excited with a linearly polarized flash of light the E-vector of which was vertical, as illustrated in Fig. 6. Linear dichroism studies on magnetically oriented chloroplasts (GEACINTOV et al., 1971) and on chloroplasts oriented as dried films (BRETON and ROUX, 1971) have revealed that the transition moments of the antennae chlorophylls absorbing above 680 nm are almost flat in the plane of the membrane. Thus, excitation of an isotropic suspension of chloroplasts with a non-saturating polarized flash at a wavelength greater than 680 nm selects for those membrane orientations which are represented by filled circles in Fig. 6. In consequence absorption changes of chlorophyll a_1 result from the same photoselected subset, which is anisotropic. These absorption changes were observed with linearly polarized measuring light the E-vector of which was directed either parallel ($\Delta A\|$) or perpendicular ($\Delta A\perp$) to the E-vector of the exciting flash (Fig. 6). We observed different extents of the absorption changes of chlorophyll a_1 depending on the polarization of the measuring beam. Dichroic ratios ($\Delta A\|/\Delta A\perp$) greater than 1.15 were observed for both major bands of chlorophyll a_1 at 700 and 430 nm, respectively (JUNGE and ECKHOF, 1973, 1974). The dichroism of the absorption change at 705 nm is documented in Fig. 7. These dichroic ratios were interpreted on the assumption of a circular degeneration of the antennae system by multiple resonance energy transfers in the plane of the membrane (JUNGE and ECKHOF, 1974). We concluded that both major transition moments of chlorophyll a_1 are inclined at less than 25° to the plane of the membrane.

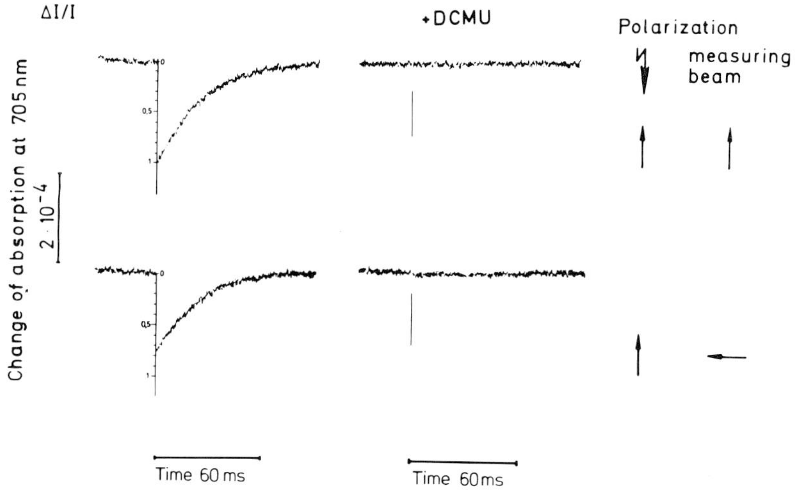

Fig. 7 Linear dichroism of the flash induced absorption changes of chlorophyll a_1 at 705 nm by photoselection (JUNGE and ECKHOF, 1974). Above: exciting beam and measuring beam polarized in parallel; below: exciting beam and measuring beam polarized perpendicularly; right: demonstration of the absence of polarization artifacts under conditions where photosynthetic electron transport is poisoned.

In order to make conclusions about the orientation of the porphyrin rings of chlorophyll a_1 to the membrane, information on the orientation of the transition moments in the coordinate system of the dimer is required. In a chlorophyll-monomer the transition moments of the blue and the red band lie approximately perpendicular to each other in the plane of the porphyrin ring (GOUTERMAN et al., 1963). Information on the geometry of the chlorophyll a_1-dimer can be inferred from the double band at 682 and 700 nm (DÖRING et al., 1968). The intensity ratio of these bands is about 1 to 4. This has been attributed to exciton interaction in the dimer. According to KASHA's treatment of dimers under exciton interaction (KASHA, 1963) the above intensity ratio implies that the red transition moments of the two chlorophyll a-components of the dimer are inclined at less than 14° to the resultant transition moment of the dimer at 700 nm. This points to a rather flat structure of the dimer.

If the transition moments of chlorophyll a_1 at 430 nm and 700 nm are more-or-less parallel to the corresponding transition moments of the two monomers then their inclination at less than 25° to the membrane implies an inclination of less than 37° of the porphyrin rings to the membrane (JUNGE and ECKHOF, 1974). Since experimental imperfections in photoselection experiments tend to decrease the apparent dichroism, 37° is an upper limit. Thus a rather flat orientation of the porphyrin rings of chlorophyll a_1 at the inner side of the membrane is probable.

The red absorption bands of chlorophyll a_1 and of its antennae pigments are shifted towards the red compared to the absorption band of chlorophyll a in organic solvents. This can be understood by exciton interaction of the respective transition moments in a head-to-tail arrangement of the porphyrin rings (HOCHSTRASSER and KASHA, 1964). Since the red transition moment of chlorophyll a_1 lies rather flat in the membrane, a head-to-tail lining up with other chlorophyll a molecules will not likely bridge the membrane. Instead, some proteinaceous structure or carotenoids might be responsible for electron conduction between excited chlorophyll a_1 at the inner side of the membrane and its electron acceptor at the outer one.

C. Acknowledgements

This work was financially supported by the Deutsche Forschungsgemeinschaft.

D. References

AUSLÄNDER, W., JUNGE, W., Biochim. Biophys. Acta, in press (1974)
BERZBORN, R.J., Z. Naturforsch. **23b**, 1096–1104 (1968)
BERZBORN, R.J., MENKE, W., TREBST, A., PISTORIUS, E., Z. Naturforsch. **21b**, 1057–1059 (1966)
BRETON, J., ROUX, E., Biochem. Biophys. Res. Commun. **45**, 557–563 (1971)
DÖRING, G., BAILEY, J.L., KREUTZ, W., WEIKARD, J., WITT, H.T., Naturwissenschaften **55**, 219–220 (1968)
EMRICH, H.M., JUNGE, W., WITT, H.T., Z. Naturforsch. **24b**, 1144–1146 (1969)
GEACINTOV, N., VAN NOSTRAND, F., POPE, M., TINKEL, J.B., Biochim. Biophys. Acta **226**, 486–491 (1971)
GOUTERMAN, M., WAGNIERE, G.H., SNYDER, L.C., J. Mol. Spectrosc. **11**, 108–127 (1963)
HAGER, A., Planta **89**, 224–243 (1969)
HAUSKA, G.A., McCARTY, R.E., BERZBORN, R.J., RACKER, E., J. Biol. Chem. **246**, 3524–3531 (1971)
HOCHSTRASSER, R.M., KASHA, M., Photochem. Photobiol. **3**, 317–331 (1964)
JACKSON, J.B., CROFTS, A.R., FEBS Letters **4**, 185–188 (1969)
JUNGE, W., AUSLÄNDER, W., Biochim. Biophys. Acta **333**, 59–70 (1974)
JUNGE, W., ECKHOF, A., FEBS Letters **36**, 207–212 (1973)
JUNGE, W., ECKHOF, A., Biochim. Biophys. Acta, in press (1974)

JUNGE, W., WITT, H.T., Z. Naturforsch. **23b**, 244–254 (1968)
KASHA, M., Radiat. Res. **20**, 55–71 (1963)
KOK, B., Biochim. Biophys. Acta **48**, 527–535 (1961)
KREUTZ, W., Adv. Bot. Res. **3**, 53–169 (1970)
MITCHELL, P., Biol. Rev. **41**, 445–502 (1966)
MÜHLETHALER, K., Chem. Phys. Lipids **8**, 259–264 (1972)
NEUMANN, J., JAGENDORF, A.T., Arch. Biochem. Biophys. **107**, 109–119 (1964)
REGITZ, G., BERZBORN, R.J., TREBST, A., Planta **91**, 8–17 (1970)
ROTTENBERG, H., GRUNWALD, T., AVRON, M., FEBS Letters **13**, 41–44 (1971)
RUMBERG, B., SIGGEL, U., Naturwissenschaften **56**, 130–132 (1969)
RUMBERG, B., WITT, H.T., Z. Naturforsch. **19b**, 693–707 (1964)
SCHLIEPHAKE, W., JUNGE, W., WITT, H.T., Z. Naturforsch. **23b**, 1571–1578 (1968)
SCHMIDT, S., REICH, R., WITT, H.T., Naturwissenschaften **58**, 414–415 (1971)
TRIBUTSCH, H., Photochem. Photobiol. **16**, 261–269 (1972)
WITT, H.T., MÜLLER, A., RUMBERG, B., Colloq. Int. Centre Nat. Rech. Sci. **p 43** (1973)
WITT, K., WOLFF, CH., Z. Naturforsch. **25b**, 387–388 (1970)
WOLFF, CH., BUCHWALD, H.-E., RÜPPEL, H., WITT, K., WITT, H.T., Z. Naturforsch. **24b**, 1038–1041 (1969)

Variable Chlorophyll Fluorescence Reflects Cation Transfer across Thylakoid Membranes

G.H. Krause

A. Introduction

Intact isolated chloroplasts suspended in an isotonic buffer medium show a characteristic induction of chlorophyll *a* fluorescence similar to the KAUTSKY phenomenon observed with algae (KAUTSKY and HIRSCH, 1931). The rise of variable fluorescence to its maximum can be understood in terms of reduction of the primary electron acceptor of photosystem II, the substance Q, which acts as a fluorescence quencher in its oxidized form (DUYSENS and SWEERS, 1963). The object of the present communication is the fluorescence decline from the maximum to the steady state. This decline cannot simply be explained as a change in the redox state of Q. In isolated chloroplasts suspended in a CO_2-free medium without artificial electron acceptors only a restricted electron transfer to oxygen through the MEHLER reaction (MEHLER, 1951) can take place during strong steady illumination, keeping Q virtually in the reduced state.

In a preceding study (KRAUSE, 1973) the fluorescence decline together with the light-induced increase in apparent absorbance at 535 nm of chloroplasts were found to reflect the high-energy state of the energy conserving apparatus. Evidence is now presented showing that the light-dependent transfer of Mg^{2+} and certain other metal cations across the thylakoid membranes is responsible for structural changes that lead to the observed changes in the fluorescence yield and apparent absorbance (KRAUSE, 1974). This agrees well with available data showing that Mg^{2+} and a number of other cations strongly increase the variable chlorophyll *a* fluorescence of thylakoids (MOHANTY et al., 1973; MURATA, 1971; HOMANN, 1969), and that the light-induced H^+ uptake is associated with efflux of cations from the thylakoids (DILLEY and VERNON, 1965).

B. Materials and Methods

Chloroplasts with intact outer double membranes were isolated from leaves of *Spinacia oleacea* L. using the method of JENSEN and BASSHAM (1966), with a slight modification (KRAUSE, 1973). For measurements the chloroplasts were suspended in 'Solution C' containing 0.33 M sorbitol, 0.5 mM KH_2PO_4, 1 mM $MnCl_2$, 1 mM $MgCl_2$, 10 mM NaCl, 2 mM EDTA, and 40 mM N-Hydroxymethylpiperazine-N-2-ethanesulfonic acid (HEPES), pH 7.6. No artificial electron acceptors were added. In order to break the chloroplast envelope, intact chloroplasts were exposed to hypotonic conditions, i.e. suspended in water and immediately brought back to the isotonic medium by adding an equal volume of double strength Solution C. In other experiments a solution of 5 mM $MgCl_2$, instead of water, and double strength Solution C with 5 mM $MgCl_2$ were used. Chlorophyll *a* fluorescence and apparent absorbance at 535 nm were measured simultaneously as described previously (KRAUSE, 1973, 1974).

C. Results and Discussion

In Fig. 1A the characteristic long-term fluorescence quenching of intact isolated chloroplasts is shown to be paralleled by a slow increase in apparent absorbance. Both changes are energy-dependent as demonstrated by the effect of FCCP (carbonyl cyanide 4-trifluoromethoxyphenylhydrazone) at a concentration that uncouples photophosphorylation but does not inhibit electron transport. Broken chloroplasts (Fig. 1B) show signals strikingly different from those of intact chloroplasts. The fluorescence yield does not rise to the same high maximum and is not quenched significantly on prolonged illumination. Consequently, uncoupling by FCCP does not exert a strong effect on fluorescence. Furthermore, no strong light-dependent change of apparent absorbance, which reflects changed light scattering, is seen. These chloroplasts have lost their envelope and soluble stroma content by osmotic rupture (see Methods). As seen from diverse photosynthetic reactions, the basic functions of the remaining thylakoids have not been impaired by the osmotic shock. Thus, it seems that certain factors are needed for the changes of fluorescence and apparent absorbance that are observed with intact chloroplasts. These substances would be retained within the chloroplast envelope and diluted by breakage.

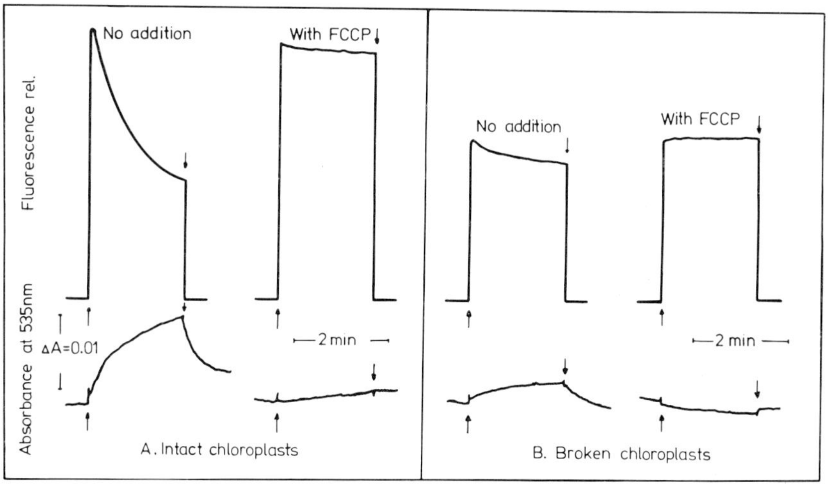

Fig. 1 Fluorescence and light-induced changes in apparent absorbance of isolated spinach chloroplasts; effect of FCCP. The chloroplasts were suspended in Solution C (see Methods; 0.1 mg chlorophyll per 1.4 ml). Intensity of exciting light 40 kerg cm^{-2} s^{-1}; half band width from 627 to 679 nm. Light on, upward arrows; light off, downward arrows; 3 min dark periods before start of illumination. Fluorescence was measured at 742 nm; absorbance at 535 nm; light path 5 mm. A, intact chloroplasts; B, broken chloroplasts. FCCP (10^{-6} M) was added in the dark to the same samples.

Fig. 2 shows that addition of $MgCl_2$ (5 mM) to broken chloroplasts results in a fluorescence signal that is very similar to that of the intact chloroplasts. Essentially the same effect is exerted by $CaCl_2$ and $MnCl_2$. The effect is saturated at about 5 mM $MgCl_2$. NaCl and KCl also restore the 'normal' fluorescence signal, but a much higher (30 to 40-fold) concentration is required. It is known that these cations drastically increase the chlorophyll fluorescence of thylakoids (see Introduction). In our experiments this increase is seen most clearly in the fluorescence maximum (Fig. 2) at the beginning of the light phase. Faster recording (Fig. 3) shows that Mg^{2+} essentially restores the characteristic fluorescence induction of intact chloroplasts. Notably, the fast (constant) fluorescence emission is the same in intact and broken chloroplasts. As the presence of cations determines the fluorescence yield, the fluorescence quenching following the maximum (not shown in Fig. 3, but in Figs. 1 and 2) can be interpreted as an indication of light-dependent efflux of cations from the thylakoids.

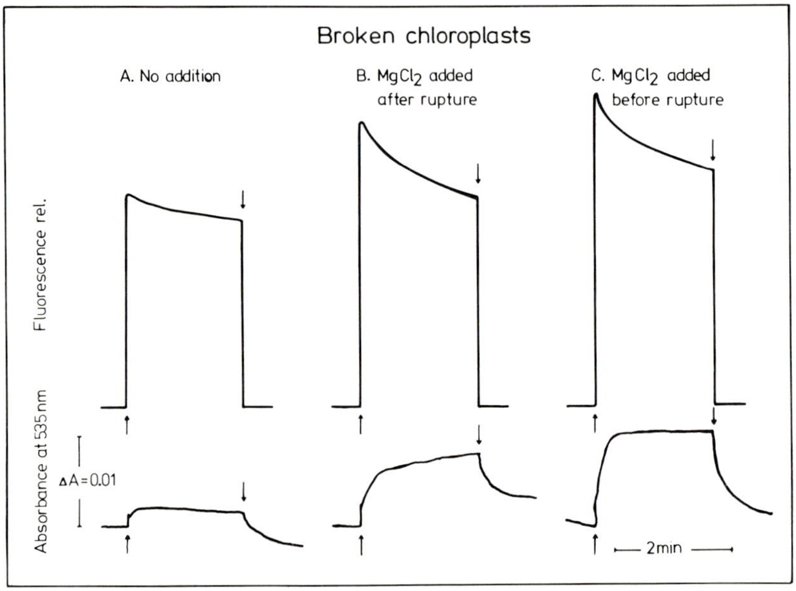

Fig. 2 Fluorescence and light-induced changes in apparent absorbance of broken chloroplasts; effects of Mg^{2+}. Conditions as for Fig. 1: A, broken chloroplasts suspended in Solution C (see Methods); B, the same chloroplast sample after addition of $MgCl_2$ (5 mM); C, broken chloroplasts suspended in Solution C, with 5 mM $MgCl_2$ added before osmotic shock.

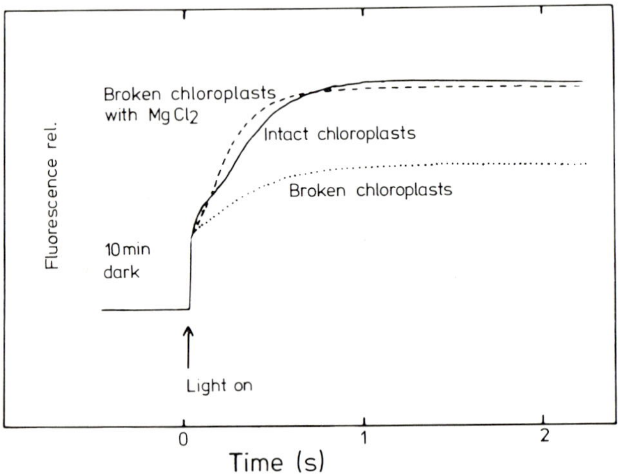

Fig. 3 Fluorescence induction of isolated chloroplasts. After dark periods of 10 min the fluorescence emission in the first seconds of illumination was recorded with an oscillograph. Intensity of exciting light 30 kerg cm^{-2} s^{-1}. Other conditions as for Fig. 1. Solid line, intact chloroplasts; dotted line, broken chloroplasts, no Mg^{2+} added; dashed line, broken chloroplasts with 5 mM $MgCl_2$, added before rupture.

As demonstrated in Fig. 2 (lower traces), also the changes in apparent absorbance seem to depend on the presence of cations. However, a larger and better reversible light-induced absorbance increase was invariably observed when, instead of adding $MgCl_2$ after osmotic rupture, the chloroplasts were broken in a 5 mM $MgCl_2$ solution (Figs. 2B, C). On the other hand, the fluorescence signal was not significantly influenced by the mode of rupture. This shows that a different type of structural change is responsible for the light scattering changes than for the long-term fluorescence quenching (MOHANTY et al., 1973). Furthermore, it indicates that some properties of the thylakoid membranes are irreversibly altered by osmotic shock in the absence of a sufficiently high cation level. As shown in Fig. 4, chloroplasts differ in light scattering (in the dark) according to the type of osmotic rupture applied. This indicates structural differences. The spectrum of apparent absorbance is closer to that of intact chloroplasts if 5 mM $MgCl_2$ is present during rupture, as compared to the case of Mg^{2+} addition after rupture.

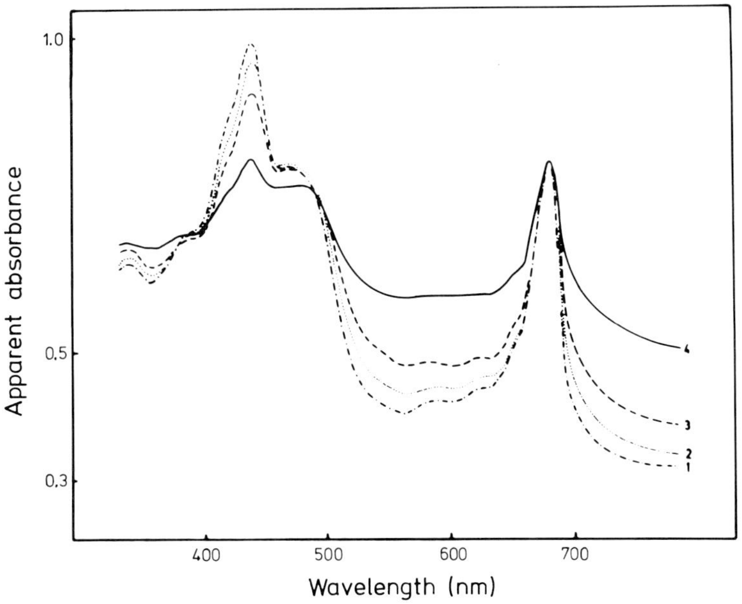

Fig. 4 Spectra of apparent absorbance of isolated chloroplasts; effects of Mg^{2+}. Chloroplasts equivalent to 10 µg chlorophyll were suspended in 3 ml Solution C (see Methods), pH 7.6. Light path 1 cm. The spectra were recorded with a Beckman Spectrophotometer Acta III. The strong apparent absorbance beside the pigment absorption bands is caused by light scattering.
1—·—·—·—, broken chloroplasts; 2······, broken chloroplasts after addition of 5 mM $MgCl_2$; 3————, broken chloroplasts with 5 mM $MgCl_2$, added before rupture; 4———, intact chloroplasts.

When the chloroplasts are broken in the presence of Mg^{2+} (5 mM), the fluorescence and apparent absorbance changes are kinetically and, with regard to light saturation, closely related to proton pumping (KRAUSE, 1974). On the other hand, chloroplasts broken in a medium of low cation strength show H^+ uptake associated with little or no changes in fluorescence and apparent absorbance. This would mean that the H^+ movement alone is not able to produce these changes. Our results lead to the conclusion that it is the light-induced cation efflux which leads to the observed changes in fluorescence and apparent absorbance when sufficiently high cation concentrations are present in the thylakoids at the beginning of the light phase. Intact chloroplasts suspended in a medium with a low concentration of free divalent cations (Solution C, see Methods) apparently retain saturating amounts of cations within their envelope. Consequently, addition of Mg^{2+} to intact chloroplasts did not significantly influence the fluorescence and absorbance changes.

Two further observations support the view that the long-term fluorescence changes indicate cation transfer.

1) As mentioned above, $MgCl_2$ (5 mM) can be replaced by 150 to 200 mM KCl or NaCl in order to restore the high fluorescence maximum followed by quenching to a low steady-state yield. When broken chloroplasts are suspended in a medium containing 0.19 M KCl (Fig. 5A), valinomycin ($2 \cdot 10^{-5}$ M), a substance that specifically facilitates permeation of biomembranes by K^+, added after about 2 minutes of illumination, increases the fluorescence yield almost to the maximum level. Obviously, the valinomycin allows back transfer of K^+ that have left the thylakoids in the preceding light period. On the other hand, the same valinomycin concentration does not, or does only slightly increase the fluorescence yield of broken chloroplasts in a NaCl medium (Fig. 5B).

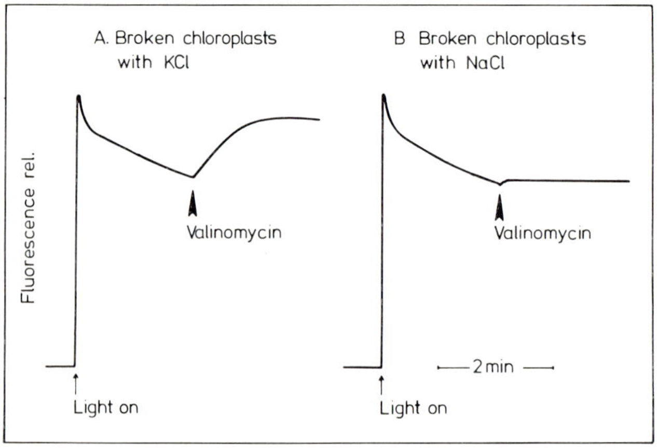

Fig. 5 Effect of valinomycin on the fluorescence of broken chloroplasts. For breakage, intact chloroplasts (0.1 mg chlorophyll) were suspended in water (1.2 ml), and then 1.2 ml Solution C, containing 0.38 M KCl (A) or NaCl (B), were added giving a final concentration of 190 mM KCl or NaCl, respectively. The suspension was placed in a cylindrical cuvette (diameter 2 cm) and agitated during measurement. Exciting light, and fluorescence recording as for Fig. 1. Valinomycin ($2 \cdot 10^{-5}$ M) was added in the light as indicated. Disturbance caused by addition has been omitted.

2) The cation effect on fluorescence requires a dark period. As depicted in Fig. 6, addition of $MgCl_2$ (10 mM) during strong illumination does not increase the fluorescence yield significantly. The restoration of the high fluorescence yield is shown by applying dark periods of increasing lengths (Fig. 7). The resulting curve that connects the fluorescence peaks recorded after different dark times (dashed line in Fig. 7) can be understood to reflect structural changes of the thylakoids caused by Mg^{2+} influx in the dark. In the light, transfer processes directed out of the thylakoids may prevent uptake of added cations. The absence of a fluorescence rise upon cation addition depends on the high-energy state of the membrane system. This can be demonstrated with uncouplers of photophosphorylation (FCCP) as well as by inhibition of electron transport with 3-(3',4'-dichlorophenyl)-1,1-dimethylurea (DCMU). Both substances lead to an immediate fluorescence rise in broken chloroplasts, when $MgCl_2$ is added in the light (KRAUSE, 1974).

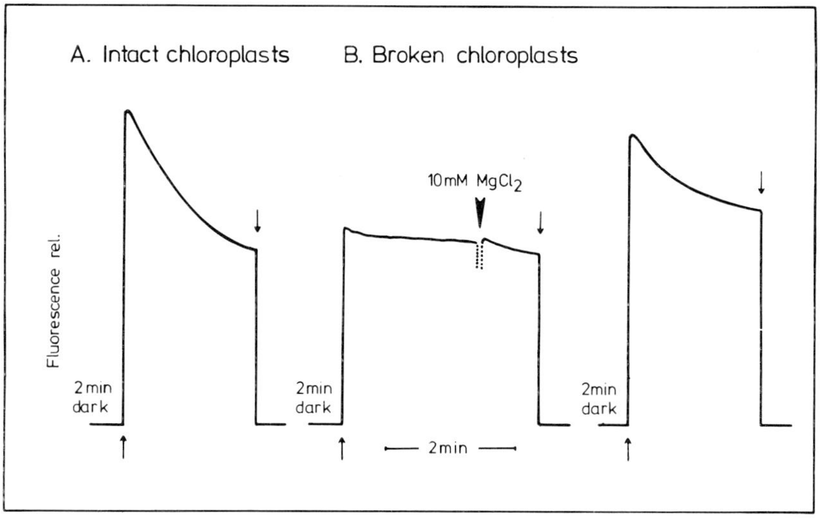

Fig. 6 Fluorescence of isolated chloroplasts; effect of $MgCl_2$ added to broken chloroplasts during illumination. Chloroplast suspensions (0.1 mg chlorophyll in 2.4 ml Solution C) were stirred during fluorescence recording (see Fig. 5). Exciting light as for Fig. 1. Light on, upward arrows; light off, downward arrows. A, intact chloroplasts; B, broken chloroplasts. $MgCl_2$ (10 mM) was added in the light as indicated. The trace to the right shows the fluorescence signal of the same sample after a dark period of 2 min.

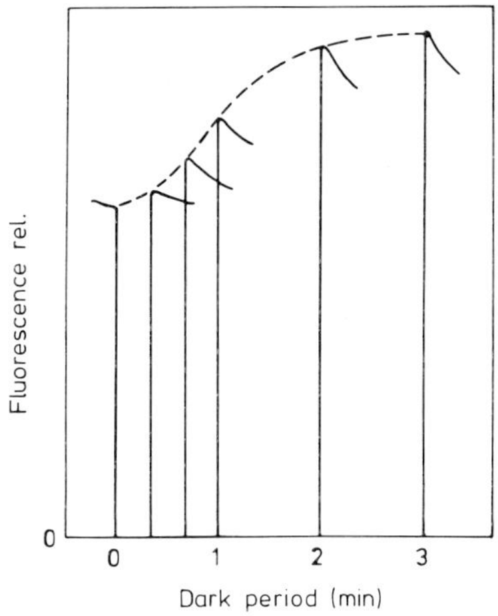

Fig. 7 Restoration of the fluorescence peak as a function of the dark time. $MgCl_2$ (5 mM) was added to broken chloroplasts in the light (see Fig. 6). Subsequently, the fluorescence signals shown were recorded after different dark periods as indicated. Experimental conditions as for Fig. 6.

D. Conclusions

Considering the experimental evidence presented by other authors (see Introduction) our results can be interpreted as follows. Cations such as Mg^{2+}, Ca^{2+}, K^+, and Na^+ influence the fluorescence yield only if they are present in the thylakoid interior. At the relatively high pH in the dark they are thought to neutralize negative charges on the inner membrane face. Under these conditions a certain structure of the membrane is maintained that is characterized by a high fluorescence yield, as observed after photochemical reduction of the quencher Q at the beginning of the light phase. On prolonged illumination H^+ uptake by the thylakoids causes protonation of negative groups on the inner side of the thylakoids. Metal ions are set free and are transferred across the membranes into the extrathylakoid space. The processes lead to different types of structural membrane alterations, e.g. to volume changes and ultrastructural changes. The former may result in an increase in apparent absorbance, the latter probably are indicated by quenched fluorescence (MOHANTY et al., 1973). As the apparent absorbance increase can also be caused by energized transport of weak organic acids out of the thylakoids (DEAMER et al., 1967), the fluorescence yield appears as the more reliable indicator of cation transfer.

We may conclude further that the intact envelope of chloroplasts provides a permeation barrier to cations. Thus cation loss from the thylakoids in the light would increase cation concentration in the stroma compartment. This may be of importance to the light-activation of certain enzymes of the carbon reduction cycle. The thylakoid membranes, on the other hand, seem to allow slow passive transfer of cations, as shown by the dark restoration of the fluorescence peak and the fact that cations responsible for the high fluorescence yield are washed out after osmotic rupture.

E. Acknowledgement

The author thanks Prof. U. Heber for critical discussions, Mrs. U. Behrend for technical assistance and the Deutsche Forschungsgemeinschaft for financial support of the work.

F. References

DEAMER, D.W., CROFTS, A.R., PACKER, L., Biochim. Biophys. Acta **131**, 81–96 (1967)
DILLEY, R.A., VERNON, L.P., Arch. Biochem. Biophys. **111**, 365–375 (1965)
DUYSENS, L.M.N., SWEERS, H.E., in Studies on Microalgae and Photosynthetic Bacteria (ed. Jap. Soc. of Plant Physiol.), pp. 353–372, The University of Tokyo Press, Tokyo 1963
HOMANN, P.H., Plant Physiol. **44**, 932–936 (1969)
JENSEN, R.G., BASSHAM, J.A., Proc. Nat. Acad. Sci. **56**, 1095–1101 (1966)
KAUTSKY, H., HIRSCH, A., Naturwissenschaften **19**, 964 (1931)
KRAUSE, G.H., Biochim. Biophys. Acta **292**, 715–728 (1973)
KRAUSE, G.H., Biochim. Biophys. Acta **333**, 301–313 (1974)
MEHLER, A.H., Arch. Biochem. **33**, 65–77 (1951)
MOHANTY, P., BRAUN, B.Z., GOVINDJEE, Biochim. Biophys. Acta **292**, 459–476 (1973)
MURATA, N., Biochim. Biophys. Acta **245**, 365–372 (1971)

Ionic Regulation in Chloroplasts as Monitored by Prompt and Delayed Chlorophyll Fluorescence

J. Barber and A. Telfer

A. Introduction

Little is known about ionic regulation in chloroplasts and to what extent these organelles exert their influence on the overall cellular ion transport processes. In this paper we wish to present evidence and discuss how certain optical phenomena, associated with the photosynthetic pigments, may help in gaining an understanding of ionic and electrical regulation within *in vivo* chloroplasts. Our investigations have involved the measurement of both prompt and delayed fluorescence and at this stage most of our data has been obtained with isolated chloroplasts retaining their outer membranes.

B. Materials and Methods

For isolation of broken spinach chloroplasts the procedure was identical to that already given (BARBER and VARLEY, 1972) while 'whole' chloroplasts were isolated by the method described by STOKES and WALKER (1971). Oxygen evolution was measured using a Rank oxygen electrode (TELFER and BARBER, 1974) and K^+ and pH stimulated 10 s delayed light was detected using a stop-flow apparatus (BARBER and VARLEY, 1972). The kinetics of prompt and ms delayed fluorescence were determined using a rotating sector phosphoroscope (BARBER, 1973). Low temperature fluorescence spectra were measured using a Perkin-Elmer spectrofluorimeter, Model MPF-3. Additional details are given in the figure legends.

C. Results and Discussion

I. Delayed Fluorescence

1. Salt and pH Effects.

All plants, after having been illuminated, emit weak light. This afterglow can be detected for several minutes after terminating the illumination, has complicated decay kinetics and has an emission spectrum corresponding to chlorophyll *a* fluorescence. There is now good evidence that this delayed fluorescence (often called delayed light emission, DLE) is a consequence of a back reaction in the photosystem II (S2) reaction centre being dependent on the concentration of the primary oxidant Z^+ and reductant Q^- (LAVOREL, 1968).

$$\text{Z Chl Q} \underset{}{\overset{h\nu}{\rightleftharpoons}} \text{Z Chl* Q} \rightleftharpoons \text{Z}^+ \text{ Chl Q}^- \qquad (1)$$

Since the back reaction giving rise to delayed light involves the chemical excitation of the chlorophyll singlet level (at rate J) then the observed emission (L) is dependent on the chl *a* fluorescence yield (ϕ)

$$L = \phi J \qquad (2)$$

where ϕ can be measured from the ratio of intensities of the prompt fluorescence (F) and the exciting light (I); that is $F = \phi I$.

Recently it has become clear that the intensity of delayed fluorescence is also controlled in some way by ionic gradients across the thylakoid membranes. This can be seen in Fig. 1A which shows that both K^+ and H^+ gradients cause a considerable increase in the intensity of the light emission. These effects have already been studied in some detail and the general conclusions are:

a) they are dependent on the permeability properties of the thylakoids (BARBER and KRAAN, 1970);

b) the stimulation by K^+ salts, and other salts, is due to the development of an electrical potential across the thylakoid membranes having a polarity which is positive on the inner side (BARBER and KRAAN, 1970; BARBER and VARLEY, 1972);

c) the H^+ effect is due to changes in the redox potentials of the primary or secondary electron donor and acceptor pools of the S2 reaction centre (KRAAN et al., 1970; WRAIGHT and CROFTS, 1971);

d) the chemical precursors of DLE are situated on opposite sides of the thylakoid membranes, Z^+ on the inner side and Q^- on the outside (BARBER and VARLEY, 1972);

e) the pH and electrical gradients act to reduce the activation barrier for the back reaction and are exponentially related to the intensity of the stimulated emission (BARBER and KRAAN, 1970; BARBER and VARLEY, 1972; FLEISCHMAN, 1971; WRAIGHT and CROFTS, 1971).

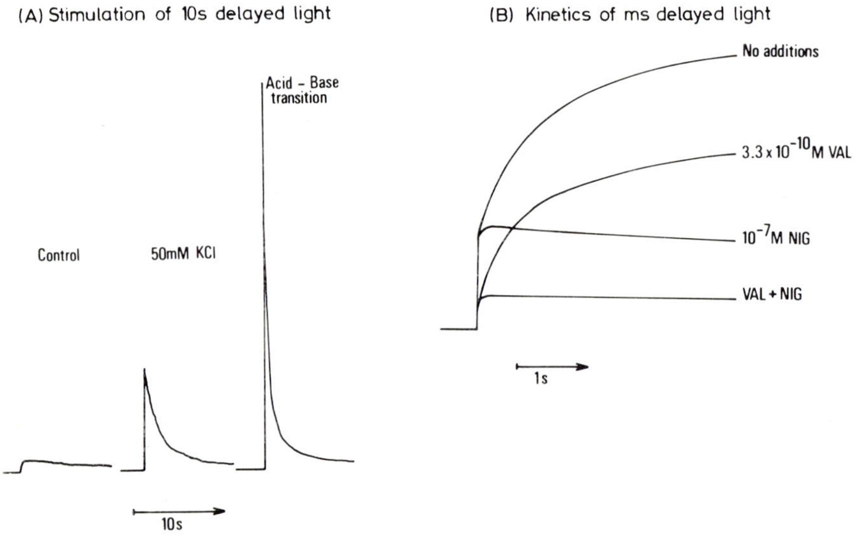

Fig. 1A Effect of a 50 mM KCl pulse and acid-base transition (pH 4.0 to 8.5) on 10 s DLE from broken chloroplasts. For the KCl induced signal the chloroplasts were treated with 10^{-7} M valinomycin. Chl conc. 20μg/ml. (B) Sensitivity of 1 ms DLE kinetics from 'whole' chloroplasts to nigericin and valinomycin. The chloroplasts were suspended in assay medium, 0.33 M sorbitol, 2 mM EDTA, 1 mM $MgCl_2$, 1 mM $MnCl_2$ and 50 mM HEPES (pH 7.6) at a chl conc. of 15 μg/ml; the beginning of the trace is the onset of illumination by blue light at an intensity of 25 kerg cm^{-2} s^{-1}.

2. High Energy State Effect

MITCHELL has suggested that electrical and pH gradients act as the high energy state (HES) between photophosphorylation and light induced electron transport (MITCHELL, 1966). If this is so, then it may be expected that the intensity of delayed fluorescence would be stimulated by the HES. Such a relationship has been reported (MAYNE, 1967) and studied in some depth with isolated broken chloroplasts (NEUMANN et al., 1973; WRAIGHT and

CROFTS, 1971). Fig. 1B shows the kinetics of 1 ms DLE measured with 'whole' chloroplasts under various conditions. The results indicate that both H^+ and K^+ are not in equilibrium between the granal and stromal compartments and that the HES consists of a rapidly produced electrogenic component and a more slowly produced chemical component. Although essentially the results are similar to earlier studies with broken chloroplasts (WRAIGHT and CROFTS, 1971) indicating that the pH component of HES is relatively significant in 'whole' chloroplasts (HELDT et al., 1973) they contrast in that very low concentrations of valinomycin inhibit, not only the fast phase, but also the extent of the slow phase. This would seem to indicate that a K^+ gradient contributes to the HES in 'whole' chloroplasts.

II. Prompt Fluorescence

1. Slow Fluorescence Quenching

At room temperature only the bulk chl ***a*** molecules of S2 fluoresce. Light energy absorbed by these S2 chl ***a*** molecules (chl a_2) has four fates: (a) used in photosynthesis, k_p; (b) radiated as fluorescence, k_f; (c) lost as heat, k_h; and (d) transferred to another non-fluorescing form of chl ***a***, k_t. Thus the fluorescence yield (ϕ) is given by:

$$\phi = \frac{k_f}{k_p + k_f + k_h + k_t} \qquad (3)$$

From Eq. 3 it can be seen that chl a_2 fluorescence can be used as an inverse measure of the efficiency of photosynthesis (DUYSENS and SWEERS, 1963). However, there are conditions when ϕ changes are not correlated with changes in k_p. This applies to slow fluorescence quenching measured with intact cells over relatively long times. These changes have been the subject of much study (GOVINDJEE and PAPAGEORGIOU, 1971), being complex and modified by the availability of O_2 and CO_2 to the cells (KRAUSE, 1973). As Figs. 2 and 3 show, we have found, in agreement with KRAUSE (1973), that a similar slow fluorescence quenching is observed with 'whole' chloroplasts but completely lost after osmotic shock (Fig. 3D). The rate of fluorescence quenching was dependent on the ability to fix CO_2 and was speeded up by increasing the rate of electron transport and/or by reducing the lag period before O_2 evolution. The chl ***a*** quenching was reversible in the dark and independent of the time taken to reach the low fluorescing state.

2. Effect of Nigericin and DCMU

Both nigericin (Fig. 2A) and DCMU (Fig. 2C) when added to a suspension of illuminated 'whole' chloroplasts relieve the low fluorescing state while at the same time inhibiting 1 ms DLE and electron transport (Fig. 2B and C). On injection of nigericin the rise in the level of fluorescence was rapid but with DCMU there was a fast and a slow phase. To clarify the involvement of electron transport in bringing about the quenched fluorescent state we have used oxaloacetate (OAA), as well as phosphoglycerate (PGA) and HCO_3^-, as electron acceptors. OAA, unlike PGA and HCO_3^-, does not require ATP for its reduction and thus allows electron transport to occur in the presence of uncouplers. As shown in Fig. 3A and associated Table (Fig. 3B) nigericin completely relieved the low fluorescent state of 'whole' chloroplasts incubated with OAA. This result, together with the slow reversal of the low fluorescent state in the dark or in the presence of DCMU and its close correlation with ms DLE, indicates that the quenching is more directly controlled by the HES than by redox changes in S2.

3. High Energy State and Cation Effects

Changes in ϕ have been induced with isolated broken chloroplasts which were independent of k_p either by creating a HES (MURATA and SUGAHARA, 1969; WRAIGHT and CROFTS, 1970) or by changing the cation content of the suspending medium (HOMMAN,

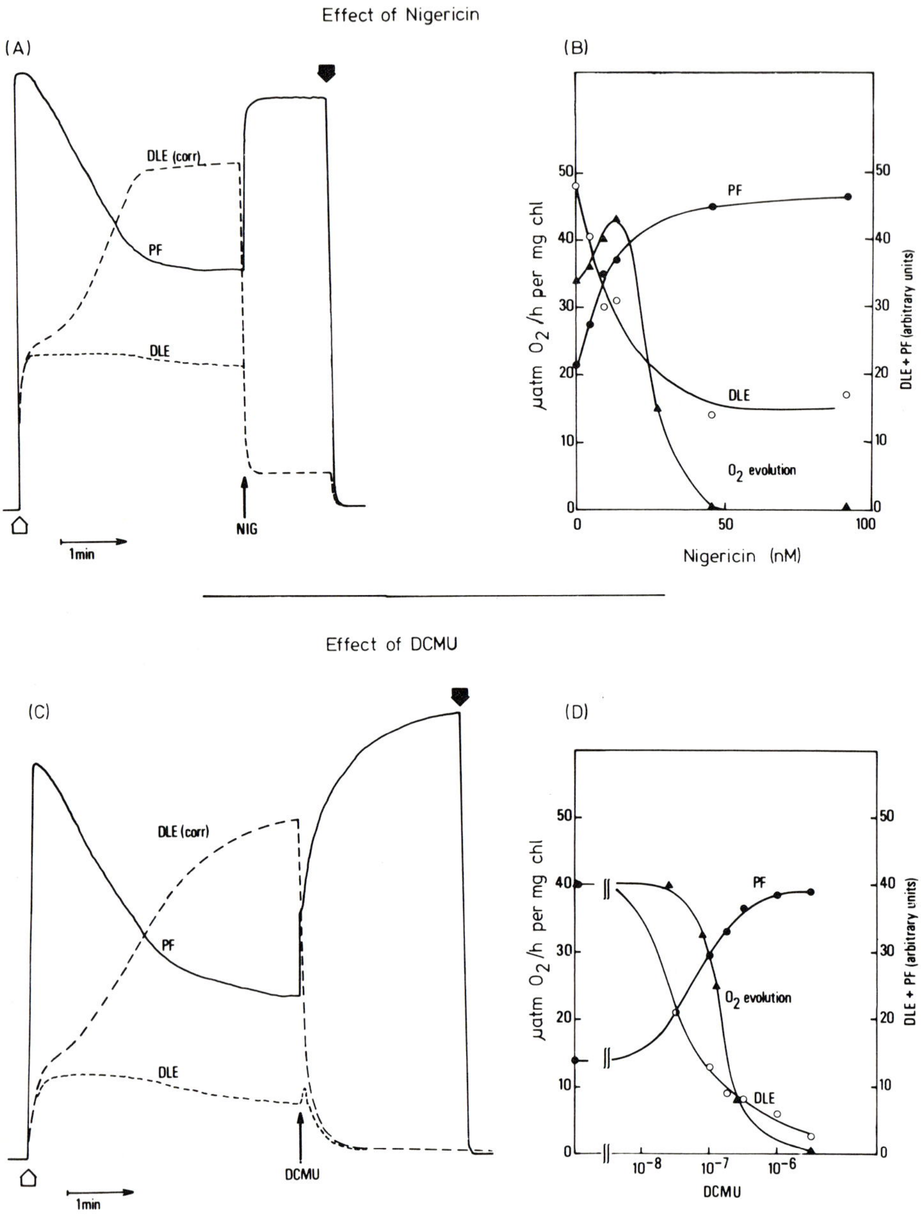

Fig. 2 Effect of nigericin and DCMU on prompt fluorescence (PF) quenching, 1 ms DLE and O_2 evolution measured with 'whole' chloroplasts in assay medium. Correction of DLE for changes of ϕ were carried out using Eq. 2. Light intensity for PF and DLE measurements was 25 kerg cm^{-2} s^{-1} and for O_2, 74 kerg cm^{-2} s^{-1}. (A) Injection of 90 nM nigericin; chl conc. 14 µg/ml. (B) Effect of various nigericin concs. on PF, DLE and O_2 evolution measured after 3 min of illumination. The suspension contained 1 mM PGA and the chl conc. was 30 µg/ml. (C) As (A) but with injection of 3.3 µM DCMU. (D) As (B) but with DCMU.

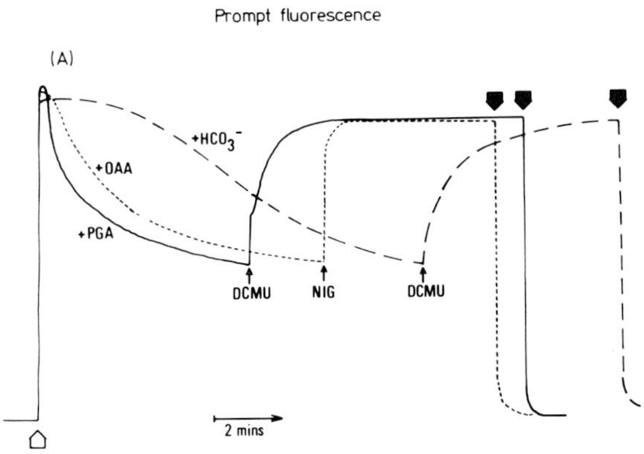

(B)

Electron Transport

	μatm O_2/h per mg chl	
	–	+ nigericin
+ HCO_3^-	22	0
+ PGA	39	0
+ OAA	18	24

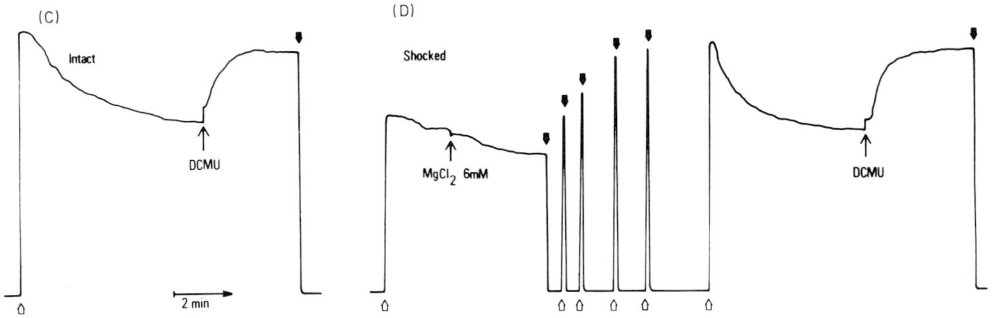

Fig. 3(A) Time course of fluorescence induced by blue light at 70 kerg cm^{-2} s^{-1} with 'whole' chloroplasts suspended in assay medium containing either 5 mM HCO_3^-, 1 mM PGA or 1 mM OAA. Injections were 90 mM nigericin and 3.3 μM DCMU. Chl conc. 31 μg/ml. (B) Electron transport data measured under the same conditions as (A) with and without 90 nM nigericin. (C) and (D) comparison of fluorescence quenching of 'whole' chloroplasts before and after osmotic shock showing effect of injecting 6 mM $MgCl_2$ into the assay medium. The open and dark arrows indicate onset and termination of illumination (90 kerg cm^{-2} s^{-1}) and the DCMU conc. was 3.3 μM; chl conc. 100 μg/ml.

1969; MURATA, 1969). It was from observations of cation control of ϕ together with the work of BONAVENTURA and MYERS (1969) and others that a model for excitation control of the two photosystems (S1 and S2) was developed (GOVINDJEE and PAPAGEORGIOU, 1971). Essentially it is visualised that S1 and S2 light can bring about net cation and water movement in such a way as to induce conformational changes of the thylakoids so that k_t varies. The mechanism operates so as to enable maximum utilization of the absorbed light energy. The high fluorescing state occurs when S2 is excited to a maximum (k_t is a minimum) and is called State 1 while the low fluorescing state is when k_t is at a maximum and is called State 2.

In our search to reconstitute osmotically broken chloroplasts in such a way as to regain the fluorescence properties of 'whole' chloroplasts we have found that the important factor is the cation level of the medium. In particular increasing the Mg^{2+} concentration to 6 mM raises the fluorescence yield and regenerates the ability of the suspension to show light induced quenching (see Fig. 3C and D). For the Mg^{2+} addition to increase ϕ there must either be a dark pretreatment period, as shown in Fig. 3D, or this divalent cation must be added to chloroplasts treated with DCMU or uncoupler. Only in the former case can the light induced fluorescence quenching occur. Ca^{2+} will mimic Mg^{2+} with maximum effect at about 6 mM while for K^+ and Na^+ concentrations in excess of 100 mM are required to induce the same effect.

4. State 1 – State 2 – Hypothesis and Low Temperature Spectra

The above results suggest that the high fluorescing state is created in someway by Mg^{2+} on the inner side of the granal membranes. The fluorescence quenching can be explained either in terms of a net efflux of Mg^{2+} from the granal interiors or as a release of Mg^{2+} from specific binding sites on the inner side of the membrane, both of which could be induced by inward primary pumping of H^+ (i.e. the establishment of the HES). The question now arises whether the high and low fluorescing levels seen with 'whole' chloroplasts correspond to State 1 and State 2 respectively.

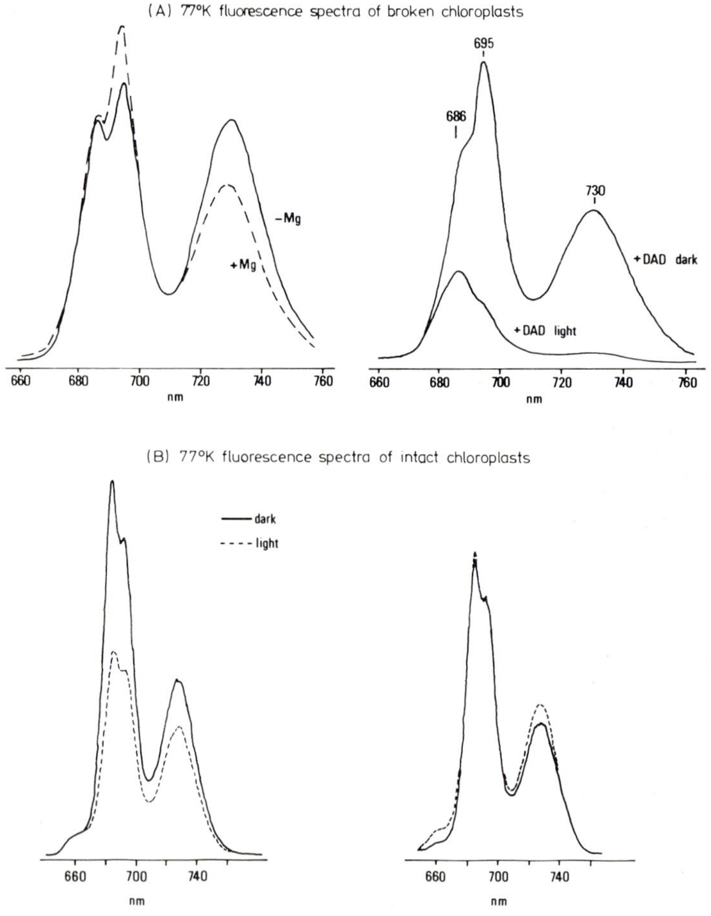

Fig. 4 Emission spectra measured at liquid N_2 temperature. Excitation monochromator was set at 440 nm with a slit width of 40 nm and emission was recorded through a slit width of 4 nm. The broken chloroplasts were treated width 3.3 μM DCMU and spectra (B) obtained with 'whole' chloroplasts have been normalised at the 695 nm peak. The light treatment to induce DAD quenching in broken chloroplasts and quenching state in 'whole' chloroplasts was for 3 and 8 min respectively with blue light at 80 kerg cm^{-2} s^{-1}.

Our preliminary investigations into this question have made use of the fact that both chl a_2 and chl a_1 fluoresce at liquid N_2 temperatures. In agreement with MURATA (1969) we found that with osmotically broken chloroplasts treated with DCMU the low temperature chl a_2 peaks at 686 nm and 695 nm were raised when the Mg^{2+} level was increased and that there was a drop in the chl a_1 peak at 730 nm indicating that there had been a change in transfer efficiency between chl a_2 and chl a_1 (Fig. 4). In contrast HES quenching of DCMU poisoned broken chloroplasts induced by diaminodurene (DAD) mediated electron flow caused an overall quenching, particularly the 695 nm and 730 nm peaks (Fig. 4). With 'whole' chloroplasts we have found it difficult to decide whether there is a change in k_t between the high and low fluorescing state. As Fig. 4 shows there is an overall light induced quenching of all three peaks and when the two spectra are normalised at the 695 nm peak there is some indication of a change in transfer efficiency.

D. Conclusions

Both delayed and prompt chl a_2 fluorescence reflect the establishment of the HES in O_2 evolving photosynthetic organisms and support the concept that the HES is an electrical and pH gradient set up across the thylakoid membranes by light induced electron flow. For DLE the HES seems to lower the activation barrier for a back reaction in the S2 reaction centre involving a specifically orientated charge transfer complex within the thylakoid membrane. In the case of prompt fluorescence the quenching is associated with secondary cation movement or displacement induced by H^+ pumping into the thylakoids. These changes in prompt fluorescence are probably due to both variations in k_h and k_t. Although the relative significance of the k_h and k_t changes are not yet known they are almost certainly due to light induced conformational changes of the thylakoid membranes which bring about changes in the physical status of the pigments (orientation and distance). Such conformational changes are likely to be induced by ionic and water movement.

Overall, from the above results it is clear that the outer chloroplast membranes play an important role in ionic regulation between granal and stromal compartments and they emphasize the possibility of using delayed and prompt fluorescence as intrinsic probes of ionic changes and regulation in *in vivo* chloroplasts.

E. Acknowledgements

The authors wish to thank Mrs. J. Nicholson for technical assistance and Dr. G.H. Krause for giving us access to his results before publication. The work was supported by grants from the Science Research Council, The Royal Society and the Central Research Fund of the University of London.

F. References

BARBER, J., in Ion Transport in Plants (ed. W.P. Anderson), pp. 191–204, Academic Press, London and New York 1973

BARBER, J., KRAAN, G.P.B., Biochim. Biophys. Acta **197**, 49–59 (1970)

BARBER, J., VARLEY, W.J., J. Exp. Bot. **23**, 216–228 (1972)

BONAVENTURA, C., MYERS, J., Biochim. Biophys. Acta **189**, 366–383 (1969)

DUYSENS, L.N.M.,SWEERS, H.E., in Studies on Microalgae and Photosynthetic Bacteria (ed. Japanese Society of Plant Physiology), pp. 353–372, University of Tokyo Press, Tokyo 1963

FLEISCHMAN, D.E., Photochem. Photobiol. **14**, 277–286 (1971)

GOVINDJEE, PAPAGEORGIOU, G., in Photophysiology (ed. A.C. Giese), Vol. VI, pp. 1–46, Academic Press, London and New York 1971

HELDT, H.W., WERDAN, K., MILOVANCEY, M., GELLER, G., Biochim. Biophys. Acta **314**, 224–241 (1973)
HOMMAN, P.H., Plant Physiol. **44**, 932–936 (1969)
KRAAN, G.P.B., AMESZ, J., VELTHUYS, B.R., STEAMERS, R.G., Biochim. Biophys. Acta **223**, 129–145 (1970)
KRAUSE, G.H., Biochim. Biophys. Acta **292**, 715–728 (1973)
LAVOREL, J., Biochim. Biophys. Acta **153**, 727–730 (1968)
MAYNE, B.C., Photochem. Photobiol. **6**, 189–197 (1967)
MITCHELL, P., Biol. Rev. Cambridge Phil. Soc. **41**, 445–502 (1966)
MURATA, N., Biochim. Biophys. Acta **189**, 171–181 (1969)
MURATA, N., SUGAHARA, K., Biochim. Biophys. Acta **189**, 182–192 (1969)
NEUMANN, J., BARBER, J., GREGORY, P., Plant Physiol. **51**, 1069–1073 (1973)
STOKES, D.M., WALKER, D.A., Plant Physiol. **48**, 163–165 (1971)
TELFER, A., BARBER, J., Biochim. Biophys. Acta **333**, 343–352 (1974)
WRAIGHT, C.A., CROFTS, A.R., Eur. J. Biochem. **17**, 319–327 (1970)
WRAIGHT, C.A., CROFTS, A.R., Eur. J. Biochem. **19**, 386–392 (1971)

Chloroplast Reflection Coefficients: Influence of Partition Coefficients, Carriers, and Membrane Phase Transitions

P.S. Nobel

A. Introduction

When the osmotic pressure in a solution surrounding cells or organelles is increased, water flows out of them. Such osmotic responses are quantitatively described by the Boyle-Van't Hoff relation, which refers to the equilibrium situation when the internal osmotic pressure equals the external osmotic pressure (π^{o}). According to the Boyle-Van't Hoff relation, the osmotic pressure times the volume of the aqueous phase within the cell or organelle (V—b), where V is the total volume and b is the volume of the non-aqueous components is constant at a given temperature, i.e. π^{o} (V—b) = constant. When the osmotic pressure of some solute in the external solution is increased, water flows out of the internal phase and the osmotic pressures on the two sides of the membrane again become equal.

The classical Boyle-Van't Hoff relation presupposes that the membranes are permeable to water only. Irreversible thermodynamics can be used to rederive the Boyle-Van't Hoff relation to include the more realistic case of penetrating solutes (NOBEL, 1974). This leads to the introduction of a reflection coefficient (σ), which can conveniently be defined as the ratio of the external osmotic pressure of an impermeant solute producing water flux equilibrium divided by the external osmotic pressure of the test solute also producing water flux equilibrium (KEDEM and KATCHALSKY, 1958). The reflection coefficient is unity for an impermeant solute and zero for freely permeant solutes, with intermediate values for the many cases between the two extremes. The modified form of the Boyle-Van't Hoff relation for penetrating solutes is $\sigma^{o}\pi^{o}$ (V—b) = constant, where σ^{o} is the mean reflection coefficient of the external solutes (NOBEL, 1974).

B. Materials and Methods

Chloroplasts were isolated from 14-day-old ***Pisum sativum*** L. cv. Blue Bantam in 0.2 M sucrose, 10 mM Tris-HCl (pH 7.9) using a gentle 2 min technique yielding 95 % class I (intact) chloroplasts (NOBEL, 1967). The chloroplasts were generally placed in 0.2 M sucrose, 10 mM Tris-HCl (pH 7.9), plus various concentrations of other solutes or additional sucrose and then centrifuged at 1°C in pre-weighed tubes for 3 min at 10 000Xg. After decanting the supernatant fluid, the weight of the pellet was determined. The weights were converted to volumes using measured densities (NOBEL and WANG, 1970). The volume of interstitial fluid trapped in a pellet was corrected for using a series of ^{14}C-dextran concentrations (NOBEL, 1968). To obtain the volume per chloroplast (V), the conversion factor of $0.93 \cdot 10^{9}$ pea chloroplasts/mg chlorophyll determined by a hemocytometer and a Coulter counter (NOBEL, 1968) was used. Osmotic pressures were measured with an Advanced Instruments Osmometer Model 31 LAS.

C. Results and Discussion

I. Osmotic Responses

Fig. 1 presents osmotic responses of pea chloroplasts when sucrose or glycerol is the sole external solute in addition to the buffer (Tris-HCl, pH 7.9, the concentration of which was 2 % of the indicated solute). The data can be analyzed using the modified Boyle-Van't Hoff relation in the following form: V = (constant/ ($\sigma^{\circ}\pi^{\circ}$)) + b. The intercept on the ordinate (b) respresents the chloroplast volume expected for an infinite external osmotic pressure; no water would then remain in the organelle, and so V then equals the non-aqueous volume b. For pea chloroplasts b is 17μm^3, the relatively large value being due to the many lamellar membranes and proteins in the chloroplasts. The osmotic pressure of the cytoplasm of pea mesophyll cells is about 7.0 bars. At this osmotic pressure the chloroplast volume would be 35 μm^3 according to Fig. 1, and therefore such chloroplasts would be only about one-half water by volume (NOBEL, 1974).

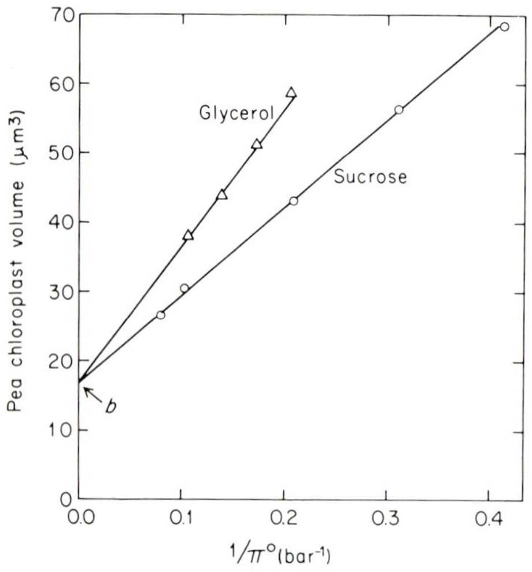

Fig. 1 Osmotic responses of pea chloroplasts to glycerol and glucose.

Experiments using ^{14}C-sucrose showed that it does not enter pea chloroplasts to any appreciable extent over the time period of these experiments, and so the reflection coefficient of sucrose can be taken as unity. Thus, comparison of the osmotic response of some solute to that of sucrose will directly give the reflection coefficient of the solute. For example, the slopes of the lines in Fig. 1 are inversely proportional to the reflection coefficients. The slope of the osmotic response to sucrose divided by that to glycerol is 0.63, and so the reflection coefficient of glycerol is here 0.63.

Having established the general validity of the modified Boyle-Van't Hoff relation for interpreting osmotic responses of pea chloroplasts, the reflection coefficients of various nonelectrolytes under different conditions were investigated. For these studies the external solution always contained 0.2 M sucrose and 10 mM Tris-HCl (pH 7.9) (plus various osmotic pressures of the compound of interest). Therefore the data could be conveniently analysed using the following relation: $1/(V-b) = a\, \sigma_j\, \pi_j + \beta$ (a and β are constants, subscript j refers to species j). When $1/(V-b)$ is plotted versus π_j, the slope is then directly proportional to the reflection coefficient of the compound added to chloroplasts in a buffered sucrose solution.

II. Polyhydroxy Alcohols and Partition Coefficients

Fig. 2 presents the osmotic responses of a series of polyhydroxy alcohols. The two-carbon alcohol, ethylene glycol, leads to a slope which is 40 % of that for sucrose, and so its reflection coefficient is 0.40. The reflection coefficient of glycerol is 0.63 (the same value as obtained in a rather different manner from Fig. 1), while the σ_j for the four-carbon polyhydroxy alcohol, erythritol, is 0.90. Thus, the permeability of these compounds becomes less as the chain length increases. This is summarized in Table 1, where the reflection coefficients of two more members of the series, methanol and adonitol, are also included.

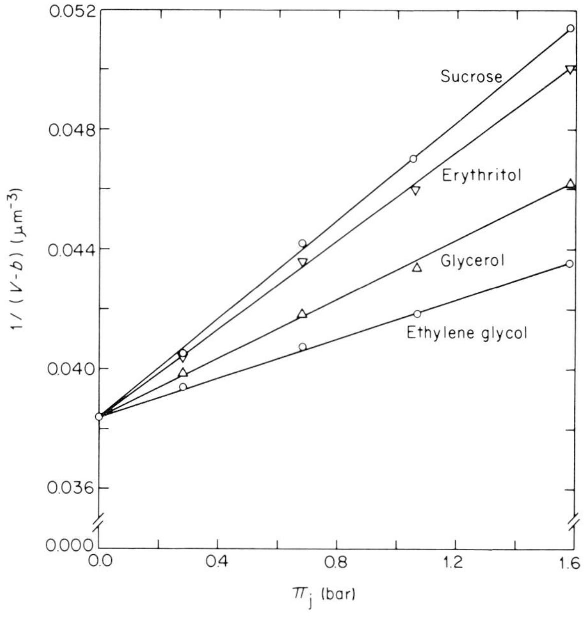

Fig. 2 Osmotic responses of polyhydroxy alcohols. Reflection coefficients are proportional to the slopes of the lines.

Except for carrier-mediated fluxes, the permeation of most nonelectrolytes through biological membranes is proportional to the partition coefficients (K_j) of the solutes (DIAMOND and WRIGHT, 1969). Table 1 indicates that the partition coefficients of alcohols decrease as hydroxymethyl groups are added, and so the increase in reflection coefficient with alcohol chain length observed for pea chloroplasts is expected. (Permeation also depends on the diffusion coefficient in the membrane which decreases about 3-fold from methanol to adonitol while the ether-water partition coefficient decreases approximately 3500-fold, and so K_j has the major influence on σ_j in this case.)

The six-carbon polyhydroxy alcohols, sorbitol, mannitol, and inositol, have reflection coefficients of unity for pea chloroplasts (WANG and NOBEL, 1971). This indication of relative impermeability for these compounds suggests that they would be suitable osmotica in which to suspend chloroplasts, as is indeed the case.

Table 1

Correlation between ether-water partition coefficients (K_j) (WRIGHT and DIAMOND, 1969) and reflection coefficients (σ_j) for pea chloroplasts.

Solute	K_j^{ether}	σ_j
Methanol	0.14	0.00
Ethylene glycol	0.0053	0.40
Glycerol	0.00066	0.63
Erythritol	0.00011	0.90
Adonitol	0.00004	1.00
Glycine	0.0000007	0.03
L-Alanine	0.0000014	0.05

III. Amino Acid Carriers

Reflection coefficients are defined only for neutral species. Since the proton which dissociates from the carboxyl group of an amino acid can be taken up by the amino group, amino acids like glycine and alanine are essentially electrically neutral at the pH's used here (both 7.9 and 6.5 were employed). Osmotic responses of such neutral amino acids can therefore be interpreted using reflection coefficients.

The ether-water partition coefficients of glycine and alanine are lower than that for adonitol, a compound with a σ_j of 1.00 (Table 1). Based on this, the reflection coefficients of glycine and alanine would be expected to be unity, instead of the very low values observed (Table 1). Apparently the movement of these amino acids across the chloroplast limiting membranes is not based on ordinary diffusion across the lipid phase of the membrane.

Fig. 3 shows that the osmotic response of pea chloroplasts to glycine depends on its osmotic pressure. Up to about 1 bar the reflection coefficient was about 0.03. At higher osmotic pressures the efflux of water for a given increment in glycine osmotic pressure increased. Above 3 bars the slope became the same as for sucrose, indicating that the local reflection coefficient then became unity. Such an increase in osmotic response as the osmotic pressure of a solute is raised is consistent with a saturation effect that can accompany a carrier-mediated uptake. The carrier that facilitates the entry of glycine and thus leads to an anomalously low reflection coefficient at the lower osmotic pressures apparently becomes saturated at high osmotic pressures, in which case a reflection coefficient of unity as predicted from the partition coefficient is obtained.

A characteristic feature of carrier-mediated transport is competition between structurally similar compounds. Fig. 3 shows that the presence of alanine (osmotic pressure of 1.1 bars) shifted the osmotic response curve of glycine. The chloroplasts responded to a given osmotic pressure of glycine plus alanine similarly to the same total osmotic pressure of glycine alone, suggesting that these two amino acids were competing equally well for a common carrier. On the other hand, the osmotic response to glycine was not affected by the presence of L-serine (even at 2.2 bars). Serine is evidently transported across pea chloroplast membranes by a carrier distinct from the one for glycine and alanine (NOBEL and CHEUNG, 1972).

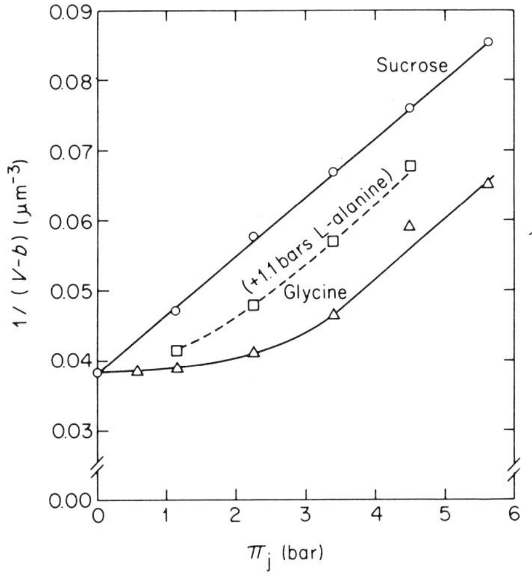

Fig. 3 Osmotic responses to glycine. Dashed line represents responses in the presence of alanine.

IV. Membrane Phase Transitions

As the temperature is increased, the passive permeabilities of nonelectrolytes generally become greater, and so the reflection coefficients should then decrease (WRIGHT and DIAMOND, 1969). To study this the usual procedure was modified by isolating chloroplasts and determining their osmotic responses at each of a series of temperatures. Fig. 4 indicates that the reflection coefficient of glycerol for pea chloroplasts monotonically decreased from 0.62 at 1°C to 0.03 at 28°C.

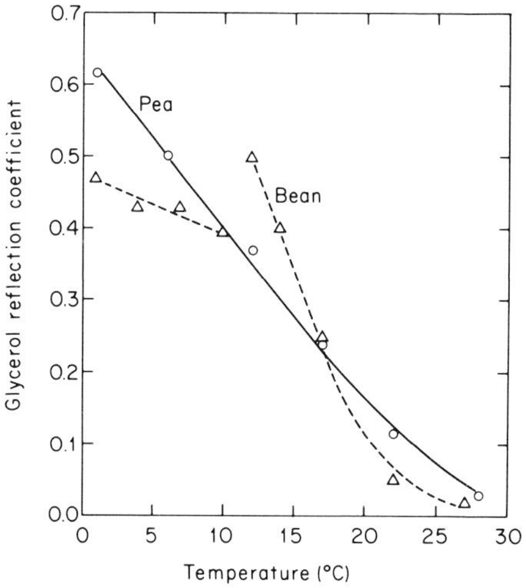

Fig. 4 Temperature dependence of the glycerol reflection coefficient for chloroplasts from chilling-resistant (pea) and chilling-sensitive (bean) plants.

Pisum sativum is a chilling-resistant plant and so does not exhibit any particular physiological dysfunction below about 12°C, as does a chilling-sensitive plant like the bean. To compare the two types, osmotic responses of chloroplasts from *Phaseolus vulgaris* L. cv. Kentucky Wonder were examined (Fig. 4). Again, the reflection coefficient for glycerol decreased as the temperature was raised. However, a break in the curve occurred between 10°C and 12°C, and the slopes were not the same on the two sides of the break.

The break may be due to a phase transition of the membrane lipids so that the chloroplast limiting membranes have different permeability characteristics above and below the transition temperature. A break in the photoreduction rate of $NADP^+$ by bean chloroplasts has been observed at 12°C, which has also been attributed to a phase transition (SHNEYOUR et al., 1973). Below the phase transition temperature, the permeability of bean chloroplasts is higher than expected based on an extrapolation of the data obtained above 12°C. The increased permeability at low temperatures could lead to a disruption of subcellular compartmentation in chilling-sensitive plants. This in turn may in part account for the detrimental effect of low temperatures on chilling-sensitive plants.

V. Other Applications of Organelle Reflection Coefficients

Ozone can reduce photosynthesis by whole plants before any external damage is visible on the leaves. One of the earliest symptoms of ozone injury is the breakdown of chloroplast limiting membranes as observed in electron micrographs (THOMSON et al., 1966). Here, pea plants were fumigated with 1.0 ppm O_3 for 4 h immediately prior to chloroplast isolation. This did not lead to any detectable net loss of chlorophyll, since such exposure to ozone led to 57.7 ± 0.9 (S.E., 12 experiments) μg chlorophyll/cm^2 of leaf, while exposure to air under the same conditions gave 57.3 ± 1.1 μg chlorophyll/cm^2. (Exposing isolated Class I or Class II chloroplasts to 30 ppm O_3 for 5 min also did not lead to any statistically significant destruction of chlorophyll compared to bubbling with air.) On the other hand, the ozone treatment of the plants caused the reflection coefficient of the chloroplasts for erythritol to be 0.22 lower than for the controls (NOBEL and WANG, 1973). This suggests that one of the important early effects of ozone on plants may be an increase in chloroplast permeability and consequent disruption of subcellular compartmentation.

The simplicity of the packed weight technique (although great precision is required) and the straightforward analysis using the modified Boyle-Van't Hoff relation suggest that the determination of osmotic responses can profitably be applied in other areas. Such studies on mitochondrial permeability have recently been undertaken (NOBEL, 1973). Changes in permeability during organelle development or effects of disease on organelle permeability could also be investigated using reflection coefficients.

D. Acknowledgements

This research was supported by Public Health Service Grant No. GM 15183 from the National Institute of General Medical Sciences.

E. References

DIAMOND, J.M., WRIGHT, E.M., Ann. Rev. Physiol. **31**, 581–646 (1969)
KEDEM, O., KATCHALSKY, A., Biochim. Biophys. Acta **27**, 229–246 (1958)
NOBEL, P.S., Plant Physiol. **42**, 1389–1394 (1967)
NOBEL, P.S., Plant Physiol. **43**, 781–787 (1968)

NOBEL, P.S., J. Membrane Biol. **12**, 287–299 (1973)
NOBEL, P.S., Introduction to Biophysical Plant Physiology. W.H. Freeman, San Francisco 1974
NOBEL, P.S., CHEUNG, Y.-N.S., Nature New Biology **237**, 207–208 (1972)
NOBEL, P.S., WANG, C.-T., Biochim. Biophys. Acta **211**, 79–87 (1970)
NOBEL, P.S., WANG, C.-T., Arch. Biochem. Biophys. **157**, 388–394 (1973)
SHNEYOUR, A., RAISON, J.K., SMILLIE, R.M., Biochim. Biophys. Acta **292**, 152–161 (1973)
THOMSON, W.W., DUGGER, W.M., JR., PALMER, R.L., Can. J. Bot. **44**, 1677–1682 (1966)
WANG, C.-T., NOBEL, P.S., Biochim. Biophys. Acta **241**, 200–212 (1971)
WRIGHT, E.M., DIAMOND, J.M., Proc. Roy. Soc. B. **172**, 227–271 (1969)

Round Table Discussion 5

Chairman: P.S. Nobel

VREDENBERG opened by saying that he thought we had reached a situation where we can measure the primary ionic events in intact chloroplasts by means of potential changes occuring at the enclosing membrane. He had a microelectrode inside the chloroplast with its tip probably in the stroma. A saturating light flash (rise time 5 μs, half width 2 ms) causes a positive going potential, of a few mV, across the chloroplast outer membrane with a rise time of the order of 0.5 ms. If the flash is given in the presence of continuous (717 nm) background illumination, the rate constant of the pseudo-first-order kinetics of the voltage response does not change but its amplitude depends on the intensity of the background illumination. (6.5, 13 and 29 kerg cm^{-2} s^{-1} were used.) VREDENBERG interprets these results according to the scheme of Grünhagen and Witt, in which the primary electron acceptors (X and Q) are at the stroma-facing side of the thylakoid membrane. The initial step in photosynthesis is charge separation followed by protonation of the electron acceptors to, say, XH and QH. For charge balance reasons these sites will then bind a negative ion, say OH^-, and it can be argued that this ion-binding process proceeds according to first-order kinetics. VREDENBERG suggests that this binding of a negative charge at the stroma face of the thylakoids causes a polarization of the chloroplast envelope membrane, which he is in fact measuring with his electrode. If this is so then he is measuring the kinetics of this binding and something proportional to the number of binding, i.e. electron acceptor, sites. Such an interpretation explains his results in the continuous background illumination, for such illumination will cause some of the sites to be already occupied and hence the voltage response, but not its kinetics, will be reduced.

He has also observed that, in the presence of DCMU, the potential response of a second flash, fired 250 ms after the first, is about one half of that of the first one. This is consistent with the experiments of Schliephake et al. (1968). The sites occupied by the primary acceptor Q of system II will be already in the reduced state, because of the presence of DCMU, when the second flash is fired. (If DCMU is not present the second response is equal to the first.) He thus thinks that we have an additional method of studying ionic events in the thylakoids associated with the primary events of photosynthesis, and in an intact system.

WALKER raised technical objections to the method, for he thought the electrode tip might well be among punctured thylakoids and he was worried about the sealing of the microelectrode shaft with the chloroplast envelope. BARBER thought well of VREDENBERG's model, even if he also appeared somewhat dubious about the technique. JAFFE was puzzled that mere redistributions of charges at the surface of a thylakoid could cause potential changes across the chloroplast envelope. VREDENBERG and JUNGE tried to persuade him that it was an induction effect, essentially, that did not violate the laws of physics.

JUNGE then made some comments on the two conflicting schemes for electrochemical potential generation in photosynthesis presented by HOPE and himself. HOPE's scheme predicts an H^+/e^- ratio of 1, whereas JUNGE's scheme predicts that H^+/e^- equals 2. JUNGE then marshalled a lot of evidence in favour of his scheme. He argued, on the basis of the electrochroic absorption changes at 515 nm under various conditions, that both light reactions contribute equal amounts to the generated electric potential across the thylakoid membrane. Using cresol red as a pH indicator in the outer phase, and by means of the different properties of benzylviologen and ferricyanide as terminal electron acceptors, and by adding FCCP when he wanted to study proton production in the inner phase, he claims to have conclusively demonstrated that $2H^+$ are produced per e^- transferred. HOPE, in reply, was somewhat sceptical about the use of P_{515} as a voltmeter measuring the PD across the thylakoid membrane; and he pointed out that the $2H^+/e^-$ ratio is found from work done with flashes and not with continuous illumination and this may reflect the differences between them. There was some argument between HOPE and JUNGE about whether the pH decay, in the ms region, after switching off continuous light was biphasic or not. JUNGE said it was and that their studies gave evidence of a diffusion potential after switching off the light, which could be influenced by changing the ratio of proton to potassium permeability of the membrane. P_{515} changes supported these results.

BARBER drew attention to the actual dimensions of the thylakoids; they are about 5000 Å in diameter and only 200 Å thick. They thus have a remarkable amount of surface area and not much volume. What do permeability, electrical neutrality, etc. mean in this kind of system? How can they maintain gradients across such large surface areas with such small volumes? JUNGE said that a lot of protons, about 20 000 per thylakoid, were accumulated in the steady state; but the inner phase had a large buffer capacity and most of these protons are buffered by the polar end groups of the lipid in the inner phase. HOPE, SMITH, and JUNGE argued for a while about the meaning of pH when only a few protons are swimming about in the inner phase volume. And SMITH also stressed the importance of BARBER's worries about the large surface area and small inner volume, particularly with respect to rates of processes across such a membrane system.

In answer to a question by BEN-AMOTZ on the influence of the outer and inner pH's on the light reactions and on whether there was any meaning to pH in the membrane itself, JUNGE spoke of the evidence for a shielding layer covering the reducing sites of the light reaction. In consequence, the pH beneath the shielding layer in the membrane will differ from the pH in the outer aqueous phase if the electron transport chain is active.

Finally KRAUSE asked: what is the counter ion in H^+ uptake? Previously HOPE had said it might be chloride, but KRAUSE believed that a more likely candidate was Mg^{2+}, for this exchange for H^+ would involve little volume change. In fact it is hard to imagine Cl^- being taken up for the thylakoid would swell; in fact, in intact systems or whole cells, the thylakoids shrink on illumination. NOBEL agreed with KRAUSE on the likelihood of Mg^{2+} (and Ca^{2+}) coming out in exchange for H^+ and suggested that more attention should be paid to this process.

The following remarks were communicated by HOPE and JUNGE. In response to a comment by JUNGE on the slowness of the response of a glass electrode which would thus underestimate the ΔH^+ signal (and thus the H^+/e^- ratio is really 2 and not 1), HOPE replied that although the response time of his glass electrode was about 0.6 s, the decay of ΔH^+ signal is approximately exponential with a decay half-time of 10 to 20 s. The speed of response does not appear to be an issue for Schröder et al. (1971) found H^+/e^- to be about 1 using fast responding cresol red to measure ΔH^+; ratios of 2 were only found when valinomycin was added. HOPE said that no doubt one day these integer values will be plausibly interpreted; perhaps, as suggested by JUNGE and others, the electrogenic effect of the H^+ influx is eliminated essentially instantaneously when the light is switched off and the driving force for passive H^+ efflux is then different from its value in the light; but this depends on the interpretation of the P_{515} changes which are controversial estimates of the PD.

JUNGE communicated the following reply.

It has been demonstrated under light flash excitation that $2H^+$ are translocated across the thylakoid membrane per electron transferred through the electron transport chain (Schliephake et al., 1968; Junge and Ausländer, 1974; Junge et al., 1974). One may argue that the H^+/e^- ratio is altered under steady illumination of photosynthetic membranes. Under these conditions the proton translocation is usually measured with glass electrodes as in HOPE's experiments. A glass electrode is a rather slowly responding device. It does not follow processes as rapid as the breakdown of the electric potential across the functional membrane of photosynthesis, which under flash excitation takes about 100 ms (Junge and Witt, 1968) and becomes even faster when continuous light is switched off (Boeck and Witt, 1971). Thus a glass electrode does not get information on the rate of proton efflux from thylakoids a few milliseconds after switching off continuous light, but at best it detects the rate 0.5 s later. Since the electric potential changes during this interval, especially if the proton permeability dominates the total conductivity of the membrane, the driving force for the proton efflux is altered, too, and hence the flux rate. The polarity of the electric potential difference being positive inside (Junge and Witt, 1968) the rate of efflux slows down after the breakdown of the electric potential which follows the switching off of continuous light. Thus glass electrode measurements necessarily underestimate the steady state H^+/e^- ratio. This argument was first stressed by Schröder et al. (1971) in order to explain their finding of an $H^+/e^- = 1$ at pH 8 in the absence, and a ratio $H^+/e^- = 2$ in the presence, of the K^+ permeability increasing valinomycin. In the latter situation the electric potential persists on switching off the light as a diffusion potential until the H^+ and K^+ concentration gradients have collapsed.

A similar argument holds even if instead of a glass electrode a pH-indicating dye is used. As evident from the contribution of Junge et al. (1974) even dyes are 'slow'indicators of pH changes in the outer phase, due to the existence of a permeability barrier for protons at the outer side of the membrane covering the core across which the electric potential exists.

References

BOECK, M., WITT, H.T., Proc. IInd Intern. Congr. Photosynthesis Res. Stresa (eds. G. Forti, M. Avron, A. Melandri), pp. 903–911, Dr. W. Junk Publ., Den Haag 1971

JUNGE, W., WITT, H.T., Z. Naturforsch. **23b**, 244–254 (1968)

JUNGE, W., AUSLÄNDER, W., Biochim. Biophys. Acta **333**, 59–70 (1974)

JUNGE, W., AUSLÄNDER, W., ECKHOF, A., in Membrane Transport in Plants (eds. U. Zimmermann, J. Dainty), Springer-Verlag, Berlin–Heidelberg–New York 1974

SCHLIEPHAKE, W., JUNGE, W., WITT, H.T., Z. Naturforsch. **23b**, 1561–1571 (1968)

SCHRÖDER, H., MUHLE, H., RUMBERG, B., Proc. IInd Intern. Congr. Photosynthesis Res. Stresa (eds. G. Forti, M. Avron, A. Melandri), pp. 919–930, Dr. W. Junk Publ., Den Haag 1971

Session 6

ATPases and Transport

Chairman: J. Willenbrink

Cation Sensitivity of the Plasma Membrane ATPase of Oat Roots

N.E. Balke, Heven Sze, R.T. Leonard, and T.K. Hodges

A. Introduction

The plasma membrane ATPase of oat roots requires Mg^{2+} and is further stimulated by salts of monovalent ions such as KCl (LEONARD and HODGES, 1973). The monovalent ion stimulation appeared to be due to the cation, however, this result was equivocal since substances such as Tris chloride and choline chloride also stimulated the ATPase. Other studies have shown that a semi-purified membrane fraction of turnip roots exhibited ATPase activity that was more sensitive to anions than to cations (RUNGIE and WISKICH, 1973), and it has also been claimed that organic cations can stimulate the ATPase activity of crude membrane preparations of cereal roots, including oats (RATNER and JACOBY, 1973). Thus, further clarification of the ion sensitivity of the plasma membrane ATPase of oat roots was sought.

In the present paper we show that the KCl stimulation is due to K^+ and not due to Cl^-. It is also shown that K^+ stimulation is an effect separate from the enhanced activity due to Tris. It has also been found that the plasma membrane ATPase is severely inhibited by Ca^{2+}. In addition, the presence of an anion-sensitive ATPase on a different, but as yet unidentified, membrane has been found.

B. Materials and Methods

All the procedures for isolating the plasma membrane from oat (*Avena sativa* L. cv Goodfield) roots and for the ATPase assays have been described (HODGES et al., 1972; HODGES and LEONARD, 1974).

C. Results and Discussion

I. Cation versus Anion Stimulation

Although previous studies indicated the plasma membrane ATPase was more sensitive to monovalent cations than to anions, the stimulation of the enzyme by salts such as Tris chloride made this interpretation uncertain (LEONARD and HODGES, 1973). Therefore, the effect of KCl on the enzyme was evaluated in the presence of high concentrations of $MgCl_2$ (Fig. 1). With increasing concentrations of $MgCl_2$, in the absence of KCl, the ATPase activity increased, reached a plateau, and then decreased. At concentrations between 3 and 12 mM $MgCl_2$, neither the Mg^{2+} nor the Cl^- stimulated the ATPase unless a stimulation by one of the ions was exactly off-set with an inhibition by the other ion. Over this concentration range of $MgCl_2$ (the chloride concentration ranged from 6 to 24 mM) the addition of 12 mM KCl caused an increase in the ATPase activity. This clearly illustrates that K^+, and not the Cl^-, is responsible for stimulating the plasma membrane ATPase of oat roots.

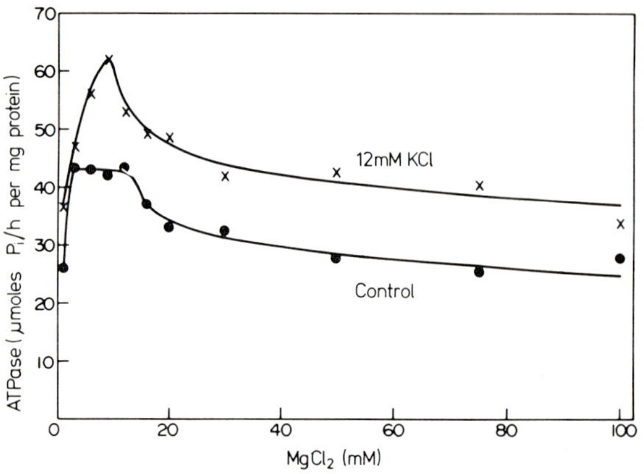

Fig.1 Effect of $MgCl_2$ concentration on plasma membrane ATPase activity in the absence and presence of 12 mM KCl. The gradient used for isolating the plasma membrane did not contain $MgSO_4$. The ATP concentration in the assay medium was 9 mM.

II. Inorganic versus Organic Cation Stimulation

Since Cl^- does not stimulate the plasma membrane ATPase, the enhanced activity in the presence of Tris chloride (LEONARD and HODGES, 1973) must have been caused by the Tris. This raises the possibility that inorganic and organic cations stimulate the ATPase at the same sites. To test this possibility, we determined the ATPase activity of the plasma membrane at various concentrations of Tris-MES buffer, pH 6.5 (Fig. 2). In the absence of inorganic ions, the buffer had no effect on enzyme activity. In the presence of Mg^{2+}, an increase in buffer concentration caused an increase in ATPase activity with the maximum velocity occurring at 250 mM. This enhanced activity was presumably due to the Tris rather than the MES since previous results showed that the anion had little, if any, effect on the ATPase (Fig. 1; LEONARD and HODGES, 1973). Thus, Tris, and presumably other organic cations, make the enzyme more reactive in the presence of Mg^{2+}.

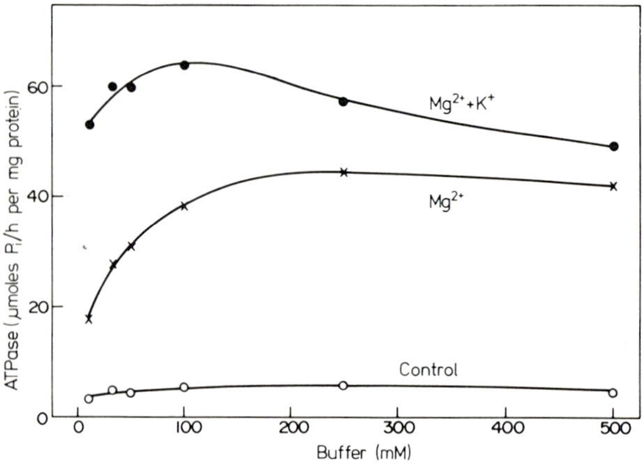

Fig. 2 Effect of Tris-MES buffer concentration on plasma membrane ATPase activity. The highest concentration of buffer was prepared by adjusting the pH of a 0.5 M Tris solution with a 0.5 M MES solution to pH 6.5. Other buffer concentrations were prepared by appropriately diluting the pH 6.5 solution.

K^+ stimulated the ATPase activity more than was observed in the presence of Mg^{2+} alone and at all buffer concentrations tested. Thus, with the maximum stimulation due to the buffer (Tris), K^+ further stimulated the ATPase. This illustrates that the stimulatory effects of K^+ and Tris on the ATPase are separate. The inhibition of K^+ stimulation at high buffer concentrations (Fig. 2) actually indicates that Tris interferes with, rather than stimulates, the site(s) on the enzyme that are sensitive to K^+. Thus, the stimulatory effect of organic cations on the plasma membrane ATPase (LEONARD and HODGES, 1973) seems to be an effect independent of the stimulation by the alkali cations.

III. Ca^{2+} Inhibition

Calcium inhibits the plasma membrane ATPase (Fig. 3). The inhibition occurs in the presence of either Mg^{2+} alone or Mg^{2+} and K^+. The K^+-ATPase is inhibited by about 50 % at 0.05 mM Ca^{2+} (Table 1). The Ca^{2+} inhibition can be observed within 5 minutes after adding Ca^{2+} to the enzyme (Fig. 4), and the inhibition is readily relieved by EDTA. Kinetic studies indicate the Ca^{2+} inhibition is uncompetitive (Fig. 5), which indicates that Ca^{2+} interferes with the enzyme-substrate complex.

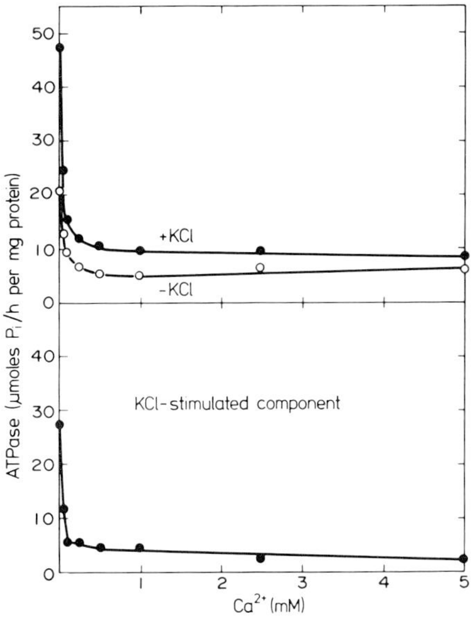

Fig. 3 ***Top.*** Effect of Ca^{2+} concentrations on plasma membrane ATPase activity in the presence of 1.5 mM $MgSO_4$ and 1.5 mM $MgSO_4$ plus 50 mM KCl. Assay pH was 6.0
Bottom. Effect of Ca^{2+} concentrations on the K^+-stimulated component of the ATPase activity.

The Ca^{2+} inhibition of the plasma membrane ATPase was surprising because Ca^{2+} is generally believed to be a promotor of monovalent ion absorption by roots, and there is substantial evidence implicating the plasma membrane ATPase in ion absorption (HODGES, 1973). There are a few examples of Ca^{2+} inhibiting ion absorption by roots (HANDLEY et al., 1965; ELZAM and HODGES, 1967), but these are exceptions rather than the rule (EPSTEIN, 1972). One explanation for the Ca^{2+} inhibition of the ATPase is that the plasma membrane vesicles are inside-out and a Ca^{2+}-sensitive site on the ATPase is exposed.

Table 1

Effect of Ca^{2+} on K^{+}-ATPase of plasma membrane of oat roots

$CaSO_4$ (mM)	% Inhibition
0.01	23.9
0.05	53.4
0.10	56.5
0.50	76.7
1.50	82.8
2.50	87.1

Reaction mixture contained 33 mM Tris-MES, pH 6.0, 3 mM ATP-Tris, pH 6.0, 1.5 mM $MgSO_4$, 50 mM KCl and various concentrations of $CaSO_4$. The % inhibition refers to the K^{+}-stimulated component of the ATPase activity.

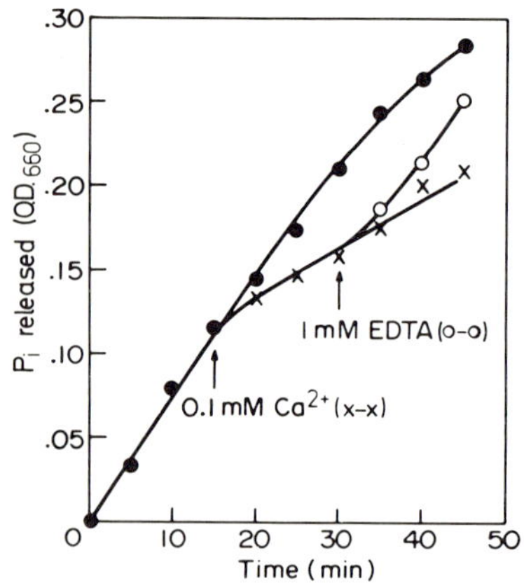

Fig. 4

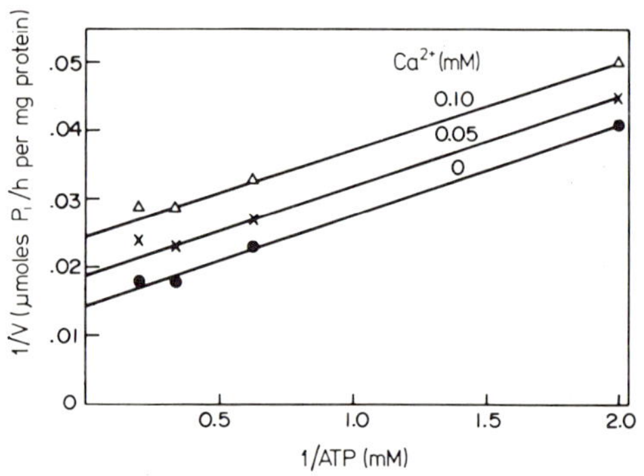

Fig. 5

Fig. 4 ATPase activity of plasma membrane vesicles as a function of time. $CaSO_4$ was added to give a final concentration of 0.1 mM at 15 minutes. EDTA was added to give a final concentration of 1 mM at 30 minutes. The assay medium contained 0.33 M Tris-MES, pH 6.5, 3 mM $MgSO_4$, 50 mM KCl and 3 mM ATP-Tris, pH 6.5.

Fig. 5 Lineweaver-Burk plot of plasma membrane ATPase activity. The assay medium was the same as shown in Fig. 4 except for varying ATP and $CaSO_4$ concentrations.

IV. An Anion Sensitive ATPase of an Unidentified Membrane

A membrane-bound ATPase of turnip roots is sensitive to anions (RUNGIE and WISKICH, 1973). A similar ATPase activity is detectable in a microsome fraction and in a semi-purified microsomal fraction of oat roots (Table 2). The ATPase activity of the microsome fraction, when assayed at pH 8.5, is stimulated by KCl, LiCl, and Kmalate, but it is inhibited by KNO_3. The similar ATPase activities in the presence of KCl and LiCl are important because these salts affect the plasma membrane ATPase differently; in the latter, Li^+ is only about 40 % as effective as K^+ (LEONARD and HODGES, 1973). The inhibitory effect of KNO_3 is particularly striking as compared to the stimulatory effect of KCl (Table 2) and clearly illustrates that this ATPase activity is sensitive to anions.

Table 2

ATPase activity of microsomes and a semi-purified microsome fraction (25/30 % interface membranes) at pH 8.5 and in the presence of various salts

	ATPase activity (μmoles/h per mg protein)			
Salts added	Microsomes	Δ	25/30 % Interface	Δ
$MgSO_4$	28.8	–	35.6	–
$MgSO_4$ + KCl	38.0	9.2	48.6	13.0
$MgSO_4$ + LiCl	37.3	8.5	46.9	11.3
$MgSO_4$ + Kmalate	38.0	9.2	48.7	13.1
$MgSO_4$ + KNO_3	17.3	-11.5	22.0	-13.6

Reaction mixture contained 33 mM Tris-MES buffer, pH 8.5, 3 mM ATP-Tris (pH 8.5), 3 mM $MgSO_4$, and 50 mM of all other salts.

A slight enrichment in the 'anion-sensitive' ATPase is obtained (Table 2) with the same discontinuous sucrose gradient procedure that is used for purifying the plasma membrane ATPase (HODGES et al., 1972). This semi-purified fraction is thought to be still heterogenous, however, since it contains many different enzyme activities (LEONARD et al., 1973).

D. General Discussion

The finding that the plasma membrane ATPase is specifically stimulated by monovalent cations (Figs. 1 and 2) and that another membrane-bound ATPase is stimulated by anions (Table 2) clarifies several heretofore puzzling results (HODGES, 1974). The plasma membrane ATPase has a pH optimum between 6.5 and 7.0, whereas the anion-sensitive ATPase has an alkaline pH optimum (unpublished results – also see Table 2 and RUNGIE and WISKICH, 1973). Thus, reports showing a stimulation of ATPase activity by KCl at neutral to low pH were undoubtedly a measure of K^+ stimulation of the plasma membrane ATPase. On the other hand, the reports showing a KCl stimulation at alkaline pH were probably the result of Cl^- stimulation of the anion-sensitive ATPase. For example, the claim by RATNER and JACOBY (1973) that organic cations are as effective as inorganic cations in stimulating the ATPase activity of microsomal fractions of roots from oats, barley, and corn is understandable since they measured the ATPase activity above pH 7.5 and only in the presence of various Cl^- salts. Under these conditions, the anion-sensitive ATPase would have been mainly responsible for the results and the cations, both inorganic and organic, would have had only a minimal effect on the ATPase activity.

The indentity of the membrane possessing the anion-sensitive ATPase is unknown. RUNGIE and WISKICH (1973) suggested such an enzyme might be on the tonoplast. In view of the properties of the enzyme, this is reasonable. For example, Cl^- (PIERCE and HIGINBOTHAM, 1970) and organic acids, especially malate, (TORII and LATIES, 1966) accumulate in vacuoles, and these anions stimulate the ATPase (Table 2). Inhibition of the ATPase by NO_3^- is also rational if the enzyme is part of the tonoplast ion transport system of root cells, since only a small fraction of the NO_3^- that is absorbed remains in the root (BEEVERS and HAGEMAN, 1969). Xylem exudates are high in nitrate (BEN ZIONI et al., 1971), and the majority of NO_3^- reduction and assimilation occurs in shoots rather than roots (BEEVERS and HAGEMAN, 1969). However, what seems on the surface to be logical may not be, and thus, the identity of the membrane containing the anion-sensitive ATPase must still be considered to be unknown.

The significance of the inhibition of the plasma membrane ATPase by Ca^{2+} is not clear, but it could represent a very important physiological control mechanism.

E. Acknowledgements

This work was supported by a grant from the National Science Foundation (GB-31052X1). Purdue University AES Journal Paper Number 5315.

F. References

BEEVERS, L., HAGEMAN, R.H., Ann. Rev. Plant Physiol. **20**, 495–522 (1969)
BEN ZIONI, A., VAADIA, Y., LIPS, S.H., Physiol. Plant. **24**, 288–290 (1971)
ELZAM, O.E., HODGES, T.K., Plant Physiol. **42**, 1438–1488 (1967)
EPSTEIN, E., 'Mineral Nutrition of Plants: Principles and Perspectives, Wiley and Sons, New York 1972
HANDLEY, R., METWALLY, A., OVERSTREET, R., Plant Physiol. **40**, 513–520 (1965)
HODGES, T.K., in Adv. in Agronomy (ed. N.C. Brady) Vol. 25, pp. 163–207, Acadmic Press, London and New York 1973
HODGES, T.K., in Short Distance Transport, Encyclopedia of Plant Physiology, New Series, (eds. M.G. Pitman and U. Lüttge), Springer-Verlag, Berlin 1975
HODGES, T.K., LEONARD, R.T., in Methods in Enzymology (eds. S.P. Colowick and N.O. Kaplan), Academic Press, London and New York, in press 1974
HODGES, T.K., LEONARD, R.T., BRACKER, C.E., KEENAN, T.W., Proc. Nat. Acad. Sci. **69**, 3307–3311 (1972)
LEONARD, R.T., HODGES, T.K., Plant Physiol. **52**, 6–12 (1973)
LEONARD, R.T., HANSEN, D., HODGES, T.K., Plant Physiol. **51**, 749–754 (1973)
PIERCE, W.S., HINGINBOTHAM, N., Plant Physiol. **46**, 666–673 (1970)
RATNER, A., JACOBY, B., J. Exp. Bot. **24**, 231–238 (1973)
RUNGIE, J.M., WISKICH, J.T., Plant Physiol. **51**, 1064–1068 (1973)
TORII, K., LATIES, G.G., Plant Cell Physiol. **7**, 395–403 (1966)

Ion Fluxes and Ion-Stimulated ATPase Activities

R.A. Leigh, R.G. Wyn Jones, and F.A. Williamson

A. Introduction

On the basis of a highly significant correlation between the rate of Rb^+ uptake and RbCl-stimulated ATPase activity, FISHER et al. (1970) postulated that ion-stimulated ATPases were involved in plant ion uptake. HODGES et al. (1972) found that the highest specific activity of KCl-stimulated ATPase from oat roots was associated with a fraction which equilibrated at high density on a sucrose gradient and was estimated to contain 75 % plasmalemma. Later LEONARD and HODGES (1973) showed that the kinetics of the KCl stimulation of this enzyme closely resembled those of ion uptake by oat roots and concluded that the dual absorption isotherm observed during ion uptake (LATIES, 1969) is the result of the activity of this plasmalemma enzyme.

In contrast, in an investigation into the subcellular distribution of ATPase activities in ***Zea mays*** roots, WILLIAMSON and WYN JONES (1971, 1972) found that, although some KCl-ATPase was associated with plasmalemma, the highest specific activity occurred in a vesiculated smooth membrane fraction equilibrating at low buoyant density on a Ficoll gradient. The membrane did not stain with the plasmalemma-specific phosphotungstic acid-chromic acid (PACP) stain (ROLAND et al., 1972). Since certain vesicular plant membranes equilibrate at higher densities on sucrose than on Ficoll gradients (COHEN et al., 1971; PITT and GALPIN, 1973) part of the activity studies by HODGES et al. (1972) may have been due to the contamination of the plasmalemma fraction by another membrane.

Since the KCl-stimulated ATPase found in the light vesicular fraction also exhibited properties which suggested a role in ion transport (WILLIAMSON and WYN JONES, 1972), the distribution of ion-stimulated ATPases in maize roots was re-examined. The results confirm our earlier findings that KCl-stimulated ATPases are associated both with the plasmalemma and with a smooth membrane of low buoyant density. Both enzymes have characteristics which may be correlated with ion transport systems. Since the results with the plasmalemma activities are very similar to those of LEONARD and HODGES (1973), this paper will be concerned principally with the enzyme found in the light fraction, with correlations with ion fluxes and with the problems of interpretation.

B. Materials and Methods

1. Definitions of ATPase Activities

Mg^{2+}-ATPase refers to ATPase activity in the presence of 2 mM Mg^{2+}. KCl-ATPase, NaCl-ATPase and (KCl + NaCl)-ATPase refer to the stimulation of Mg^{2+}-ATPase caused by the relevant salt (KCl, NaCl or mixture) and is calculated by the difference between the Mg^{2+}-ATPase activity with and without the monovalent ions.

2. Membrane Fractionation and Isolation

Root tips of ***Zea mays*** seedlings (WF9 x M14, Crow Hybrid Corn Co., Milford, Illinois, USA) were excised, chopped into 1 mm segments and homogenised in 0.3 M sucrose, 4 mM mercaptoethanol, 25 mM Tris-HCl, pH 7.8 (1 ml medium to 1 g tissue) with a Polytron homogeniser. After filtering through

Miracloth, the filtrate was centrifuged at 6000 X g for 10 min and the supernatant further fractionated by density gradient centrifugation (Fig. 1). All solutions were buffered with 25 mM Tris-HCl, 4 mM mercaptoethanol at pH 7.2 and, in addition, all Ficoll solutions contained 0.3 M sucrose. Fractions were removed from the gradients with a Pasteur pipette, pelleted by centrifugation at 70 000 X g for 1 h and resuspended in buffered 0.3 M sucrose.

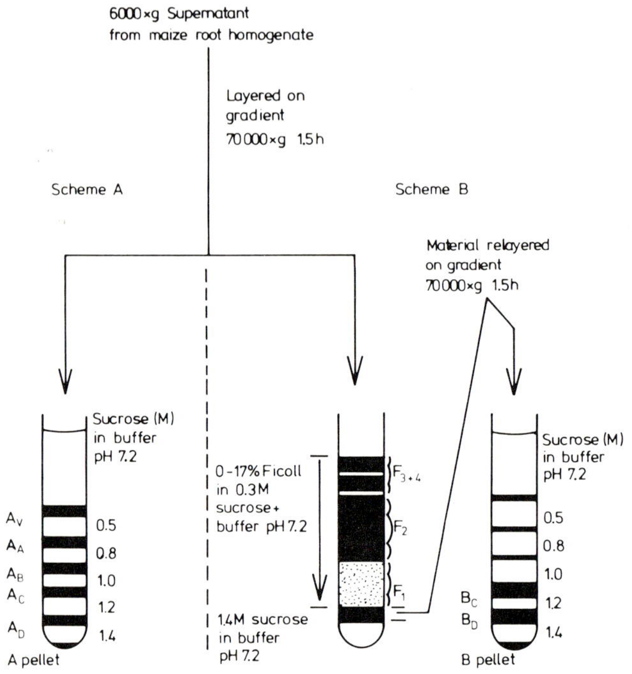

Fig. 1 Fractionation Schemes

3. Tracer Influx Measurements

In measuring tracer influxes into maize roots two uptake-wash regimes were used. A 10 min loading period followed by a 5 min wash in 5 mM KCl and 0.5 mM $CaCl_2$ (10/5 uptake-wash regime) was considered to measure influx across the plasmalemma. A 40 min uptake period followed by a 30 min wash (40/30 uptake-wash regime) was also employed and probably measured influx across an internal cell membrane. The rationale for these regimes, which are based on the studies of CRAM and LATIES (1971), is discussed in detail by LEIGH et al. (1973).

C. Results and Discussion

I. Fractionation of ATPase Activities

The distribution of ATPase and some other enzyme activities was examined in a single batch of corn roots fractionated by both schemes A and B (Fig. 1). That is, a comparison was made of enzyme distribution on sucrose only and on Ficoll and sucrose gradient systems (Table 1). When roots were fractionated by scheme A, the highest specific activities of KCl-ATPase were found in the A_C and A_D fractions. PACP staining showed these two fractions to contain 30—50 % and 50—70 % plasmalemma, respectively. Thus the results are in broad agreement with those of HODGES et al. (1972). However, there was no correlation between the plasmalemma content and KCl-ATPase since the A_C fraction had the highest activity of the enzyme but the A_D fraction had the greater plasmalemma content. This suggested that KCl-ATPase activity might be associated with at least one other membrane. Little KCl-ATPase was associated with the light A_V and A_A fractions equivalent in buoyant density to the light vesicular fraction previously described by WILLIAMSON and WYN JONES (1972).

Table 1

Distribution of enzyme activities between fractions from maize roots prepared by Schemes A and B

	Enzyme activity (μmoles/h per mg)			
Fraction[a]	Mg^{2+}-ATPase	KCl-ATPase	NPPase	INTase[b]
Homogenate	3.17	0.53	2.06	0.17
6 000 X g pellet	2.43	0.59	0.93	0.46
70 000 X g supernatant	3.65	0.20	2.30	0.08
A_V	3.29	0.11	1.71	0.03
A_A	3.67	0.05	1.43	0.47
A_B	3.97	0.82	1.33	0.57
A_C	4.35	2.04	1.20	0.41
A_D	2.49	1.26	0.90	0.66
A pellet	2.10	0.25	1.10	0.16
F_{3+4}	3.00	0.34	1.28	0.24
F_2	3.52	1.60	1.06	0.35
F_1	2.93	1.66	0.96	0.44
B_C	3.72	1.48	0.76	0.17
B_D	1.45	1.04	0.58	0.80
B pellet	2.42	1.12	0.72	0.19

a Fraction nomenclature as Fig. 1
b INTase = INT-succinate reductase

A different picture emerged when the same batch of roots was fractionated by scheme B. A relatively high KCl-ATPase activity was again found at high densities in the B_C and B_D fractions on the sucrose gradient. However, the fractions equivalent to the A_V, A_A, and A_B fractions were absent; the material being retained on the 0 to 17 % Ficoll gradient. In contrast to the light sucrose fractions, these Ficoll fractions had relatively high KCl-ATPase specific activities suggesting the material equivalent to that found in the A_C and possibly the

A_D fractions had also been retained. This postulate was reinforced by calculating the percentage of the total protein and the total KCl-ATPase recovered in the individual fractions on the gradient systems (Table 2). In roots fractionated by scheme A 80 % of the KCl-ATPase was found in the A_C and A_D fractions whereas using scheme B the major part of this activity was found in the Ficoll fractions.

Table 2

Percentage of total KCl-ATPase and protein, recovered from gradients, which was found in individual fractions of maize roots prepared by Schemes A and B

	% Total recovered from gradient			% Total recovered from gradient	
Fraction	KCl-ATPase	Protein	Fraction	KCl-ATPase	Protein
A_V	2	38	F_{3+4}	13	15
A_A	0	24	F_2	40	17
A_B	13	18	F_1	33	14
A_C	52	6	B_C	9	22
A_D	30	10	B_D	11	21
A pellet	3	4	B pellet	4	11

Electron microscopy showed the F_2 fraction to contain smooth membranes which did not stain with PACP and were therefore not plasmalemma. The B_D and A_D fractions both contained plasmalemma and mitochondria (Table 1) but further evidence showed the pH 5.5 KCl-ATPase activity to be associated with the plasmalemma and not the mitochondria. These results therefore substantiate the findings of WILLIAMSON and WYN JONES (1972) that KCl-ATPase activity is associated both with the plasmalemma and with a smooth intracellular membrane which is not plasmalemma. The identity of this latter membrane is unknown. Freeze-etch micrographs (LEIGH et al., 1973) revealed structures in this fraction very similar to the provacuoles isolated by MATILE (1968). A tonoplast fraction from yeast has a similar buoyant density (VAN DER WILDEN et al., 1973). Thus it is possible that the membrane is tonoplast but since tonoplast is ontogenetically related to smooth endoplasmic reticulum (MATILE and MOOR, 1968; BERJAK, 1972), this membrane may also be present.

II. Properties of KCl-ATPase Activities

The plasmalemma and smooth internal membrane KCl-ATPase may be distinguished by a number of characteristics: their pH response curves; their inhibitor sensitivities; their response to MES, which in maize is a specific stimulator of plasmalemma KCl-ATPase activity. The activities also differ in their response to the salt status of the roots as will be discussed later. It is therefore probable that two different KCl-stimulated ATPases are associated with the two membranes. Certain properties of these enzymes were studied to assess their possible involvement in ion transport and we report here principally on the activity of the F_2 fraction enzyme (see Fig. 1).

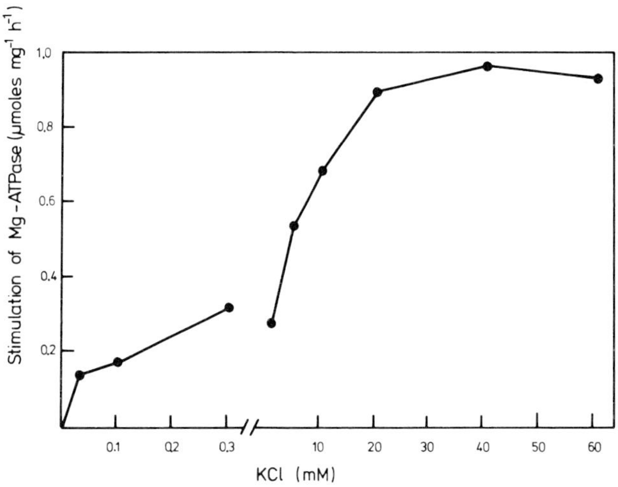

Fig. 2 Stimulation of Mg^{2+}-ATPase activity of the F_2 fraction by KCl over the range 0.03–60 mM KCl.

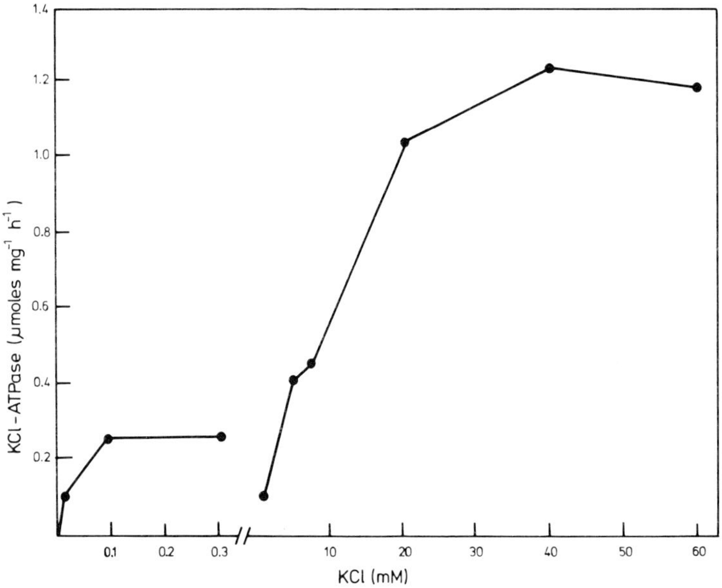

Fig. 3 Stimulation of Mg^{2+}-ATPase activity of the B_D fraction by KCl over the range 0.03–60 mM KCl.

Substantial monovalent ion stimulation was only found in the presence of Mg^{2+}. Although mixtures of NaCl and KCl enhanced Mg^{2+}-ATPase activity, no synergistic (i.e. $(K^+ + Na^+)$-Mg^{2+}-ATPase) activity was detected. Similarly when the total salt concentration was held constant over a wide range of K^+/Na^+ ratios (HANSSON and KYLIN, 1969), no indication of significant synergistic stimulation was obtained. A study of the dependence of the KCl-stimulation of ATPases on KCl concentration produced kinetics very similar to the dual isotherm of ion uptake with both the F_2 and the plasmalemma enzyme activities (Figs. 2 and 3), although stimulation of the plasmalemma enzyme at low concentrations was variable. The observations on the plasmalemma enzyme confirm those of LEONARD and HODGES (1973) using oats but the presence of similar kinetics in another enzyme found on an internal membrane poses many problems which will be discussed in detail later.

High internal salt status inhibited influxes into corn roots (LEIGH and WYN JONES, 1973). Therefore the effect of salt status on the KCl-ATPase of both fractions was studied (Fig. 4). No consistent effect was detectable in the plasmalemma fraction but high salt status was found to decrease KCl-ATPase activity of the F_2 fraction. In contrast the Mg^{2+}-ATPase was elevated in agreement with KYLIN and KÄHR (1973). As observed with influxes, the greatest inhibition occurred at low KCl concentrations.

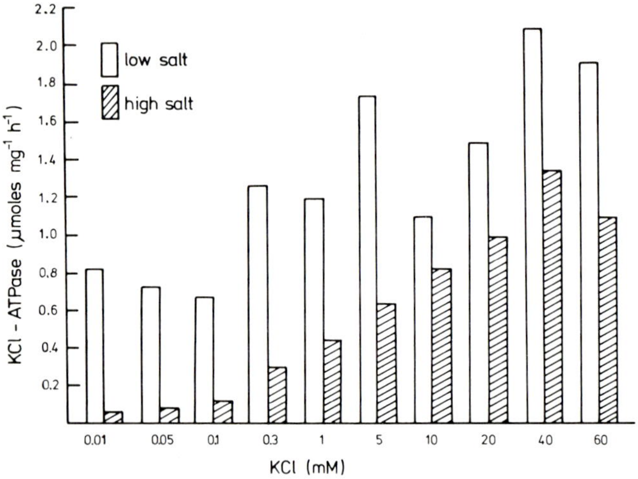

Fig. 4 Effect of high salt status on F_2 KCl-ATPase activity.

The influence of the accompanying anion on both the ion-stimulated activities has been examined (Table 3). With the F_2 enzyme the substitution of either MES or SO_4^{2-} for the Cl^- decreased the monovalent cation stimulated activity substantially. In contrast no clear picture emerged from a study of the plasmalemma enzyme activity as MES strongly stimulated the $(K^+ + Na^+)$-Mg^{2+}-ATPase activity.

Table 3

Effect of accompanying anion upon ATPase activity of F_2 and B_D fractions

Accompanying anion (X)	F_2 fraction Mg^{2+}-ATPase	KX-ATPase	NaX-ATPase	B_D fraction Mg^{2+}-ATPase	KX-ATPase	NaX-ATPase
Cl^-	2.38	1.53	0.97	2.52	0.16	-0.45[1]
SO_4^{2-}	2.98	0.98	0.29	2.36	-0.08	-0.53
MES	3.13	0.77	0.18	1.89	1.72	1.40

[1] negative value indicates inhibition of Mg^{2+}-ATPase by salt

III. Correlations of Fluxes and Ion-Stimulated ATPase

The significant correlation obtained by FISHER et al. (1970) between RbCl-stimulated ATPase activity and Rb^+ influx is remarkable in that an impure membrane fraction was used and no special precautions were taken to relate the activity to influx data at a specific membrane. It is probable that active fluxes occur across both plasmalemma and tonoplast (ANDERSON, 1972) and this study has shown the presence of, at least, two KCl-stimulated ATPases. Therefore the interpretation of correlation of FISHER et al. (1970) must be weighed carefully. Clearly a correlation between a single membrane bound ATPase and the flux across that same membrane would be more desirable. It follows that the enzymes, one associated with the plasmalemma and the other with an internal smooth membrane, must be considered separately.

A technique for measuring separate tracer Cl^- fluxes across the plasmalemma and tonoplast was reported by CRAM and LATIES (1971) and rigorously tested by CRAM (1973). Using this technique, the plasmalemma and internal membrane fluxes of $^{86}Rb^+$ were assessed using the 10/5 and 40/30 uptake-wash regimes. However the intracellular location of the 40/30 flux is unknown as, unlike Cl^- influx, it may not be confidently assigned to the tonoplast (LEIGH et al., 1973). It is suggested that at high external ion concentrations the flux measured is at a membrane internal to the plasmalemma and may be associated with either endoplasmic reticulum or tonoplast. It will be referred to as influx across an internal membrane. Therefore possible correlations could be tested between KCl-stimulated ATPase and rates of ion flux for both plasmalemma and the internal membrane i.e. between the B_D KCl-ATPase and 10/5 fluxes and between the F_2 KCl-ATPase and 40/30 fluxes.

A highly significant correlation was observed ($P < 0.001$; $R = 0.93$) between plasmalemma enzyme activity and influx over the range 0.03 to 60 mM KCl but the 95 % confidence limits were very wide. Using $^{86}Rb^+$ the 10/5 uptake-wash regime overestimates influx across the plasmalemma since a considerable amount of tracer is retained by the cell wall. If a correction was made for this, there was little change in the correlation coefficient but there was a marked decrease in the 95 % confidence limits (Fig. 5).

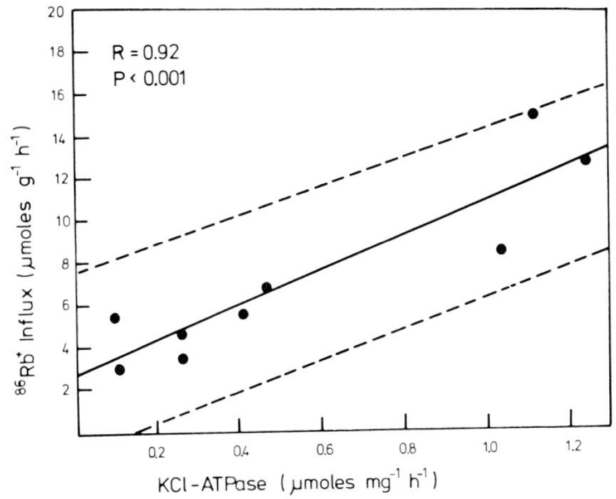

Fig. 5 Correlation between B_D KCl-ATPase and influx of $^{86}Rb^+$ measured by 10/5 regime over the range 0.03–60 mM KCl corrected for inadequate washing of cell wall. ------: 95 % confidence limits.

A good correlation was also observed between the F_2 KCl-ATPase and influx across an internal membrane (40/30 regime) (Fig. 6). An excellent correlation was obtained between 1 and 20 mM but became poorer as higher concentrations were included. Flux experiments indicated that a non-metabolic component contributes significantly to 40/30 influx at higher concentrations. If a correction was applied to the flux data for this non-metabolic activity, an excellent correlation could be obtained up to 60 mM with very narrow 95 % confidence limits ($R = 0.998$, $P < 0.001$). These experiments were conducted with low salt tissues and, as previously mentioned, high salt status decreased both ion influxes and F_2 KCl-ATPase activity. The correlation between F_2 KCl-ATPase activity and 40/30 influx was therefore also examined in high salt tissue and a highly significant result ($P < 0.001$; $R = 0.94$) obtained.

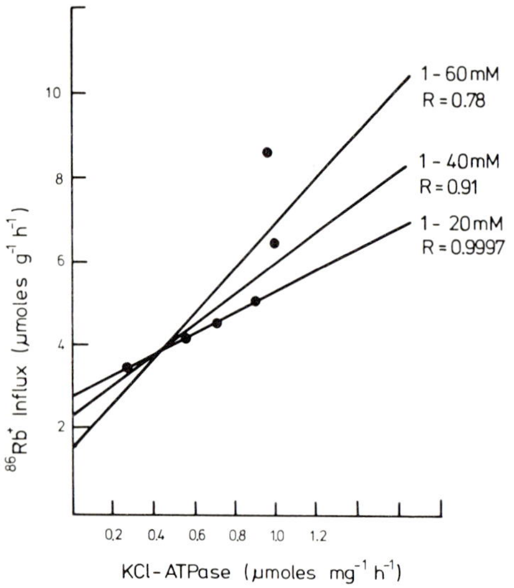

Fig. 6 Correlation between F_2 KCl-ATPase activity and $^{86}Rb^+$ uptake measured by 40/30 uptake-wash regime over the range 1–60 mM KCl.

Three strands of evidence therefore point to the involvement of the F_2 KCl-ATPase in ion transport: 1) presence of dual isotherm of ion-stimulation of activity; 2) similar response to high salt status; 3) correlation of KCl-ATPase activity with $(^{86}Rb^+)K^+$ influx.

Our results also support the conclusion of HODGES and his colleagues concerning the plasmalemma enzyme. However these conclusions are too facile as correlations could also be obtained between F_2 ATPase activity and 10/5 influx, and B_D-ATPase on an internal cell membrane, and ***vice versa***, although in these cases the 95 % confidence limits were wide. These unpredicted correlations explain the good correlation reported by FISHER et al. (1970) while using an impure membrane preparation but raise doubts as to the validity of the other, apparently highly significant, correlations discussed above. Clearly a number of factors may contribute to the appearance of unpredicted correlations. In the first place, the similarity of the KCl-stimulation kinetics of the two enzymes lead to cross correlations. Also the fluxes measured represent at best only approximations to the true fluxes across the membranes involved. There is also the possibility that the correlations have a real physiological basis in ion regulation in plants and the elucidation of these possibilities must await further study.

D. Conclusion

While subject to obvious limitations, the present data suggest that at least two KCl-stimulated ATPases are present in maize roots and that each is involved in ion influx, one at the plasmalemma and the other at some internal membrane. Further, the results indicate that uptake at both membranes, at all external concentrations, is an active process. However, previous models involving the series arrangements of the concentration-dependent uptake isotherms have suggested that uptake of ions across the plasmalemma at concentrations above 1 mM is a non-active, diffusive process (TORII and LATIES, 1966; CRAM and LATIES, 1971). This appears not to be so and in agreement with the results of GERSON and POOLE (1972), we suggest that this uptake is active and may be mediated by an ATPase. However, since the rate-limiting step at higher external concentration resides at an internal membrane (CRAM and LATIES, 1971; CRAM, 1973) it is necessary to propose that the ATPase on the plasmalemma delivers ions to the cytoplasm faster than they are removed by the ATPase on the internal membrane. This model is similar to that proposed by NISSEN (1973) which envisaged multiphasic mechanisms on both the plasmalemma and tonoplast.

The ATPases described above and by HODGES and his group differ radically from the ion transport ATPases found in animals since there is no evidence that they are synergistically stimulated by Na^+ and K^+. The fact that K^+-stimulated Na^+ efflux can be detected in certain higher plants (JESCHKE, 1973) suggests that further work in this field is urgently required. Nevertheless a number of pieces of evidence now point to the association of KCl-stimulated ATPase with ion fluxes.

Although the results with the F_2 enzyme indicate that Cl^- was not the stimulating ion, the stimulation in the presence of either K^+ or Na^+ was affected by replacing Cl^- with $SO_4{}^{2-}$ or MES. Somewhat similar results were obtained with the B_D fraction although other complicating factors were noted. A possible explanation for the anion sensitivity is that the F_2 enzymes are salt-stimulated and act as salt transporting systems rather than monovalent-ion transporting systems. This postulate is further supported by the correlation ($R = 0.984$; $P < 0.2$) found between KCl-stimulated ATPase activity and Cl^- influx. Influx data also support the existence of salt-transporting systems both in *Nitella* and *Hydrodictyon* (MacROBBIE, 1970) and in maize cortices (CRAM, 1973). This proposal may also partly explain the inconsistency of the enzymic data and electrochemical measurement, as the latter do not indicate an active K^+ influx at higher external ion concentrations (HIGINBOTHAM, 1973).

E. Acknowledgements

R.A.L. is grateful to the Science Research Council for a postgraduate studentship.

F. References

ANDERSON, W.P., Ann. Rev. Plant Physiol. **23**, 51–72 (1972)
BERJAK, P., Ann. Bot. **36**, 73–81 (1972)
COHEN, E., SHAIN, Y., BEN-SHAUL, Y., MAYER, A.M., Can. J. Bot. **49**, 2053–2057 (1971)
CRAM, W.J., Aust. J. Biol. Sci. **26**, 757–779 (1973)
CRAM, W.J., LATIES, G.G., Aust. J. Biol. Sci. **21**, 633–646 (1971)
FISHER, J.D., HANSEN, D., HODGES, T.K., Plant Physiol. **46**, 812–814 (1970)
GERSON, D.F., POOLE, R.J., Plant Physiol. **50**, 603–607 (1972)
HANSSON, G., KYLIN, A., Z. Pflanzenphysiol. **60**, 270–275 (1969)
HIGINBOTHAM, N., Bot. Rev. **39**, 15–70 (1973)

HODGES, T.K., LEONARD, R.T., BRACKER, C.E., KEENAN, T.W., Proc. Nat. Acad. Sci. **69**, 3307–3311 (1972)
JESCHKE, W.D., in Ion Transport in Plants (ed. W.P. Anderson), pp. 285–296, Academic Press, London and New York 1973
KYLIN, A., KÄHR, M., Physiol. Plant. **28**, 452–458 (1973)
LATIES, G.G., Ann. Rev. Plant Physiol. **20**, 89–116 (1969)
LEIGH, R.A., WYN JONES, R.G., J. Exp. Bot. **24**, 787–795 (1973)
LEIGH, R.A., WYN JONES, R.G., WILLIAMSON, F.A., in Ion Transport in Plants (ed. W.P. Anderson), pp. 407–418, Academic Press, London and New York 1973
LEONARD, R.T., HODGES, T.K., Plant Physiol. **52**, 6–12 (1973)
MacROBBIE, E.A.C., Quart. Rev. Biophys. **3**, 251–294 (1970)
MATILE, PH., Planta **79**, 181–196 (1968)
MATILE, PH., MOOR, H., Planta **80**, 159–175 (1968)
NISSEN, P., in Ion Transport in Plants (ed. W.P. Anderson), pp. 539–553, Academic Press, London and New York 1973
PITT, D., GALPIN, M., Planta **109**, 233–258 (1973)
ROLAND, J.-C., LEMBI, C.A., MORRE, D.J., Stain Technol. **47**, 195–200 (1972)
TORII, K., LATIES, G.G., Plant Physiol. **41**, 863–870 (1966)
VAN DER WILDEN, W., MATILE, PH., SCHELLENBERG, M., MEYER, J., WEIMKEN, A., Z. Naturforsch. **28c**, 416–421 (1973)
WILLIAMSON, F.A., WYN JONES, R.G., Biochem. J. **123**, 4P (1971)
WILLIAMSON, F.A., WYN JONES, R.G., in Isotopes and Radiation in Soil-Plant Relationships Including Forestry, pp. 69–80, International Atomic Energy Agency, Vienna 1972

On the Reversible ATPase of Plant Mitochondria

J.B. Hanson

A. Introduction

MITCHELL (1973) has recently hypothesized on the H^+ and cation translocating properties of membrane ATPase. With respect to mitochondria he visualizes a direct relationship between H^+ translocation and ATP or its hydrolysis products as carriers. Although he does not completely reject the possibility of a role for intermediates (e.g. IOH and XH) in the H^+ efflux accompanying ATP hydrolysis, he favours direct coupling of group translocation and H^+ (cation) translocation.

Recent investigations with corn and cauliflower mitochondria in our laboratory make us question if a H^+ translocating ATPase as visualized by MITCHELL actually functions in plant mitochondria. It appears that H^+ translocation may be indirectly accomplished through the P_i^--OH^- antiporter (or P_i^--H^+symporter). The electrical component of the H^+ motive force could arise from operation of the AdN (adenine nucleotides) translocator which is coupled by a transfer mechanism of the F_1-ATPase (coupling ATPase of the inner mitochondrial membrane). This paper attempts to show that the reversible H^+-conducting ATPase proposed by MITCHELL (1961) is probably the concerted result of two separate exchange reactions and one transfer reaction in addition to the hydrolysis of ATP.

B. Results

In phosphate-loaded, mersalyl-blocked corn mitochondria, the addition of ADP produces ATP formation commensurate with the amount of accumulated phosphate (HANSON et al., 1972). Concurrently, the mitochondria contract and respiration declines. In effect, there has been efflux of the K^+ and P_i^- initially accumulated. Atractyloside prevents this removal.

When the same experiments are done by substituting arsenate for phosphate, it can be deduced that the ADP-As formed is not released to the aqueous matrix (BERTAGNOLLI and HANSON, 1973). Rather, ADP-As is transferred to the AdN transporter without leaving the hydrophobic membrane phase. Hydrolysis occurs on exit into the intermembrane phase.

Studies of the respiratory 'priming' of the DNP-stimulated ATPase of cauliflower mitochondria (JUNG and HANSON, 1973a, b; JUNG, 1974) indicate that the AdN translocator only functions when a transmembrane potential exists, possibly as a H^+ gradient although this is difficult to prove. Respiratory priming is absolutely dependent on added Mg^{2+}, and added phosphate (or arsenate) is required for optimum priming. Since the K_m for phosphate in respiratory priming is quite low (17 μM), and since some endogenous phosphate seems to be transferred from the intermembrane phase to the matrix through a mersalyl-sensitive site when respiration starts (EARNSHAW et al., 1973), it is likely that the requirement for phosphate is absolute as well. JUNG (1974) concludes that part of the requirement for a transmembrane potential may lie with maintaining adequate Mg^{2+} and phosphate concentrations, but that the potential is needed over and above this for the complex of ATP/ADP, P_i^-/OH^- and H^+ transport to operate as an ATPase.

Uncoupler-stimulated ATPase is inhibited by both mersalyl and atractyloside (JUNG and HANSON, 1973a, b).

C. Discussion

These observations can be assembled to give a schematic representation of the reversible ATPase and its associated H^+ (OH^-) transport. ATPase activity involves the concerted reaction of 4 distinct processes (Fig. 1A).

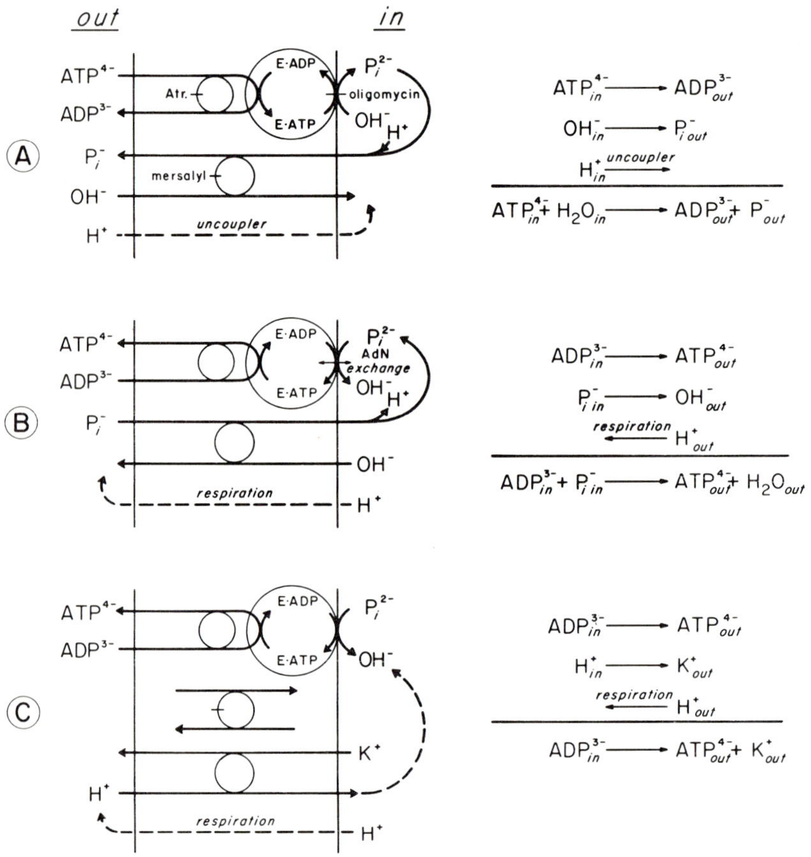

Fig. 1 The reversible H^+-conducting ATPase of plant mitochondria viewed as a membrane complex. A: Function in uncoupler-stimulated ATP hydrolysis. B: Function in normal oxidative phosphorylation. C: Function in oxidative phosphorylation with phosphate-loaded, mersalyl-blocked mitochondria.

1) Exchange transfer of ATP^{4-}/ADP^{3-} with the F_1-ATPase through the atractyloside-sensitive AdN transporter. The exchange as depicted is electrogenic creating an electrical potential, positive outside.
2) Hydrolysis by the F_1-ATPase with release of phosphate to the matrix, an electrogenic act linked with that of ATP^{4-}/ADP^{3-} exchange. In plant mitochondria, as in animal, the pH

optimum is high (9.0–9.5), Mg^{2+} is required, and the process is oligomycin-sensitive (JUNG and HANSON, 1973a, b).

3) Exchange of P_i^- for OH^-. To simplify illustration of the stoichiometry, the dominant divalent phosphate of the alkaline matrix is pictured as being protonated to monovalent phosphate for exchange transport. It is in this electrically neutral exchange that the H^+ gradient is created (or sustained).

4) Completion of the cycle by H^+ influx via an uncoupler, which simultaneously collapses both the membrane potential and the H^+ gradient.

Figs. 1B and C simply illustrate the action of the reversible ATPase in ATP formation. The change from hydrolysis to synthesis of ATP lies with reversal of the H^+ current through linkage with respiratory electron transfer. Fig. 1C applies to the special case of phosphate-loaded, mersalyl-blocked mitochondria (HANSON et al., 1972; BERTAGNOLLI and HANSON, 1973). Here it is necessary to assume that an active K^+/H^+ exchange mechanism exists in plant mitochondria. MITCHELL and MOYLE (1969) have suggested a comparable Na^+-H^+ antiporter for rat liver mitochondria, and MITCHELL (private communication) suggests a K^+-H^+ antiporter might explain the observed uncoupling properties of valinomycin on plant mitochondria (HANSON et al., 1972; WILSON et al., 1972; KIRK and HANSON, 1973). Lack of an active K^+-H^+ antiporter might account for the failure of phosphate-loaded, mercurial-blocked liver mitochondria to form ATP other than a small amount from endogenous nucleotides (McGIVAN et al., 1971).

Comparison of this hypothesis with that of MITCHELL (1973) shows several points of difference.

The ATPase depicted here is a complex of reactions based on the properties of the P_i^--OH^- transporter (CHAPPELL, 1968), the F_1-ATPase (RACKER, 1970) and the ADP^{3-}-ATP^{4-} transporter (KLINGENBERG et al., 1969), to which has been added the hypothetical transfer site between the AdN translocator and the F_1-ATPase (BERTAGNOLLI and HANSON, 1973). For the ATPase to function as illustrated the free energy of hydrolysis must drive an electrogenic ADP^{3-}/ATP^{4-} exchange and create a P_i^- gradient which drives an electrically neutral P_i^-/OH^- exchange. KLINGENBERG et al. (1969) make essentially the same point. The total H^+ motive force is thus derived from both transporters. ATP and ADP react from the outside and do not cross the membrane (except as they may exchange between the F_1-ATPase and the matrix). Hence, these components could not be directly linked to H^+ transport. However, phosphate is transferred and is the critical component in the transmembrane transport of charge. Indeed, the entire process of creating the H^+ motive force ($\Delta p = \Delta\psi + 2.3\frac{RT}{F}\Delta pH$) lies with transmembrane phosphate transport.

A single H^+, or its OH^- equivalent, is transported. Dissociation of $P_i^- \rightarrow P_i^{2-}+H^+$ externally could produce an extra H^+ giving $\Delta H^+/ATP$ approaching 2, but this is not a stoichiometric element of the membrane H^+ conduction. MITCHELL and MOYLE (1968) found corrected values for $\Delta H^+/ATP$ to approach 2.0 in the presence of valinomycin, and deduced from control experiments with added oligomycin that in the entry of ATP and the exit of ADP+P_i there was no net H^+ translocation. Although we have not done these same experiments, work on K^+ transport in corn mitochondria gives a ratio of K^+ to high energy intermediate approaching one (KIRK and HANSON, 1973), half of what would be expected from the chemiosmotic hypothesis. Mersalyl, which inhibits the P_i^--OH^- antiporter and thereby generation of the H^+ gradient, inhibits the DNP-ATPase of plant mitochondria (JUNG and HANSON, 1973a).

It will be noted that this formulation of what constitutes the reversible ATPase does not mitigate against its function in oxidative phosphorylation as required by the chemiosmotic hypothesis. The H^+ motive force arising from respiration would have to furnish both a H^+ gradient for P_i^- transport and adequate electrical potential to pull the ADP^{3-}/ATP^{4-}

exchange. This is surely as feasible as in MITCHELL's formulation of the ATPase. However, neither is the hypothesis inconsistent with our concept that the coupling act somehow produces both a H^+ gradient and a 'squiggle', and analogous schemes can be devised incorporating intermediates (Fig. 6, BERTAGNOLLI and HANSON, 1973).

D. Acknowledgements

I am indebted to the Unites States Atomic Energy Commission (AT-11-1-790) for support.

E. References

BERTAGNOLLI, B.L., HANSON, J.B., Plant Physiol. **52**, 431–435 (1973)
CHAPPELL, J.B., Brit. Med. Bull. **24**, 150–157 (1968)
EARNSHAW, M.J., MADDEN, D.M., HANSON, J.B., J. Exp. Bot. **24**, 828–840 (1973)
HANSON, J.B., BERTAGNOLLI, B.L., SHEPHERD, W.D., Plant Physiol. **50**, 347–354 (1972)
JUNG, D.W., HANSON, J.B., Arch. Biochem. Biophys. **158**, 139–148 (1973a)
JUNG, D.W., HANSON, J.B., Biochim. Biophys. Acta **325**, 189–192 (1973b)
JUNG, D.W., The 2,4-Dinitrophenol-Stimulated Adenosine Triphosphatase Activity in Cauliflower and Corn Mitochondria. Ph.D. Thesis, University of Illinois, 1974
KIRK, B.I., HANSON, J.B., Plant Physiol. **51**, 357–362 (1973)
KLINGENBERG, M., WULF, R., HELDT, H.W., PFAFF, E., in Mitochondria-Structure and Function (eds. L. Ernster, Z. Drahota), pp. 59–77, Academic Press, London and New York 1969
McGIVAN, J.D., GREBE, K., KLINGENBERG, M., Biochem. Biophys. Res. Commun. **45**, 1533–1541 (1971)
MITCHELL, P., Nature **191**, 144–148 (1961)
MITCHELL, P., FEBS Letters **33**, 267–274 (1973)
MITCHELL, P., MOYLE, J., Eur. J. Biochem. **4**, 530–539 (1968)
MITCHELL, P., MOYLE, J., Eur. J. Biochem. **9**, 149–155 (1969)
RACKER, E., in Membranes of Mitochondria and Chloroplasts (ed. E. Racker), pp. 127–171, Van-Nostrand-Reinhold, New York 1970
WILSON, R.H., DEVER, J., HARPER, W., FRY, R., Plant Cell Physiol. **13**, 1103–1111 (1972)

Effects of Divalent Cations and Oligomycin on Membrane ATPases from Roots of Wheat and Oat in Relation to Salt Status and Cultivation

M. Kähr and A. Kylin

A. Introduction

In a previous article (KYLIN and KÄHR, 1973) the ATPases from wheat roots were studied. The pH-curves and the effects of Mg^{2+} and Ca^{2+} did not correspond to those reported by FISHER and HODGES (1969) from oat roots. For instance, ATPase stimulated by Ca^{2+} dominated over Mg^{2+}-stimulated activity in wheat (KYLIN and KÄHR, 1973) whereas the reverse had been reported for oat (FISHER and HODGES, 1969). Our investigation was extended to oat and it was established that the differences were primarily due to the species used (KYLIN and KÄHR, 1973).

The activity of the wheat ATPase was also affected by the nutritional level of the culture medium (KYLIN and KÄHR, 1973). The present work will present data on how oat roots react to the same treatments as were applied to wheat, and discuss the possible relationship between the stimulation of ATPases by different divalent cations and the nutritional requirements of the two species. Data on the oligomycin sensitivity are also included.

B. Materials and Methods

The methods were the same as used in the wheat investigation (KYLIN and KÄHR, 1973), and only the outlines will be indicated here. Oats (*Avena sativa* L. cv Brighton) were cultured under low and high salt conditions. From the roots a microsomal fraction was prepared (FISHER and HODGES, 1969). The resuspended pellet (1 h at 10 000—30 000 X g) was stored at $-86°C$.

ATPase activity was assayed in a 1 ml reaction system containing 1 mM ATP, 37.5 mM histidine buffer (Tris for pH values above 7), enzyme and $MgCl_2$ or $CaCl_2$ as indicated for each case. After 60 min at 30°C the reaction was stopped by adding 0.1 ml 33 % TCA.

The liberated inorganic phosphate was determined spectrophotometrically (LINDEMAN, 1958) and the specific activity was calculated as μmoles P_i released per hour by 1 mg protein. All figures given are the means of duplicate samples, the single samples not differing by more than 10 % from each other. The background activity without any divalent cation added was subtracted.

The protein contents were measured according to the standard micro— Kjeldahl method, with a conversion factor of 6.3 from nitrogen to protein.

C. Results

Ca^{2+} and Mg^{2+} each stimulated the ATPases from oat roots. The activation depended on the pH, on the cation in the reaction system and on the nutritional conditions of the roots. At 1 mM Mg^{2+}, pH was optimal at a value of 7—8, whereas the stimulating effect of 1 mM Ca^{2+} was increased by increasing the pH over the whole range tested (Fig. 1A, B). On the whole,

Mg^{2+} stimulated more than Ca^{2+}. Especially with Mg^{2+} as the stimulating cation, the ATPase from the low salt roots was more active than the one from the high salt roots (Fig. 2A, B).

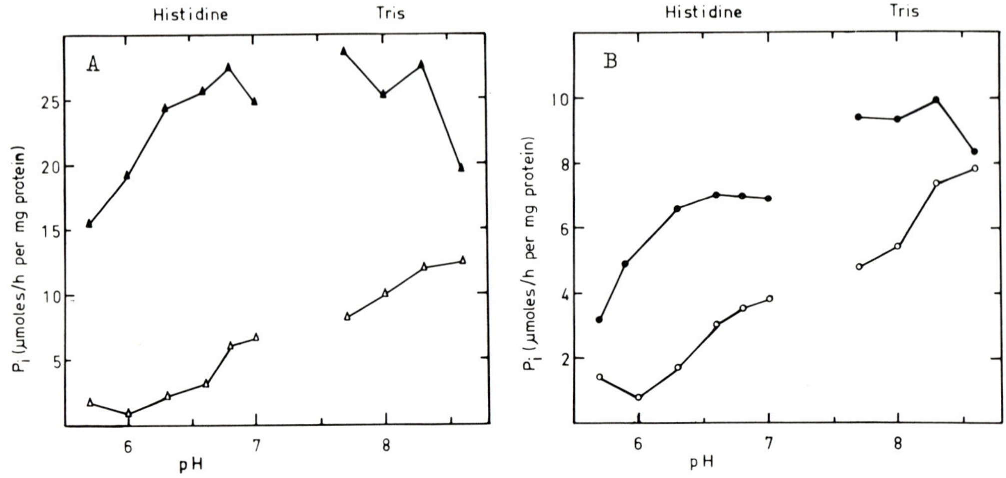

Fig. 1 Effect of pH on the Ca^{2+} and Mg^{2+} dependent ATPase activity of a microsomal fraction from oat roots.
A) Preparation from low salt roots.
B) Preparation from high salt roots.
Open symbols give the activation due to 1 mM Ca^{2+} over the background values. Closed symbols give the same for activation by 1 mM Mg^{2+}. Histidine and Tris buffers used as indicated.

Table 1

The effect of a range of Ca^{2+} and Mg^{2+} mixtures on the ATPase activity of a microsomal fraction from low salt oat roots. Data in µmoles P_i/h per mg protein.

$CaCl_2$ (mM)	pH 6.6, $MgCl_2$ (mM)							pH 8.3, $MgCl_2$ (mM)							
	0	0.05	0.2	0.5	1	2	3	0	0.05	0.2	0.5	1	2	3	5
0	...	2	8	16	21	23	23	...	4	11	17	16	12	11	9
0.05	0	2	7	14	18	19	21	0	5	11	16	15	11	10	9
0.2	0	2	6	11	13	14	15	3	7	11	14	13	10	9	8
0.5	2	3	7	11	14	11	12	7	9	12	12	11	9	9	8
1	3	3	6	8	9	9	9	10	10	9	9	9	8	8	7
2	4	5	6	6	7	8	8	13	10	8	7	7	7	7	7
3	4	5	5	6	6	6	7	...	...	7	7	6	7	6	6
5	...	...	...	...	...	...	...	13	9	7	7	6	6	5	5

At pH 6.6 and pH 8.3, the influence of different concentrations of Ca^{2+} and Mg^{2+} on the ATPase activity was analyzed. At the lower pH (Fig. 2A), Ca^{2+} (open symbols) activated the ATPases slightly and nearly independently of the cultivation condition. Mg^{2+} (filled symbols) stimulated the ATPases from low salt roots (broken line) much more than the ATPases from high salt roots (full line). At pH 8.3 the differences were similar although less marked (Fig. 2B).

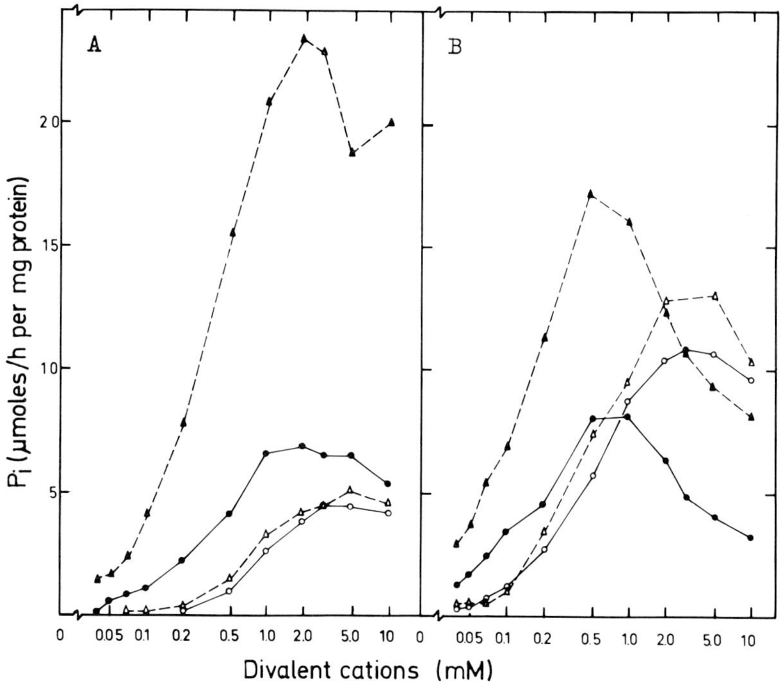

Fig. 2 Activation of the ATPase activity of the microsomal fraction of oat roots as a function of the concentration of Ca^{2+} (open symbols) and Mg^{2+} (closed symbols).
Broken line: low salt roots. Full line: high salt roots. A) pH 6.6. B) pH 8.3.

Combinations of the two cations showed both cooperation (at low concentrations) and competition in their ability to activate the ATPases (Tables 1, 2). The results were complex and depended on both the salt status of the roots and the pH of the reaction system. Further experiments are needed to analyse the kinetics of the system completely.

Oligomycin inhibited the cation-dependent ATPases. At pH 8.3 the activation due to Mg^{2+} addition was inhibited 70–90 %. The Ca^{2+} dependent ATPase was somewhat less sensitive to oligomycin, especially at high concentrations of Ca^{2+} (Table 3).

Table 2

The effect of a range of Ca^{2+} and Mg^{2+} mixtures on the ATPase activity of a microsomal fraction from high salt oat roots. Data in µmoles P_i/h per mg protein.

$CaCl_2$ (mM)	pH 6.6 MgCl₂ (mM)								pH 8.3 MgCl₂ (mM)							
	0	0.05	0.1	0.2	0.5	1	2	5	0	0.05	0.1	0.2	0.5	1	2	5
0	...	1	1	2	4	7	7	6	...	2	3	5	8	8	6	4
0.05	0	0	1	2	4	5	6	5	0	2	3	5	8	7	5	4
0.2	0	1	1	2	3	4	4	4	3	3	4	5	7	6	4	3
0.5	1	2	2	3	3	3	3	3	6	5	6	6	6	5	4	3
1	3	3	4	4	3	3	2	2	9	7	6	6	5	4	4	3
2	4	4	4	3	3	2	2	2	10	7	6	5	4	4	3	3
5	4	4	4	4	3	2	1	1	11	7	6	6	4	3	3	3

Table 3

The effect of oligomycin on the divalent cation-dependent ATPase of a microsomal fraction from low salt oat roots at pH 8.3. Data in % inhibition of the specific activation.

Divalent cation (mM)		Oligomycin (µgml⁻¹) 1	2	4	8	16
None		0	0	0	0	0
Mg^{2+}	0.05	90	90	90	90	90
	0.5	70	80	80	90	90
	1	70	60	...	...	80
	5	70	70	80	80	80
Ca^{2+}	0.2	70	70	70	70	60
	1	60	70	70	70	70
	10	40	40	40	40	30

D. Discussion

A soil with high amounts of Ca^{2+} has normally a relatively high (near neutral) pH value. The high pH decreases the rate of decomposition of minerals, thus preventing the mineral nutrients from being leached out. Instead, the ions are attached to colloids, where they are easily available for the plants (JANSSON, 1970). This is necessary for a good crop of wheat. Oats can more easily utilize complex, bound minerals and give a good crop even if the pH in the soil is as low as 4.5 (ÅBERG, 1970). At the same time, oats are indicator plants for Mg^{2+} deficiency (LUNDBLAD, 1955), which may easily occur in acid soils.

Our results from ATPase experiments may be compared with this knowledge from agriculture. Oats that give a good crop with a low Ca^{2+} supply in a soil exposed to leaching, have ATPases which are stimulated more by Mg^{2+} than by Ca^{2+}, especially if the roots are cultured in a low salt solution and the ATPases have to react at low pH (Fig. 2A). On the other hand, wheat has an ATPase that is stimulated more by Ca^{2+} than by Mg^{2+} (KYLIN and KÄHR, 1973), especially when the roots are cultured in a complete culture medium, which corresponds to the need for high levels of nutrition and Ca^{2+} by wheat. Since the ATPases from oat and wheat roots show such different responses to Mg^{2+} and Ca^{2+} in relation to the salt level of the medium, it is tempting to look upon the root ATPases as expressions of basic functions of the membranes, showing both genotypic and phenotypic adaptations to different ecological circumstances.

The Ca^{2+}- and Mg^{2+}-dependent ATPases from oat roots are inhibited by oligomycin (Table 3), while ATPases from wheat are not affected (KYLIN and KÄHR, 1973). As ATPases from mitochondria are sensitive to oligomycin (FISHER and HODGES, 1969), one could explain the differences between oats and wheat by assuming that the mitochondria of oats are more fragile than those of wheat. This would be an interesting difference between oats and wheat; at the same time caution in interpretating the data is needed. Systems other than mitochondria may be affected by oligomycin, for the antibiotic affects ion uptake without changes in the respiration of intact oat roots (HODGES, 1966). Furthermore, JACOBY and PLESSNER (1970) reported that Cl^- uptake by excised barley roots was more inhibited by oligomycin if the Cl^- concentration was low ($<$ 0.2 mN) than if the Cl^- concentration was high ($>$1 mN), indicating two different places for the inhibition.

E. Acknowledgements

Grants from the Swedish and from the Danish Natural Sciences Research Foundations are gratefully acknowledged.

F. References

ÅBERG, E., in Växtodlingslära (ed. O. Hammar), pt. 2, pp. 157–206, LT:s förlag LTK, Borås, Sweden 1970

FISHER, J., HODGES, T.K., Plant Physiol. **44**, 385–395 (1969)

HODGES, T.K., Nature **209**, 425–426 (1966)

JACOBY, B., PLESSNER, O.E., Planta **90**, 215–221 (1970)

JANSSON, S.L., in Växtodlingslära (ed. O. Hammar), pt. 1, pp. 213–324, LT:s förlag LTK, Borås, Sweden 1970

KYLIN, A., KÄHR, M., Physiol. Plant. **28**, 452–457 (1973)

LINDEMAN, W., Proc. IInd. Int. Conf. of the UN on the Peaceful Uses of Atomic Energy **24**, 8–15 (1958)

LUNDBLAD, K., Växtnäringsnytt **11**(5), 1–13 (1955)

Session 7

Kinetics of Transport

Chairman: F. Führ

Uptake of Sulphate by Barley Roots: Separate Sites for Uptake and Phase Transitions

M.S. Vange, K. Holmern, and P. Nissen

A. Introduction

Uptake of inorganic ions by higher plants can be represented by single, multiphasic isotherms, i.e. by a series of phases following Michaelis-Menten kinetics and separated by discontinuous transitions (NISSEN, 1971, 1973a-d, 1974). These kinetics indicate mediation by a structure which changes characteristics at certain discrete salt concentrations. Uptake of glucose and 3-O-methylglucose (LINASK and LATIES, 1973; NISSEN, 1974) and at least some amino acids (NISSEN, 1974; SHTARKSHALL and REINHOLD, 1974) are similarly mediated by multiphasic mechanisms.

The concept of two uptake mechanisms in parallel at the plasmalemma (WELCH and EPSTEIN, 1968; EPSTEIN, 1972, 1973) has been shown to be invalid by a comprehensive reanalysis of concentration-dependence data (NISSEN, 1973a-d, 1974). The series model of ion uptake (TORII and LATIES, 1966; LATIES, 1969) has been amended and combined with the concept of multiphasic uptake mechanisms in a multiphasic series model (NISSEN, 1973b, 1974; LATIES, 1974). This model involves a single, multiphasic mechanism at the plasmalemma and a single, probably multiphasic mechanism at the tonoplast which may become rate limiting at high salt concentrations and after 'filling' of the cytoplasm.

Discontinuous phase transitions are a striking and hitherto unrecognized feature of ion uptake. As yet the phenomenon cannot be explained in molecular terms, but it will be shown in this contribution that transitions are not caused by binding of ions to the uptake site.

B. Materials and Methods

Seedlings of barley, ***Hordeum vulgare*** cv Herta, were grown in the dark at room temperature in 40-litre aquaria using $5 \cdot 10^{-4}$ M $CaCl_2$ (EPSTEIN, 1961). Roots were excised from 7-day-old plants. Uptake experiments were done with about 1 g roots in 75 ml of solution on a gyratory shaker in the light at room temperature. The 20 min uptake period was followed by a 5 min rinse in water at about 4°C. This provides a good estimate of influx across the plasmalemma (CRAM and LATIES, 1971; NISSEN, 1973a). Uptake of radiosulphate was performed in the presence of $5 \cdot 10^{-4}$ M $CaCl_2$, and pH was adjusted as described in Table 1. The root samples were assayed for radioactivity by liquid scintillation counting.

C. Results

In the experiment shown in Fig. 1, roots were allowed to take up radiosulphate from a 10^{-5} M solution in the presence of various concentrations of unlabelled sulphate or tungstate. Tungstate competitively inhibited uptake of radiosulphate, but more weakly than sulphate itself. The addition of sulphate or tungstate also caused transition from phase 1 to phase 2 at $2.9 \cdot 10^{-5}$ M or $3.0 \cdot 10^{-5}$ M, respectively.

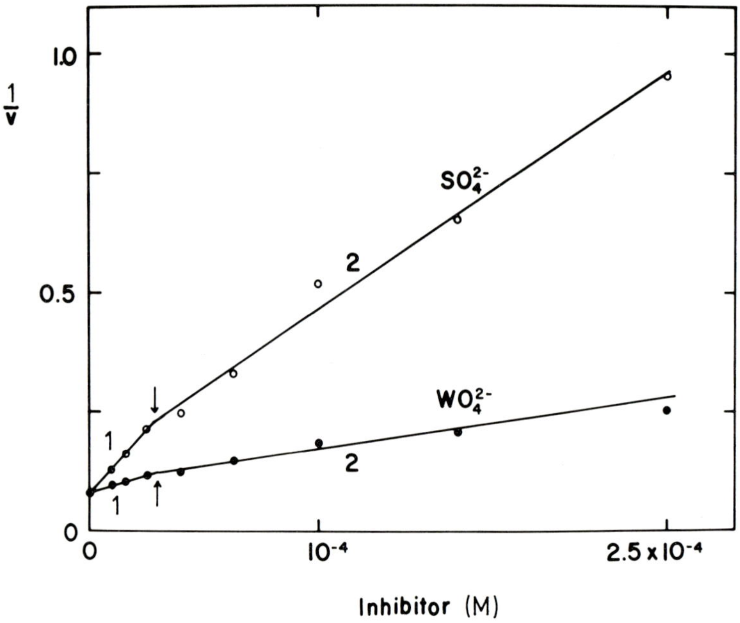

Fig. 1 Dixon plot for uptake of sulphate-^{35}S (10^{-5} M $K_2{}^{35}SO_4$) as a function of K_2SO_4 or Na_2WO_4 added. No buffer added, pH 5.8. Uptake expressed as nmoles/g fresh weight per 20 min. Lines determined by the method of least squares, transition points indicated by arrows. Average S.E. 5.2 % (n=4).

Table 1

Sulphate uptake: interactions of anions with uptake and phase transitions

Anion	pH	Inhibition of uptake	Effect on transition
Divalent sulphate analogues	5.8	+	+
$HMoO_4^-$	3.0	±	±
$H_2PO_4^-$	5.8	—	—
HPO_4^{2-}	8.15	—	+
$H_2P_2O_7^{2-}$	3.0	—	+
$HP_2O_7^{3-}$	6.9	—	+

No buffer added for divalent sulphate analogues and $H_2PO_4^-$. For HPO_4^{2-} and $HP_2O_7^{3-}$, pH adjusted with 10^{-2} M Tris-HCl buffer (H_3PO_4 : $pK_1 = 2.12$, $pK_2 = 7.21$, $pK_3 = 12.67$). 10^{-2} M Tris has no deleterious effects on sulphate uptake. For $HMoO_4^-$ and $H_2P_2O_7^{2-}$, pH adjusted with HCl (H_2MoO_4 : $pK_2 = 5.3$; $H_4P_2O_7$: $pK_2 = 1.49$, $pK_3 = 5.77$, $pK_4 = 8.22$ (c.f. TWEEDIE and SEGEL, 1970; WEAST, 1970).

Other divalent analogues ($S_2O_3{}^{2-}$, $SO_3{}^{2-}$, $SeO_4{}^{2-}$, $MoO_4{}^{2-}$, $CrO_4{}^{2-}$) also competitively inhibit sulphate uptake. The degree of inhibition varies, but the analogues are equally effective in causing transitions, i.e. transitions occur at the same ionic concentrations (these and other findings will be documented elsewhere).

Monovalent molybdate (Table 1) may inhibit uptake by the higher but possibly not by the lowest phase. It causes fewer phase transitions than sulphate or divalent molybdate.

Monovalent phosphate neither inhibits sulphate uptake nor causes transitions. Uptake from 10^{-5} M and 10^{-3} M solutions of radiosulphate was not affected by a 100-fold excess of unlabelled phosphate.

Divalent phosphate similarly does not inhibit sulphate uptake, but does cause phase transitions (Fig. 2). Phase transitions occurred at $9 \cdot 10^{-5}$ M and $3.4 \cdot 10^{-4}$ M when sulphate was added and in the intervals $6.3 \cdot 10^{-5}$ M – $1 \cdot 10^{-4}$ M and $2.5 \cdot 10^{-4}$ M – $4 \cdot 10^{-4}$ M when phosphate was added.

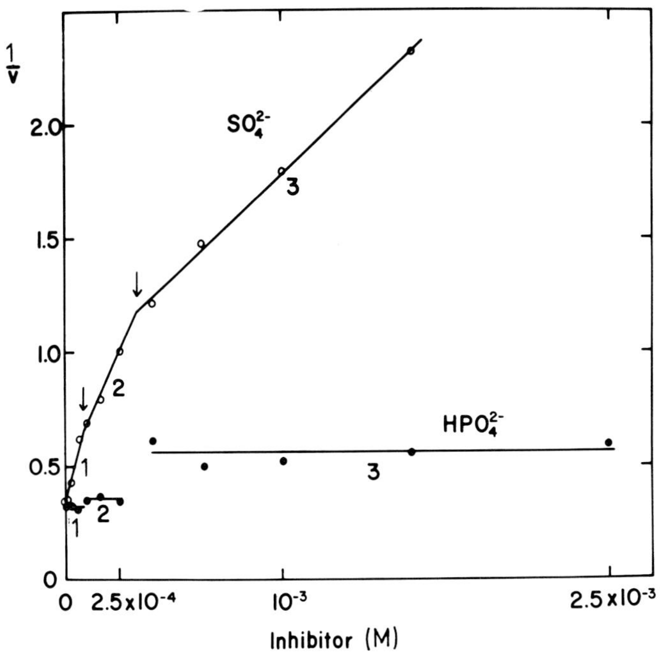

Fig. 2 Dixon plot for uptake of sulphate-^{35}S (10^{-5} M $K_2{}^{35}SO_4$) as a function of K_2SO_4 or KH_2PO_4 added. Tris-HCl buffer (10^{-2} M), pH 8.15 (cf. Table 1). Uptake expressed as nmoles/g fresh weight per 20 min. Lines determined by the method of least squares ($SO_4{}^{2-}$) or represented by mean values ($HPO_4{}^{2-}$); transition points indicated by arrows. Average S.E. 3.9 % (n=4).

Similar results were obtained with di- and trivalent pyrophosphate (Table 1). Pyrophosphate does not inhibit sulphate uptake, but the phase transitions correspond well to those caused by sulphate.

D. Discussion

Sulphate analogues can both compete with sulphate for uptake and cause phase transitions, but they do so by interacting with different sites as shown by differences in specificity. Their affinity for the uptake site may vary considerably, but they are equally effective in shifting phases (Fig. 1). These differences become still clearer in experiments with phosphate and pyrophosphate. $H_2PO_4^-$ is without effect on sulphate uptake, while HPO_4^{2-}, $H_2P_2O_7^{2-}$, and $HP_2O_7^{3-}$ cause phase transitions (Fig. 2, Table 1). Only sulphate analogues interact with the uptake site, while di- and trivalent anions are equally effective in causing transitions, presumably by interacting with a 'transition site'. Divalent anions may normally be required for phase transitions, cf. $H_2PO_4^-$ vs. HPO_4^{2-} and the finding that $HMoO_4^-$ causes no or at least fewer transitions than MoO_4^{2-}.

The existence of separate transition sites has also been inferred from experiments with other solutes. Reanalysis of data for alkali cations, chloride, and boric acid reveals remarkably similar multiphasic patterns (NISSEN, 1973d, 1974). The transition points for uptake of K^+ and Cl^- by roots of tall wheatgrass (ELZAM and EPSTEIN, 1969) may in fact be identical (NISSEN, 1973d). The patterns for uptake of phosphate and sulphate are also similar in a variety of plants and tissues (NISSEN, 1973c). This suggests that multiphasic patterns may be caused by transitions in a structure common to ions exhibiting the same pattern. There may, alternatively, be separate but very similar structures for the various ions as indicated by the fact that $H_2PO_4^-$ is without effect on SO_4^{2-} uptake.

Recent findings indicate that the postulated common structures may be transport ATPases. There is very good correlation between K^+ uptake and KCl stimulated ATPase activity in plasma membranes from oat roots (LEONARD and HODGES, 1973). The ATPase data may in fact be represented by a single, multiphasic isotherm (NISSEN, 1974). This provides good correlative evidence for the involvement of ATPases in ion uptake in plants and indicates that multiphasic kinetics of ion uptake result from phase transitions in transport ATPases.

Transport ATPases are membrane-bound, and there are as yet no clues as to whether the structures undergoing discontinuous transitions are proteins, lipids, or lipoproteins. The ATPase data have been discussed (LEONARD and HODGES, 1973) in terms of negative cooperativity between enzyme subunits (KOSHLAND, 1970; DALZIEL and ENGEL, 1968; ENGEL and FERDINAND, 1973), but it is not clear whether this model can adequately explain the clearly discontinuous kinetics of ion uptake (Fig. 2 and NISSEN, 1974).

The multiphasic kinetics of glucose uptake (LINASK and LATIES, 1973) are discussed in terms of the concept of cardinal adsorbents (LING, 1970). Interacting protein or lipoprotein structures are envisaged as undergoing all-or-none transitions as the external glucose concentration is raised. Whether these concepts apply to ion uptake is not clear (NISSEN, 1974).

E. References

CRAM, W.J., LATIES, G.G., Aust. J. Biol. Sci. **24**,633–646 (1971)
DALZIEL, K., ENGEL, P.C., FEBS Letters **1**, 349–352 (1968)
ELZAM, O.E., EPSTEIN, E., Agrochimica **13**, 196–206 (1969)
ENGEL, P.C., FERDINAND, W., Biochem. J. **131**, 97–105 (1973)
EPSTEIN, E., Plant Physiol. **36**, 437–444 (1961)
EPSTEIN, E., Mineral Nutrition of Plants: Principles and Perspectives, John Wiley and Sons, New York–London–Sydney–Toronto 1972
EPSTEIN, E., Int. Rev. Cytol. **34**, 123–168 (1973)
KOSHLAND, D.E., Jr., in The Enzymes (ed. P.D. Boyer), Vol. 1, ed.,3, pp. 342–396, Academic Press, London and New York 1970

LATIES, G.G., Ann. Rev. Plant Physiol. **20**, 89–116 (1969)
LATIES, G.G., in F.C. Steward Commemorative Plant Physiology Symposium, Cornell University Press, Ithaca, in press, 1974
LEONARD, R.T., HODGES, T.K., Plant Physiol. **52**, 6–12 (1973)
LINASK, J., LATIES, G.G., Plant Physiol. **51**, 289–294 (1973)
LING, G.N., Proc. Nat. Acad. Sci. **67**, 296–301 (1970)
NISSEN, P., Physiol. Plant. **24**, 315–324 (1971)
NISSEN, P., Physiol. Plant. **28**, 113–120 (1973a)
NISSEN, P., in Ion Transport in Plants (ed. W.P. Anderson), pp. 539–553, Academic Press, London and New York 1973b
NISSEN, P., Physiol. Plant. **28**, 304–316 (1973c)
NISSEN, P., Physiol. Plant. **29**, 298–354 (1973d)
NISSEN, P., Ann. Rev. Plant Physiol. Vol. 25, in press, 1974
SHTARKSHALL, R.A., REINHOLD, L., in Membrane Transport in Plants (eds. U. Zimmermann, J. Dainty), Springer-Verlag, New York–Heidelberg–Berlin 1974
TORII, K., LATIES, G.G., Plant Physiol. **41**, 863–870 (1966)
TWEEDIE, J.W., SEGEL, I.H., Biochim. Biophys. Acta **196**, 95–106 (1970)
WEAST, R.C. (ed).; Handbook of Chemistry and Physics, ed. 51, The Chemical Rubber Co., Cleveland 1970
WELCH, R.M., EPSTEIN, E., Proc. Nat. Acad. Sci. **61**, 447–453 (1968)

Influx Isotherms – their Interpretation and Use

W.J. Cram

A. Introduction

Since the shapes of influx isotherms are being accepted as evidence for two (or more) transport processes involved in influx to plant cells, and since parameters of curves fitted to isotherms are being used as fundamental characteristics of transport systems, it is becoming important to examine critically whether these interpretations of influx data are valid. The necessity for such a reappraisal is most strikingly apparent when one considers the possible ways of identifying isolated transport machinery. If the shape of the isotherm is not sufficient to distinguish or characterise transport processes it will not be profitable to attempt to identify transport machinery amongst microsomal fractions by a response to changing concentrations of the transported substance.

B. Two Basic Questions Concerning Influx Isotherms

I. What Flux is Being Measured?

The flux measured most frequently is the quasi-steady influx from the external solution to the vacuole (CRAM, 1969, 1973a). This flux can only be interpreted as the flux across a single membrane if one of the several steps involved is rate limiting. The only other flux that can be measured with any certainty in higher plant cells is the influx across the plasmalemma, based on the idea that the initial influx to the cell must equal the influx across the plasmalemma. This can be compared with the quasi-steady influx to the vacuole to see if there is a single rate limiting step at all external concentrations.

The experimental difficulty is that the initial influx to the cell must be estimated from the cell content after a short influx period; but in a tissue most of the substance is in the extracellular spaces after a short loading period, and only a fraction is in the cells. This difficulty can be avoided in three ways.

First, one can wash out the extracellular space content, but this also leads to loss from the cells and hence to an underestimation of the plasmalemma influx (CRAM, 1969; CRAM and LATIES, 1971; CRAM, 1973a). Nevertheless, this at least gives a minimum estimate of the plasmalemma influx, and can show if it is greater than the quasi-steady influx to the vacuole.

Secondly, one can wash out the extracellular space content and also estimate the concomitant loss from the cells by extrapolation of the cellular wash-out curve. This is another way of describing one aspect of what is sometimes called 'efflux analysis'. This method gives unambiguous (CRAM, 1973a, 1973b) and accurate results, but can only be used near a steady state, and has some inherent technical traps. Some misapprehensions concerning the assumptions and restrictions involved in this method have arisen (EPSTEIN, 1973).

Thirdly, instead of washing out, one can estimate the extracellular space content by including in the solution a labelled substance which does not penetrate the cells as well as the penetrating substance under investigation. After a short uptake period the amount of the non-penetrating substance in the tissue gives the extracellular space content. This can be subtracted from the total amount of the penetrating substance in the tissue to give the amount of the latter in the cells. This is potentially a rapid and accurate method of measuring the plasmalemma influx even under non-steady conditions. The problem is to pick a non-penetrating substance occupying the same extracellular volume at the same concentration as the substance under investigation. Sulphate nearly fulfils these requirements for anions, but a faster diffusing and slower penetrating anion is being sought.

Results from both methods in a range of tissues (e.g. Figs. 1A, 1B) agree in showing that the quasi-steady influx is not limited by the plasmalemma influx, except possibly at low external concentrations, in agreement with the conclusion of TORII and LATIES (1966). Hence the flux generally used in constructing isotherms cannot be analysed as though it were the flux across a single membrane.

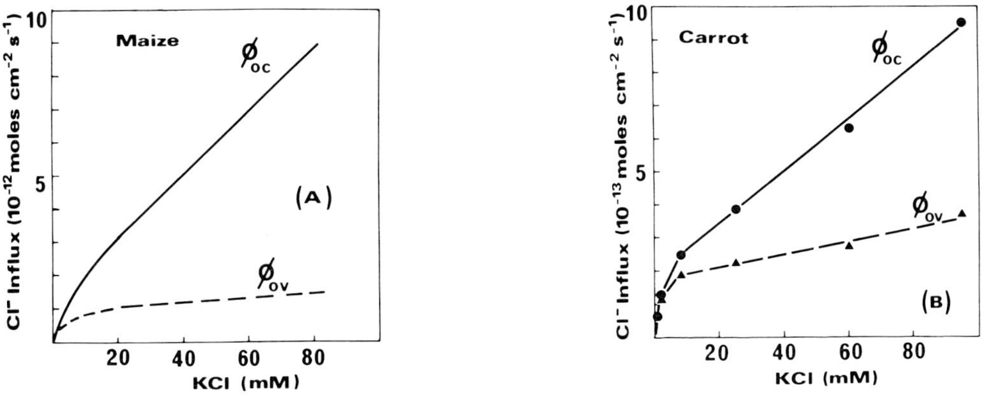

Fig. 1 Cl^- influx across the plasmalemma (ϕ_{OC}) and the quasi-steady influx to the vacuole (ϕ_{OV}) at various external KCl concentrations.
(A) Maize (***Zea mays*** L.) root cortical cells. From CRAM (1973a)
(B) Carrot (***Daucus carota*** L.) storage root cells. From CRAM (1968).

II. What Theoretical Basis is there for Interpreting Influx Isotherms?

The only bases for predicting isotherms of transport processes are kinetic models, which show what might occur, and observations on artificial membrane systems. Passive movement of ions might be as independent ions and/or in combination with natural carriers. (EISENMAN et al., 1967; EISENMAN, 1968; PAGANO and THOMPSON, 1968; HANAI et al., 1965.)

Non-carrier mediated passive movements of ions could increase more than proportionately to external concentration (LATIES et al., 1964) or less than proportionately (BRIGGS, 1964; SMITH, 1973) depending on the relative permeabilities of ions and the presence or absence of concomitant active transport of other ions. Under certain circumstances a fit to a rectangular hyperbola may obtain. Carrier-mediated passive movements of ions might rise more than proportionately or less than proportionately, or fall, with increasing external concentration; and might even rise, plateau, and rise again as external ion concentration increases (SZABO et al., 1969 – Fig. 2; STARK and BENZ, 1971; GERSON and POOLE, 1971; CIANI et al., 1973). Active transport isotherms are likely to be intrinsically at least as complex as carrier mediated passive movements (HOSHIKO and LINDLEY, 1967; THELLIER, 1973). There seems to be no basis for the frequent implicit assumption that all active transport processes obey a rectangular hyperbolic relationship to concentration.

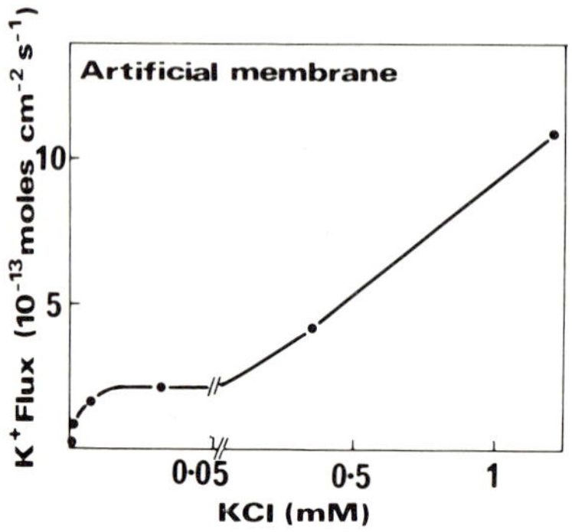

Fig. 2 K^+ influx across an artificial phospholipid bilayer membrane mediated by nonactin (10^{-7} M in the external solution) acting as a carrier. In 10 mM KCl the K^+ flux in the absence of nonactin is 10^{-16} moles cm^{-2} s^{-1}. From SZABO et al. (1969) with permission of the authors.

These results show that passive fluxes may 'saturate', that carrier-mediated fluxes may rise linearly or even quasi-exponentially with external concentration, and that the isotherms of both could have non-monotonically changing slopes. Influx isotherms therefore appear to provide no basis for distinguishing or identifying active or passive transport processes.

C. Discussion

Further questions may also be asked (for instance, as to the proportion of a measured flux which is active) but the above considerations alone make it difficult to interpret the observed relationships of quasi-steady influx to the external concentration of an ion or molecule.

Two main conclusions can be drawn about influx isotherms in general.

1. A fit to a rectangular hyperbola is not evidence for carrier-mediation or active transport. The parameters of a fitted rectangular hyperbola cannot ***a priori*** be taken to characterise a transport system, and an apparent 'K_m' cannot necessarily be equated with the 'affinity' of a carrier for the transported substance.
2. A transport system may fit a rectangular hyperbola in one range of concentrations but not in another. There is therefore no justification for extrapolating a rectangular hyperbola fitted to influx points in one concentration range to concentrations outside that range; and hence the operation of two transport processes is not established by observing a deviation from extrapolations of a rectangular hyperbola.

There therefore appears to be no basis for attaching a mechanistic interpretation to a quasi-steady influx isotherm. In particular, the existence of the transport processes commonly referred to as 'System I' and 'System II', as based on the kinetic parameters fitted to quasi-steady influx isotherms, is not convincingly established; and hence differences between the properties of quasi-steady influx at high and low external concentrations cannot be equated with properties of these 'systems'. Parameters of a fitted curve may be a convenient summary of experimental results (THELLIER, 1973), but the curve cannot prove, or ***a priori***, characterise, mechanism. All that an isotherm can be used for is to differentiate between kinetically distinguishable alternatives. The insuperable limitation to the use of isotherms is that different transport mechanisms are, in general, almost kinetically indistinguishable.

D. References

BRIGGS, G.E., J. Exp. Bot. **14**, 191–197 (1964)
CIANI, S., LAPRADE, R., EISENMAN, G., SZABO, G., J. Membrane Biol. **11**,255–292 (1973)
CRAM, W.J., Biochim. Biophys. Acta **163**, 339–353 (1968)
CRAM, W.J., Plant Physiol. **44**,1013–1015 (1969)
CRAM, W.J., Aust. J. Biol. Sci. **26**, 757–779 (1973a)
CRAM, W.J., in Ion Transport in Plants (ed. W.P. Anderson), p. 354, Academic Press, London and New York 1973b
CRAM, W.J., LATIES, G.G., Aust. J. Biol. Sci. **24**, 633–646 (1971)
EISENMAN, G., Fed. Proc., Fed. Am. Soc. Exp. Biol. **27**, 1249–1251 (1968)
EISENMAN, G., SANDBLOM, J.P., WALKER, J.L., Science **155**, 965–974 (1967)
EPSTEIN, E., Int. Rev. Cytol. **34**, 123–168 (1973)
GERSON, D.F., POOLE, R.J., Plant Physiol. **48**, 509–511 (1971)
HANAI, T., HAYDON, D.A., TAYLOR, J., J. Theor. Biol. **9**, 433–443 (1965)
HOSHIKO, T., LINDLAY, B.D., J. Gen. Physiol. **50**, 729–758 (1967)
LATIES, G.G., MacDONALD, I.R., DAINTY, J., Plant Physiol. **39**, 254–262 (1964)
PAGANO, R., THOMPSON, T.E., J. Mol. Biol. **38**, 41–57 (1968)
SMITH, P.G., J. Theor. Biol. **41**, 269–286 (1973)
STARK, G., BENZ, R., J. Membrane Biol. **5**, 133–153 (1971)
SZABO, G., EISENMAN, G., CIANI, S., J. Membrane Biol. **1**, 346–382 (1969)
THELLIER, M., in Ion Transport in Plants (ed. W.P. Anderson), pp. 47–63, Academic Press, London and New York 1973
TORII, K., LATIES, G.G., Plant Physiol. **41**, 863–870 (1966)

Multiphasic Amino Acid Transport in Leaf Cells

R.A. Shtarkshall and L. Reinhold

A. Introduction

We have previously brought forward evidence (SHTARKSHALL et al., 1970; REINHOLD et al., 1970) for the operation of an amino acid transport system in leaf cells, and demonstrated the usefulness of the unnatural analogue α-aminoisobutyric acid (AIB) in characterizing it. In the present communication we report on further studies on this system: a more extensive examination of the uptake versus concentration curve and its possible explanation; the sensitivity of the system to pH; and its structural specificity.

B. Materials and Methods

The materials and methods used were essentially as described in our earlier investigation (SHTARKSHALL et al., 1970). Briefly, narrow strips of barley leaf tissue which had been 'aged' overnight in 10^{-4} M $CaSO_4$ were incubated in labelled AIB solutions for 45 min. They were then washed for 25 min to remove AIB from the free space, after which their AIB content was determined by liquid scintillation counting.

C. Results and Discussion

I. The Relationship between Uptake and External Concentration

The kinetic data which we published previously (REINHOLD et al., 1970) for AIB uptake by barley leaf strips suggested that the curve relating uptake to external concentration was diphasic over the concentration range examined. We have now extended our observations to a wider range of external concentrations, and have noted additional phases in the uptake curve. Fig. 1A shows the results of two experiments A and B carried out at different times over the concentration range 0.0025 mM to 15 mM. It seems clear that 4 'steps' in the uptake versus concentration curve can be distinguished; these are labelled I, II, III, IV in Fig. 1A. The different phases are brought out more clearly if, as in our earlier paper, v (rate of uptake) is plotted against v/s (s = external concentration). A straight line would be expected for a system conforming with Michaelis-Menten kinetics, the slope of the line giving the K_m. Figs. 1B, C show these plots for the successive concentration ranges covered by the two curves shown in Fig. 1A. Inspection shows that the slopes of the lines (fitted statistically through the points) differ substantially from phase to phase.

If the system effecting uptake within each successive concentration range operates exclusively, without contribution from the systems operating in the neighbouring concentration ranges, then the true parameters of each system may be deduced from the solid lines in Figs. 1B, C. However, all systems may operate to a certain extent over the whole range, and the dashed lines in Figs. 1B, C give the corrected line for each phase after allowance for the maximal contribution of other phases according to the method suggested by HAGEN and

HOPKINS (1955). NISSEN (1973) has strongly objected to this procedure, but we present both the solid and the dashed lines here because, as we shall make clear below, we think that the true parameters may lie between these two limits.

Two – or multiphase-uptake curves have been reported for the uptake of many ions by plant cells, and were earlier interpreted as indicating two or more separate uptake mechanisms

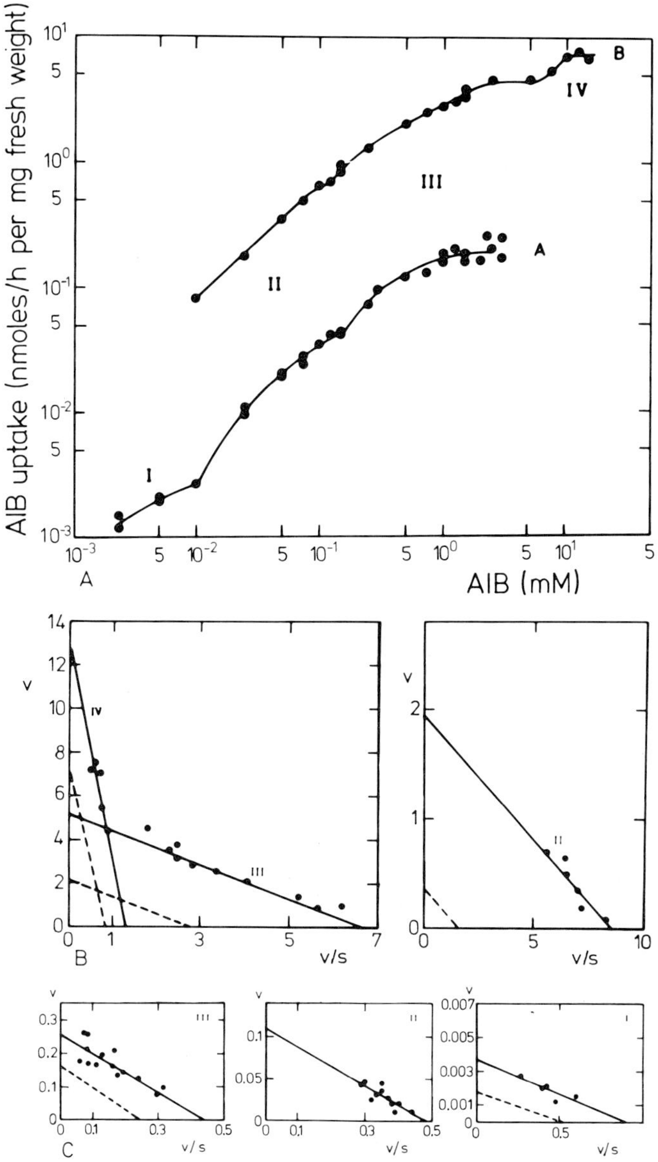

Fig. 1 Kinetics of AIB uptake by 'aged' strips of barley leaves. A (Top): relation between uptake velocity (v) and external concentration (s). Curves A and B are two separate experiments. B (Middle): analysis of kinetic data for upper curve in Fig. 1A. C (Bottom): analysis of kinetic data for lower curve in Fig. 1A. For explanation of dashed lines see text.

operating either in parallel or in series (TORII and LATIES, 1966; WELCH and EPSTEIN, 1968). More recently the curves have been taken to indicate the involvement in uptake of a structure which changes characteristics (kinetic constants) at certain discrete concentrations (NISSEN, 1971, 1973). On the basis of work by LING (1962), NISSEN (1971) has suggested that the changes are 'all—or—none' transformations, a proposal taken up by LINASK and LATIES (1973) in connection with sugar uptake by potato discs. These workers consider that the all—or—none hypothesis is a necessary consequence of the fact that the 'stepped' empirical curves cannot be generated by summing two or more Michaelis-Menten type mechanisms. For this reason NISSEN (1973) disapproves of extrapolation and subtraction of any one phase from adjacent phases, as suggested by HAGEN and HOPKINS (1955).

We wish to point out, however, that all—or—none changes do not constitute the only possible explanation of the multiphase uptake curve. We should like to suggest at least one mechanism in which non-interacting sites, each displaying Michaelis-Menten kinetics and acting in parallel, can account for a stepped curve. This is the case of excess substrate inhibition. For example, if 2-point attachment of the substrate to a receptor site is essential for the formation of the complex which leads to transport, then with increasing concentration there is an increasing probability that two substrate molecules will each combine at a single point. Each thus prevents the other from consummating the 2-point attachment. The rate of transport, v, will be given by

$$v = \frac{V_{max}}{1 + \frac{K_m}{s} + \frac{s}{K'_s}}$$

where V_{max} is the maximum rate, s is the substrate concentration, K_m the Michaelis constant and K'_s the dissociation constant for the complex with two substrate molecules (DIXON and WEBB, 1958). The curve relating v to s is bell-shaped. We have set up a model involving 2-point attachment to each site on a multiple site molecule, the sites being characterized by different constants. Two-point attachment is in our case quite probable in view of our conclusions about the structural requirements for optimal uptake (see below). Using constants suggested by our data we have been able to generate stepped curves.

II. The Effect of pH on Uptake

Fig. 2 shows that uptake is highly sensitive to pH, and it is interesting that the direction of this effect is the opposite of that observed in amino acid transport systems in animal cells (OXENDER and CHRISTENSEN, 1963). The pH effect is evident at both low (0.075 mM) and high (15 mM) AIB concentrations. (Only the low concentration curves are shown in Fig. 2.)

The uptake versus pH curve is shifted in various buffers (Fig. 2) suggesting that the buffer ions themselves exert effects. This possibility has been confirmed by experiments in which the concentration of the buffer ions was varied. Analysis of the inhibition caused by citrate ion according to the method of CLELAND (1970) (which involves replotting the slope, S, or intercept, I, obtained in reciprocal plots against inhibitor concentration) results in parabolic curves. The inhibition thus falls under his category 'S-parabolic, I-parabolic', indicating that at least two molecules of inhibitor combine with the carrier, which again suggests more than one site for attachment.

We believe that the effect of pH is not primarily an effect on the concentration of the species of ion transported, and that the latter is likely to be the zwitterion. Our reasons are as follows.

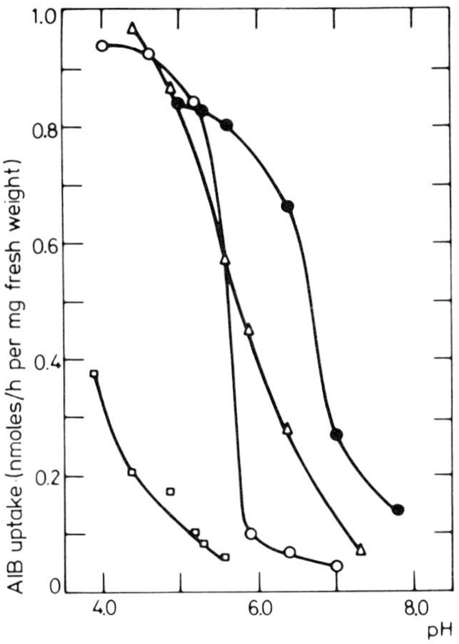

Fig. 2 The effect of pH on AIB uptake by 'aged' strips of barley leaves in 4 buffer systems. External AIB concentration 0.075 mM. Buffer concentration 0.05 M. ○ Phosphate-citrate; △ Phosphate; □ Citrate; ● Tris maleate.

1) The pK's of AIB are well above and below the experimental range (pK_1 = 2.35; pK_2 = 9.8). Over the experimental range more than 98 % of the AIB was present as the zwitterion.
2) Lowering the pH from 5.8 to 5.2 in phosphate-citrate buffer brought about an 8-fold increase in uptake. The only ion showing a concentration change in the same direction is the positive ion H_3N^+RCOOH, and this shows only a 3-fold increase in concentration.
3) If this were the species of ion taken up then it can be calculated from our data that the K_m would have to be approximately 10^{-7} M, a value a thousand times lower than any value in the amino acid transport literature.
It appears likely that protonation or deprotonation of the transport site is mainly responsible for the pH effect on uptake.

III. Structural Analogues as Competitors for Uptake

Table 1 shows that alanine is a very efficient inhibitor at both the pH's investigated. β-amino-isobutyric acid is less inhibitory at both pH's. Isopropylamine (which lacks the carboxylic group) and isobutyric acid (which lacks the amino group) are both somewhat inhibitory at pH 4.6, but at pH 7.2 isobutyric acid enhances AIB uptake. It is likely that this effect is related to that of dicarboxylic amino acids, to be described in a further communication. Dicarboxylic amino acids enhance uptake at the pH where they are in their positive ionic form. At the pH where they are near their zwitterion form they compete like neutral amino acids. Our preliminary conclusion is that for optimal uptake by the system studied both α-amino and α-carboxylic groups are required in their ionic forms.

Table 1

The effect of structural analogues of AIB on uptake of the latter by barley leaf strips. AIB concentration (1 mM). Buffer system, phosphate-citrate.

	Concentration	Uptake of AIB as % control	
Analogue	(mM)	pH 7.2	pH 4.6
β-amino-isobutyric acid	2	55	42
	10	40	27
Isobutyric acid	2	142	40
	10	164	27
Isopropylamine	2	48	65
	10	17	83
Alanine	2	18	19
	10	13	6

IV. Loss of Membrane Transport Ability as a Result of Osmotic Shock

As we have recently reported (AMAR and REINHOLD, 1973) osmotic shock severely reduces the ability of aged leaf strips to take up AIB. About 3.5 % of the total cell protein is released as a result of the shock treatment. The fact that uptake is reduced to near zero by treatment which allows the cells to retain over 95 % of their protein suggests that the small fraction of protein lost may be closely involved in the transport mechanism.

D. References

AMAR, L., REINHOLD, L., Plant Physiol. **51**, 620–625 (1973)

CLELAND, W.W., in The Enzymes (ed. P.D. Boyer), Vol. 2, pp. 1–65, Academic Press, London and New York 1970

DIXON, M., WEBB, E.C., Enzymes. Longmans, Green and Co., London 1958

HAGEN, C.E., HOPKINS, H.T., Plant Physiol. **30**, 193–199 (1955)

LINASK, J., LATIES, G.G., Plant Physiol. **51**, 289–294 (1973)

LING, G.N., Fed. Proc. **25**, 958–970 (1962)

NISSEN, P., Physiol. Plant. **24**, 315–324 (1971)

NISSEN, P., in Ion Transport in Plants (ed. W.P. Anderson), pp. 539–553, Academic Press, London and New York 1973

OXENDER, D.L., CHRISTENSEN, H.N., J. Biol. Chem. **238**, 3686–3699 (1963)

REINHOLD, L., SHTARKSHALL, R.A., GANOT, D., J. Exp. Bot. **21**, 926–932 (1970)

SHTARKSHALL, R.A., REINHOLD, L., HAREL, H., J. Exp. Bot. **21**, 915–925 (1970)

TORII, K., LATIES, G.G., Plant Physiol. **41**, 863–870 (1966)

WELCH, R.M., EPSTEIN, E., Proc. Nat. Acad. Sci. **61**, 447–453 (1968)

The Cellular Electropotential Isotherm as Related to the Kinetic K^+ Absorption Isotherm in Low-salt Barley Roots

S.M. Mertz, Jr. and N. Higinbotham

A. Introduction

The results of our first attempt to determine if there was any correlation between membrane electrical potential difference PD and rate of K^+ uptake by low-salt barley roots were inconclusive (PITMAN et al., 1971). Then, we cut the root into 1-cm segments in order to expose cortex cells for the measurement of PD. The cutting led to changes in the PD with time, and we did not know whether the PD of cells in freshly-cut or aged segments was equivalent to those found in whole roots. Therefore we developed new techniques and improved the instrumentation in order to routinely measure the electrical PD of cells in whole roots. In this report we compare data of the PD with the rate of K^+ influx as a function of the external concentration of KCl.

B. Materials and Methods

Barley seeds (***Hordeum vulgare*** L. cv. Himalaya) were surface sterilized and soaked in aerated deionized water for 24 h. Germinated seedlings were dark-grown on a plastic screen with their roots in aerated 0.5 mM $CaSO_4$ at 20°C. Roots from 4- or 5-day old seedlings were used for measurements of electrical PD, and from 6-day old seedlings for measurements of K^+ uptake.

The PD of cells in the cortex in the region 1–2 cm from the root tip was measured using pairs of 3 M KCl-filled glass micropipettes, Ag^+/AgCl wire electrodes, a high impedance voltmeter, and a recorder (MERTZ, 1973). A plastic chamber held a whole excised root horizontally on the stage of a microscope. The impaling micropipette was positioned perpendicular to the axis of the root. The micropipette was driven first into an epidermal cell, then continuously into underlying cortical cells. The driving unit was a piezoelectric bimorph linear motor mounted on a micromanipulator which was in turn mounted on the microscope stage. The tissue chamber was continuously perfused with solution. While the PD of a cell was continuously recorded, small stepwise increases were made in the KCl concentration. After the PD reached a steady level, another increase was made. The cells were exposed for less than 30 min to various concentrations of KCl during measurement of the PD.

Because the PD became more negative radially from the epidermis into the cortex (MERTZ, 1973), the PD's were averaged according to the radial position of the cell. The electrical PD reported here is that between the cell vacuole and the solution bathing the root.

Using $^{42}K^+$ as a label for K^+, short term uptake rates were determined at many KCl concentrations in order to duplicate the absorption isotherm characteristic for low-salt roots. The procedures used were those of EPSTEIN (1961) and associates (EPSTEIN et al., 1963) with some modifications (MERTZ, 1973). Excised roots were pretreated 30 min in 0.5 mM $CaSO_4$. Then they were transferred to labeled KCl solutions in beakers for 20 min. The temperatures of the pretreatment and uptake solutions were the same (20 or 30°C). Uptake was followed by a 30 min wash in 5 mM KCl at 4°C. All solutions were aerated, stirred, and contained 0.5 mM $CaSO_4$. After blotting and weighing, the roots were frozen, thawed, dried in planchettes and the radioactivity counted.

C. Results

When cells of low-salt barley roots were bathed in various concentrations of KCl, their electrical PD varied with the KCl concentration. This is shown in Fig. 1 for the outermost

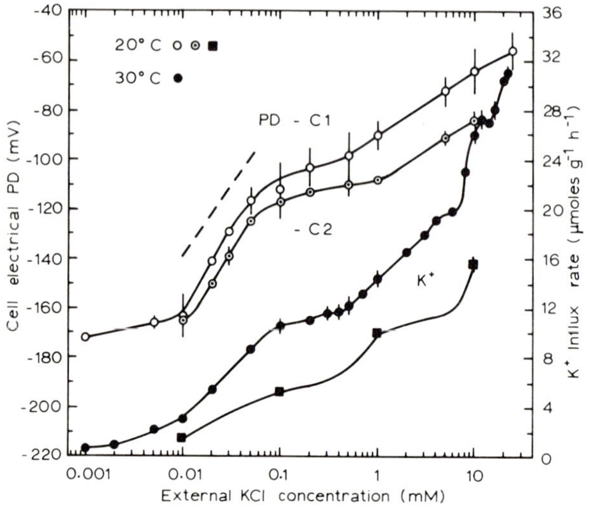

Fig. 1 The electrical PD of cortical cells and the rate of K^+ influx in excised, low-salt barley roots as a function of temperature and of the external concentration of KCl from 0.001 to 25 mM. All solutions contained 0.5 mM $CaSO_4$. The PD is shown for the first (C1) and second (C2) layers of cells in the cortex adjacent to the epidermis. The dashed line represents a change in PD of 58.2 mV per ten-fold change in ion concentration. The data are the means of duplicate samples for K^+ influx, 16 cells for C1 PD's, and 10 cells for C2 PD's. The vertical bar through a point indicates one standard deviation about the mean. Bars are shown only where they exceed the symbol size.

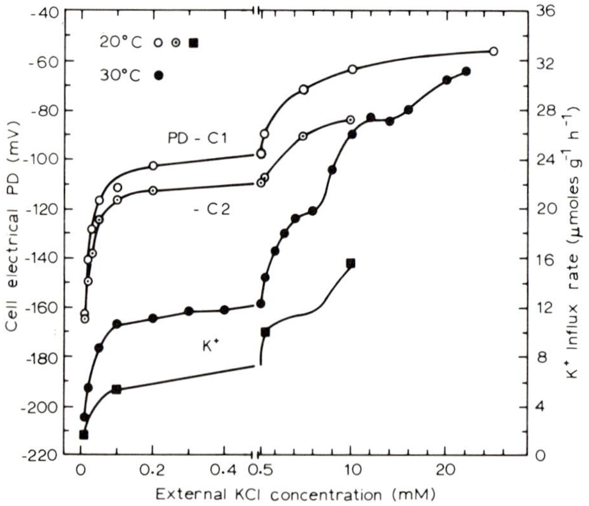

Fig. 2 The electrical PD of cortical cells and the rate of K^+ influx in excised, low-salt barley roots as a function of the temperature and of the external concentration of KCl from 0.01 to 25 mM. Other information is the same as in the legend of Fig. 1.

two layers of cells in the cortex. The KCl concentration is plotted on a logarithmic scale. Note that even though cells in the layers C1 and C2 are adjacent and interconnected via plasmodesmata they had different PD's. At all concentrations studied, the PD became more negative from the epidermis into the third layer of cells in the cortex (MERTZ, 1973). The PD isotherm for C2 is most likely characteristic of the majority of the cortex tissue. The isotherm for C1 is intermediate between that of C2 and the epidermis. The PD's found in roots attached to the shoot are the same as those reported here for excised roots.

Increasing the KCl concentration resulted in depolarization. The amount of depolarization varied depending on the concentration range. The depolarization per ten-fold increase in concentration was 8–9 mV in the 0.001–0.01 mM range, 58–62 mV in the 0.01–0.1 mM range, 9–18 mV in the 0.1–1 mM range, and 24 mV in the 1–25 mM range. This pattern of depolarization found in cells in the cortex of whole roots is different than that found in cells in the cortex of freshly-cut or aged segments of roots (PITMAN et al., 1971).

The logarithmic plot demonstrates those concentration ranges where the change in the PD is an exponential function of ion concentration. If the PD is the result of independent diffusion of K^+ only, then the change in PD per ten-fold change in K^+ concentration (ΔPD_{10}) would be 58.2 mV at 20°C. This is shown by the dashed line in Fig. 1. The experimental data fit this theoretical maximum value only in the range 0.01 to 0.05 mM K^+; values less than the maximum suggest significant permeability to Cl^-.

The rate of K^+ influx was determined at 20 and 30°C over the same time period as the measurements of PD. A temperature of 30°C was used in order to be able to compare our results with those of many other workers who have used 30°C. At 20°C the rates of K^+ uptake are less than those at 30°C by 1.4- to 2.0-fold, but the overall shape of the isotherms is thought to be similar.

In Fig. 1, one can observe that there is a correspondence in the overall pattern of the isotherms of PD and K^+ influx. The correspondence at low external K^+ concentrations is shown more strikingly in Fig. 2. Fig. 2 is a double-scale plot. The two linear scales of KCl concentration (0.01–0.5 mM and 0.5–25 mM) have the ordinate axes and the concentration of 0.5 mM in common. The PD isotherm parallels that of the rate of K^+ influx. At this time we have not yet obtained enough data to show whether or not there is any correlation between the inflections in the isotherm of K^+ influx (which occur here at 0.5, 6, and 14 mM) and changes in the membrane PD.

D. Discussion

One criterion used by some workers to distinguish between active and passive transport processes is the Q_{10} value. Most chemical reactions have Q_{10}'s between 2 and 3, whereas physical processes such as ion diffusion often have Q_{10}'s slightly greater than 1. The Q_{10} values (30°C/20°C) for the short term rate of K^+ influx were determined from the data of Fig. 1. They were 2.0 at 0.01 mM, 2.0 at 0.1 mM, 1.4 at 1.0 mM, and 1.7 at 10 mM. Based upon the Q_{10} criterion, K^+ influx is coupled to an active process at concentrations less than 1 mM and may have a passive component at 1 mM and above.

The electrical PD became more negative from the epidermis into at least the third layer of cells in the cortex. This needs to be explored further as a driving force for transport of ions from cell to cell.

In the regions 0.001–0.01 and 0.1–1 mM where there was little change in the rate of K^+ uptake with a change in K^+ concentration, there was also little change in the PD. Conversely, in the regions 0.01–0.1 and 1–25 mM where there was a significant change in the rate of K^+ uptake with a change in K^+ concentration, there was also a significant change in the PD. Thus, there is a definite correlation between membrane PD and the rate of K^+ uptake.

The electrical PD across membranes is primarily a function of membrane permeability, ion activity, and active (electrogenic) transport processes. A brief analysis of the data in terms of these parameters may be instructive.

1) The ΔPD_{10} does attain the theoretical maximum value possible for diffusion in the range 0.01–0.05 mM. This suggests a high ratio of the membrane permeability coefficient for K^+ (P_K) to that for Cl^- (P_{Cl}) in this range. Calculations using the Goldman voltage relationship yield a cytoplasmic K^+ concentration (activity) of 5–10 mM.

2) Only at concentrations of 1 mM or less can the magnitude of the PD be explained in terms of the classical Goldman voltage relationships for a PD generated by ion diffusion.

3) At higher external K^+, the PD exceeds values possible for diffusion. One must thus invoke electrogenic ion transport to account for the extra charge separation. The influx of K^+ is in the wrong direction for it to be the subject of an electrogenic ion pump, however, K^+ could be driven inward by the PD. Also in this region, the ΔPD_{10} of 24 mV was significantly less than the maximum value of 58 mV. This suggests a lower ratio of P_K to P_{Cl} than that found in the 0.01–0.05 mM range.

4) The inflections in the influx isotherm do not appear to be directly related to the membrane PD. It is likely that such inflections in general are associated with changes in membrane resistance (SPANSWICK, 1972), possibly as a result of changes in ionic strength of the solution at the surface of membranes.

E. Acknowledgements

We gratefully acknowledge the technical assistance of R. McKenzie, Yee Ping Wong, and J. Mertz. This work was supported in part by National Science Foundation Grants GB-5117X and GB-19201 to N.H.

F. References

EPSTEIN, E., Plant Physiol. **36**, 437–444 (1961)

EPSTEIN, E., SCHMID, W.E., RAINS, D.W., Plant Cell Physiol. **4**, 79–84 (1963)

MERTZ, S.M., JR., Electrical Potentials and Kinetics of Potassium, Sodium, and Cloride Uptake in Barley Roots. Doctoral thesis, Washington State University, Pullman (1973)

PITMAN, M., MERTZ, S.M., JR., GRAVES, J.S., PIERCE, W.S., HIGINGBOTHAM, N., Plant Physiol. **47**, 76–80 (1971)

SPANSWICK, R.M., Biochim. Biophys. Acta **288**, 73–89 (1972)

Round Table Discussions 6 and 7

Chairmen: M.G. Pitman and W. Tanner

PITMAN opened the proceedings by saying that we should bring together a number of topics previously discussed and see whether we can relate them to the mechanism of ion uptake. First, in ***Neurospora***, as discussed by SLAYMAN, there is a K^+ influx which can be coupled with a K^+, or Na^+, or H^+ efflux; this transport system is associated with an electrogenic potential, which correlates closely with the ATP levels in the cell; and the electrogenic potential appears to be due to active extrusion of H^+. Secondly, TANNER and KOMOR have shown that in ***Chlorella*** the kinetics of glucose uptake, because it is coupled with H^+ uptake, are affected by pH; pH also affects the uptake of K^+ in ***Neurospora***. Both examples show that the kinetics may be multiphasic and affected by pH and that uptake (and may be ATP) seem to be correlated with an electrogenic potential. PITMAN asked how these results may be applied to roots and he pointed to complex uptake kinetics, different transport sites for different ions, electrogenic potentials, H^+ active efflux, etc. all in barley roots. He continued: is transport due to operation of an ATPase? What is the coupling between the ATPase and the actual ion movement? And do we have transport systems in roots that show multiple site control, or multiple site kinetics?

By invitation, SLAYMAN spoke about the kinetics of cotransport systems. He pointed that even the simplest model of such a system is pretty complicated, kinetically; it has 12 rate constants, for instance. Assumptions can be made, such as that the transit reactions are slow, which reduce the effective number of constants. And one can of course assume steady state conditions. But even then the system remains very complex, and this is with no charge or voltage term introduced! He had to conclude that if one is, for instance, looking at the binding constants in such a transport system, then these constants depend on all the rest of the kinetic parameters of the system. They are not independent of each other, either. SLAYMAN also pointed out that in such systems, because of this mutual dependence of the measured parameters, one could get, say, correlations between potassium sensitive ATPase activity and potassium transport which were quite false. He was just stressing how complicated even an apparently simple cotransport system is and the dangers of reading too much into kinetic data.

TANNER pointed out that in his ***Chlorella*** system, nice biphasic curves are obtained depending on the pH; they arise because the protonation of the carrier changes the K_m of the sugar transporter. CRAM said, very definitely, that he does not believe there are biphasic curves. He also pointed out that in most of the published isotherms the variability of the individual points has not been established. And, furthermore, even if a single rectangular hyperbola could be fitted to an isotherm, he would not know what it meant. THELLIER doubted very much whether from purely macroscopic measurements of uptake against concentration, from which one supposedly gets V_{max}'s and K_m's, one could obtain the molecular data about numbers of carriers, affinities, transit velocities, etc. that people claim. He wondered whether with this data a flow-force 'black box' approach is not the best, even if there are doubts about what to take for the force. If the logarithm of the force is taken, there is a domain of linearity between flow and force and from the relationship a conductance and a kind of internal contribution to the force can be derived. This, for THELLIER, is the only kind of thing one can say.

NISSEN then explained how his multiphasic curves were obtained. In his own experiments he ran 4 or 5 replicates, resulting in S.E.'s of 4 to 5 %. His own, and other people's, data are then analysed by a FORTRAN programme in which the sum of the squares of deviations in log v is minimised; this is used to minimise the weighting errors associated with the linear transformation of the Michaelis-Menten equation. Bias and error are mainly introduced into the choice of the transition points. He always runs an F test to see whether the data fit better 1 or 2 phases. He thinks that the data, whether his own, or Epstein's, Laties's, etc. are consistent. WALKER did not think it legitimate to split, subjectively, a set of data into two phases and then do statisties on the two phases. He said that the only legitimate thing is to do an objective test which tells you whether you have cheated by splitting into two phases or not; almost anything fits two lines better than one! NISSEN denied that this latter statement was true with respect to his data.

HODGES now spoke of his result on the potassium stimulation, as a function of $[K^+]$ from 1 to 50 mM, of his ATPases. It showed multiphasic behaviour similar to influx data. Only the alkali metal cations showed this typical deviation from Michaelis-Menten kinetics. He then described Koshland's model of a multi-sub-unit enzyme in which, when the first ligand binds to a sub-unit it puts a constraint on the other sub-units and either increases or decreases their affinity to subsequent ligands. The negative cooperativity type, in which the affinity decreases, seems to him a reasonable model to account for multiphasic kinetics. JESCHKE next said that, in spite of the remarks discrediting Michaelis-Menten constants, at low concentrations the K_m's do give the relative affinities of the various ions being compared and these are in line with competitive inhibitor constants. HIGINBOTHAM was disturbed by the fact that absorption isotherms are never carried out at constant ionic density; Mertz had studied K^+ uptake in high concentrations of $CaCl_2$ and it showed no inflections. PITMAN said that K^+ uptake by leaf slices is very sensitive to Ca^{2+}, which is therefore another variable to be kept in mind.

At this point MERTZ introduced the audience to a class of toxic compounds secreted by the fungus responsible for Southern corn leaf blight disease. They are steroid-like compounds at least one of which is a glycoside. It (the effective compound) may be the ouabain of the plant world! In maize it inhibits K^+ uptake and H^+ efflux, and depolarizes the membrane potential, all with the same time course. It inhibits the K^+ stimulated (microsomal) ATPase. It inhibits Cl^- influx, but has no effect on Na^+ and phosphate influx. It does not affect mitochondrial transport, and appears to act only at the surface of the plasmalemma. It also has some effects on water transport, e.g. stomata are closed because K^+ transport is blocked. MERTZ stressed what an exciting inhititor it is. Not to be outdone, SLAYMAN then spoke of the phytoalexins which are plant antifungal toxins. Many of them are steroid-like and some of these are glycosides. In ***Neurospora*** one of them, phaseolin, partially depolarizes the membrane potential, inhibits K^+ influx and H^+ efflux, and clots the cytoplasm in a manner suggesting that it has become acid in pH.

PITMAN changed the discussion to ecological aspects; he asked KYLIN whether in his work changes in repressor compounds or in enzyme content were involved. KYLIN had no idea, but said that external conditions must change the ATPases in a cell. He stressed that ecology, physiology and biochemistry must closely work together to find out how plants respond to various external conditions. LÜTTGE thought that the prototype of a biochemical adaptation to ecological conditions was the inducible salt-excretion mechanism in the salt glands of ***Limonium***; there seems to be an inducible Cl^--ATPase. KYLIN agreed but warned: don't believe it is always ATP and ATPases that are involved. LÄUCHLI asked KYLIN whether he had found any specific differences between the ATPases of high salt and low salt plants. KYLIN thought that looking at two concentrations was not enough. In his mind when an ATPase functions, there should be some balance point on an optimum curve where there is minimal ATPase activity so as to save energy.

Finally WYN-JONES spoke of the problem of the biochemists. They get an enzyme out which may be involved in ion uptake. How does one correlate its activity with an ion flux across a specific membrane? He accepted that the short-uptake short-wash method gave a reasonable estimate of the plasmalemma flux, but what about the tonoplast flux?

CRAM communicated firstly that the short-uptake short-wash technique for the plasmalemma influx for cations is inherently non-quantitative unless one can be sure what is happening to the cell wall cations. Secondly, there are great difficulties in measuring tonoplast fluxes. If one is sure that the cytoplasm and vacuole are in series, then from an analysis of the exchange kinetics one can get the tonoplast fluxes. But there are some indications of a 'straight-through' pathway from the work of MacRobbie, and there is also the possibility of exchange between the cytoplasm and the organelles in the cell. Both these phenomena raise doubts about the simple series model using which tonoplast fluxes can be calculated.

Session 8

Transport in Organs of Higher Plants

Chairman: M. Thellier

Co-Operation of Organs in Intact Higher Plants: A Review

U. Lüttge

A. Introduction

The need for co-operation by organs is well known as an evolutionary consequence of increasing specialization. Specialization in plants usually implies involvement of various systems of supply and demand[1]) of particular substances. For instance, the supply of products of photosynthesis from leaves is related to demand from the non-green parts, especially from storage tissues and rapidly growing organs. Similarly, the supply of ions and water from the soil is determined by activity of the roots and demand from the aerial parts of the plant. In both of these examples phytohormones can be shown to be involved in signalling demand. Let us consider the second example, i.e. supply of water and ions, to be more specific and to introduce a system which will be used more extensively below. The phytohormone abscisic acid (ABA) is produced in leaves in response to water stress. It affects stomatal regulation, i.e. the resistance to, and hence the driving force on, transpiration (WRIGHT and HIRON, 1969; MIZRAKI et al., 1970; MILBORROW and NODDLE, 1970; MOST, 1971; KRIEDEMANN et al., 1972; HIRON and WRIGHT, 1973). As a consequence, the supply of water and ions is altered.

In each of these examples a prerequisite of co-operation is traffic of goods, or transport of matter. The simplified structural models of Fig. 1 appear suitable to demonstrate that maintenance of co-operation of organs in higher plants requires three principally distinguished modes of transport of matter, i.e. membrane transport, transcellular transport, and long distance transport. It might be attractive to illustrate the role of these modes of transport by drawing analogies to the diversity of traffic systems in the equally complex structure of modern industrial society. It will,' however, be more revealing to proceed without delay to specific model cases selected from plant biochemistry and plant physiology, where we shall consider the concerted action of membrane transport, transcellular transport in the symplasm, and long distance transport in the xylem and phloem.

B. Model Cases

I. Membrane Transport and Transcellular Transport in C_4-Plants

An understanding of the involvement of membrane transport in the regulation of cellular metabolism has come from investigations of transport at chloroplast and mitochondrial membranes interrelating photosynthesis, respiration and glycolytic reactions. Concerted action of such membrane transport processes with transcellular transport is implied in models of the so-called C_4-photosynthesis (HATCH et al., 1971; LÜTTGE, 1973), where different tissues of a leaf co-operate in CO_2 fixation and eventual formation of starch as depicted in Fig. 2. Malate formed as the primary product of CO_2 fixation in the so-called

[1]) The terms 'supply' and 'demand' are used throughout this review in the same sense as the more widely used but perhaps less descriptive terms 'source' and 'sink'.

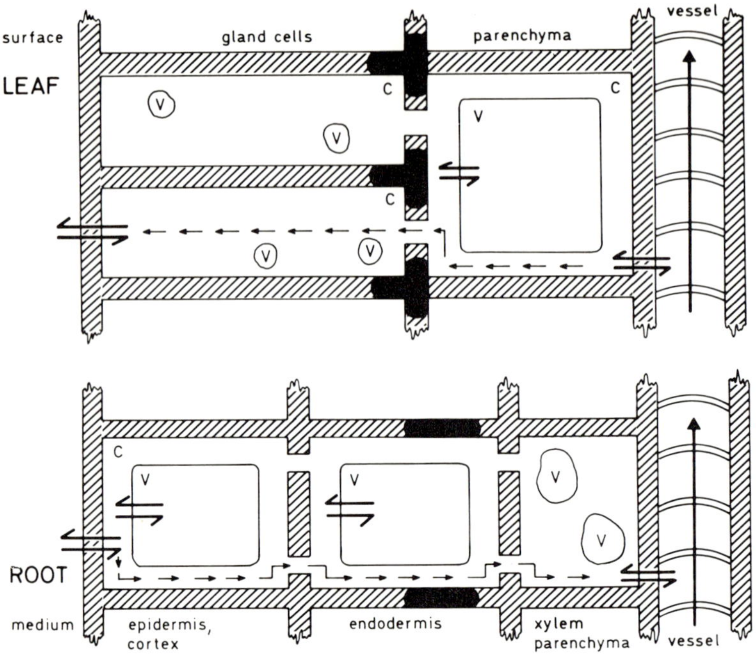

Fig. 1 Simplified structural models of a plant root and of a leaf with a salt gland, demonstrating the basic similarity in the concerted action of membrane transport, transcellular transport and long distance transport in the overall function of the respective organ.
Arrows: transport processes. C = cytoplasm, V = vacuole. Casparian strip-like incrustations of cell walls of leaf gland cells and Casparian strip of the root endodermis are drawn black.

mesophyll tissue is supplied via the symplasm as a transport metabolite transferring CO_2 to the bundle sheath cells, where the CO_2 is liberated by malic enzyme and demanded for refixation by RudP-carboxylase. One of the properties of this transport process reveals two principal features of symplasmic transport: 1) continuous metabolic control on the particles transported, and 2) efficiency. The first point is self-evident; release of a metabolite from the cytoplasm during transport would imply release from metabolic control. The second point is made clear by the fact that the flow of CO_2 through malate and into the carbohydrate eventually deposited in the bundle sheath cells is 100–200 μmoles $h^{-1} g^{-1}$ fresh weight, whereas malate transport across the plasmalemma or tonoplast would be only of the order of 1–2 μmoles h^{-1} g^{-1} (OSMOND, 1971).

Fig. 2 Simplified scheme of C_4-photosynthesis and C_4-leaf anatomy. M = mesophyll, B = bundle sheath, Chl = chloroplast, Pj = inorganic phosphate, PP_i = pyrophosphate, RudP = ribulose-1,5-diphosphate, NADP, ATP, and AMP have the usual meanings. Thick black arrows: chloroplast membrane transport plus symplasmic transcellular transport. The reaction scheme (after LÜTTGE, 1973, Fig. 6.18, p. 196) aims to demonstrate the need for concerted action of these two modes of transport. This would be maintained in principle if some of the reactions shown were cytoplasmic rather than chloroplastic, which is not yet unequivocally clear. Leaf anatomy is shown by a photograph of a colour poster original drawn by Mrs. D. Schäfer for teaching purposes in our Institute.

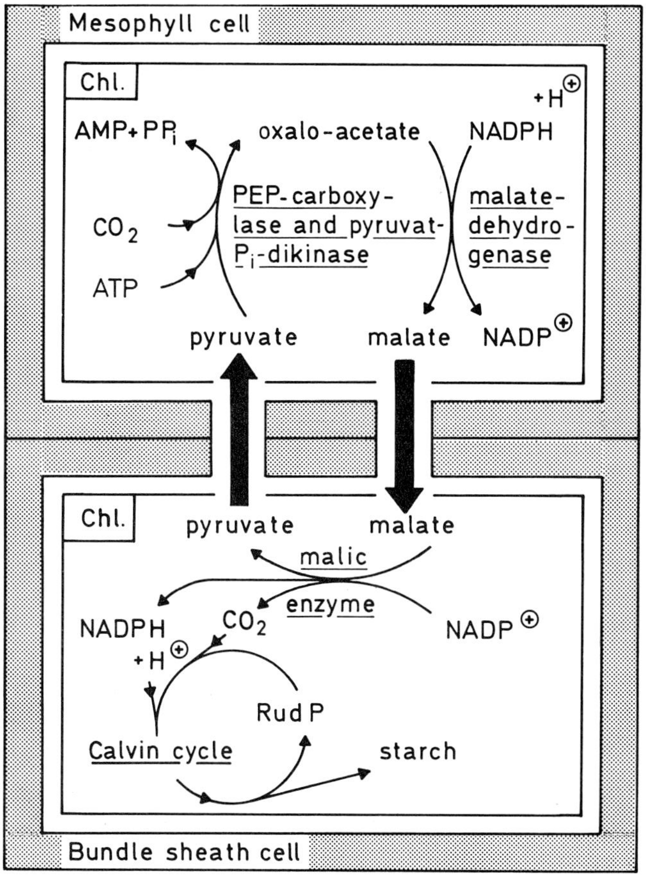
Mesophyll cell
Chl.
+H⊕
AMP+PP$_i$
oxalo-acetate
NADPH
PEP-carboxy-
lase and pyruvat-
P$_i$-dikinase
malate-
dehydro-
genase
CO$_2$
ATP
pyruvate
malate
NADP⊕
Chl.
pyruvate
malate
malic
enzyme
NADPH
CO$_2$
NADP⊕
+H⊕
Rud P
Calvin cycle
starch
Bundle sheath cell

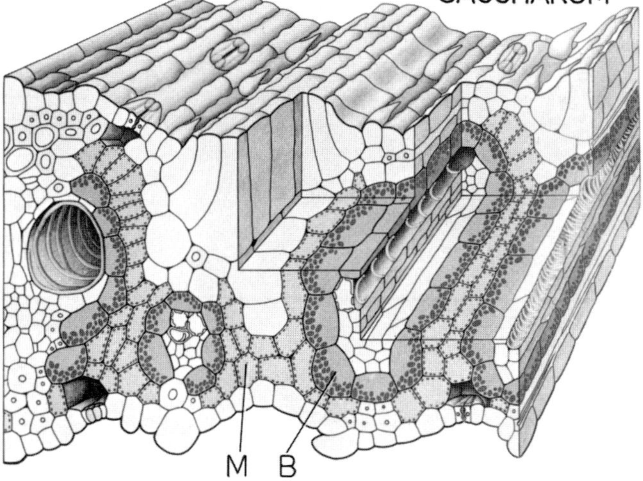
SACCHARUM
M
B

II. Membrane Transport, Transcellular Transport, and Long Distance Transport in Gland Systems

From considering co-operation of different tissues of a leaf as in C_4-photosynthesis mediated by membrane transport and symplasmic transport of transport metabolites we may proceed to structurally much more complex, though biochemically perhaps more simple, systems by adding the third component: long distance transport.

We can do this by an evaluation of nectary or salt gland systems. Supply now originates from more or less distant organs: salts in salt excretion by salt glands come from the roots via the xylem (Fig. 1), sugars in nectar secretion by nectary glands are supplied via the phloem. The gland cells and the supporting tissue never contain more than a fraction of the total amount of material which they excrete or secrete (FINDLAY et al., 1971; LÜTTGE and SCHNEPF, 1975). In the case of the glands we may prefer to use the term 'elimination' (FREY-WYSSLING, 1970, 1972) instead of 'demand'; however, this does not change the model situation in principle. In C_4-metabolism the metabolic supply of malate in the mesophyll and the metabolic demand of CO_2 carried by the malate in the bundle sheath appears to be the driving force of transport. However, the driving force becomes rather difficult to discern in the more complex gland systems. In nectary secretion, in addition to chemical reactions such as sugar metabolism in the gland cells and sucrose inversion at the gland surface, active membrane transport processes must be considered as possible driving forces, e.g. active unloading of sieve tubes, active accumulation of sugars from the parenchyma into the secretory cells, active elimination at the gland surface (reviews: LÜTTGE, 1971; LÜTTGE and SCHNEPF, 1975).

Although metabolism of the particles transported is not involved in excretion of salt, the situation is not much less complex in salt gland systems. Again possible driving forces are salt accumulation from the routes of long distance transport (xylem) into the non-specialised parenchyma, accumulation from the parenchyma by the specialised gland cells, and active elimination (Fig. 1; LÜTTGE, 1971). Nevertheless, since for salt excretion the well advanced theory and methodology of the electro-chemical approach is applicable, somewhat more is known here. Excretion in most instances is clearly active in that overall transport is against a gradient of electrochemical potential.

In some cases accumulation by the gland cells from the parenchyma can be ruled out as a possible site of active membrane transport driving excretion. Using a cytological precipitation technique ZIEGLER and LÜTTGE (1967) have shown the numerous plasmodesmata between the cells of the ***Limonium*** salt gland and the embedding leaf parenchyma to be readily accessible for ions. Thus plasmodesmata appear to mediate a transport between the co-operating tissues which is symplasmic, i.e. which does not require the passage of membrane barriers.

Salt excretion into the epidermal bladders of ***Atriplex*** and ***Chenopodium*** leaves (Fig. 3) can be used nicely to demonstrate such co-operation. Fig. 4 shows that light greatly stimulates the excretion of Cl^- into the bladder vacuoles. This light stimulation involves photosynthetic energy, but the gland-like stalk cells and the bladder cells are photosynthetically inactive. This means that either the energy-consuming driving force (ion pump) is located in the photosynthetically active mesophyll tissue or that energy is transported from there to the bladder to drive active excretion or that just a photosynthesis dependent signal is transferred.

Electrophysiological measurements with the mesophyll-epidermal bladder system of chenopods are very suitable to further elucidate the ready contact within the symplasmic continuum. Transient light-dependent membrane potential oscillations (such as those shown in Fig. 5) are generated by switching on or off photosynthesis in the green parts of the

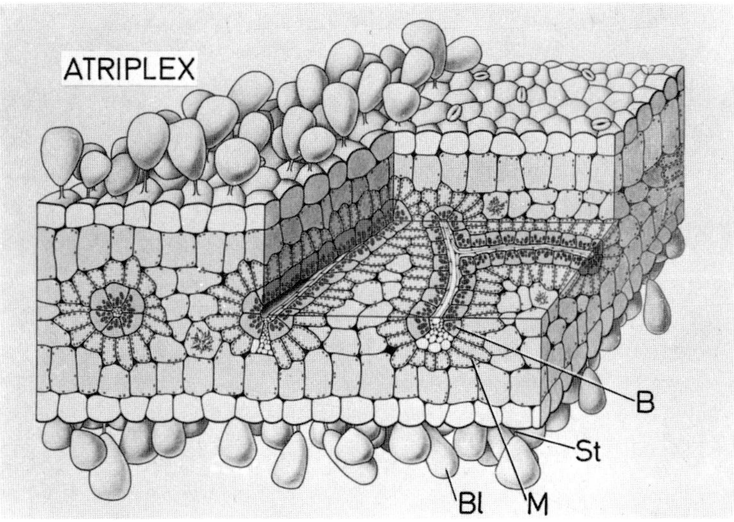

Fig. 3 ***Atriplex spongiosa*** leaf anatomy. M = mesophyll, B = bundle sheath, St = stalk cell, Bl = bladder cell. Photograph of a colour poster original drawn by Mrs. D. Schäfer for teaching purposes in our Institute.

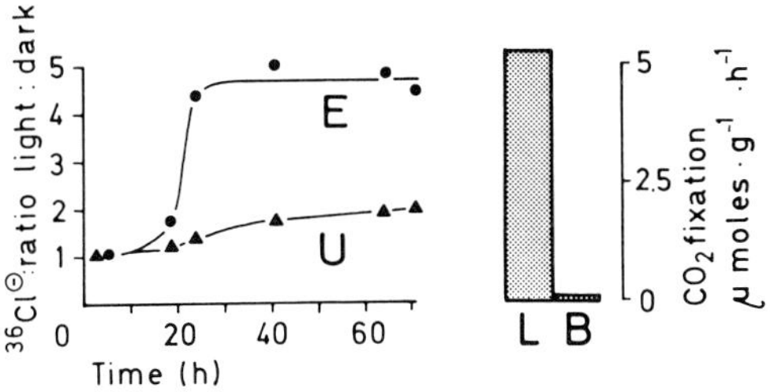

Fig. 4 Light stimulation of Cl^- uptake (U) by ***Atriplex spongiosa*** leaf slices (3 mm wide) and excretion (E) to bladders; and photosynthetic CO_2 fixation of leaf lamina (L) and stalk plus bladder cells (B). Drawn from Fig. 2 in OSMOND et al. (1969) and from Table 3 in LÜTTGE and OSMOND (1970).

leaves. They are also picked up in the photosynthetically inactive bladder cells when the green mesophyll is in place but not when it is removed (LÜTTGE, 1973, p. 221, Fig. 7.2). Cutinization of the bladder, stalk and epidermal cell walls prevents apoplasmic electrical contact. BRINCKMANN (1973) demonstrated electrical contact in the symplasm more elegantly by using variegated photosynthesis mutants of ***Oenothera*** (Fig. 5). Photosynthesis-dependent membrane potential oscillations are observed with the electrode in white mutated cells when the tissue sample also contains green, non-mutated and photosynthetically fully active, cells but not when there are white cells alone. Apoplasmic contact or contact from cell to cell via the plasmalemma membranes and the cell walls is ruled out here by short-circuiting in the apoplast, since the whole leaf sample is immersed in the experimental solution. The signal triggered by switching on or off photosynthesis in the green cells is transferred to an electrode in a white cell without discernable lag-phase. In one experiment the distance between the electrode in a white cell and the next green cell was 0.89 mm. If

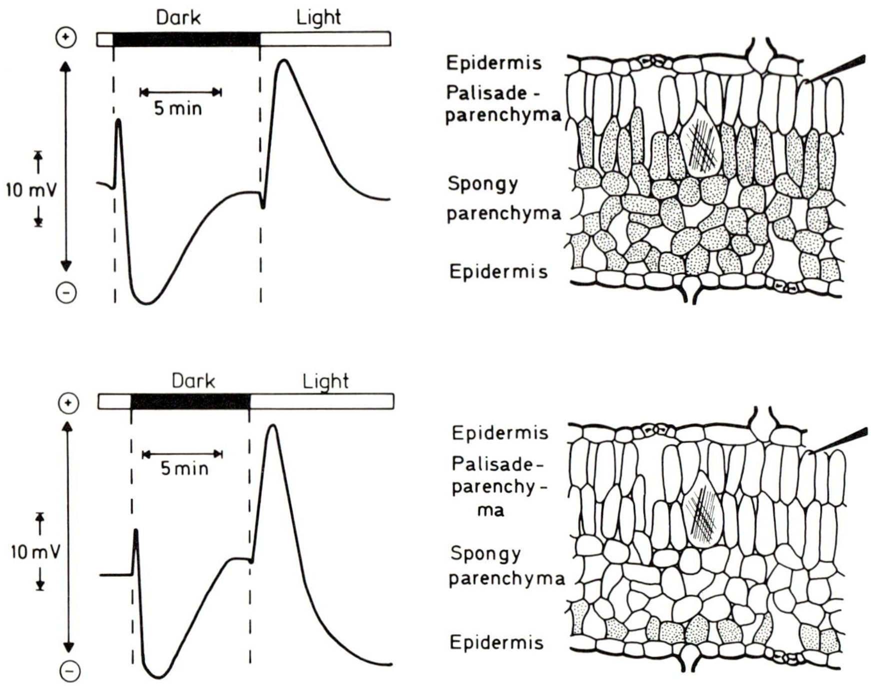

Fig. 5 Transfer of the electrical signal of photosynthesis-dependent membrane potential oscillations from photosynthetically active cells to an electrode in a mutated cell in an ***Oenothera*** leaf. In the leaf section diagrams the electrode picking up the signal in a mutated cell is shown in the upper right; mutated cells are left white, photosynthetically active green cells are dotted. Comparing the two membrane potential tracings with the corresponding leaf section diagrams it can be seen that the size of the signal obtained with the electrode in a mutated cell is rather independent of the amount of green cells, which alone can generate the signal, present in the close vicinity of the electrode. (Redrawn from Fig. 13 in BRINCK-MANN, 1973).

transfer of the signal had occurred at the usual velocity of symplasmic transport of about 5–10 μm s^{-1} (ARISZ and WIERSEMA, 1966), a lag-phase should have been observed. Hence, electrical contact within the symplasm is possible and most probably it is not associated with the actual transport of ions[2]).

III. Membrane Transport, Transcellular Transport, and Long Distance Transport in the Root

The scheme of Figs. 1A and B already suggests that a model of ion transport in the root must be principally similar to a model of salt excretion, and *vice versa*. Membrane transport, symplasmic transcellular transport, and long distance transport in the xylem are involved. In the root, supply is via the cortex from the medium and 'demand' is excretion to the transpiration stream in the xylem.

---

[2]) Incidentally, the approach described here, using the internally generated signal of photosynthesis-dependent membrane potential oscillations is experimentally much easier than attempts to investigate resistance from cell to cell by introducing an external signal with an additional electrode (SPANS-WICK, 1972; GOLDSMITH et al., 1972).

An experimental set up introduced independently by WEIGL (1969, 1970, 1971) and by PITMAN (1971, 1972a) allows empirical distinction between total ion uptake 'U' to the root, accumulation 'A' of ions in the root, and export 'E' from the root via the xylem of the stele. Membrane transport is involved in all three of these processes, i.e. in uptake at the plasmalemma of epidermis and cortex cells, in accumulation at the plasmalemma and tonoplast of epidermis and cortex cells, and in export at the plasmalemma membranes of epidermal and stelar parenchyma cells. In addition to membrane transport, export (E) involves transcellular transport in the root (see also Fig. 1). The use of chemical agents, which differentially affect export (E), demonstrates that E has other, additional, properties than the membrane transport resulting in uptake (U) and accumulation (A). These additional properties must be the features of transcellular transport or particular characteristics of transport across the stelar plasmalemma, or both. Chemical agents differentially affecting export (E), while uptake (U) remains unchanged for long periods and accumulation (A) is increased (or decreased) as uptake decreases (or increases), are the phytohormones abscisic acid (ABA) and benzyladenine (BA) (CRAM and PITMAN, 1972; PITMAN et al., 1974a). Similarly the inhibitor of protein synthesis, cycloheximide, selectively inhibits export (E) under conditions where it has no secondary effects on energy metabolism, i.e. respiratory O_2 uptake and availability of ATP within the root cells (LÄUCHLI et al., 1973; LÜTTGE et al., 1974). On the basis of the latter findings it has been hypothezised, that membrane-protein turnover of the endoplasmic reticulum might be a salient property of transcellular transport. Finally resolving the problem will, however, be considerably more difficult.

C. The Intact Plant in its Environment

We have proceeded above from co-operation of tissues in a leaf mediating a metabolic event, i.e. C_4-photosynthesis, to the structurally more complex gland and root systems. However, in these systems we have considered co-operation leading to a one-way net transfer of material, i.e. to elimination of salt or sugar from the glands, and to export of ions from the root. Considering the organs in place, in the intact plant, we will often obtain results different from those with isolated organs, and we must unavoidably operate with feedback models. Regarding transport of water and ions out of the root (i.e. export E) we will then have to consider three components: 1) transport itself, and in addition two components acting on it, namely 2) the energy status of the plant, and 3) a highly sensitive signalling system. The energy status comprises factors residing within the root, but also feedback from the shoot by way of establishing a water potential gradient along the 'soil-root-leaves-atmosphere', or by transport of substrates originating from photosynthesis to the root. The high-sensitivity signalling system which is ready to respond to small physiological and environmental changes consists of interacting hormone systems and possibly involves also electrical signals (Fig. 6).

It might be interesting to extend this by surveying all known phenomena on transport-mediated co-operation in intact higher plants and by pressing the emerging ideas into a few model sketches. However, it seems more rewarding for the moment to dwell upon the specific problem already introduced above.

After discovery of the effects of the senescence and wilting hormone ABA on ion export from the root, CRAM and PITMAN (1972) (PITMAN, 1972b) have already speculated that this may be of ecological relevance. This was further suggested by experiments showing that the effect of ABA depended on environmental factors such as nutrition and temperature (PITMAN et al., 1974a).

The ideas emerging from this are sketched in the feedback model of ionic relations of a higher plant described previously (LÜTTGE, 1973, p. 266). The basic principle of the model

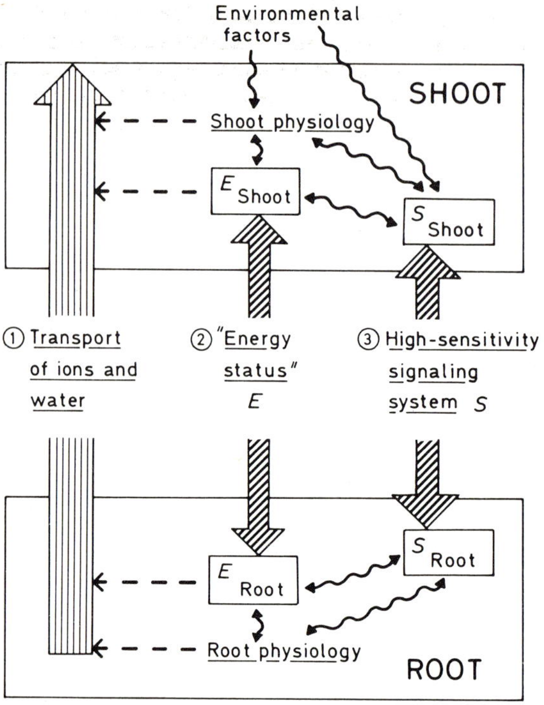

Fig. 6 Three components of a feedback system regulating transport of ions and water from root to shoot: 1) transport itself, 2) 'energy status' (E), 3) high sensitivity signalling system (S). Wavy arrows: interactions between E and S, and with general physiological activities; dashed arrows: action upon transport.

is that uptake by the roots determines overall supply to the plant and that the shoot and the leaves can tell the root about their physiologically and ecologically regulated demand. The facts inherent in the model are the following.

1) The root needs substrate (sugars) from the shoot to perform ion uptake (PITMAN et al., 1971; PITMAN, 1972b; HATRICK and BOWLING, 1973).

2) Hormones affect export of ions from the root (see above).

3) Hormones affecting export of ions from the root such as ABA are formed in the leaves in response to physiological and ecological conditions, e.g. water stress and wilting (refs. see Introduction).

4) Substrates and hormones (ABA: HOCKING et al., 1972) are transported from the leaves to the root in the phloem.

The model can be tested in various way. Coarse regulation by substrate levels has been demonstrated by PITMAN and coworkers (PITMAN, 1972b; PITMAN et al., 1971, 1974b). For instance, ion uptake by the root and export from the root are much more sensitive to reduced supply of substrate by a drastically shortened photoperiod than ion uptake to leaf cells.

A test of fine regulation by hormone levels has been attempted in experiments summarized in Fig. 7 (PITMAN et al., 1974c). Leaves of intact barley plants were carefully wilted so that after a very short period of recovery from water stress physiological activities of leaf and root cells such as photosynthesis, respiration and protein synthesis, and ion uptake by leaf

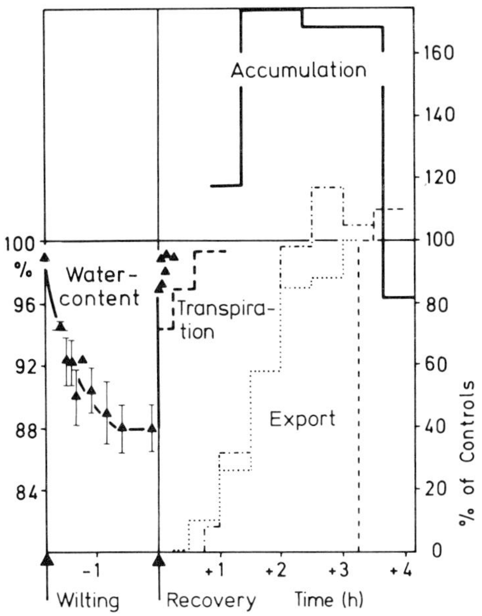

Fig. 7 Water content (▲) of barley leaves during careful wilting and during recovery; transpiration of plants during recovery; and accumulation and export of ions in roots harvested after various times of recovery (i.e. 15 min:.........., 45 min:–.–.–.–, and 195 min:–––––––––). Water content is given as %

$$\% = \frac{\text{water content of leaves} \cdot 100}{\text{water content of fully turgid leaves}} .$$

Transpiration, accumulation, and export are given as % of non-wilted controls. Redrawn from Figs. 1, 2 and 3 of PITMAN et al. (1974c)

cells were indistinguishable from non-wilted control plants. It was assumed, however, that wilting caused formation of a hormonal factor (ABA) in the leaves which was transported to the roots, and it was hoped that the half-life of this factor in the roots was longer than the half-life of discernable wilting effects during recovery. Indeed this was borne out by the experimental results. An appreciable time after obvious physiological recovery in roots of pre-wilted plants export (E) was selectively inhibited, accumulation (A) was enhanced much like in the ABA treated roots mentioned above. Thus a hormonal factor (ABA) produced during wilting in the leaves may not only lead to the well-known regulatory action of closure of stomata (KRIEDEMANN et al., 1972) but may in addition act at a second regulatory site at the level of the roots.

D. Acknowledgements

Our experimental contributions to this review have been supported by Deutsche Forschungsgemeinschaft. New data presented were obtained in co-operation with Profs. M.G. Pitman and A. Läuchli, and with Dr. E. Brinckmann. I thank Prof. M.G. Pitman for stimulating discussion of this paper. Miss E. Ball and Mrs. G. Zirke are acknowledged for valuable technical assistance and Mrs. D. Schäfer for drawing the figures.

E. References

ARISZ, W.H., WIERSEMA, E.P., Proc. Kon. Ned. Akad. Wetensch. C **69**, 223–241 (1966)
BRINCKMANN, E., Zur Messung des Membranpotentials und dessen lichtabhängigen Änderungen an Blattzellen höherer Landpflanzen. Dissertation, Darmstadt, 1973
CRAM, W.J., PITMAN, M.G., Aust. J. Biol. Sci. **25**, 1125–1132 (1972)
FINDLAY, N., REED, M.L., MERCER, F.V., Aust. J. Biol. Sci. **24**, 665–675 (1971)
FREY-WYSSLING, A., Ber. Schweiz. Bot. Ges. **80**, 454–466 (1970)
FREY-WYSSLING, A., Saussurea **3**, 79–90 (1972)
GOLDSMITH, M.H.M., FERNANDEZ, H.R., GOLDSMITH, T.H., Planta **102**, 302–323 (1972)
HATCH, M.D., OSMOND, C.B., SLATYER, R.O., Photosynthesis and Photorespiration, John Wiley and Sons, New York–London–Sydney–Toronto 1971
HATRICK, A.A., BOWLING, D.J.F., J. Exp. Bot. **24**, 607–613 (1973)
HIRON, R.W., WRIGHT, S.T.C., J. Exp. Bot. **24**, 769–781 (1973)
HOCKLING, T.J., HILLMAN, J.R., WILKINS, M.B., Nature New Biology **235**, 124–125 (1972)
KRIEDEMANN, P.E., LOVEYS, B.R., FULLER, G.L., LEOPOLD, A.C., Plant Physiol. **49**, 842–847 (1972)
LÄUCHLI, A., LÜTTGE, U., PITMAN, M.G., Z. Naturforsch. **28c**, 431–434 (1973)
LÜTTGE, U., Ann. Rev. Plant Physiol. **22**, 23–44 (1971)
LÜTTGE, U., Stofftransport der Pflanzen, Springer-Verlag, Berlin–Heidelberg–New York 1973
LÜTTGE, U., OSMOND, C.B., Aust. J. Biol. Sci. **23**, 17–25 (1970)
LÜTTGE, U., SCHNEPF, E., in Encyclopedia of Plant Physiology, New Series (eds. U. Lüttge, M.G. Pitman), Springer-Verlag, Berlin–Heidelberg–New York 1975
LÜTTGE, U., LÄUCHLI, A., BALL, E., PITMAN, M.G., Experientia (Basel), in press (1974)
MILBORROW, B.V., NODDLE, R.C., Biochem. J. **119**, 727–734 (1970)
MIZRAKI, Y., BLUMENFELD, A., RICHMOND, A.E., Plant Physiol. **46**, 169–171 (1970)
MOST, B.H., Planta **101**, 67–75 (1971)
OSMOND, C.B., Aust. J. Biol. Sci. **24**, 159–163 (1971)
OSMOND, C.B., LÜTTGE, U., WEST, K.R., PALLAGHY, C.K., SHACHER-HILL, B., Aust. J. Biol. Sci. **22**, 797–814 (1969)
PITMAN, M.G., Aust. J. Biol. Sci. **24**, 407–421 (1971)
PITMAN, M.G., Aust. J. Biol. Sci. **25**, 243–257 (1972a)
PITMAN, M.G., Aust. J. Biol. Sci. **25**, 905–919 (1972b)
PITMAN, M.G., MOWAT, J., NAIR, H., Aust. J. Biol. Sci. **24**, 619–631 (1971)
PITMAN, M.G., LÜTTGE, U., LÄUCHLI, A., BALL, E., J. Exp. Bot., in press (1974a)
PITMAN, M.G., LÜTTGE, U., LÄUCHLI, A., BALL, E., Z. Pflanzenphysiol., in press (1974b)
PITMAN, M.G., LÜTTGE, U., LÄUCHLI, A., BALL, E., Aust. J. Plant Physiol., in press (1974c)
SPANSWICK, R.M., Planta **102**, 215–227 (1972)
WEIGL, J., Planta **84**, 311–323 (1969)
WEIGL, J., Planta **91**, 270–273 (1970)
WEIGL, J., Planta **98**, 315–322 (1971)
WRIGHT, S.T.C., HIRON, R.W., Nature **224**, 719–720 (1969)
ZIEGLER, H., LÜTTGE, U., Planta **74**, 1–17 (1967)

Ultrastructure and Ion Localization in Xylem Parenchyma Cells of Roots

A. Läuchli, D. Kramer, and R. Stelzer

A. Introduction

Among the examples of cooperation of organs in intact higher plants (see the review by LÜTTGE, 1974), the function of the root in supplying inorganic ions to the shoot is of particular importance. Recent evidence from our laboratories indicates that ion transport through the root to the xylem and eventually to the shoot is different in its basic mechanism from active uptake by the root, as is shown by the differential effects of cycloheximide (LÄUCHLI et al., 1973), abscisic acid and benzyl adenine (PITMAN et al., 1974a), and of previous water stress (PITMAN et al., 1974b). This suggests that in the root there are sites of control for regulation of ion transport to the shoot. Most probably, the xylem parenchyma cells adjacent to the vessels represent these sites of control (LÄUCHLI et al., 1971; ANDERSON, 1972; EPSTEIN, 1972; LÄUCHLI, 1972a; PITMAN, 1972a; BIELESKI, 1973; LÄUCHLI et al., 1974a).

The present work was undertaken to examine structure and ion localization in xylem parenchyma cells of roots in relation to transport of ions through the root. Some results on the ultrastructure of xylem parenchyma in barley roots are described in detail elsewhere (LÄUCHLI et al., 1974a).

B. Materials and Methods

Experiments were done with seedlings of barley, ***Hordeum vulgare***, cv 'Kocherperle', and with seedlings of soybean, ***Glycine max***, varieties 'Lee' and 'Jackson'.

For ultrastructural studies of barley roots, seedlings were germinated and grown in the dark at 25°C on an aerated nutrient solution until they were 6 days old (PITMAN, 1972b). Roots from seedlings grown in the light did not appear to differ in their ultrastructural features (LÄUCHLI et al., 1974a). Root segments approximately 7 mm long were cut with a razor blade about 1.5 to 2.2 cm and between 10 and 15 cm (8-day-old roots) behind the tip, respectively, and fixed immediately in buffered glutardialdehyde (1 %, pH 7.2) for 2 h, postfixed for 2 h with OsO_4 (1 %, pH 7), and dehydrated in a graded ethanol series followed by propylene oxide. Spurr's medium (SPURR, 1969) was used for embedding. Ultrathin sections were cut with glass knives on a Reichert ultramicrotome OM U 3. The sections were poststained with methanolic uranyl acetate and lead citrate and examined in a Zeiss EM 9 electron microscope.

For ultrastructural studies of soybean roots, seeds were germinated for 6 days in the dark at 25°C. The seedlings were grown for 3 weeks on an aerated nutrient solution of the following composition: 1.2 mM KNO_3, 0.8 mM $Ca(NO_3)_2$, 0.2 mM $NH_4H_2PO_4$, 0.2 mM $(NH_4)_2HPO_4$, 0.2 mM $MgSO_4$, 4μM Fe-EDTA, 0.2 ml/l of a micronutrient stock solution (see p. 39 in EPSTEIN, 1972). For studies of possible salt effects, NaCl was added to the nutrient solution after 1 week of growth at an initial concentration of 25 mM which was increased gradually up to 75 mM NaCl during the remaining growth period. The growth conditions were 16 h photoperiod, ca. 10 000 lux light intensity, ca. 25–30°C day temperature and ca. 20–25°C night temperature. Root segments approximately 5 mm long were cut about 5, 10, and 15 cm behind the tip, respectively, and prepared for electron microscopy as described for barley.

For localization of Cl^- in barley roots, seeds were germinated for 24 h and then grown for 2 days in the dark at 25°C on an aerated 0.2 mM solution of $CaSO_4$. Thereafter, the seedlings were grown for 11 days in the light on an aerated, $^{1}/_{10}$ concentration nutrient solution (see p. 39 in EPSTEIN, 1972); Fe-EDTA, however, was supplied at $^{1}/_{2}$ concentration. NaCl between 0 and 10 mM was either added to the nutrient solution after 7 days of growth in the light or 5 mM NaCl was supplied as a pulse up to 120 min at the end of the growth period. The growth conditions were as described for soybean.

A Ag^+ precipitation technique was used for electron microscopic localization of Cl^- (KOMNICK, 1962; KOMNICK and BIERTHER, 1969). Root-tip segments approximately 15 mm long were washed for 1 min with ice-cold 0.2 mM $CaSO_4$, cut to 1 mm segments and fixed immediately for 2 to 3 h at +2°C with either 1 % OsO_4 or 1 % $KMnO_4$ in 0.1 M cacodylate- acetate buffer pH 6.4, containing either 0.5 % Ag-acetate or 0.8 % Ag-lactate, under red light. The fixed specimens were rinsed briefly with cold buffer, washed for 15 min with H_2O in the cold, and dehydrated in acetone. Dehydration was started for 15 min each in 5, 10, and 20 % acetone, followed by a 15 min wash with 0.05 N HNO_3 in 20 % acetone to dissolve nonspecific Ag-precipitates. The specimens were then transferred into 30 % acetone and dehydrated continuously according to SITTE (1962), transferred into propylene oxide and embedded in Spurr's medium (SPURR, 1969). 'Silver-sections' were cut and examined in the electron microscope either unstained or poststained with methanolic uranyl acetate.

C. Results and Discussion

I. Ultrastructure of Xylem Parenchyma in Barley Roots

It was shown by LÄUCHLI et al. (1974a) that xylem parenchyma cells at 1 cm behind the tip, where ion uptake and transport was going on at high rates, were rich in cytoplasm with extensive rough endoplasmic reticulum and well-developed mitochondria. Their cell walls contained numerous plasmodesmata establishing the possibility of a symplasmic pathway across the stele up to the vessels. The vessel walls regularly contained primary pits. A distinct plasmalemma lined the cytoplasm of the xylem parenchyma cells, including the areas which faced the primary pits of the vessel walls. In this paper, the structural studies were expanded to include the xylem parenchyma of the basal regions of the root.

Examinations of xylem parenchyma cells about 2 cm behind the tip confirmed the structural features of these cells at 1 cm behind the tip. Fig. 1 shows parts of two xylem parenchyma cells adjacent to a vessel. The main features are dense cytoplasm with several organelles and a secondary vacuole in one cell and rough endoplasmic reticulum, dictyosomes and mitochondria in the other. The mitochondria are well-developed and exhibit numerous sacculi (Fig. 2). Xylem parenchyma cells in the basal region of the root appear similar in structure (Fig. 3). In comparison with the region about 1 to 2 cm behind the tip, the main cytological differences in the xylem parenchyma cells are the relatively thick cell walls and the great number of cytoplasmic ribosomes (Fig. 3). Plasmodesmata were also found in the cell walls of the xylem parenchyma of the basal region.

We consider that these structural features of xylem parenchyma cells in roots make them suitable to play an active role in the stele. In the younger, apical region of the root, the structure of xylem parenchyma appears to be suitable for involvement in ion transport from the stele to the xylem. PITMAN et al. (1974a) suggested from a transport study that ion transport through the root is under hormonal regulation, the regulation occuring at the level of RNA activity and membrane turnover. The location of this transport process remained uncertain, though symplasmic transport within the endoplasmic reticulum and secretion into the vessels were advanced as possibilities (LÄUCHLI et al., 1973). The special structures of the xylem parenchyma cells, particularly their rough endoplasmic reticulum, may in fact indicate that regulation of symplasmic transport and secretion of ions into the vessels takes place in the xylem parenchyma cells of the younger part of the root. In the older, basal region of the root, however, xylem parenchyma cells may function as sites of ion re-absorption from the xylem sap. Such a conclusion is supported by the findings of LÄUCHLI et al. (1974a) showing that the rate of $^{86}Rb^+$ export from the basal end of barley roots was

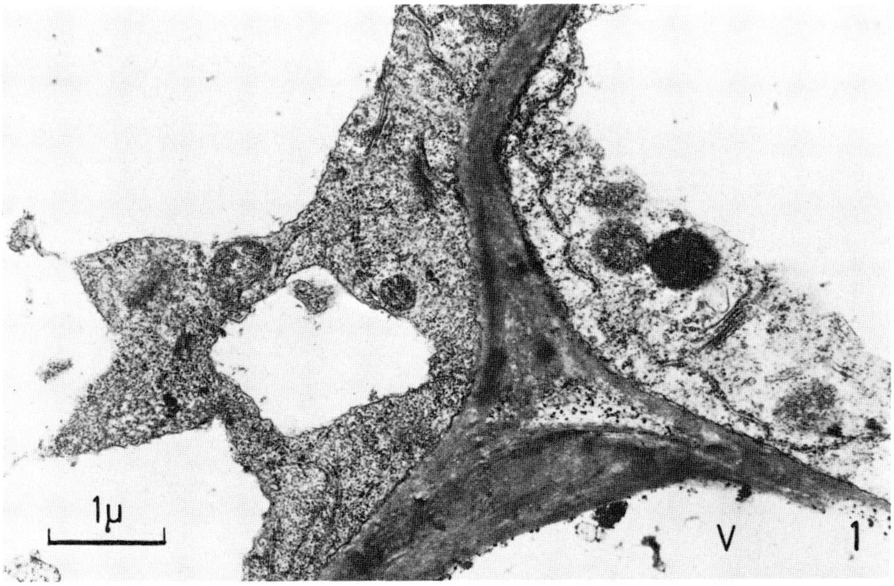
1μ
V
1

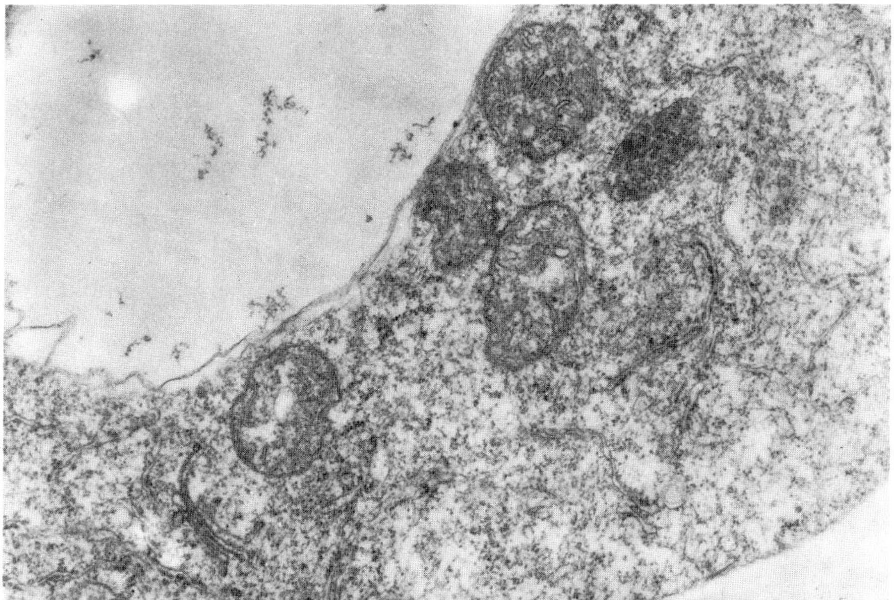
1μ
2

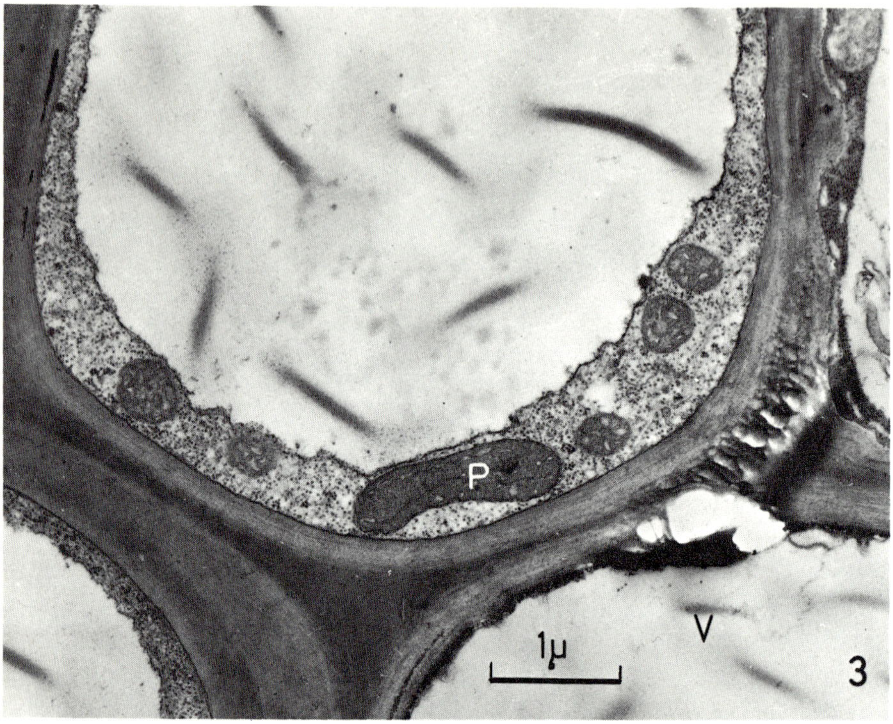

Figs. 1–3 Xylem parenchyma of barley roots.
Fig. 1 Parts of two xylem parenchyma cells adjacent to a vessel (v), about 2 cm behind the tip. Note the rough endoplasmic reticulum and dictyosomes. X 16 000. Fig. 2 Mitochondria, rough endoplasmic reticulum, and dictyosome in a xylem parenchyma cell about 1.5 cm behind the tip. X 23100. Fig. 3 Xylem parenchyma cell about 10 to 15 cm behind the tip, featuring a proplastid (p), mitochondria, rough endoplasmic reticulum, and cytoplasmic ribosomes. Note the relatively thick cell wall. Vessel (v). X 16 900.

only about 50 % of the rate of total export from the $^{86}Rb^+$-labelled tip-region of the root. Ion re-absorption from the xylem sap appears to be particularly important for cations. EPSTEIN and NORLYN (1973) determined the velocities of longitudinal transport within the xylem of excised corn roots to be 103 cm/h for $^{82}Br^-$ but only 35 cm/h for $^{86}Rb^+$. They suggested that the cation gets involved in a chromatographic exchange migration on negative-

---

Figs. 4–7 Chloride localization in xylem parenchyma of barley roots about 1 cm behind the tip.
Fig. 4 Cross-sectioned plasmodesmata in the cell wall between an endodermis and a xylem parenchyma cell (xyp). Silver deposits associated with plasmodesmata. Root exposed to 5 mM NaCl for 20 min prior to fixation. OsO_4 fixation, precipitation of Cl^- *in situ* with Ag-lactate. Section unstained. X 38 600. Fig. 5 Endoplasmic reticulum and mitochondrion (M). Silver deposits associated with the endoplasmic reticulum and the cytoplasm. Root grown in nutrient solution with 10 mM NaCl for the last 4 days of growth. $KMnO_4$ fixation, precipitation of Cl^- *in situ* with Ag-acetate. Section unstained. X 29 000. Fig. 6 Endoplasmic reticulum and mitochondrion (M). Silver deposits mainly in the endoplasmic reticulum and the cytoplasm. Root exposed to 5 mM NaCl for 20 min prior to fixation. OsO_4 fixation, precipitation of Cl^- *in situ* with Ag-lactate. Section poststained with uranyl acetate. X 34 400. Fig. 7 Cisterna of the endoplasmic reticulum containing silver deposits. Root exposed to 5 mM NaCl for 120 min prior to fixation. OsO_4 fixation, precipitation of Cl^- *in situ* with Ag-lactate. Section poststained with uranyl acetate. X 171 000.

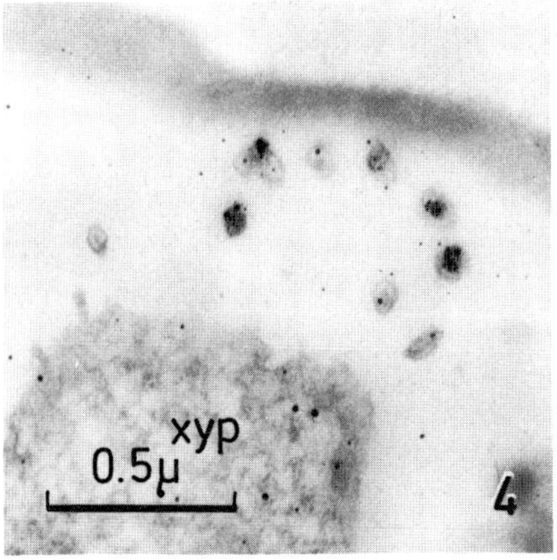
xyp
0.5µ
4

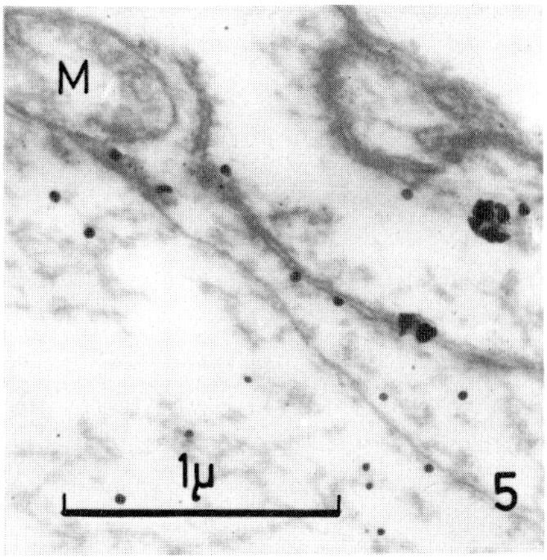
M
1µ
5

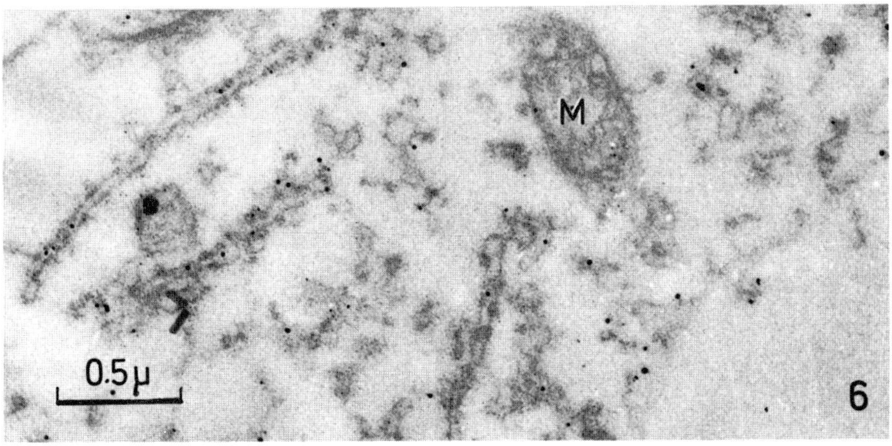
M
0.5µ
6

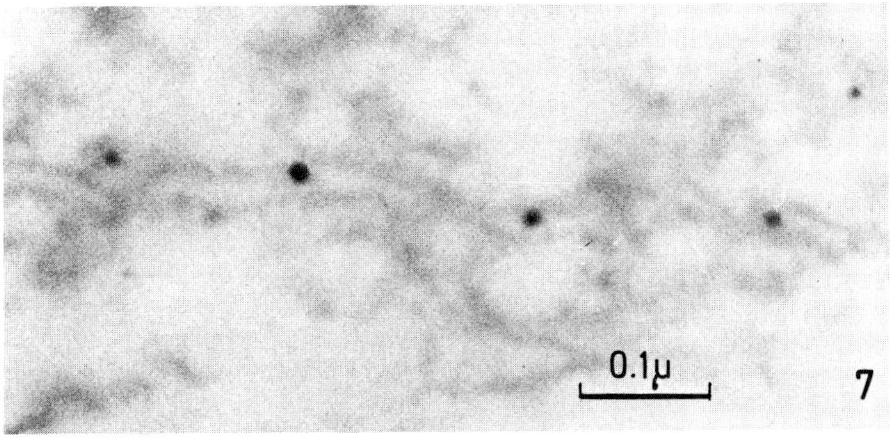
0.1µ
7

ly charged sites of the cell walls of vessels. In view of the special structures of the xylem parenchyma, predominant cation re-absorption by these cells may have been an additional cause for the slow longitudinal movement of $^{86}Rb^+$. Selective ion re-absorption from the xylem sap may also affect $Na^+ - K^+$ selectivity in the plant (BOWLING and ANSARI, 1972). Notwithstanding the possibility of ion re-absorption by xylem parenchyma cells in the older region of the root, this appears to be the normal function of these cells in leaves (SRIVASTAVA and SINGH, 1972; PALLAGHY, 1973).

II. Intracellular Localization of Chloride in Xylem Parenchyma of Barley Roots

Intracellular localization of Cl^- in plant cells by means of a Ag^+ precipitation technique has only been used in a few laboratories (ZIEGLER and LÜTTGE, 1967; VAN STEVENINCK and CHENOWETH, 1972; VAN STEVENINCK et al., 1973; VAN STEVENINCK et al., 1974). A detailed methodical study and critical review showed, however, that this method is valid for studies on Cl^- transport in plant roots (LÄUCHLI et al., 1974b). In particular, elemental analysis of the silver deposits by electron probe analysis (LÄUCHLI, 1972b) confirmed the specificity of the method for Cl^-.

Cl^- was found to be associated with plasmodesmata in the walls of xylem parenchyma cells. Fig. 4 exhibits this for plasmodesmata in the wall common to a xylem parenchyma and an endodermis cell. Within the cytoplasm of xylem parenchyma cells, Cl^- is mainly associated with the endoplasmic reticulum (Figs. 5 and 6). Silver deposits are also found in the cytoplasm but mitochondria contain only few deposits (Figs. 5 and 6). Furthermore, Cl^- appears to be localized partly within the cisternae of the endoplasmic reticulum (Figs. 6 and 7).

Thus, plasmodesmata and endoplasmic reticulum, presumably forming a functional unit (LÄUCHLI, 1972a), appear to be the pathway of intercellular Cl^- transport through barley roots. Since xylem parenchyma cells are shown to be a part of this transport pathway (Figs. 4–7), they are most probably the sites of Cl^- secretion into the xylem vessels, as was concluded earlier from a structural investigation (LÄUCHLI et al., 1974a). These results further emphasize the importance of the symplasmic pathway in Cl^- transport through the endodermis and the stele to the vessels. In recent reviews (EPSTEIN, 1972; LÄUCHLI, 1972a; BIELESKI, 1973; LÜTTGE, 1973), there was general agreement as to the symplasm forming the predominant pathway of ion transport through the root to the vessels. Ultrastructural studies apparently show that the pores of the plasmodesmata are in open continuity with the endoplasmic reticulum (review: LÄUCHLI, 1972a). In terms of symplasmic transport, this paramount conclusion is in line with TYREE (1970) who interpreted the plasmodesmata as constituting the pathway of least resistance for transport of small solutes through the symplasm. Furthermore, measurements of fluxes across plasmodesmata also led to such an interpretation (CLARKSON et al., 1971).

The localization of Cl^- at the endoplasmic reticulum and most probably within its cisternae (Figs. 5–7) is particularly interesting, since LÄUCHLI (1972a) suggested that the endoplasmic reticulum may represent an integral part of the symplasm. Our observation, exhibited particularly in Figs. 6 and 7, represents the first experimental indication of the involvement of the endoplasmic reticulum in intercellular ion transport in plant cells. (Reports by

---

Figs. 8–10 Xylem parenchyma of roots of soybean, variety 'Lee', about 15 cm behind the tip. ▶
Fig. 8 Xylem parenchyma adjacent to a vessel with a primary pit (pp). Note the wall ingrowths (arrows) in the xylem parenchyma cell exclusively in the area of the primary pit. X 11 700. Fig. 9 Wall ingrowths facing a primary pit (pp) of a vessel. Note the distinct plasmalemma, which follows the contours of the wall ingrowths, and some mitochondria. Proplastid (p). X 19 300. Fig. 10 Part of a xylem parenchyma cell of a root grown in nutrient solution with 75 mM NaCl. Wall ingrowths and many mitochondria in the area facing a primary pit (pp). X 15 400.

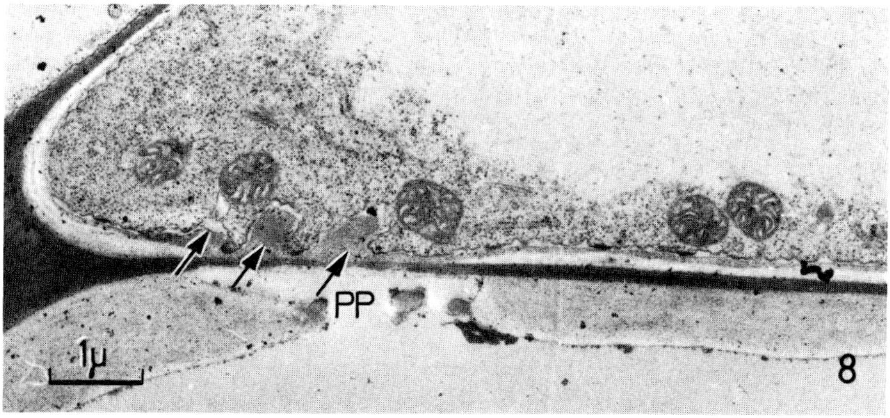
PP
1µ
8

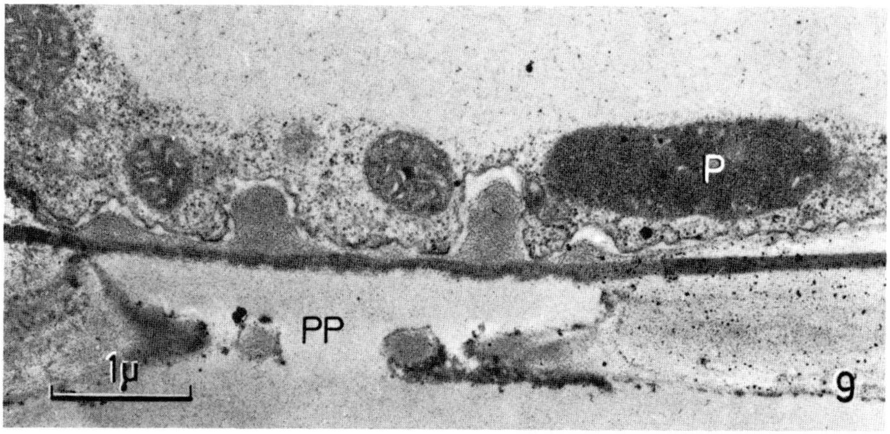
P
PP
1µ
9

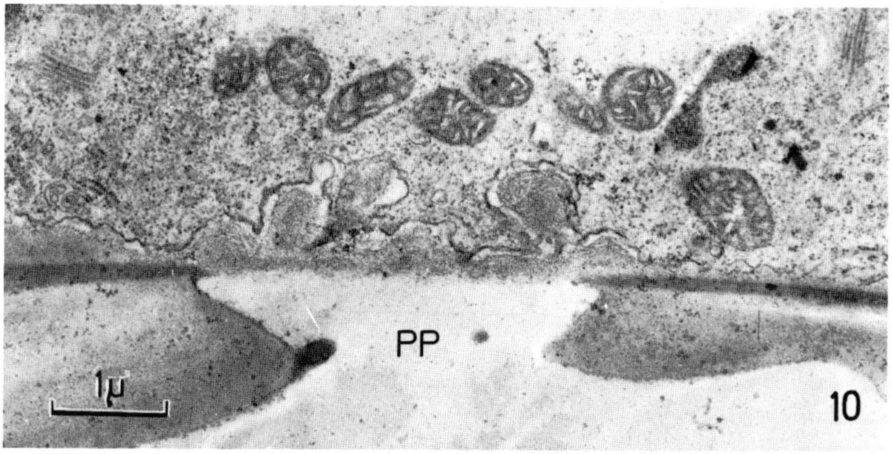
PP
1µ
10

PICKETT-HEAPS (1967) and FRANKE and KARTENBECK (1971) suggest it is also implicated in transport of organic materials.) Thus, our earlier interpretation that cycloheximide inhibits intercellular ion transport through an effect on membrane turnover of such compartments as the cisternae of the endoplasmic reticulum (LÄUCHLI et al., 1973; LÜTTGE et al., 1974) is possibly correct. It gains force by the observation that cycloheximide prevents or delays the production of long endoplasmic reticulum membranes in aging beetroot (VAN STEVENINCK and VAN STEVENINCK, 1971). Nonetheless, Cl^- appears localized in the endoplasmic reticulum of the xylem parenchyma cells, which are extremely rich in rough endoplasmic reticulum (LÄUCHLI et al., 1974a). Whether the endoplasmic reticulum in xylem parenchyma cells is furthermore involved in ion secretion to the vessels, by a process like that suggested for the fast transport of Cl^- through the cytoplasm of *Nitella* cells (MacROBBIE, 1973), remains to be elucidated.

III. Ultrastructure of Xylem Parenchyma in Soybean Roots

There is increasing evidence that ion transport processes are under genetic control (EPSTEIN, 1972). It was therefore deemed to be a fruitful approach to examine structure in relation to ion transport in roots of varieties that possess the characteristics of transport mutants. The soybean varieties 'Lee' and 'Jackson' are salt tolerant and salt sensitive, respectively, the differential response to salt being a function of the extent to which Cl^- is transported from the root to the shoot (ABEL and MacKENZIE, 1964). A genetic study of the inheritance of Cl^- transport to the shoot showed that the varietal differences are controlled by a single gene pair (ABEL, 1969).

A preliminary examination of the soybean roots, which were about 25 cm in length, showed that vessels and xylem parenchyma appeared completely differentiated at approximately 15 cm behind the tip. In Fig. 8, part of a differentiated xylem parenchyma cell adjacent to a vessel can be seen. The most evident structural features are wall ingrowths, exclusively in the area of the primary pit, and mitochondria. At higher magnification (Fig. 9), it is evident that the plasmalemma follows the contours of the wall ingrowths. Furthermore, the mitochondria are well-developed with sacculi and the cytoplasm is dense with numerous ribosomes and some rough endoplasmic reticulum. Fig. 10 shows a comparable picture of the xylem parenchyma from a salt-stressed root of 'Lee'. ('Lee' soybeans exhibited no visible, macroscopic signs of salt damage under mild salt stress of 75 mM NaCl in the medium.) Apparently, salt stress did not alter the basic structures of the xylem parenchyma cells. One may emphasize that these cells in particular contained many mitochondria (HECHT-BUCHHOLZ et al., 1971). Comparative studies of the salt-sensitive variety 'Jackson' are in progress. They have confirmed thus far the occurence of wall ingrowths in xylem parenchyma cells and demonstrated structural salt damage of the xylem parenchyma under conditions at which these cells in 'Lee' appeared normal.

Wall ingrowths, clearly shown in Figs. 8–10, eştablish the xylem parenchyma cells in soybean roots as transfer cells with increased surface area of the plasmalemma and, presumably, facilitated transmembrane flux of solutes (PATE and GUNNING, 1972). Transfer cells are situated at the beginning or end of symplasmic pathways, are common in the xylem parenchyma of leaf minor veins and nodes but are very rarely found in roots (PATE and GUNNING, 1972). In fact, xylem parenchyma transfer cells in roots have only been described in the vascular bundles of the nodules of legumes (PATE et al., 1969). Thus, xylem parenchyma cells in soybean roots may function as transfer cells for active secretion of ions into or active re-absorption from the xylem sap. The fact that their structure appears undamaged in salt-stressed roots of the salt-tolerant variety 'Lee', while there is damage in the salt-sensitive variety 'Jackson' under comparable conditions, makes these single gene mutants most promising in further elucidating the transport function of xylem parenchyma cells.

D. Acknowledgements

This work was supported by the Deutsche Forschungsgemeinschaft. Seed of the soybean varieties 'Lee' and 'Jackson' were kindly provided by Dr. G.H. Abel, United States Department of Agriculture, Mesa, Arizona 85201. We are grateful to Mrs. R. Brandtner and Miss G. Weber for technical assistance. Stimulating discussions with Professors U. Lüttge and M.G. Pitman are gratefully acknowledged.

E. References

ABEL, G.H., Crop Sci. **9**, 697–698 (1969)
ABEL, G.H., MacKENZIE, A.J., Crop Sci. **4**, 157–161 (1964)
ANDERSON, W.P., Ann. Rev. Plant Physiol. **23**, 51–72 (1972)
BIELESKI, R.L., Ann. Rev. Plant Physiol. **24**, 225–252 (1973)
BOWLING, D.J.F., ANSARI, A.Q., J. Exp. Bot. **23**, 241–246 (1972)
CLARKSON, D.T., ROBARDS, A.W., SANDERSON, J., Planta **96**, 292–305 (1971)
EPSTEIN, E., Mineral Nutrition of Plants: Principles and Perspectives. J. Wiley, New York–London–Sydney–Toronto 1972
EPSTEIN, E., NORLYN, J.D., Plant Physiol. **52**, 346–349 (1973)
FRANKE, W.W., KARTENBECK, J., Protoplasma **73**, 35–41 (1971)
HECHT-BUCHHOLZ, CH., PFLÜGER, R., MARSCHNER, H., Z. Pflanzenphysiol. **65**, 410–417 (1971)
KOMNICK, H., Protoplasma **55**, 414–418 (1962)
KOMNICK, H., BIERTHER, M., Histochemie **18**, 337–362 (1969)
LÄUCHLI, A., Ann. Rev. Plant Physiol. **23**, 197–218 (1972a)
LÄUCHLI, A., in Microautoradiography and Electron Probe Analysis: Their Application to Plant Physiology (ed. U. Lüttge), pp. 191–236, Springer-Verlag, Berlin–Heidelberg–New York 1972b
LÄUCHLI, A., KRAMER, D., PITMAN, M.G., LÜTTGE, U., Planta, in press (1974a)
LÄUCHLI, A., LÜTTGE, U., PITMAN, M.G., Z. Naturforsch. **28c**, 431–434 (1973)
LÄUCHLI, A., SPURR, A.R., EPSTEIN, E., Plant Physiol. **48**, 118–124 (1971)
LÄUCHLI, A., STELZER, R., GUGGENHEIM, R., HENNING, L., in Microprobe Analysis of Cells and Tissues (eds. T. Hall, P. Echlin, R. Kaufmann), pp. 107–118, Academic Press, London and New York 1974b
LÜTTGE, U., Stofftransport der Pflanzen, Springer-Verlag, Berlin–Heidelberg–New York 1973
LÜTTGE, U., in Membrane Transport in Plants (eds. U. Zimmermann, J. Dainty), Springer-Verlag, Berlin–Heidelberg–New York 1974
LÜTTGE, U., LÄUCHLI, A., BALL, E., PITMAN, M.G., Experientia, in press (1974)
MacROBBIE, E.A.C., in Ion Transport in Plants (ed. W.P. Anderson), pp. 431–446, Academic Press, London and New York 1973
PALLAGHY, C.K., Aust. J. Biol. Sci. **26**, 1015–1034 (1973)
PATE, J.S., GUNNING, B.E.S., Ann. Rev. Plant Physiol. **23**, 173–196 (1972)
PATE, J.S., GUNNING, B.E.S., BRIARTY, L.G., Planta **85**, 11–34 (1969)
PICKETT-HEAPS, J.D., Dev. Biol. **15**, 206–236 (1967)
PITMAN, M.G., Aust. J. Biol. Sci. **25**, 243–257 (1972a)
PITMAN, M.G., Aust. J. Biol. Sci. **25**, 905–919 (1972b)
PITMAN, M.G., LÜTTGE, U., LÄUCHLI, A., BALL, E., J. Exp. Bot., in press (1974a)
PITMAN, M.G., LÜTTGE, U., LÄUCHLI, A., BALL, E., Aust. J. Plant Physiol., in press (1974b)
SITTE, P., Naturwissenschaften **49**, 402–403 (1962)
SPURR, A.R., J. Ultrastruct. Res. **26**, 31–43 (1969)
SRIVASTAVA, L.M., SINGH, A.P., Can. J. Bot. **50**, 1795–1804 (1972)
TYREE, M.T., J. Theor. Biol. **26**, 181–214 (1970)
VAN STEVENINCK, M.E., VAN STEVENINCK, R.F.M., Protoplasma **73**, 107–119 (1971)
VAN STEVENINCK, R.F.M., CHENOWETH, A.R.F., Aust. J. Biol. Sci. **25**, 499–516 (1972)
VAN STEVENINCK, R.F.M., CHENOWETH, A.R.F., VAN STEVENINCK, M.E., in Ion Transport in Plants (ed. W.P. Anderson), pp. 25–37, Academic Press, London and New York 1973
VAN STEVENINCK, R.F.M., VAN STEVENINCK, M.E., HALL, T.A., PETERS, P.D., Histochemie, in press (1974)
ZIEGLER, H., LÜTTGE, U., Planta **74**, 1–17 (1967)

What Do We Know about the Function of Plasmodesmata in Transcellular Transport?

H. Ziegler

Plasmodesmata were first described by TANGL (1879) who called them 'Protoplasmafortsätze'. The generally accepted name 'plasmodesmos' was given by STRASBURGER (1901). Plasmodesmata are protoplasmic connections between cells which originated from a common mother cell by cell division. They are formed by inclusion of either microtubular elements from the spindle apparatus or of elements from the endoplasmic reticulum in the newly formed cell wall. 'Ectodesmata' are non-plasmatic structures (SCHÖNHERR and BUKOVAC, 1970a, b) and have, therefore, nothing in common with true plasmodesmata. There are many indications that plasmodesmata play a role in transcellular transport of substances and of stimuli, even if there is little direct proof to date for this assumption.

The following facts are especially significant (MEEUSE, 1957).

1) The walls between cells which transfer substances in substantial amounts have a high density of plasmodesmata in strategic positions. Examples are the cell walls between storage tissues and conducting tissues in seeds or seedlings (Fig. 1), between mesophyll cells and bundle sheath cells in plants which belong to the C_4-type of photosynthetic CO_2-fixation

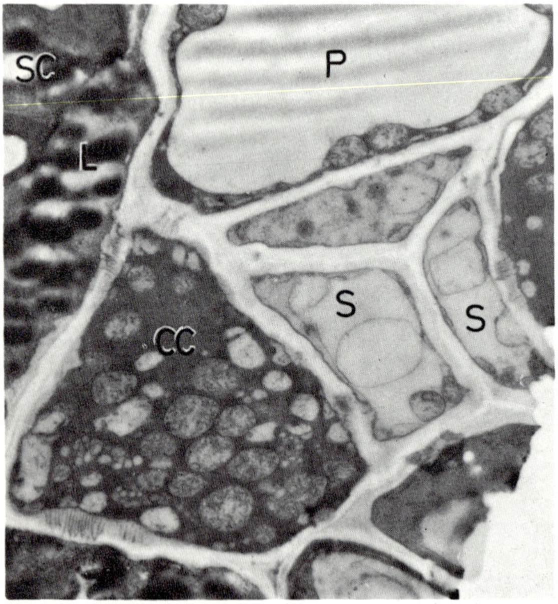

Fig. 1 Electron microscope picture of a cross-section through the storage tissue and the phloem of a cotyledon of ***Sesamum indicum***. SC storage cell with lipids L, CC companion cell, S sieve tube, P phloem parenchyma cell. Note the numerous, branched plasmodesmata between these cells. Fixation OsO_4, embedding vestopal. X 7500.

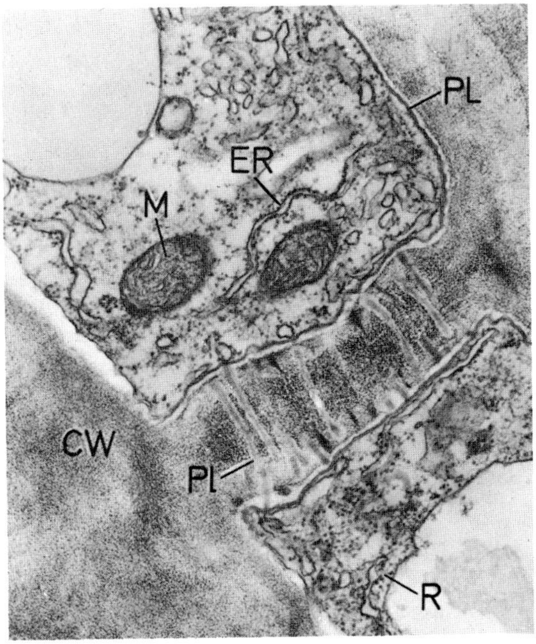

Fig. 2 Plasmodesmata Pl between two parenchyma cells in the petiole of ***Ficus bengalensis***. No cross-membrane in the inner part of the plasmodesmata. PL plasmalemma, M mitochondrion, ER endoplasmic reticulum, R ribosomes, CW cell wall. Preparation as in Fig. 1. X 25 000.

(OSMOND, 1971), between companion cells and sieve tubes (Figs. 1 and 8) and between the companion cells themselves (Figs. 2 and 3).

2) There are large numbers of plasmodesmata in the cross-walls of the 'trumpet-hyphae' in some brown algae (Fig. 4); these cells are known to carry out an effective transport of assimilates (SCHMITZ et al., 1972). In ***Laminaria saccharina*** (ZIEGLER and RUCK, 1967) one cross wall between two trumpet-hyphae has about 20 000–30 000 plasmodesmata, i.e. 50–60/μm^2 or 2000 times the normal density in parenchyma cells (e.g. ***Viscum album***, KUHLA, 1900).

3) The most intensive transport of assimilates in large brown algae (Fig. 5) as well as in vascular plants (Figs. 6A, B) proceeds through the sieve pores of the sieve plates, structures which could be considered as modified plasmodesmata, or at least derived from plasmodesmata. In the sieve cells of the pteridophytes and of the gymnosperms and in the longitudinal walls between sieve tube members of angiosperms (Fig. 7), they may still have the dimensions of plasmodesmata.

4) STRASBURGER (1901) found that ***Mnium*** plants which were plasmolysed and subsequently thoroughly washed and allowed to become turgid again, did not survive the treatment very long, even under the most favourable conditions. For strong plasmolysis causes disruption of plasmodesmata which may well produce the lethal effect of plasmolysis. (There are, of course, also other possibilities.)

5) SPANSWICK and COSTERTON (1967) showed that plasmodesmata have a much lower electrical resistance than membranes, even though higher than that of an open pore.

6) ARISZ (1969) showed with leaves of ***Vallisneria*** that transcellular transport was not influenced by membrane-transport inhibitors, thus showing that the moving substances do not go through the plasmalemma on leaving one cell and entering the next one. He also

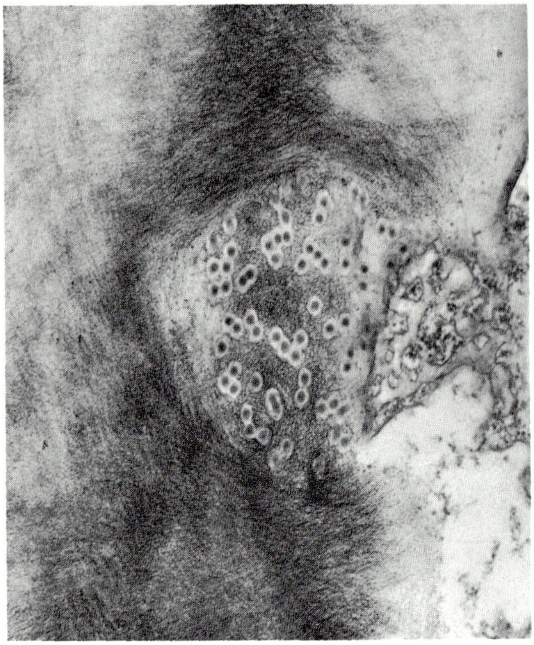

Fig. 3 Cross-section through a field of plasmodesmata in the wall between two companion cells in the shoot of ***Hevea brasiliensis*** seedling. Preparation as in Fig. 1. X 25 000.

showed that labelled substances, once taken up through the plasmalemma into the symplast, did not leach out into the external solution, again indicating that they did not move in the apoplast.

7) Longitudinal transport of auxin in coleoptiles follows the longitudinal rows of cells from apex to base. When a permanent torsion is brought about in the upper part of an ***Avena*** coleoptile, the phototropic and geotropic stimuli are no longer transmitted longitudinally but follow the direction of the twisted cells (TAMMES, 1931; ARISZ, 1969). It is well established that plasmodesmata are numerous in the transverse walls between the cells of one row, but scarce on the longitudinal walls (JUNIPER and BARLOW, 1969).

There are two direct proofs that material can go through plasmodesmata.

1) By the electron microscope viruses have been demonstrated in the plasmodesmata between parenchyma cells and sieve elements of the small veins in a ***Beta vulgaris*** leaf infected with beet yellow (ESAU et al., 1967).

2) The precipitation of Cl^- by Ag^+ in the phloem-plasmodesmata of ***Limonium*** which were supplied with Cl^-, has been demonstrated by ZIEGLER and LÜTTGE (1967) (Fig. 8).

A few facts also indicate a role for plasmodesmata in transmitting stimuli (which may be chemical in nature, and this would mean that there is no difference in their transcellular transport from normal substance transport).

1) HESLOP-HARRISON (1966) showed that, in the developing meiocytes of several plant species, the number of protoplasmic connections increased when synchronised meiotic divisions began. The same author also found (1968) that the persistence of these connections in the massulae of some orchids allowed the mitotic divisions of the pollen nucleus to be synchronised.

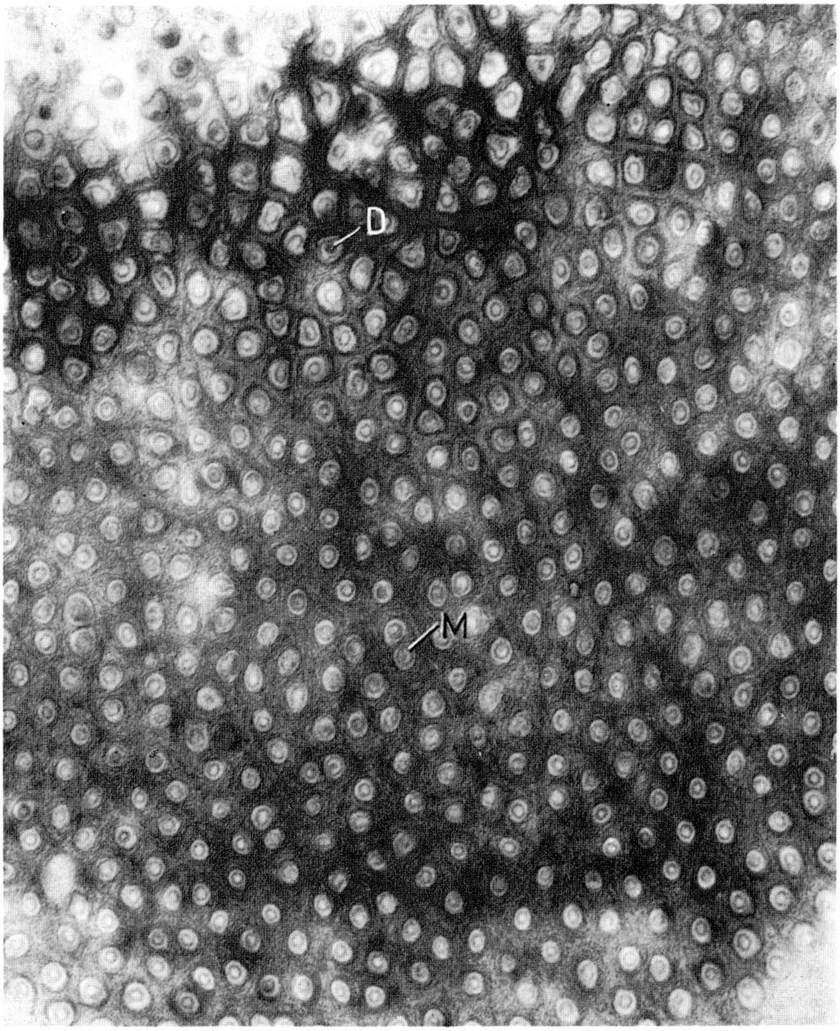

Fig. 4 Cross-section through a trumpet cell of the brown alga ***Laminaria saccharina*** with surface view of the cross-wall. D central rod, M limiting double membrane. X 42 300. (ZIEGLER and RUCK, 1967.)

2) It is well known that the ordered function of the flagellae on the surface of ***Volvox*** is maintained because the single cells are joined by protoplasmic bridges; thus ***Volvox*** has to be considered as a multicellular organism and not as a colony.

Symplasmic transport has one great advantage compared to apoplasmic transport in the intact plant. In the latter substances can only move down thermodynamic potential gradients, but the direction of movement in the symplasm can be regulated according to physiological needs. It therefore seems probable that assimilates and especially growth regulators are translocated exclusively in the symplast. This would seem especially true in submerged plant parts (e.g. ***Vallisneria*** leaves or rice coleoptiles), where substances in the apoplast would be leached out.

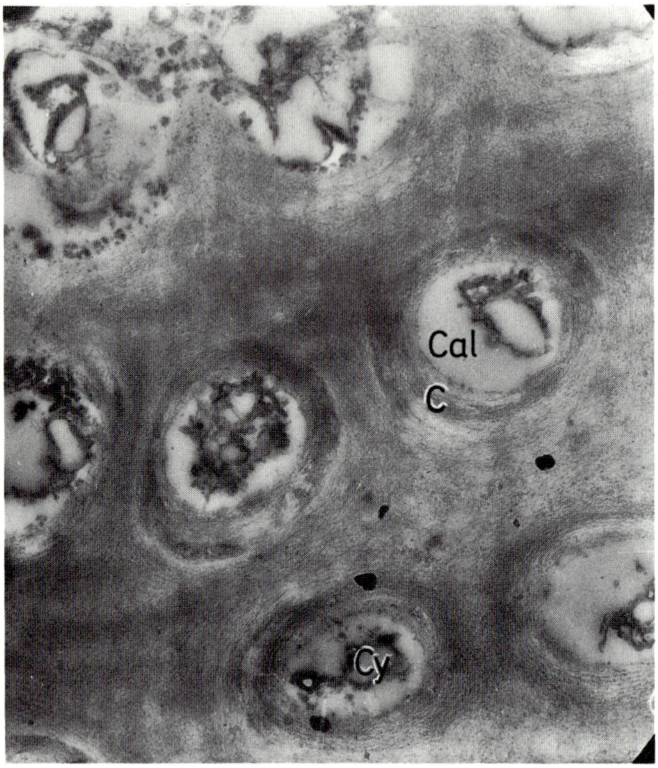

Fig. 5 Sieve plate in the sieve filament of the brown alga ***Macrocystis pyrifera***. C circular texture of the microfibrils around the sieve pore, Cal callose, Cy cytoplasm of the sieve pore. X 12 000. (Compare with Fig. 4) (ZIEGLER, 1963).

There are probably some exceptions to the rule that transcellular, parenchyma, transport always takes place in the symplasm. For instance where there is an intensive unidirectional transfer of material, transfer cells can help to pump the substances rapidly through the cell wall from one cell to the next (GUNNING and PATE, 1969). Another exception, perhaps, is the transfer of ions (e.g. K^+) from the subsidiary epidermal cells to the guard cells and ***vice versa*** during the movement of the stomata; plasmodesmata are scarce or absent between these cells (but cf. PALLAS and MOLLENHAUER, 1972).

Since translocation of substances, synthesized in the symplasm, through the plasmodesmata does not, as far as we know, involve the passage through cytoplasmic membranes (e.g. plasmalemma), experiments in which substances (assimilates, growth hormones) normally produced in the symplasm, are supplied to sections of organs or tissues (leaf slices, hypocotyl or coleoptile sections) from outside cannot give a true picture of the transport processes between cells in the intact plant, which are connected by plasmodesmata.

Since transcellular auxin transport in parenchyma tissues presumably takes place in the symplasm, then the question arises as to whether the polarity can be localised in the plasmodesmata. At present we do not have any evidence for this idea from ultrastructural studies, and it would be excluded if transport through the plasmodesmata proceeds by diffusion only, as many research workers assume (TYREE, 1970). Polarity in plasmodesmata is

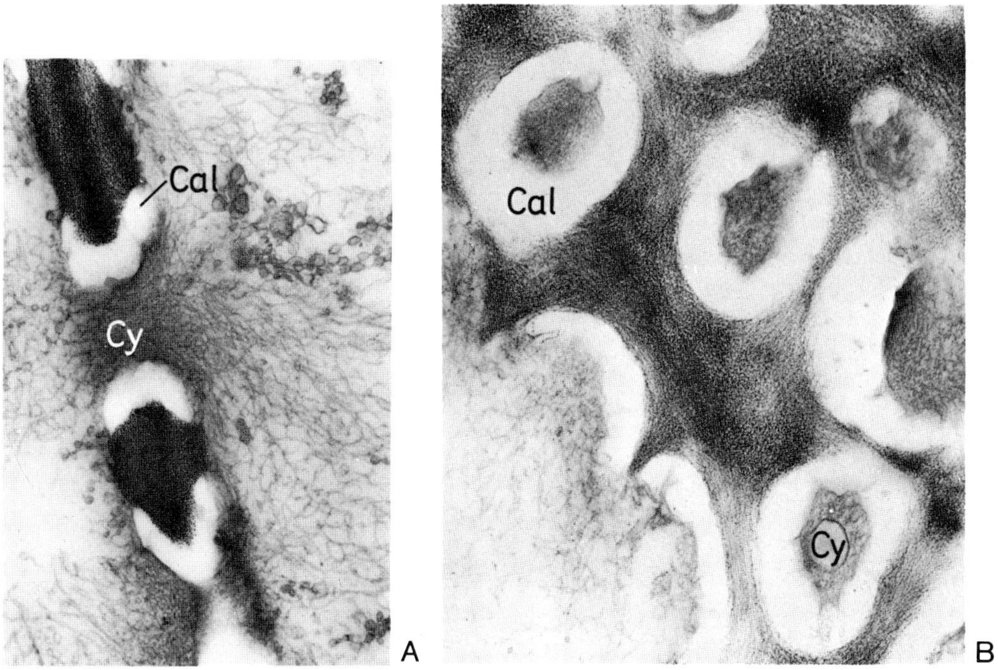

Fig. 6 Sieve plate from the petiole of ***Heracleum mantegazzianum.*** (A) cross section, (B) oblique section (surface view). Cal callose, Cy cytoplasm of the sieve pore. Fixation chrome-osmium, embedding vestopal. X 29 200.

indirectly indicated by the findings of WENT (1941) that in shoot sections an inversion of the original polarity can arise, if new cells are formed by the cambium under the inverse influence of gravity.

To summarize: our knowledge of the function of plasmodesmata in transcellular transport is at present very poor and circumstantial. The main purpose of this paper is to draw attention to this point.

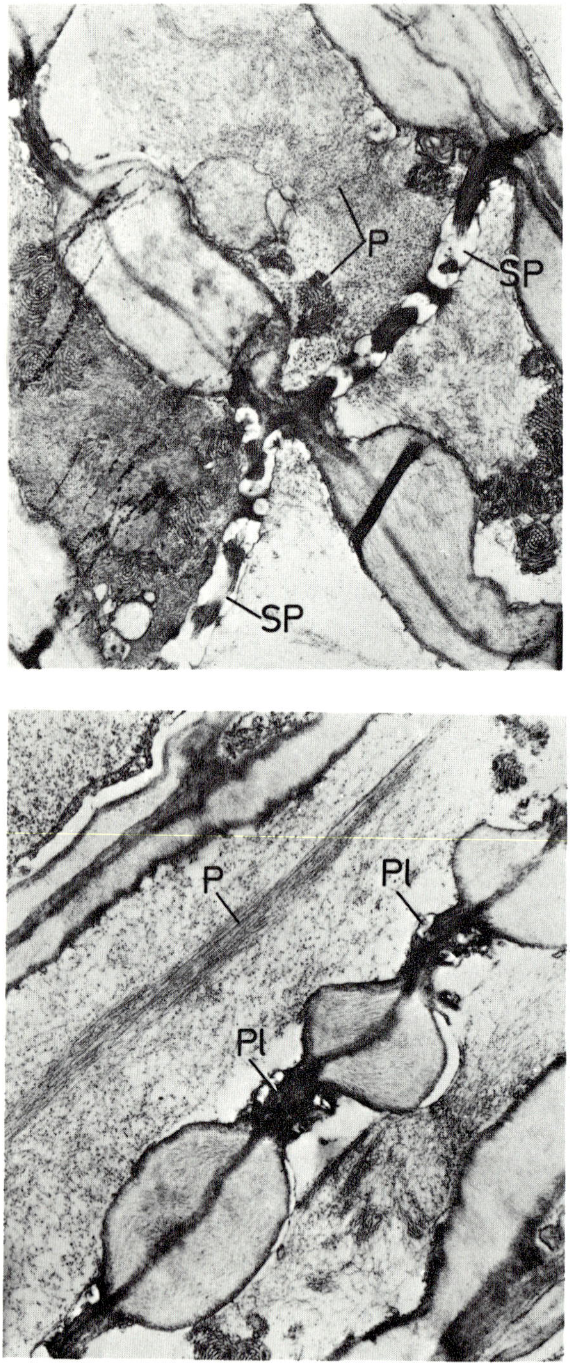

A

B

Fig. 7 Sieve plates (A) and plasmodesma-like connections (B) (on the longitudinal walls) in the sieve tubes of petioles from ***Ficus bengalensis***. SP sieve plate, Pl plasmodesmata, P P-protein. Fixation OsO_4, embedding epon. X 10 500.

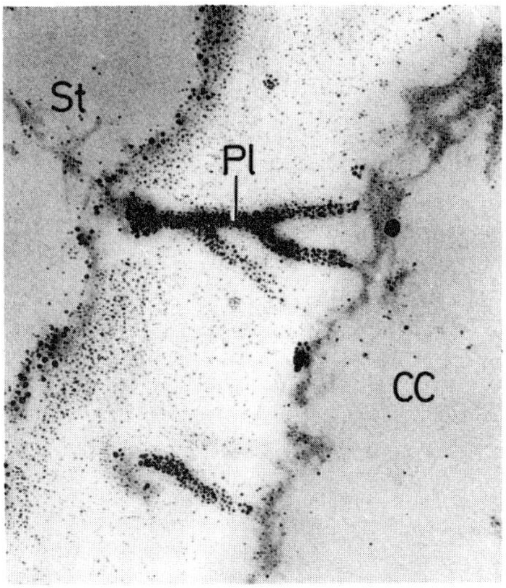

Fig. 8 Plasmodesmata Pl in the cell wall between sieve tube St and companion cell CC in the phloem of the petiole from ***Limonium vulgare***. The chloride ions are precipitated as AgCl (dark spots). X 55 000. (ZIEGLER and LÜTTGE, 1967.)

Acknowledgements

I thank Mrs. G. Lange and Mrs. K. Blase for skilful technical assistance in the electron microscopic work.

References

ARISZ, W.H., Acta Bot. Neer. **18**, 14–38 (1969)
ESAU, K., CRONSHAW, J., HOEFFERT, L.L., J. Cell Biol. **32**, 71–87 (1967)
GUNNING, B.E.S., PATE, J.S., Protoplasma **68**, 107–133 (1969)
HESLOP-HARRISON, J., Endeavour **25**, 65–72 (1966)
HESLOP-HARRISON, J., J. Cell Sci. **3**, 457–466 (1968)
JUNIPER, B.E., BARLOW, P.W., Planta **89**, 352–360 (1969)
KUHLA, F., Bot. Ztg. **58**, 29–58 (1900)
MEEUSE, A.D.J., Plasmodesmata (Vegetable Kingdom). Protoplasmatologia II A 1c. Springer-Verlag, Wien 1957
OSMOND, C.B., Aust. J. Biol. Sci. **24**, 159–163 (1971)
PALLAS, J.E., MOLLENHAUER, H.H., Am. J. Bot. **59**, 504–514 (1972)
SCHÖNHERR, J., BUKOVAC, M.J., Planta **92**, 189–201 (1970a)
SCHÖNHERR, J., BUKOVAC, M.J., Planta **92**, 202–207 (1970b)
SCHMITZ, K., LÜNING, K., WILLENBRINK, J., Z. Pflanzenphysiol. **67**, 418–429 (1972)
SPANSWICK, R.M., COSTERTON, J.W.F., J. Cell Sci. **2**, 451–464 (1967)
STRASBURGER, E., Jb. wiss. Bot. **36**, 493–601 (1901)
TAMMES, P.M.L., Rec. Trav. Bot. Neer. **28**, 77–81 (1931)
TANGL, E., Jb. wiss. Bot. **12**, 170–190 (1879)
TYREE, M.T., J. Theor. Biol. **20**, 181–214 (1970)
WENT, F.W., Bot. Gaz. **103**, 386–390 (1941)
ZIEGLER, H., Protoplasma **97**, 786–799 (1963)
ZIEGLER, H., LÜTTGE, U., Planta **74**, 1–17 (1967)
ZIEGLER, H., RUCK, I., Planta **73**, 62–73 (1967)

H^+ Fluxes, Cytoplasmic pH and the Control of Salt Accumulation in Plants

F.A. Smith and J.A. Raven

A. Introduction

Regulation of intracellular (cytoplasmic) pH in plant cells growing under a wide range of conditions involves exchange of H^+ or OH^- with the surrounding medium. We have previously suggested (RAVEN and SMITH, 1973, 1974) that the occurrence of fluxes of H^+ or OH^- in association with salt accumulation by plant cells can be regarded as a particular expression of this pH-regulating role. The aim of the present paper is to extend this approach, with reference to the ionic relations of cells of higher plants.

B. Results and Discussion

I. Carboxylation and Salt Accumulation in Non-Green Cells

The accumulation of salts of organic (carboxylic) acids by excised roots and storage tissues has been studied in detail, with considerable emphasis on regulation by internal factors. These include cytoplasmic concentrations of HCO_3^- or carboxylate anions (e.g. malate), or transport of cations plus anions to the vacuole (JACOBY and LATIES, 1971). We believe that this approach is one-sided. Accumulation of carboxylate salts (as opposed to the free acid) requires concomitant cation uptake and loss of H^+ either by H^+ efflux, or possibly by HCO_3^- influx, when the external pH is high. Both H^+ efflux and HCO_3^- influx are thermodynamically active processes (POOLE and POEL, 1965; PITMAN, 1970; POOLE, 1973). We have previously suggested (RAVEN and SMITH, 1973) that in the many situations where H^+ extrusion occurs, this process is the primary mechanism whereby metabolic energy is coupled to solute transport. It was proposed that, as well as being regulated by cytoplasmic pH, the H^+ pump can respond to other signals which may affect the ionic balance within the cell. In the present context, we suggest that increased H^+ efflux coupled to cation influx will tend to increase cytoplasmic pH and stimulate organic acid synthesis. This latter effect is explicable in terms of the following sequence of reactions, suggested by DAVIES (1973) to act as an internal 'pH stat':

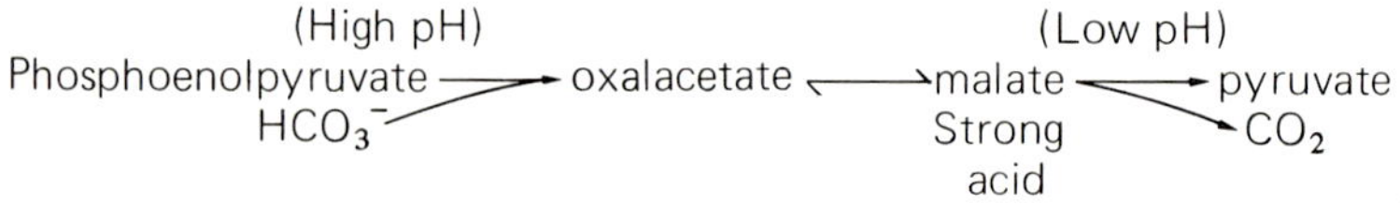

Transport of carboxylate plus inorganic cations (and probably H^+) to the vacuole is necessary for this situation to last (i.e. to prevent decarboxylation). When the cell reaches equilibrium, the whole complex of reactions may be regarded as being under the control of a pH stat involving not only carboxylation and decarboxylation, but also membrane transport of ions.

Accumulation of carboxylate by excised roots grown under low-salt conditions is essentially a conversion of the cells from a 'high-sugar' to a 'high-carboxylate' stage. Under these

Table 1

Influx of K^+ (as $^{86}Rb^+$) into excised 'low-salt' barley roots: effects of removal of external CO_2

	Influx (μmoles $g^{-1} h^{-1}$)		
Conditions	Air	CO_2-free air	N_2
KCl	6.75	7.00	—
K_2SO_4	6.50	5.75	1.70

Solutions contained 0.5 mM $CaSO_4$ plus either 0.5 mM KCl or 0.25 mM K_2SO_4, labelled with $^{86}Rb^+$. Influx at 20°C for 2 h, followed by 15 min wash in tap water.

conditions respiration is rapid and CO_2 availability within the cytoplasm must be high. It appears that in these cells refixation of endogenous CO_2 from sugar breakdown is sufficient to allow carboxylation to proceed rapidly. This is indicated by rapid K^+ influx from K_2SO_4 solutions bubbled with CO_2-free air (Table 1). The presence of O_2 would be necessary for the respiratory disposal of Acetyl CoA produced during glycolysis, as well as for the provision of respiratory energy for K^+/H^+ exchange. This type of experiment supports the body of physiological evidence that the substrate for PEP carboxylase is HCO_3^- rather than CO_2, as the latter would be expected to diffuse rapidly from the cell under external CO_2-free conditions, and so reduce carboxylation. This point is explored further by JACOBY and LATIES (1971), with especial reference to the stoichiometric similarities between K^+/H^+ exchange and influx of HCO_3^- plus K^+.

II. Accumulation of Inorganic Salts by Non-Green Cells

In most studies of inorganic ion transport internal control, other than that exerted by energy supply (e.g. ATP or reducing power), has received little attention. This is unfortunate, in view of the ability of many higher plant cells (e.g. in roots and storage tissues) to accumulate cations plus either organic or inorganic anions (see Table 1), or to switch from the former to the latter. CRAM and PITMAN (see references in PITMAN and CRAM, 1973) have shown that influx of inorganic cations and anions is reduced when the internal concentrations of these ions are high, i.e. there is a feedback effect on membrane transport. CRAM (1973) showed that Cl^- influx into barley and carrot root cells was reduced not only when the cells were high in Cl^-, but also when they were high in NO_3^-. When the cells were pretreated in K_2SO_4 plus $KHCO_3$, giving high internal carboxylate (malate) content, Cl^- influx remained high. Similar effects were found on net NO_3^- influx into barley roots (SMITH, 1973). As shown in Table 2, roots high in KCl absorbed NO_3^- at about 55 % of the control rates in low-salt roots, although rather more rapidly than roots pretreated in KNO_3. Pretreatment in $KHCO_3$ also reduced net NO_3^- influx, but when roots were pretreated in K_2SO_4 there was a decrease only after several hours in NO_3^-. Pretreatment in $KHCO_3$ or K_2SO_4 should both produce roots high in carboxylate. The differential effects of these two pretreatments have been usually (but not invariably) found in experiments on both NO_3^- and Cl^- influx (see Table 2), and may reflect differences in levels and types of carboxylation products.

Net NO_3^- influx into excised roots pretreated in $KHCO_3$ is insensitive to removal of external K^+, and results in increased external pH (SMITH, 1973). This 'excess anion accumulation' must represent exchange of NO_3^- with OH^- (or HCO_3^-?), resulting from respiration of

Table 2

Effects of pretreatment in K^+ salt solutions on uptake of NO_3^- and Cl^- into roots grown under low salt conditions

Pretreatment conditions	(a) Net NO_3^- influx (μmoles $g^{-1} h^{-1}$)	(b) Cl^- influx (μmoles $g^{-1} h^{-1}$)
0.2 mM $CaSO_4$	3.55	4.65
+ KNO_3	1.25	1.50
+ KCl	1.95	1.60
+ $KHCO_3$	2.35	2.85
+ K_2SO_4	3.54	4.85

(a) Net NO_3^- influx measured with NO_3^- electrode. Roots were grown in 0.2 mM $CaSO_4$, excised and pretreated overnight in solutions containing 0.2 mM $CaSO_4$ and additions as shown. K^+ was 1 mM in all cases. Net influx from 1 mM KNO_3 + $CaSO_4$ measured over 5 h.

(b) Cl^- influx measured with $^{36}Cl^-$-labelled solutions. Roots grown as in (a), pretreated in solutions containing K^+ at 2 mM. Influx from 1 mM $K^{36}Cl$ + $CaSO_4$, measured over 2 h.

carboxylates synthesized during the pretreatment period. This experiment separates in time cation influx (K^+/H^+ exchange or K^+ plus HCO_3^- influx) from anion influx (NO_3^-/OH^- exchange). It provides support for the proposal that simultaneous influx of inorganic cations and anions involves two ion-exchange sites, e.g. K^+/H^+ exchange and NO_3^-/OH^- exchange, as discussed previously (RAVEN and SMITH, 1973). The links between these proposed ion-exchange sites and metabolic energy are at present a matter for debate.

CRAM (1974) has discussed why, given the choice, low-salt roots placed in KCl accumulate Cl^- rather than carboxylate. One possibility is that Cl^-/OH^- exchange lowers cytoplasmic pH and hence HCO_3^- concentration. Furthermore, in terms of the DAVIES pH stat, lower pH would reduce PEP carboxylase activity and increase decarboxylation. This would also account for the rapid influx of Cl^- into 'high-carboxylate' roots.

III. Ion Transport in Leaf Tissues

The ionic relations in leaf tissues are conveniently studied using thin leaf slices. As yet few species have been investigated, but there appears to be considerable excess cation influx (associated with carboxylation) when the leaf slices are placed in KCl (NOBEL, 1969; KHOLDEBARIN and OERTLI, 1970; SMITH and ROBINSON, 1971). There is the added complication of light effects on ion transport, which are variable (LÜTTGE et al., 1971) and, as in non-green tissues, effects of 'ageing' and 'salt status' on ion fluxes are to be expected. The influx of both K^+ (measured with $^{86}Rb^+$) and Cl^- into ***Citrus*** (orange) leaf slices increases when the slices are aged by aeration in $CaSO_4$ solutions (SMITH, unpublished results). The increases are greater at low external concentrations, and would be ascribed conventionally to the development of 'high affinity' transport systems. They are

Table 3
Influx of K^+ (as $^{86}Rb^+$) into ***Citrus*** leaf slices

Expt. No.	Conditions	Influx (μmoles $g^{-1}h^{-1}$) Light	Dark
1.	KCl, fresh slices	0.80	0.75
	K_2SO_4, fresh slices	0.75	0.65
	KCl	5.40	5.55
	K_2SO_4	5.40	5.45
2.	K_2SO_4	4.45	3.60
	K_2SO_4, N_2	3.90	0.30
	K_2SO_4, CO_2-free air	–	4.15
	K_2SO_4, 2°C	0.35	–
3.	K_2SO_4	4.85	–
	K_2SO_4, N_2	3.20	–
4.	KCl	4.15	–
	KCl, + 5μM CCCP	2.05	–
	K_2SO_4	3.70	–
	K_2SO_4, + 5 μM CCCP	2.15	–

Experimental treatments as for Table 1, see also SMITH and ROBINSON (1971). Unless otherwise stated, experiments were with slices aged 24 h in light, with air bubbled through experimental solutions at 20°C.

accompanied by increases in respiration and photosynthesis. Table 3 shows the relatively small effects of light on K^+ ($^{86}Rb^+$) influx in both freshly-cut slices and in slices aged in the light. However, ageing in the dark produces subsequent light-stimulation of ion transport (results not shown). The insensitivity of K^+ influx to removal of Cl^- is also shown. In comparable experiments, Cl^- influx into aged slices from 0.5 mM KCl was only 0.4–0.8 μmoles per gram fresh weight per hour.

Effects of CO_2-free air and N_2 on K^+ influx in the dark are similar to those in non-green tissue (Table 3; cf. Table 1), but in the light there is only slight reduction in K^+ influx under N_2. These results are substantially in agreement with those obtained by other workers (e.g. RAINS, 1968) and the aim here is to consider if the ideas developed in Sections I and II can be applied to these green cells. If the capacity for green cells to accumulate carboxylate rather than Cl^- is of widespread occurrence, then this is of considerable interest. Teleologically, a low Cl^- influx can be regarded as a first defence against salinity damage, and in particular inhibition of PEP carboxylase. This enzyme appears to be more sensitive to NaCl in leaves than in roots (OSMOND and GREENWAY, 1972), and is involved in the removal of HCO_3^- resulting from NO_3^- reduction (DIJKSHOORN, 1969).

In the light, CO_2 levels within green cells must be low due to removal by photosynthesis but, if excess K^+ influx can be taken as a guide, reduction in CO_2 is not accompanied by reduction in PEP carboxylase activity (Table 3). In fact K^+ influx under N_2 in the light is inhibited by only about 15—30 %. It should be stressed that the influx values shown in Table 3 do not indicate net uptake of K^+, but estimates obtained by flame photometry showed that net influx was about 80 % of the influx values obtained with $^{86}Rb^+$-labelled solutions. There are a number of possible explanations for this light-promoted carboxylation under apparently anaerobic conditions. One is that, as in the dark, carbohydrate breakdown provides cytoplasmic CO_2 plus HCO_3^- for carboxylation, and that photosynthetic fixation of CO_2 also occurs, resulting in O_2 production. This in turn would allow the Krebs cycle to dispose of Acetyl CoA, as in the dark. Efficient recycling of O_2 would be necessary for this to proceed, and this might seem unlikely if there is competition for CO_2 between ribulose-1,5-diphosphate carboxylase and PEP carboxylase. However, it is also possible that glycollate, produced under limiting CO_2 levels, might be oxidized to glyoxylate, which could then react with Acetyl CoA to give further malate (ZELITCH, 1964). A direct stimulation of K^+/H^+ exchange in the light due to availability of photosynthetic energy could be an important additional factor in promoting carboxylation under 'anaerobic' conditions.

It is interesting to compare briefly the mechanisms outlined above in relation to stomatal physiology. In the past, the main stumbling block to hypotheses including CO_2 fixation (either into sugars or carboxylate) as a means of turgor generation is that stomates open under CO_2-free conditions. Nevertheless, there is now no doubt that K^+ absorption into guard cells is an important part of the stomatal opening process, and that in some of the plant species so far examined K^+ is balanced by carboxylate (malate). This apparent anomaly is explained if the guard cells retain and refix HCO_3^- from starch breakdown, and the pathways described above are applicable. Thus, the role of the so-called 'K^+ pump' in stomates should realistically be considered in relation to H^+ exchange, i.e. as part of a pH regulating mechanism.

Cytoplasmic pH may be involved in the control of competition between PEP carboxylase and ribulose-1,5-diphosphate carboxylase in stomatal and other green cells. If the substrate for PEP carboxylase is HCO_3^- rather than CO_2, then this enzyme will be favoured if the pH is high enough to reduce CO_2 to low levels. In fact, the balance between CO_2 and HCO_3^- would be dramatically affected by relatively small pH changes on either side of pH 6.4 (pK for CO_2/HCO_3^-). This aspect of pH regulation could have important implications for plants with the C_4 pathway of photosynthesis and those with Crassulacean acid metabolism.

In the intact leaf, the volume of solution surrounding the cells is very small, and extracellular pH changes could be physiologically disadvantageous. However, it is feasible that processes resulting in net H^+ production (e.g. excess cation uptake) can be balanced by those resulting in net OH^- production (e.g. NO_3^- reduction). The overall balance must depend on control of ions entering the xylem stream. Intracellular disposal of H^+ or OH^- and export of carboxylate through the phloem may have an important pH-regulating role in this situation. In certain water plants, HCO_3^-/OH^- exchange takes place in the light, resulting in large external pH increases (RAVEN, 1970). This process is thus essentially a 'CO_2 pump', enabling photosynthesis to take place when the external pH is high, and appears to be quite distinct from influx of cations plus HCO_3^- resulting in malate synthesis.

C. Conclusion

We believe that fluxes of H^+ and OH^- play a key role in the regulation of ion transport within plant cells. In particular, control of the synthesis of carboxylate anions is possible via the regulation of cytoplasmic pH. Where inorganic ions are accumulated (e.g. KCl) it is suggested that increased cytoplasmic pH resulting from H^+ extrusion favours Cl^-/OH^- exchange, with at least part of the energy for Cl^- influx coming from the downhill OH^- efflux (RAVEN and SMITH, 1973, 1974). In photosynthetic cells, changes in cytoplasmic pH may affect the balance between accumulation of carboxylate anions (involving PEP carboxylase) and carbohydrate (involving ribulose-1,5-disphosphate carboxylase). In the present paper we have not discussed possible mechanisms for H^+ or OH^- transport. Nevertheless, we feel that the ideas presented above may help to bridge the gap between biophysical aspects of H^+ or OH^- transport and the biochemistry of solute accumulation in plants.

D. Acknowledgements

We are grateful to DR. W.J. CRAM for making available papers prior to publication and for helpful discussions. Experimental work (shown in Tables 1–3) was supported by the Australian Research Grants Committee.

E. References

CRAM, W.J., J. Exp. Bot. **24**, 328–342 (1973)
CRAM, W.J., J. Exp. Bot., in press (1974)
DAVIES, D.D., Symp. Soc. Exp. Biol. **27**, 513–529 (1973)
DIJKSHOORN, W., in Ecological Aspects of the Mineral Nutrition of Plants (ed. I.H. Rorison), pp. 201–213, Blackwell Scientific Publications, Oxford 1969
JACOBY, B., LATIES, G.G., Plant Physiol. Lancaster **47**, 525–531 (1971)
KHOLDEBARIN, B., OERTLI, J.J., Z. Pflanzenphysiol. **62**, 237–244 (1970)
LÜTTGE, U., BALL, E., VON WILLERT, K., Z. Pflanzenphysiol. **65**, 336–350 (1971)
NOBEL, P.S., Plant Cell Physiol. **10**, 597–605 (1969)
OSMOND, C.B., GREENWAY, H., Plant Physiol. Lancaster **49**, 260–263 (1972)
PITMAN, M.G., Plant Physiol. Lancaster **45**, 787–790 (1970)
PITMAN, M.G., CRAM, W.J., in Ion Transport in Plants (ed. W.P. Anderson), pp. 465–481, Academic Press, London and New York 1973
POOLE, R.J., in Ion Transport in Plants (ed. W.P. Anderson), pp. 129–134, Academic Press, London and New York 1973
POOLE, R.J., POEL, L.W., J. Exp. Bot. **16**, 453–461 (1965)
RAINS, D.W., Plant Physiol. Lancaster **43**, 394–400 (1968)
RAVEN, J.A., Biol. Rev. **45**, 167–221 (1970)
RAVEN, J.A., SMITH, F.A., in Ion Transport in Plants (ed. W.P. Anderson), pp. 271–278, Academic Press, London and New York 1973
RAVEN, J.A., SMITH, F.A., Can. J. Bot., in press, 1974
SMITH, F.A., ROBINSON, J.B., Aust. J. Biol. Sci. **24**, 861–871 (1971)
SMITH, F.A., New Phytol. **72**, 769–782 (1973)
ZELITCH, I., Ann. Rev. Plant Physiol. **15**, 121–142 (1964)

Measurement of Intracellular pH in Roots Using a H^+ Sensitive Microelectrode

D.J.F. Bowling

A. Introduction

HOAGLAND and BROYER (1940) were among the earliest workers to study the effect of the pH of the external environment on the uptake of salts by roots. They obtained no evidence for a relationship between pH and salt uptake by barley roots. The work of OLSEN (1953), however, showed that pH could have a marked effect on salt uptake by rye grass. He found that potassium uptake was reduced at pH values below 5.0. Conversely anion uptake was considerably inhibited at pH values over 6.0, a finding which was later confirmed by VAN DEN HONERT and HOOYMANS (1955).

The pH of the vacuolar sap appears to remain fairly constant in the face of wide fluctuations in the external pH (HOAGLAND and BROYER, 1940). Root cells are normally electrically negative inside relative to their environment and vacuolar potentials of over 100 mV can often be recorded. RAVEN and SMITH (1973) have pointed out that this implies an internal pH which is several pH units more acid than the external solution assuming that H^+ is in thermodynamic equilibrium. To maintain the internal pH close to its usual value of 6.0–7.0 when the external pH is below 7.0 an active efflux of H^+ is required.

Direct evidence for active H^+ has been obtained by SPEAR et al. (1969). They found that in the light the cell surface of ***Nitella clavata*** has distinct areas of acid and base secretion. In the dark the pH of the cell surface quickly becomes uniform. Indirect evidence for a H^+ efflux pump in barley roots was obtained by PITMAN (1970). He observed a proton flux which was sensitive to metabolic inhibitors, strongly suggesting active transport.

To understand the role of pH in salt transport it is obviously important to know the pH gradient across the cell membranes and it is now possible to determine this using a pH microelectrode. In this paper some preliminary experiments are described in which the vacuolar pH of sunflower root cells has been measured with this technique.

B. Materials and Methods

pH measurements were carried out on young roots of sunflower ***(Helianthus annuus)***. The plants were grown in culture solution (ANSARI and BOWLING, 1971) at 25°C under constant artificial illumination at 10000 lux. The finest branch roots with a diameter of approximately 500 µm were used in the experiments. The root was excised from the plant approximately 2 cm from the apex and mounted between two coverslips fixed together at one edge with Araldite adhesive. The coverslips were placed on a Leitz hanging drop chamber filled with culture solution. The root was illuminated under the microscope using a long focal length condenser. All the cells between the epidermis and protoxylem could usually be clearly seen in optical section.

Vacuolar pH was determined using an antimony coated glass microelectrode (type 801) manufactured by Transidyne General Corporation, Ann Arbor, Michigan, USA. This electrode has a tip diameter of 1 µm and proved to be very stable although its response time tended to increase with use. The following electrode arrangement was employed:

Calomel electrode	Salt bridge 3 M KCl	pH micro-electrode	Cell vacuole	Cytoplasm and Membranes	Calomel reference electrode in external solution

The calomel electrodes were connected to an E.I.L. electrometer model 33.B2 and a chart recorder.

The pH microelectrode was calibrated using standard pH buffer solutions and the response was linear over the pH range 4–9. Approximately 50 mV was obtained for a change of pH of 1 unit.

The microelectrode was implanted in the cell using a Zeiss sliding micromanipulator under a microscopic magnification of X 640. A measurement was acceptable if the reading was steady for 1 minute or longer. In some experiments the trend in vacuolar pH across the whole root was determined by moving the microelectrode from cell to cell and readings were only accepted where the electrode tip could be clearly seen to be in the vacuole. Several runs from epidermis to protoxylem were often made on the same root.

Three external pH levels were employed in the experiments: pH 3.6 obtained by using glycine-HCl buffer in the culture solution; pH 8.0 using Tris-HCl buffer in the culture solution; and the pH of the unbuffered culture solution, 6.4.

The pH microelectrode gives a signal in mV and for the pH of the vacuolar sap to be calculated the cell transmembrane electrical potential has to be subtracted. The transmembrane PD was determined in the usual way using glass microelectrodes (tip diameter 1 μm) filled with 3 M KCl.

All the measurements were carried out in the root zone approximately 1 cm from the root tip at room temperature (20°C).

C. Results

To investigate the effect of the external pH, plants were left for 20 h in culture solution at the three pH levels employed in the experiments.

Before the vacuolar pH could be determined the effect of the pH of the external solution on the transmembrane potential had to be determined. The results for the epidermal cells are shown in Fig. 1. The cell PD tended to become more negative as the external pH was

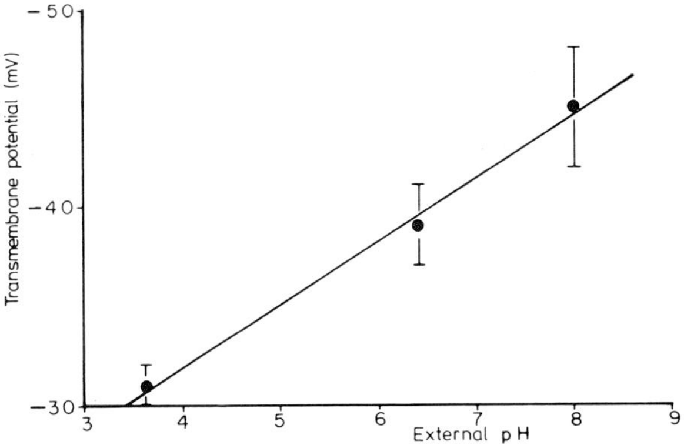

Fig. 1 The relationship between external pH and the transmembrane potential of the epidermal cells of sunflower roots. Points are means of 20 determinations. Vertical bars represent standard errors of the mean.

increased. The effect, however, was not very great. A change of 1 pH unit altered the PD by approximately 3 mV over the pH range investigated.

The effect of the external pH on the pH of the vacuole of the epidermal cells is shown in Fig. 2. Increasing the external pH caused a rise in the internal pH which was exactly half that in the external solution. Thus the vacuolar pH was not kept constant but there was obviously some degree of control being exercised by the cell. At pH 5.0 there was no pH gradient between the vacuole and the external solution.

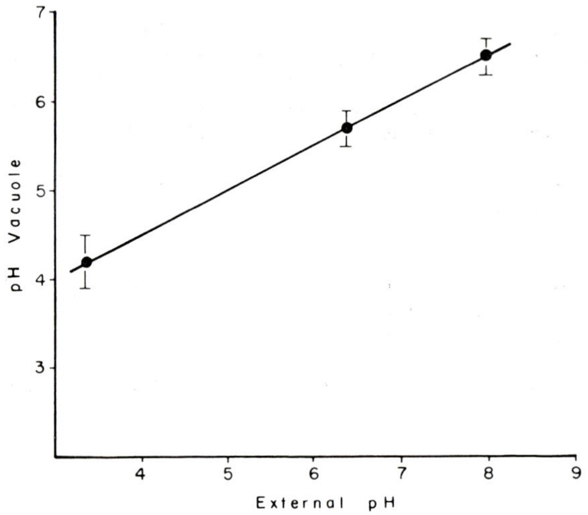

Fig. 2 The relationship between external pH and vacuolar pH for epidermal cells of sunflower roots. Points are means of 15 measurements. Vertical bars represent 95 % confidence limits.

The electrochemical potential gradient for H^+ was calculated from the data in Figs. 1 and 2 and is shown in Fig. 3. It can be seen that at external pH's below 6.4 the direction of the driving force on H^+ is outwards indicating that there is a H^+ efflux pump in operation.

Fig. 3 Driving force on H^+ between the vacuole and the external solution in relation to the external pH for epidermal cells of sunflower roots.

The pH profile across the root was determined for roots in unbuffered culture solution (pH 6.4). The profile of electrical potential has been previously described for these roots (BOWLING, 1972, 1973). There is no significant gradient in vacuolar PD between the epidermis and the pericycle. With the culture solution used in the present experiments which contained 7 mM K^+ the transmembrane potential of all the living cells was approximately -35 mV and the protoxylem -23 mV. The pH, however, was found to rise progressively from cell to cell as the microelectrode was moved inwards to the xylem. Results from 20 runs are shown in Fig. 4. There was an overall rise in pH of 1.2 units from epidermis to protoxylem giving an approximate increment from cell to cell of 0.15 units.

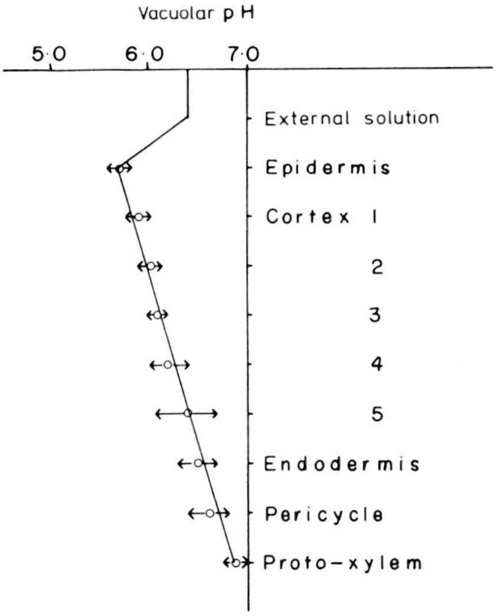

Fig. 4 Profile of vacuolar pH across the root of sunflower. Data were obtained from seven roots and each point is the mean of 20 measurements. The bars represent 95 % confidence limits.

D. Discussion

In absolute terms the effect of the external pH on the cell membrane potential in sunflower roots appears at 3 mV/pH unit to be almost negligible compared with the large effects found in *Nitella clavata* (KITASATO, 1968) and *Nitella translucens* (SPANSWICK, 1972). In relative terms, however, a change from -31 mV to -45 mV between pH 3.6 and 8.0 compares well with the change from -120 mV to -170 mV obtained by SPANSWICK for *N.translucens* over the range pH 5–8.

The vacuolar pH of these cells is obviously protected to some degree from wide fluctuations in the external pH. There appears to be a H^+ extrusion pump raising the pH of the vacuole when the cell is in acid solution. This pump is most likely to be located at the plasmalemma thus controlling the pH of the cytoplasm, the enzymes of which are generally believed to function best at pH 6.0–7.5 (RAVEN and SMITH, 1973). The pump, however, is not particularly successful at maintaining the cellular pH at a constant level. Its inefficiency may explain the lack of tolerance of most plants to wide changes in the external pH. Above pH 7.0 in the medium there is a lower pH in the vacuole than predicted by the Nernst equation and H^+ appears to be retained by the cell against the electrochemical potential gradient. Is there a H^+ influx pump operating under these conditions or is the cell actively extruding OH^- when the external pH is high?

The observation that there is a steady gradient of pH across these roots was unexpected, but perhaps it should not have been as SCOTT and PRIESTLEY (1928) postulated the stelar tissues to be more acid than the cortex in order to explain root exudation. However, the present results indicate the opposite trend. The gradient persists even when the external pH is altered although the absolute values for the cell sap change in the manner shown in Fig. 2.

The pH gradient across the cortex is particularly interesting. Measurements of apparent free space indicate that the external solution pervades the cell walls of the root as far as the endodermis. The presence of the pH gradient indicates that this does not lead to a metabolic uniformity of the cortex as might perhaps be expected. Obviously other forces are involved and it would be interesting to know how the pH gradient is developed and maintained.

We have for the first time a clear cut steady activity gradient of an ion across the root. Obviously we need to know if the gradient is of general occurrence in roots. If it is we are faced with the interesting question: is the pH gradient responsible for centripetal ion transport? The probability is quite high because the pH gradient across the membrane is thought to play a major role in salt transport in cell organelles (ROBERTSON, 1968). If the H^+ efflux pump is at the outer membrane of the cell then the symplasm will show a similar trend in pH to that of the vacuoles. Fig. 4 indicates that the magnitude and direction of the pH gradient across the membranes at the outer surface of the symplasm are quite different from those at the xylem surface. The activity of anion pumps at the outer and inner surfaces of the symplasm is likely to be quite different due to the differing pH gradients prevailing especially if the pumps are powered by a chemiosmotic system like those postulated by MITCHELL (1968). Therefore it is not difficult to see how the pH gradient could bring about unequal ion pumping at the two symplasmic surfaces resulting in centripetal salt flux from the external solution to the xylem. Monovalent cations such as K^+ and Na^+ could move into the symplasm in exchange for H^+ on the H^+ efflux pump. There is increasing speculation that the K^+- Na^+ pump in animal cells might be replaced by a K^+- H^+ pump in plant cells (RAVEN and SMITH, 1973). Divalent cations appear to move across the root passively in response to the transroot electrical potential (BOWLING, 1973).

E. References

ANSARI, A.Q., BOWLING, D.J.F., Planta (Berl.) **98**, 323–329 (1971)

BOWLING, D.J.F., Planta (Berl.) **108**, 147–151 (1972)

BOWLING, D.J.F., in Ion Transport in Plants (ed. W.P. Anderson), pp. 483–491, Academic Press, London and New York 1973

HOAGLAND, D., BROYER, T.C., Am. J. Bot. **27**, 173–185 (1940)

van den HONERT, T.H., HOOYMANS, J.J.M., Acta Bot. Neer. **4**, 376–384 (1955)

KITASATO, H., J. Gen. Physiol. **52**, 60–87 (1968)

MITCHELL, P., Chemiosmotic Coupling in Oxidative and Photosynthetic Phosphorylation, Glynn Research, Bodmin 1968

OLSEN, C., Physiol. Plant. **6**, 848–858 (1953)

PITMAN, M.G., Plant Physiol. **45**, 787–790 (1970)

RAVEN, J.A., SMITH, F.A., in Ion Transport in Plants (ed. W.P. Anderson), pp. 221–228, Academic Press, London and New York 1973

ROBERTSON, R.N., Protons, Electrons, Phosphorylation and Active Transport, Cambridge University Press 1968

SCOTT, L.I., PRIESTLEY, J.H., New Phytol. **27**, 125–140 (1928)

SPANSWICK, R.M., Biochim. Biophys. Acta **288**, 73–89 (1972)

SPEAR, D.G., BARR, J.C., BARR, C.E., J. Gen. Physiol. **54**, 397–414 (1969)

Effect of Abscisic Acid on Fluxes of Ions in Barley Roots

M.G. Pitman, N. Schaefer, and R.A. Wildes

A. Introduction

In certain circumstances abscisic acid (ABA) leads to inhibition of transport of ions through the root to the xylem in maize and barley, without any prior reduction in influx to root cells. The purpose of this paper is to investigate the location of this inhibition in the root, and its relation to other ion fluxes in root cells. The results are considered in relation to the similar effect of cycloheximide on transport through the root (LÄUCHLI et al., 1973).

B. Materials and Methods

Barley seedlings were germinated on moist blotting paper and then grown on stainless steel gauze in nutrient solution for 5 days in the dark at 25°C. The nutrient solution contained 10 mM KNO_3 and other nutrients as described elsewhere (PITMAN, 1971).

Fluxes of chloride were determined following the procedure described by PITMAN (1971).

Fluxes (ϕ) and specific activities (S) are designated by subscripts: o=solution; c = cytoplasm; v = vacuole; x = xylem; thus ϕ_{oc} is the flux from solution into the cytoplasm, ϕ_{cv} is the flux from cytoplasm to vacuole etc. Fluxes and amounts of uptake or transport are referred to fresh weight.

For uptake measurements roots were removed from tracer, rinsed 1 min in ice cold KCl solution then blotted, weighed and dried on planchettes for counting.

Transport of tracer through the root was measured as described by PITMAN (1971).

C. Results and Discussion

I. Effect of ABA on Transport

An example of the response to ABA is shown in Fig. 1. Roots of barley seedlings were excised and set up in apparatus to measure tracer transported through the root. In 5 mM KCl alone (a), roots transported $^{36}Cl^-$ through the cut end of the root at a steady rate of about 1.5 μmoles $g^{-1} h^{-1}$ over the 12 hours of the experiment. When 10^{-5} M ABA was present throughout (b) transport was inhibited after 3 hours and remained low for the duration of the experiment. Transport was measured at 28°C since it was shown previously that the inhibition developed consistently and strongly only above 25°C. There was little inhibition at 22°C and no detectable reduction in transport at 15°C (PITMAN et al., 1974).

Fig. 1 also shows the extent of recovery from 10^{-5} M ABA. (c) Samples of roots were set up in the same solution as in (b), but after 2 hours the solution was changed to 5 mM KCl labelled with $^{36}Cl^-$ and the same specific activity but without ABA. At the changeover, the roots were rinsed several times to remove ABA from the free space and ABA was estimated

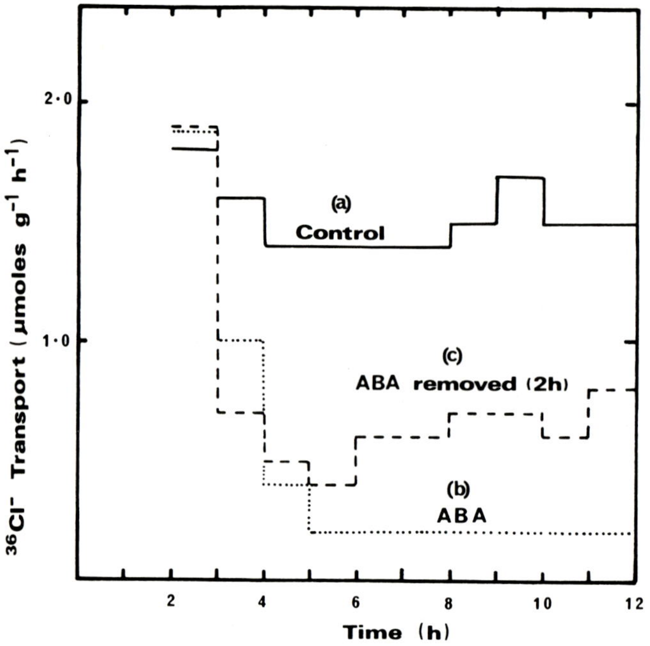

Fig. 1 Effect of ABA on transport of Cl^- through barley roots at 28°C. (a) Control, (b) 10^{-5} M ABA throughout, (c) 10^{-5} M ABA only during first two hours, but then removed.

to be less than 10^{-8} M in the solution from 2 hours onwards. This concentration does not inhibit transport. Despite removal of ABA the response of the roots had apparently been started within 2 hours and substantial inhibition occurred with only slow recovery over the subsequent 8 hours.

Although reduced transport is referred to as 'ABA induced inhibition' this may be an indirect effect, and the action of ABA may be to set in train production of other compounds that act as inhibitors of transport. The time for development of inhibition was variable, being evident in some experiments within 1 hour but in others 3 to 4 hours elapsed before transport was reduced. Once initiated, inhibition recovered only slowly following removal of ABA (Fig. 1c).

Another feature of the response to ABA is that reduction in transport is initially accompanied by increased accumulation of tracer in cells of the root. For example, at the end of the experiment shown in Fig. 1 the roots in ABA contained 22 μmoles g^{-1} compared with 16 μmoles g^{-1} tracer Cl^-. Again, this response was variable. In some cases the increase in content of roots in ABA relative to controls persisted for many hours. In others there was an increase followed by a decline as the influx of tracer to the root as a whole decreased. This decreased influx prevented any observed increase in accumulation in extreme cases. Despite the variability it was clear that inhibition of transport (ϕ_{cx}) could and usually did occur without accompanying reduction in ϕ_{oc}.

II. Flux Measurements

Estimation of Cl^- fluxes was made using roots treated with ABA for various periods in order to investigate how the fluxes changed during inhibition. Fig. 2 shows elution of tracer from the surface of the root expressed as μmoles $g^{-1}h^{-1}$. The control set of roots (a) was kept in 5 mM KCl throughout. Samples of roots were taken from 5 mM KCl + 10^{-5} M ABA and put into unlabelled 5 mM KCl + 10^{-5} M ABA at 80 and 255 minutes. Each set of data is the mean of 3 replicates. Extrapolation to the starting time for the elution estimates ϕ_{co} since the specific activity of Cl^- was the same in the tissue as in the labelling solution at the start of elution. Estimates for ϕ_{co} from these and other experiments have been combined in Fig. 3, which shows ϕ_{oc} constant at about 2.0 μmoles $g^{-1}h^{-1}$ up to 2 hours and then increasing. This increase in ϕ_{oc} runs parallel with the increase in efflux from 0.45 μmoles $g^{-1}h^{-1}$ in Fig. 2 evident from 200 minutes onwards. (Also shown in Fig. 3.)

Loss of tracer from the cut end decreased as expected from the effect on transport through the root shown in Fig. 1. In an elution starting 20 minutes after addition of ABA the efflux from the cut end had fallen by nearly 50 % within 60 minutes. The flux to the xylem (ϕ_{cx}) was calculated from ϕ_{co} (Fig. 3) and the ratio of efflux from the cut end to efflux from the surface (ϕ_{cx}/ϕ_{co}) at various times after equilibration of the cytoplasmic phase.

It follows from the changes in efflux from surface and cut end that the total efflux from the root in ABA decreases at about the time ϕ_{cx} decreases. Thus in control roots the total efflux was 0.9 μmoles $g^{-1}h^{-1}$ between 1 and 2 hours but in ABA only 0.56 μmoles $g^{-1}h^{-1}$; this reduction persisted for several hours. The reduced efflux can be explained by the inhibition

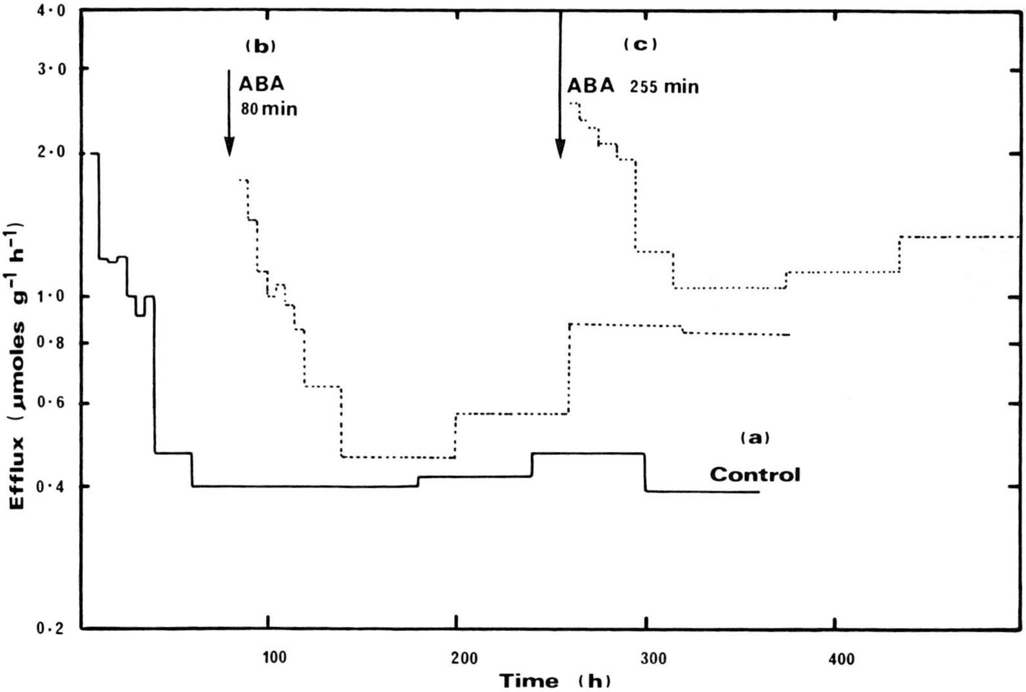

Fig. 2 Elution of $^{36}Cl^-$ from barley roots — surface only. (a) starting t = 0, Control; (b) starting after 80 minutes in 10^{-5} M ABA; (c) starting after 255 minutes in 10^{-5} M ABA.

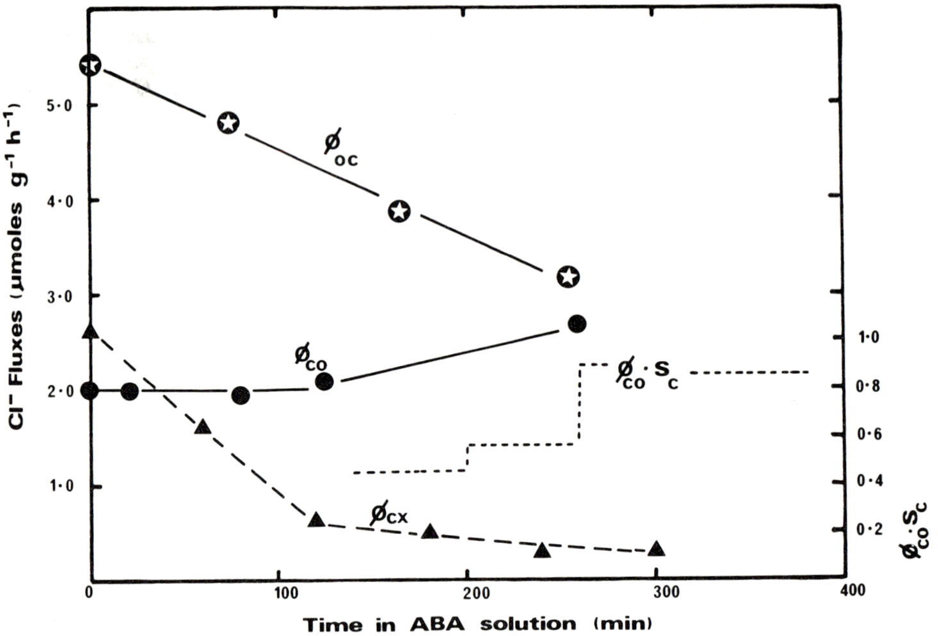

Fig. 3 Estimated values of ϕ_{co}, ϕ_{oc} and ϕ_{cx}, and observed values of surface efflux into ABA solution, $\phi_{co} \cdot S_c$.

of ϕ_{cx}. At this stage of efflux, $S_v > S_c$ and loss of tracer is limited by diffusion from vacuole to cytoplasm. However, tracer diffusing from the cytoplasm passes to solution, vacuole and xylem in the ratio $\phi_{co} : \phi_{cv} : \phi_{cx}$. As ϕ_{cx} becomes smaller, the fraction $(\phi_{co} + \phi_{cx})/(\phi_{co} + \phi_{cv} + \phi_{cx})$ becomes smaller and only increases again as ϕ_{co} increases relative to ϕ_{cv} at later times.

Influx to the vacuole was determined from ϕ_{cx} (Fig. 3) and the ratio ϕ_{cx}/ϕ_{cv}, estimated from measurements of tracer uptake (the rate of increase of tracer in the roots is $\phi_{cv} \cdot S_c$ $(t > 2h)$) and transport out of the xylem $\phi_{cx} \cdot S_c$. The value of ϕ_{vc} can then be calculated since efflux from the vacuole through the surface (Fig. 2) is $\phi_{co} \cdot \phi_{vc}/(\phi_{co} + \phi_{cv} + \phi_{cx})$. For the set of data in Fig. 3 values were:

time in ABA (h)	0	1-2	2-3	3-4
ϕ_{cv} (μmoles $g^{-1} h^{-1}$)	2.9	3.4	2.0	1.6
ϕ_{vc} (μmoles $g^{-1} h^{-1}$)	1.2	1.0	0.9	0.9

Note that there appears to be a small increase in ϕ_{cv} at 1–2 hours, which is the time when reduction in ϕ_{cx} was observed.

In other experiments larger increases in tracer influx to the tissue were observed that could only be accounted for by an increase in ϕ_{cv}.

The response of the fluxes to ABA is difficult to reduce to a simple pattern of operational interactions since it is so complex. It seems that the prime effect is on ϕ_{cx}, but the start of inhibition may be variable. Changes in tracer uptake to the vacuoles then result partly from an increase in specific activity, S_c, but also due to real increases in ϕ_{cv}. Such an increase is not unreasonable if blocking ϕ_{cx} leads to increased content of 'cytoplasm' but in the absence of detailed knowledge about regulation of ϕ_{cv} little more can be said. It is thought that the reduction in ϕ_{cx} is not due to stimulation of ϕ_{cv} so that in effect the vacuoles compete for tracer with the stele, as in low salt roots. Thus the reduction in ϕ_{cx} (2.5 μmoles $g^{-1} h^{-1}$) is much larger than any increase in ϕ_{cv} (about 0.5 μmoles $g^{-1} h^{-1}$).

It is clear that ABA did not produce its prime effect on ϕ_{cx} by reduction in ϕ_{oc}. Although reduction in ϕ_{oc} may occur it seems to be a later consequence of ABA action than inhibition of ϕ_{cx}. There was no effect of ABA on the passive flux ϕ_{co}, which eventually rises as net uptake to the vacuole becomes negligible. In other experiments ABA had no effect on tracer uptake, although transport across the root was inhibited.

III. Transport from Cortex to Stele

A possible mode of action of ABA is in stopping symplasmic movement across the root, or at least, preventing transport across the endodermis. To test this view, roots pretreated with ABA were put into labelled 5 mM KCl with and without ABA then separated into stele and cortex at various times. Both barley and maize were used, but as the cortex of the barley roots was damaged during stripping it was discarded and tracer measured only in the stele. Cortical content was estimated as the difference between whole roots and the steles. The steles were rinsed and measured so that the results could be expressed relative to root lenght.

Table 1 gives results of these experiments showing that ABA did not inhibit transport to the stele. The maize examples show that there was increased uptake both to stele and cortex.

Table 1

Distribution of $^{36}Cl^-$ between cortex and stele of barley and maize roots

	Time (min)		Content in Control roots (nmoles cm^{-1})		Content in ABA roots (nmoles cm^{-1})	
	ABA	$^{36}Cl^-$	Cortex	Stele	Cortex	Stele
Barley	180	34	2.8	0.64	3.3	0.67
	180	74	4.4	1.20	5.7	1.27
Maize	30	30	3.5	0.56	7.5	0.88
	120	120	11.2	1.80	22.8	3.0

D. Conclusion

The above results confirm that the primary action of ABA seems to be on the release of ions from the stele to the xylem, i.e. on ϕ_{cx}. What is not established from these results is whether inhibition of ϕ_{cx} is due to a decrease in membrane permeability reducing a passive flux or reduction in an active transport process.

E. Acknowledgements

We are grateful to J. Le Claire, M. Corderoy and D. Wellfare for their assistance and to the Australian Research Grants Commission for support of the project.

F. References

LÄUCHLI, A., LÜTTGE, U., PITMAN, M.G., Z. Naturforsch. **28c**, 431–434 (1973)
PITMAN, M.G., Aust. J. Biol. Sci. **24**, 407–421 (1971)
PITMAN, M.G., LÜTTGE, U., LÄUCHLI, A., BALL, E., J. Exp. Bot., in press (1974)

The Effect of Inhibitors on the K^+-Dependent Na^+ Efflux and the K^+-Na^+ Selectivity of Barley Roots

W.D. Jeschke

A. Introduction

The K^+-Na^+ selectivity of ion uptake by higher plants has been attributed to the presence of a Na^+ efflux pump at the plasmalemma of their root cells (PITMAN and SADDLER, 1967; ETHERTON, 1967; JESCHKE, 1970). Recently, a transient K^+-dependent Na^+ efflux at the plasmalemma of the cortical cells of barley roots, induced by an addition of K^+ to the external medium, has been measured quantitatively (JESCHKE, 1972, 1973). This Na^+ efflux was followed by a substantial drop in the Na^+ transport and the beginning of K^+ transport through the xylem of these roots. I proposed that the efflux was due to a K^+-dependent Na^+ pump in the cortical cells and that this was the basis of the K^+-Na^+ selectivity of transport in barley roots.

It could be objected that the effect of K^+ is unspecific and due to a depolarization of the root cells by K^+ in the external medium (MEARES, 1973). Although the high effectiveness of K^+ (JESCHKE, 1973) and to some extent also the ineffectiveness of Li^+ in producing a Na^+ efflux argue against unspecific effects, it seemed desirable to prove the metabolic linkage of the K^+-dependent Na^+ efflux.

The inhibitor of animal Na^+-K^+ ATPases, ouabain, was ineffective, as it is in the salt-stimulated ATPase of plant roots (FISHER and HODGES, 1969). In this paper the effect of three inhibitors was studied. CCCP (carbonyl cyanide m-chlorophenylhydrazone) is a potent uncoupler of oxidative phosphorylation (HEYTLER and PRITCHARD, 1962); NEM (N-ethylmaleimide) reacts with sulphhydryl groups (WEBB, 1966), inhibits the Na^+-K^+ transport ATPase, and reacts specifically with different sulphhydryl groups in this enzyme (HART and TITUS, 1973); DCCD (N,N'-dicyclohexyl carbodiimide) has been shown to inhibit the ATPase and the K^+ transport of ***Streptococcus*** (HAROLD and BAARDA, 1969) and the Na^+/K^+ exchange in ***Chlorella*** (SHIEH and BARBER, 1971).

B. Materials and Methods

1. Abbreviations

ϕ_{OC}, ϕ_{CO} influx and efflux at the plasmalemma of the cortical cells (µmoles/h per g fresh weight); ϕ_{CV}, ϕ_{VC} influx and efflux at the tonoplast; R' transport from the cytoplasm (of the xylem parenchyma cells) through the xylem of the roots; ϕ'_{CO} apparent efflux from the symplasm of the entire excised roots: $\phi'_{CO} = \phi_{CO} + R'$; Q_C,Q_V cytoplasmic or vacuolar content (µmoles/g fresh weight); ϕ out steady, quasi-steady $^{22}Na^+$ efflux at the end of the elution (cpm/h per g fresh weight); S_O,S_C specific activity of $^{22}Na^+$ (cpm/µmole) in the external solution or in the cytoplasmic phase (symplasm); k_V rate constant of $^{22}Na^+$ exchange between the vacuole and the external solution (h^{-1}).

2. Material and Solutions

Hordeum distichon L. cv Müllers Sommergerste was grown for 7 days in the greenhouse. The culture solutions contained in mequiv/l: low-salt roots: $CaSO_4$ 1.0; 'Na-roots': 'Na-solution' with NaH_2PO_4/Na_2HPO_4 (pH 5.8) 1.0, $Ca(NO_3)_2$ 6.0, $MgSO_4$ 1.0, Fe-EDTA 0.05 and 1 ml/l of a trace element solution (Hoagland) but without K^+; 'K-roots': 'K-solution' with the same composition but with K^+ instead of Na^+.

The same Na- and K-solutions were used in the experiments, but here Fe and the trace elements were omitted. Where indicated, 0.2 mM KCl was added to the Na-solution.

3. Influx of Na^+ and K^+

Low-salt roots were exposed for 20 min at 20°C to Na-solution labelled with $^{22}Na^+$ or to K-solution labelled with $^{86}Rb^+$. After repetitive washing in inactive solutions for 8 min the radioactivity in the roots was measured. The washing procedure removed the radioactivity from the free space as was seen from a linear time course of influx with a zero ordinate intercept. As the half-time of cytoplasmic $^{22}Na^+$ exchange is about 20–30 min and that of K^+ is even longer (determined at 5°C, PITMAN and SADDLER, 1967) the measured influxes probably are limited mainly by ϕ_{oc}.

4. Determination of Na^+ Fluxes

Individual fluxes were determined by a modified compartmental method (PITMAN, 1971); the methods have been described in detail (JESCHKE, 1972, 1973). Briefly low-salt roots were mounted in a two-chambered vessel with the apical 3 cm in a chamber S and the cut ends (3–4 mm) in a chamber X. After a period of 24 h of $^{22}Na^+$ uptake from labelled Na-solution, inactive Na-solutions were introduced into S and X, exchanged at frequent intervals, and the radioactivity lost to these wash-out solutions was determined by liquid scintillation counting. In this way the rates of $^{22}Na^+$ efflux through the root cortex (S) and through the xylem (X) were obtained. They were related to the fresh weight of the roots in S (cortical efflux) or to the total fresh weight of the roots (xylem efflux), and their logarithms were plotted against the time of elution. From the efflux curves thus obtained, the fluxes, transport and contents of Na^+ in the roots were estimated.

5. K^+-Dependent Na^+ Efflux in Re-Elution Experiments

When barley roots have been washed out for more than 3 h, the $^{22}Na^+$ efflux through the cortex and the xylem reach quasi-steady state values $\phi_{\text{out steady}}$ (S or X), which drop slowly with the rate constant k_v of vacuolar Na^+ exchange. In this state the cytoplasmic specific activity of $^{22}Na^+$ is given by

$$S_c(t) = S_o \exp(-k_v \cdot t) \cdot \phi_{vc} \cdot (\phi'_{co} + \phi_{cv})^{-1} \qquad (1)$$

When K^+ is added to the wash-out solution in this state (= re-elution), the cortical $^{22}Na^+$ efflux is transiently increased (see Fig. 5, control) due to an increased plasmalemma efflux ϕ_{co}. This K^+-stimulated ϕ_{co} can be estimated by the use of $S_c(t)$, Eq. (1); subtraction of the value of ϕ_{co} before addition of K^+ yields the K^+-dependent Na^+ efflux.

6. Transport of Na^+ or K^+

Na-roots were placed into the two-chambered vessels containing Na-solution in S and 6 mequiv/l $Ca(NO_3)_2$ plus 1 mequiv/l $MgSO_4$ in X; the transport of Na^+ was estimated from the Na^+ appearing in X as determined by flame photometry. Determination of K^+ in X yielded a transport of endogenous K^+, i.e. of K^+ contained in the cytoplasm or the vacuoles of the root cells. K^+ transport and an endogenous Na^+ transport were correspondingly obtained using K-roots and K-solution in S. This method yielded higher transport rates than those in which Na^+ or K^+ were also present in X.

7. Inhibitors

NEM was added in aqueous solution; DCCD was added in ethanol, the final concentration of 0.5 % ethanol had only a negligible effect on the steady state of K^+-stimulated Na^+ efflux; CCCP was dissolved in methanol, the final concentration was 0.1 % or less.

C. Results

I. Effects of CCCP on Na^+ and K^+ Influx and Transport

CCCP inhibited the initial Na^+ and K^+ influxes of low-salt barley roots (Fig. 1) (RAINS and EPSTEIN, 1967; PITMAN and SADDLER, 1967). The K^+ influx was somewhat more sensitive than the Na^+ influx. Also the transport of K^+ and Na^+ through the roots was inhibited and again the Na^+ transport was less sensitive (Fig. 1).

The endogenous K^+ transport in Na-roots (see Methods B. 6.) was as sensitive to CCCP as the K^+ transport through K-roots. As the endogenous transport essentially is a transport of

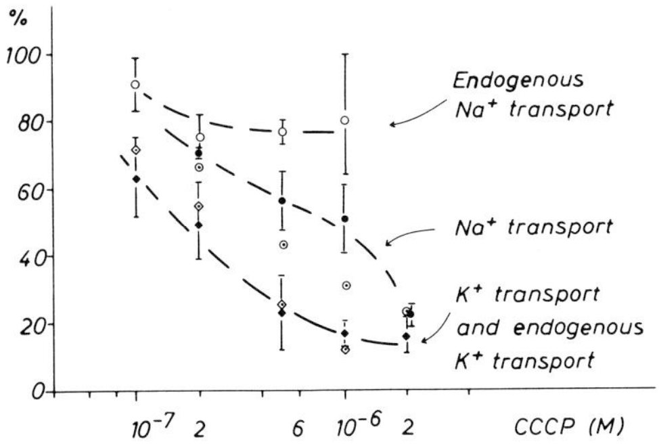

Fig. 1 Effect of CCCP on influx and transport of Na^+ and K^+ in barley roots. ◇ K^+($^{86}Rb^+$) and ⊙ Na^+($^{22}Na^+$) influx; ● transport of Na^+ by Na-roots, ○ transport of endogenous, predominantly vacuolar Na^+ by K-roots (see methods, 6); ◆ transport of K^+ by K-roots or of endogenous K^+ by Na-roots. The control rates were in µmoles/h per g fresh weight: Na^+ transport 7.1 ± 1.3; endogenous Na^+ transport 0.6 ± 0.3; K^+ transport 14.6 ± 2.9; endogenous K^+ transport 1.1 ± 0.4; Na^+ influx (1 mM Na^+) 6.5; K^+ influx (0.2 mM K^+) 11.5.

vacuolar K^+ via the cytoplasm through the xylem vessels, its inhibition could indicate that two active steps are involved in the overall K^+ transport, one at the plasmalemma of the cortical cells and one in the transfer from the symplasm to the xylem as suggested for Cl^- by PITMAN (1972). Remarkably, the endogenous Na^+ transport was rather insensitive to CCCP (Fig. 1), which suggests that the second step is predominantly passive for Na^+.

II. Effect of CCCP on Accumulation and Individual Fluxes of Na^+

Measurements of individual Na^+ fluxes in roots in the presence of CCCP require that the roots are loaded with $^{22}Na^+$ almost to flux equilibrium also in the presence of the inhibitor. Na^+ accumulation by low-salt barley roots from Na-solution with or without 0.2 mM K^+ proceeds for about 24 h before flux equilibrium is attained (Fig. 2). The accumulation is inhibited in the presence of K^+.

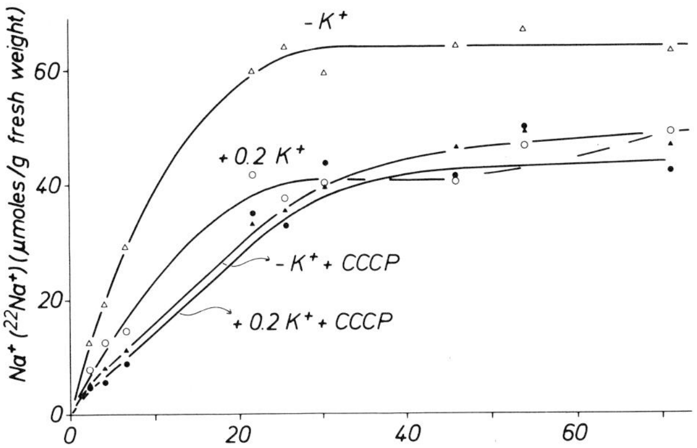

Fig. 2 Time course of the accumulation of Na^+($^{22}Na^+$) by low-salt barley roots. △, ▲ Accumulation from Na-solution (1 mM Na^+) without △ or with ▲ 0.5 µM CCCP; ○, ● accumulation from Na-solution (1 mM Na^+ + 0.2 mM K^+) ○ control and ● in the presence of 0.5 µM CCCP.

When CCCP is added, the Na^+ accumulation, essentially in the vacuoles, is strongly inhibited but it continues when the accumulation in the control roots has reached saturation (Fig. 2). This shows that CCCP apparently does not interfere as strongly with the regulatory mechanisms (CRAM, 1973; ZIMMERMANN and STEUDLE, 1974) which control the amount of ion accumulated by the cells. Since ϕ_{oc} is decreased by this regulation (PITMAN, 1971; JESCHKE, 1972) CCCP effects an apparent increase in ϕ_{oc} after 24 h compared to the control (Fig. 2).

In the presence of CCCP 0.2 mM K^+ hardly inhibits the Na^+ accumulation (Fig. 2), indicating that CCCP interferes also with the K^+-Na^+ selectivity of uptake.

Although in the presence of CCCP 50 h of loading would be needed to comply with the requirement of flux equilibrium (Fig. 2) flux measurements with 24 h loading were conducted in order to avoid prolonged exposures to the inhibitor. In this way regular exponential efflux curves were obtained and the fluxes and contents of Na^+ (Table 1) were calculated after corrections were applied for a small endogenous Na^+ content in the roots and for the net uptake occurring in the presence of CCCP.

Table 1

Effect of CCCP on the estimated values of fluxes, transport (μmoles/h per g fresh weight) and contents (μmoles/g fresh weight) of Na^+ in barley roots, measured in Na-solution in the presence of 0.2 mM KCl at room temperature (21°C).

Na^+	1 mM	1 mM
K^+	0.2 mM	0.2 mM
CCCP	–	0.5 μM
ϕ_{vc}	0.18	1.46
ϕ_{co}	0.85	1.6
ϕ_{co}/Q_c	0.84	0.59
R'	0.63	0.57
R'/Q_c	0.62	0.22
Q_c	1.0	2.7
Q_v	66	69

As is seen the effluxes ϕ_{vc} and ϕ_{co} were increased by CCCP; the greatest change occurred in ϕ_{vc} which was increased by a factor of 4. Also the cytoplasmic Na^+ content Q_c was strongly increased by CCCP, possibly as a consequence of the change in ϕ_{vc}. These effects of CCCP on ϕ_{vc} and Q_c are comparable to those of DNP in ***Chaetomorpha*** (DODD et al., 1966). R' was hardly affected by CCCP. This corresponds to the relatively small inhibition of the endogenous Na^+ transport since R', too, measures a vacuole-xylem transport out of the roots via the xylem vessels.

III. Effects of CCCP, NEM, and DCCD on the K^+-Stimulated Na^+ Efflux in Re-Elution Experiments

The time course of the $^{22}Na^+$ efflux through the root cortex and the xylem of barley roots in the absence of inhibitors was very similar in all experiments, it is shown only in Fig. 5. In these re-elution experiments an addition of 0.2 mM K^+ effected a stimulation of the cortical $^{22}Na^+$ efflux and a subsequent strong inhibition of the $^{22}Na^+$ xylem transport (Fig. 5).

To observe the effect of inhibitors, these were added to the roots 2.5 h prior to the addition of K^+. As is seen in Fig. 3, the steady state cortical Na^+ efflux is tripled within 1 h after the addition of CCCP, pointing to an increase in ϕ_{co} and ϕ_{vc} also in the absence of K^+. When 0.2 mM K^+ was added to induce a K^+-stimulated Na^+ efflux, only a small stimulation occurred, which was equivalent to a K^+-dependent Na^+ efflux of 1.5 μmoles/h per g fresh weight, i.e. 27 % of the control.

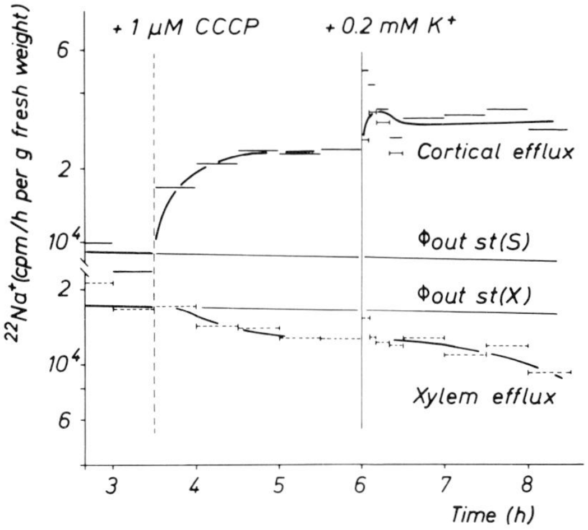

Fig. 3 Steady state Na^+ efflux through the root cortex and the xylem of barley roots. CCCP was added at 3.5 h and was present in all wash-out solutions, also when 0.2 mM K^+ was added at 6 h. The cortical efflux rates ⊢⊣ were corrected for the effect of short wash-out intervals.

The steady state $^{22}Na^+$ xylem efflux on the other hand was slowly and scarcely affected by CCCP (Fig. 3), and the subsequent addition of K^+ failed to inhibit the xylem efflux selectively. In an other experiment a 35 % inhibition of the xylem efflux occurred 4 h after the addition of K^+ in the presence of CCCP. So there was only a residual K^+-Na^+ selectivity of transport in the presence of CCCP.

Similarly NEM (Fig. 4) and DCCD (Fig. 5) produced an increase in the steady state $^{22}Na^+$ efflux through the root cortex. But additionally these compounds inhibited the steady state $^{22}Na^+$ efflux through the xylem vessels, indicating an inhibition of the vacuole-xylem transport in addition to an increase in ϕ_{vc} and ϕ_{co}. When K^+ was added to the roots in the presence of NEM or DCCD there was only a small, if any, K^+-stimulated Na^+ efflux and no further decrease in $\phi_{out\ steady\ (X)}$ (Figs. 4 and 5). So these inhibitors, too, prevented the selective sodium transport apparently by an inhibition of the K^+-dependent Na^+ efflux at the plasmalemma of the cortical cells.

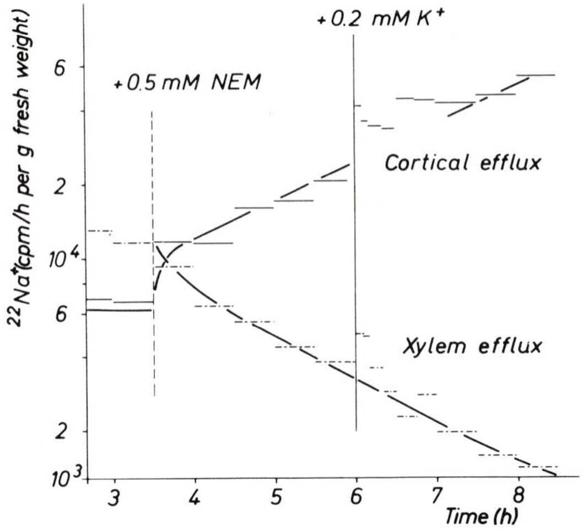

Fig. 4 The same as Fig. 3, save that 0.5 mM NEM was added.

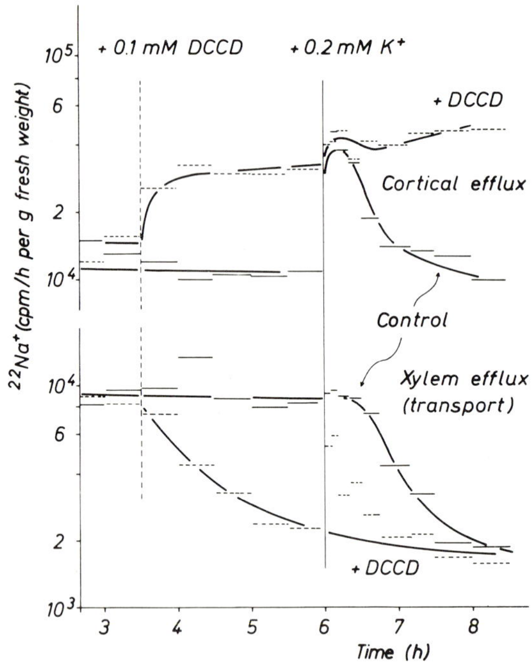

Fig. 5 Steady state Na^+ efflux through the root cortex and through the xylem of barley roots. In the control 0.2 mM K^+ was added to the wash-out solution (1 mM Na^+) at t=6 h, resulting in a transiently increased cortical efflux and an inhibited xylem efflux. In the '+DCCD' roots the inhibitor was added at 3.5 h and was present in all wash-out solutions, also during the re-elution, when 0.2 mM K^+ was added at 6 h.

D. Discussion

I. Inhibition of Na^+ Uptake by CCCP

This inhibition is seemingly at variance with electrochemical considerations (PITMAN and SADDLER, 1967). However, in vacuolated plant cells a salt uptake and a discrimination between K^+ and Na^+ as the major univalent cations occur. When no K^+ or only low concentrations are present Na^+ is accumulated by barley roots and an inhibition of its uptake may be due to inhibited uptake of the anion to which the cation uptake is linked. Further, at 1 mM Na^+ the sodium in the cells is not at a much lower electrochemical potential than in the external medium. Taking -110 mV as the electrical potential of barley root cells at 0.2 mM K^+ and 1 mM Na^+ (PITMAN and SADDLER, 1967; PITMAN et al., 1971) the expected internal concentrations would be 80 mM Na^+ and 16 mM K^+. Experimentally these concentrations were estimated to be 75 mM (efflux analysis) or 93 mM (flame photometry) Na^+ and 40 mM K^+. Thus K^+ is at a somewhat higher electrochemical potential within the roots while Na^+ appears to be close to equilibrium. Assuming further that the cytoplasm occupies 3 % of the cell volume, a cytoplasmic $[Na^+]$ of 33 mM would follow from Q_c = 1 μmole/g fresh weight. Hence the transport of Na^+ across the plasmalemma into the cytoplasm would be a passive process. But if the overall transfer of Na^+ to the vacuole occurs close to equilibrium ($\Delta G = O$) then the transport across the tonoplast to the vacuole would have to be an active process. So the effect of CCCP on the accumulation of Na^+ in the vacuoles (Fig. 2) could also be influenced by an active step occuring at the tonoplast.

At any rate, the lower sensitivity of the uptake and transport of Na^+ to CCCP (Fig. 1) is consistent with a lower energy requirement compared to the uptake and transport of K^+.

II. Action of CCCP on the Near to Steady State Na^+ Fluxes

CCCP increased the effluxes ϕ_{vc} and ϕ_{co} and the cytoplasmic content Q_c; the transport R' was hardly affected (Table 1). The increase in ϕ_{vc} and ϕ_{co} occurred also in the absence of K^+ (see $\phi_{out\ steady\ (S)}$ in Fig. 3). From the energetic considerations in the last paragraph the increase in ϕ_{co} is unexpected while the change in ϕ_{vc} could be due to an increased tonoplast permeability, if ϕ_{vc} were a passive flux. This would be consistent with the energetic considerations. An active transport of Na^+ across the tonoplast to the vacuole has been suggested also by JENNINGS (1968) and is possibly present in ***Chaetomorpha*** (DODD et al., 1966).

The increase in Q_c could result then from the larger Na^+ efflux out of the vacuole. Certainly an increased Q_c will reflect on the fluxes out of the cytoplasm ϕ_{co}, ϕ_{cv} and R'. Therefore the ratios ϕ_{co}/Q_c and R'/Q_c have been included in Table 1. As is seen, ϕ_{co}/Q_c decreased in the presence of CCCP indicating that the change in ϕ_{co} (Table 1) resulted from the increase in Q_c and that the process transporting Na^+ across the plasmalemma to the external solution was inhibited rather than increased by CCCP. When the fluxes were measured in the absence of K^+, however, ϕ_{co} as well as ϕ_{co}/Q_c were increased by CCCP. It may be concluded therefore, that CCCP inhibits the K^+-dependent part of ϕ_{co} in agreement with the direct measurements (Fig. 3) and the energetic considerations.

A similar consideration applies to R' which measures a vacuole-xylem transport like the endogenous Na^+ transport (Fig. 1). The small effect of CCCP in both cases (Fig. 1 and Table 1) appears to result from the increased Q_c, since R'/Q_c was decreased in the presence of CCCP. It cannot be excluded, therefore, that Na^+ ions are transferred to the xylem by a similar process to K^+, involving an active step or metabolic control (PITMAN, 1972). But again the energy requirement of the overall transport of Na^+ appears to be lower than that of K^+ (Fig. 1).

The effects of CCCP on the Na^+ fluxes and contents of barley roots in the presence of K^+, especially the increase in ϕ_{vc} and Q_c are in good agreement with the effects of DNP found with cells of ***Chaetomorpha*** (DODD et al., 1966) indicating that similar processes are involved in the fluxes and in the distribution of Na^+ between the cytoplasm and the vacuole in the two organisms. The effects of CCCP would be consistent with a model in which Na^+ ions are actively exported from the cytoplasm in the presence of K^+.

III. Inhibition of the K^+-Dependent Na^+ Efflux

In a previous paper it was shown that a K^+-dependent Na^+ efflux pump at the plasmalemma of the cortical cells of barley roots maintains the low cytoplasmic Na^+ content Q_c which in turn decreases ϕ_{cv} and the transport of Na^+ through the xylem of the roots (JESCHKE, 1973). CCCP as well as DCCD and NEM inhibited the K^+-dependent Na^+ efflux and the subsequent decrease in the sodium transport. As shown in the previous paragraph, the inhibition of the K^+-dependent part of ϕ_{co} by CCCP was apparent also from the decrease in ϕ_{co}/Q_c in the presence of K^+ (Table 1). From the effect of CCCP it may be inferred that the K^+-dependent Na^+ efflux is linked to oxidative phosphorylation and derives its energy from ATP.

The inhibition of the K^+-dependent Na^+ efflux by DCCD agrees with its effect on the K^+/Na^+ exchange in ***Chlorella*** (SHIEH and BARBER, 1971) and with the inhibition of the ATPase and the K^+ transport of ***Streptococcus*** (HAROLD et al., 1969). In both organisms, as in barley, ouabain had no effect on the K^+/Na^+ exchange or the ATPase. The action of DCCD suggests that an ATPase is involved in the K^+-Na^+ selectivity of barley roots, too. It should be emphasized, however, that DCCD (as CCCP) has effects other than inhibiting the K^+-dependent Na^+ efflux. In the absence of K^+ it increased ϕ_{vc} and ϕ_{co} and inhibited the Na^+ transport through the xylem (Fig. 5). The same was true for NEM, but as its effects on the steady state Na^+ effluxes increased progressively (Fig. 4) these secondary effects may be more severe. So the inhibition of the K^+-dependent Na^+ efflux by NEM should be considered with caution, although it is not inconsistent with the participation of an ATPase in the K^+-Na^+ selectivity.

To conclude, all inhibitors used in this study affected more than one Na^+ transport process in barley roots, if only because of the interdependence of fluxes, ion content and transport through the xylem. The results suggest that an ATPase participates in the K^+-dependent Na^+ efflux at the plasmalemma of the cortical cells of barley roots. Whether it is similar to the salt-stimulated ATPase isolated from oat roots (HODGES et al., 1972) is not known. A difference between this ATPase and the Na^+ efflux pump in barley appears to be in their affinities to K^+. While the ATPase had an apparent K_m of about 10 mM K^+ (FISHER and HODGES, 1969), half-saturation of the K^+-dependent Na^+ efflux occurred at 0.02 mM K^+ (JESCHKE, 1973).

The finding that CCCP also prevents the decrease of the Na^+ uptake in the presence of K^+ (Fig. 2) is consistent with the suggestion (PITMAN and SADDLER, 1967) that the Na^+ efflux pump at the plasmalemma is involved not only in the selective K^+ and Na^+ transport but also in the selective accumulation of K^+ and Na^+ by barley roots.

E. Acknowledgements

This investigation was supported by the Deutsche Forschungsgemeinschaft. Thanks are due to Miss H. Eschenbacher for excellent technical assistance.

F. References

CRAM, W.J., J. Exp. Bot. **24**, 328–341 (1973)
DODD, W.A., PITMAN, M.G., WEST, K.R., Aust. J. Biol. Sci. **19**, 341–354 (1966)
ETHERTON, B., Plant Physiol. **42**, 685–690 (1967)
FISHER, J. HODGES, T.K., Plant Physiol. **44**, 385–395 (1969)
HAROLD, F.M., BAARDA, J.R., J. Biol. Chem. **244**, 2261–2268 (1969)
HART, W.M., TITUS, E.O., J. Biol. Chem. **248**, 4674–4681 (1973)
HEYTLER, P.G., PRITCHARD, W.W., Biochem. Biophys. Res. Commun. **7**, 272–275 (1962)
HODGES, T.K., LEONARD, R.T., BRACKER, C.E., KEENAN, T.W., Proc. Nat. Acad. Sci. **69**, 3307–3311 (1972)
JENNINGS, D.H., New Phytol. **67**, 899–911 (1968)
JESCHKE, W.D., Planta **94**, 240–254 (1970)
JESCHKE, W.D., Planta **106**, 73–90 (1972)
JESCHKE, W.D., in Ion Transport in Plants (ed. W.P. Anderson), pp. 285–296, Academic Press, London and New York 1973
MEARES, P., in Ion Transport in Plants (ed. W.P. Anderson), p. 355, Academic Press, London and New York 1973
PITMAN, M.G., Aust. J. Biol. Sci. **24**, 407–421 (1971)
PITMAN, M.G., Aust. J. Biol. Sci. **25**, 243–257 (1972)
PITMAN, M.G., SADDLER, H.D.W., Proc. Nat. Acad. Sci. **57**, 44–49 (1967)
PITMAN, M.G., MERTZ, S.M., JR., GRAVES, J.S., PIERCE, W.S., HIGINBOTHAM, N., Plant Physiol. **47**, 76–80 (1971)
RAINS, D.W., EPSTEIN, E., Plant Physiol. **42**, 314–318 (1967)
SHIEH, Y.J., BARBER, J., Biochim. Biophys. Acta **233**, 594–603 (1971)
WEBB, J.L., Enzyme and Metabolic Inhibitors, Vol. 3, Academic Press, London and New York 1966
ZIMMERMANN, U., STEUDLE, E., J. Membrane Biol. **16**, 331–352 (1974)

Potassium Uptake with Respect to Cation-Anion Balance in Pea Epicotyl Segments

N. Higinbotham and W.S. Pierce

A. Introduction

There are three basic ways in which cation-anion balance in living systems may be maintained: 1) absorption rates for cations equal absorption rates for accompanying anions; 2) cations are exchanged for internal cations and anions are exchanged for internal anions; and 3) excess accumulated ions are balanced by endogenous organic ions of opposite charge. Each mechanism has been reported at various times to be the principal mechanism in barley roots. Furthermore, JACOBSON and ORDIN (1954) reported that roots of young barley seedlings respond to unbalanced cation uptake through the production of malate while older roots respond to unbalanced cation uptake by exchanging previously absorbed cations. These reports suggest that perhaps subtle differences in conditions may determine which of these balancing mechanisms is functioning in a given tissue. The present investigation indicates that in etiolated pea epicotyl segments cation uptake is balanced by mineral anion uptake and the organic acid content remains essentially constant at a low level under the conditions tested.

B. Materials and Methods

Epicotyls were obtained from ***Pisum sativum*** cv Alaska seedlings grown in darkness for 7 days at 25°C. The substrate was vermiculite moistened with nutrient solution of the following composition in mM: KCl, 1.0; $Ca(NO_3)_2$, 1.0; $MgSO_4$, 0.25; NaH_2PO_4, 0.904; and Na_2HPO_4, 0.048. This is designated as 1X solution; 10X solution is 10-fold more concentrated. In preparing segments the apical 1.5 cm of the third internode was discarded, then two 1-cm segments were excised. Twenty segments (about 0.75 g) were used for each sample. Each sample was placed in 100 ml of solution in a 250 ml flask and placed on a reciprocating shaker. During absorption periods K^+ and Na^+ were assayed by a flame emission spectrophotometer (Jarrell-Ash Model 52-700). Phosphate was determined by the Fiske and Subbarrow method. NO_3^- was assayed using an ion-specific electrode (Orion Research, Inc.). Cl^- was measured using an Aminco-Cotlove titrator. The uptake of SO_4^{2-} was measured by labelling with $^{35}SO_4^{2-}$. Extraction of tissue samples was performed in boiling water as described previously (HIGINBOTHAM et al., 1967).

Organic acids were extracted in hot 80 % ethanol with successive aliquots. Amino acids were removed by passing the extract through a cation exchange column, then the organic acids were placed on a column of AG-1 formate resin. Fractions of organic acids were removed by eluting with 4 N formic acid. After driving off the formic acid, each fraction was assayed by titration with standardized NaOH.

Cell electropotential measurements were performed as previously described (HIGINBOTHAM et al., 1970).

C. Results and Discussion

I. Organic Acids

According to a number of reports, beginning with that of ULRICH (1941), plant tissues may selectively absorb cations and maintain cation-anion balance by an increased synthesis of organic acids. Differential cation uptake may be induced by high pH or high CO_2 concentration (POOLE and POEL, 1965) or by use of K^+ salts having a slowly absorbed anion, e.g. $SO_4{}^{2-}$ (TORII and LATIES, 1966). Pea epicotyl fails to show an enhanced K^+ uptake at higher pH unlike beet root (POOLE and POEL, 1965); the effect of increased CO_2 concentration was not tested.

In pea epicotyl segments we found considerable K^+ uptake in 1X or 10X solution or in 5 mM K_2SO_4 + 1 mM $CaSO_4$ solution. During the 48-hour period there was no change in the organic acid content, and it seems clear that inorganic anion absorption balanced K^+ uptake. The average concentration of organic acids, about 13 mM, approximately balanced the sum of Na^+, Ca^{2+}, and Mg^{2+}; uptake of the latter three ions during the period (~ 4 mM) was quite closely matched by Cl^- absorption. HIATT's (1967) results with barley roots in K_2SO_4 solution showed an increase of organic acids to 53 μequiv/g.

II. Cation-Anion Balance in Nutrient Solution

In 1X solution a small excess of anion uptake occurred (Table 1). In 10X solution uptake of cations and anions nearly balanced one another. It may be seen that K^+ absorption matches the sum of NO_3^-, $H_2PO_4^-$, and $SO_4{}^{2-}$; the net influx of Na^+, Ca^{2+}, and Mg^{2+} is not shown since in charge balance it was quite small and closely matched by Cl^- accumulation (3–4 μequiv/g tissue in 48 h). Compared to beet root which shows a cation excess of 34–130 μequiv in uptake over that of Cl^- (in 2 mM KCl) (POOLE and POEL, 1965) the

Table 1

Net uptake of cations and anions by pea epicotyl segments during the experimental period[a]

Solution	Time from excision (h)	Net uptake, μequiv/g tissue					
		K^+	Total cations	NO_3^-	$SO_4{}^{2-}$	$H_2PO_4^-$	Total anions
1X	12	4	4	3	0	4	7
	24	17	17	12	4	8	24
	36	30	30	20	7	13	40
	48	43	43	29	10	18	57
10X	16	19	19	10	(2)[b]	6	18
	24	33	33	18	(3)	9	30
	32	47	47	25	(5)	13	43
	40	61	61	33	(6)	17	56
	48	75	75	41	(7)	20	68

a Data on Na^+, Ca^{2+}, and Mg^{2+}, and on Cl^- are excluded since the sum of the net uptake of these cations rather closely matched Cl^- net uptake.

b Numbers in parentheses were estimated from data of HIGINBOTHAM et al. (1967).

cation excess reported here is quite small. It could be accounted for readily by NO_3^- reduction or conversion of $H_2PO_4^-$ to organic phosphates.

The time course of uptake in 1X solution is shown in Fig. 1. There is a lag in uptake during a period of about 12 h after excision, then a steady rate of absorption extending over many hours (MACKLON and HIGINBOTHAM, 1968; PIERCE, 1971). This lag phase may be prolonged by low temperature and maintained indefinitely by cycloheximide which inhibits protein synthesis (HENDRIX, 1973). During net uptake the cation-anion balance is as shown in Table 1.

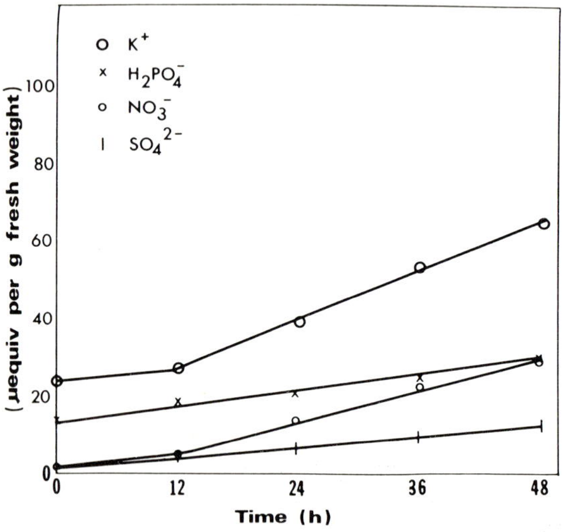

Fig. 1 Content of K^+, $H_2PO_4^-$, NO_3^-, and SO_4^{2-} during the first 48 h following excision.

III. Cation-Anion Balance in Various Salt Mixtures

In addition to evaluating charge balance in essentially complete nutrient solutions (1X and 10X), single salts and various combinations were used; in all cases some Ca^{2+} was present. The results of these experiments show that K^+ uptake is limited by the anions present. With SO_4^{2-} only, net K^+ uptake is markedly reduced compared to a solution containing NO_3^- + SO_4^{2-} (Fig. 2); greatest K^+ uptake is in 1X solution (or 10X). Furthermore, in a mixture containing NO_3^- and SO_4^{2-} a close balance of cation and anion uptake occurred (Table 2). It appears that in pea epicotyl there is a specific rate of uptake for each anion and that the sum of anion uptake determines the pace of K^+ net flux. With respect to uptake rates the sequence is $NO_3^- > H_2PO_4^- > SO_4^{2-}$ (Fig. 1); Cl^- uptake, not shown, is smallest.

Another feature shown in Fig. 2 is that the shortest lag period occurred in the nutrient mixture. In the presence of only some of the anions the lag phase was prolonged. This is consistent with the hypothesis that a transport system is induced during the lag phase and that it involves specific sites for each anion.

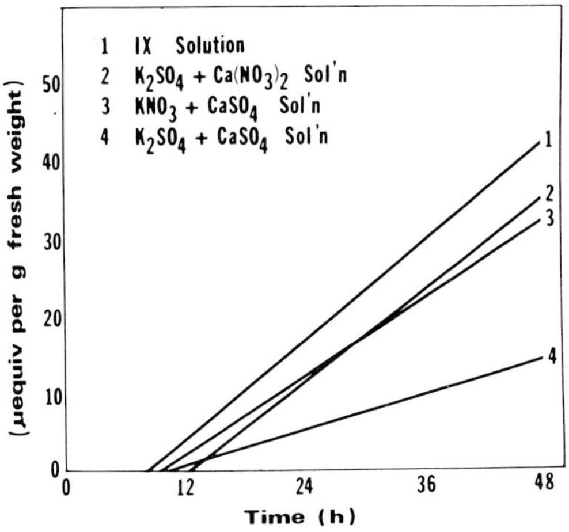

Fig. 2 Net uptake rates of K^+ in each of several solutions. Initial values of K^+ content were subtracted and the slopes calculated by the least squares method.

Table 2
Net uptake of cations and anions by pea epicotyl segments

Solution	Time from excision (h)	Total cations (µequiv/g)	Total anions (µequiv/g)
$KNO_3 + CaSO_4$	12	2	3
	24	13	13
	36	23	23
	48	33	34
$K_2SO_4 + Ca(NO_3)_2$	12	0	1
	24	12	14
	36	24	26
	48	36	38

IV. The Relationship of Cell Electropotentials to K^+ Uptake

Immediately following excision the cell electropotentials, PD's, are quite low but after a few hours of immersion, approximately coincident with the lag phase, the PD's increase rapidly (Fig. 3) (MACKLON and HIGINBOTHAM, 1968; PITMAN et al., 1971). Thus, the increase in cell PD is associated with the induction of ion transport mechanisms, not with the increase in K^+ internally; however, after the lag phase, when the tissue enters a period of constant absorption, there is a less rapid but steady increase in cell PD. The latter is

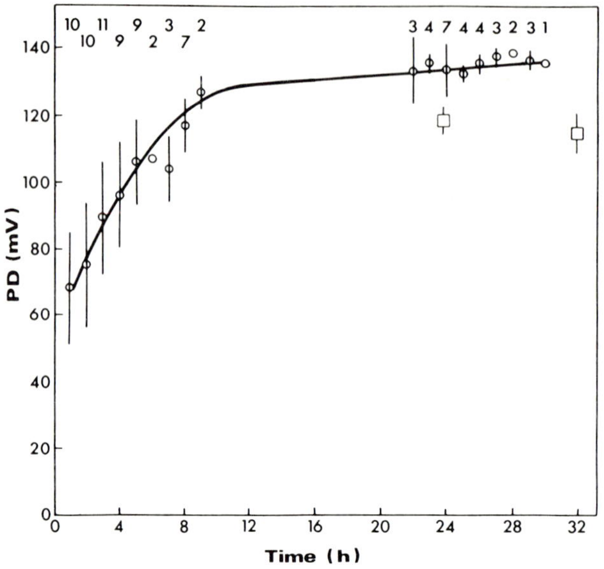

Fig. 3 The change in cell electropotential following excision; the number of cells measured at each time interval is indicated. The squares represent cells in a solution containing 0.5 mM K_2SO_4 plus 1 mM $CaSO_4$. Vertical bars represent the standard deviation.

apparently correlated with the increase in internal K^+ and explicable in terms of the Goldman equation as a passive diffusion with K^+ influx being driven by the electropotential gradient.

A large portion of the cell PD (40–60 %) is electrogenic in origin and may be blocked by inhibitors, e.g. CN^- or dinitrophenol (HIGINBOTHAM et al., 1970). Whether the induced anion influx pumps directly generate this portion of the PD or are coupled with another generating mechanism is not known. The properties of the electrogenic pump have been summarized recently (HIGINBOTHAM and ANDERSON, 1974). At present the anion transport system cannot be excluded as an electrogenic mechanism.

An interesting feature of the induced ion uptake mechanism is that it will develop in distilled water, i.e. in the absence of external ions (MACKLON and HIGINBOTHAM, 1968). Further, tissue in 0.5 mM K_2SO_4 plus 1 mM $CaSO_4$, in which there is relatively little K^+ uptake, develop a cell PD almost as great as that in 1X solution (Fig. 3). Thus, the magnitude of the PD is not dependent directly on the rate of ion influx; this is consistent with the model of electrogenesis now envisaged (HIGINBOTHAM and ANDERSON, 1974).

D. Acknowledgements

The authors wish to acknowledge the support of National Science Foundation Grants GB 5117X and GB 19201 to N. Higinbotham.

E. References

HENDRIX, D.L., The Effect of Selected Inhibitors on Ion Transport in ***Pisum sativum*** Epicotyl Tissue. Doctoral thesis, Washington State University, Pullman, WA., USA (1973)
HIATT, A.J., Plant Physiol. **42**, 294–298 (1967)
HIGINBOTHAM, N., ETHERTON, B., FOSTER, R.J., Plant Physiol. **42**, 37–46 (1967)
HIGINBOTHAM, N., GRAVES, J.S., DAVIS, R.F., J. Membrane Biol. **3**, 210–222 (1970)
HIGINBOTHAM, N., ANDERSON, W.P., Can. J. Bot., in press (1974)
JACOBSON, L., ORDIN, L., Plant Physiol. **29**, 70–75 (1954)
MACKLON, A.E.S., HIGINBOTHAM, N., Plant Physiol. **43**, 888–892 (1968)
PIERCE, W.S., Ion Accumulation, Charge Balance, and Cell Electropotentials in Pea Epicotyls and Roots. Doctoral thesis, Washington State University, Pullman, WA., USA (1971)
PITMAN, M.G., MERTZ, S.M., JR., GRAVES, J.S., PIERCE, W.S., HIGINBOTHAM, N., Plant Physiol. **47**, 76–80 (1971)
POOLE, R.J., POEL, L.W., J. Exp. Bot. **16**, 453–461 (1965)
TORII, K., LATIES, G.G., Plant Cell Physiol. **7**, 395–403 (1966)
ULRICH, A., Am. J. Bot. **28**, 526–537 (1941)

Heavy Metals and Sulphhydryl Reagents as Probes of Ion Uptake in Pea Stem

D.L. Hendrix and N. Higinbotham

A. Introduction

Is K^+ actively or passively absorbed? Although there are proponents for an active uptake mechanism in many cells, the electrical gradient is of a magnitude which would permit a passive influx sufficient to explain K^+ accumulation (HIGINBOTHAM 1973a, 1973b). About 60 % of the cell electropotential in pea epicotyls is inhibitor sensitive and is thus thought to result from a metabolically driven ion pump, i.e. an electrogenic pump (HIGINBOTHAM et al., 1970). Since the cell interior is negative the net effect of such an electrogenic pump would be to drive cations inward or anions outward.

Using microelectrodes with higher plant cells (in solutions containing more than 1 mM K^+) ETHERTON and HIGINBOTHAM (1960) and ETHERTON (1963) found (1) that the cell electropotentials, PD's, (vacuole to exterior) exceed any values predicted by the Nernst equation and (2) that metabolic inhibitors, e.g. CN^- and 2,4-dinitrophenol, rapidly and reversibly depolarized the cells (HIGINBOTHAM et al., 1970). While metabolic inhibitors can be used to depolarize plant cells and also impair the uptake of mineral ions and oxygen, they have the disadvantage of having a mitochondrial site of action; thus the inhibitor enters the cytoplasm and disrupts metabolic processes required by a wide range of systems besides ion uptake. It is appropriate, therefore, to examine the effects of reagents which might have a selective effect on the ion uptake process while not affecting respiration.

Heavy metals might accomplish this objective by blocking cation recognition sites (HIGINBOTHAM, 1973a) in the plasmalemma. Of course, heavy metals also interfere with cytoplasmic processes but if added in low concentrations or in conjunction with ionic organic groups their actions may be primarily at the limiting membrane of the cell (MADSEN, 1963; ROTHSTEIN, 1970). An 'ionic organic group' also increases the specificity of reactions in which the heavy metal atom can participate (MADSEN, 1963; ROTHSTEIN, 1970).

Several heavy metal ions react selectively with sulphhydryl groups to form mercaptides; at low ratios of inhibitor to sulphhydryl, the actions of these reagents are specific for thiol groups and disulphide bonds (ROTHSTEIN, 1970; BOYER, 1954). These reagents have long been known to interfere with cation movement in intact cells and membrane-bound enzymes (ROTHSTEIN, 1970; WEBB, 1966a).

The N-substituted maleimides also react with sulphhydryl groups but in a more specific and less reversible manner than heavy metals. Being small and nonpolar, they penetrate membranes more readily than the larger ionic inhibitors and form double bond addition products with thiols. Unlike the heavy metals they do not react with disulphide bonds (WEBB, 1966a, 1966b).

B. Materials and Methods

Pea seedlings (***Pisum sativum*** L. cv Alaska) were grown in darkness for 7 days at 25°C on vermiculite watered with a solution termed IX which has the following composition in mM: KCl, 1.0; NaH_2PO_4, 0.904; Na_2HPO_4, 0.048; $Ca(NO_3)_2$, 1.0; $MgSO_4$, 0.25. The pH of this solution is 5.8. Whenever silver was used, the above solution was modified by substituting KNO_3 for KCl. Epicotyl segments were taken 2 cm below the apical hook and pretreated in nutrient solution for 10 to 14 h before use. All work after excision was carried out at 20°C and all work with silver was performed in the dark or in dim light.

The uptake of K^+ and Cl^- by the stem segments was followed essentially as described elsewhere (HENDRIX and HIGINBOTHAM, 1973) using $^{42}K^+$ and $^{36}Cl^-$, respectively. Aliquots of tissue extracts were dried in planchettes and counted by conventional gas flow techniques. The $^{36}Cl^-$ extracts were made basic with Tris before drying to prevent the loss of $^{36}Cl^-$ as $H^{36}Cl$. Other extract aliquots were analyzed for K^+ by flame emission spectrophotometry.

The respiration of similar stem segments was measured with a Gilson respirometer. In most cases the respirometer flasks were lined with a plastic film (Beckman Desicote) to prevent adsorption of the inhibitor on the glass surface. Plastic containers were utilized wherever possible to minimize adsorption.

C. Results

I. Effect of Sulphhydryl Reagents on Ion Uptake and Respiration

Both mercury and silver have similar effects on ion uptake and respiration in pea stem cells (Figs. 1A, 1B). With both of these reagents, low concentrations stimulate K^+ uptake. Slightly higher concentrations strongly depress K^+ uptake and stimulate respiration, and still higher concentrations either depress respiration or stimulate less strongly. The relative stimulation of respiration by mercury is almost as great as that caused by the uncoupling agent CCCP (HENDRIX, unpublished); indeed, it appears that a high concentration of either Ag^+ or Hg^{2+} induces an uncoupling of respiration from K^+ uptake. Also, there appears to be a concentration of either Ag^+ or Hg^{2+} at which K^+ uptake is strongly depressed while respiration is only slightly affected. It is hypothesized that at these concentrations (3 μM Hg^{2+} and 0.6 μM Ag^+) the heavy metal ion is acting primarily at the plasma membrane, supposedly at the site of the K^+ uptake mechanism.

To test the effect of mercury in a form having limited penetration, p-chloromercuribenzoate (PCMB) and p-chloromercuribenzene sulphonic acid (PCMBS) were used (Figs. 1C, 1D). Both of these reagents blocked K^+ uptake but required concentrations roughly an order of magnitude higher than that of mercuric chloride to produce a similar depression of K^+ uptake. They had little effect on respiration, even at concentrations exceeding 100 μM.

Both PCMB and PCMBS appear to have less effect on the uptake of Cl^- by pea stem cells than on the uptake of K^+. This is consistent with the idea that they are blocking cation uptake sites, although other explanations are possible. The fact that K^+ accumulation and $^{42}K^+$-labelled K^+ influx in the presence of PCMBS decrease with time (Figs. 2A, 2B), while control tissue exhibits a linear K^+ uptake for several days (PIERCE, 1971) suggests that PCMBS impairs K^+ influx in pea stem cells.

N-ethylmaleimide (NEM) was found to have an effect on K^+ uptake similar to that of PCMB (Fig. 2C) but unlike PCMB or PCMBS, it appears to stimulate respiration at concentrations which depress K^+ uptake. Here again, there appears to be a concentration at which respiration is not altered but K^+ uptake is depressed, as if NEM is having an extra-mitochondrial effect.

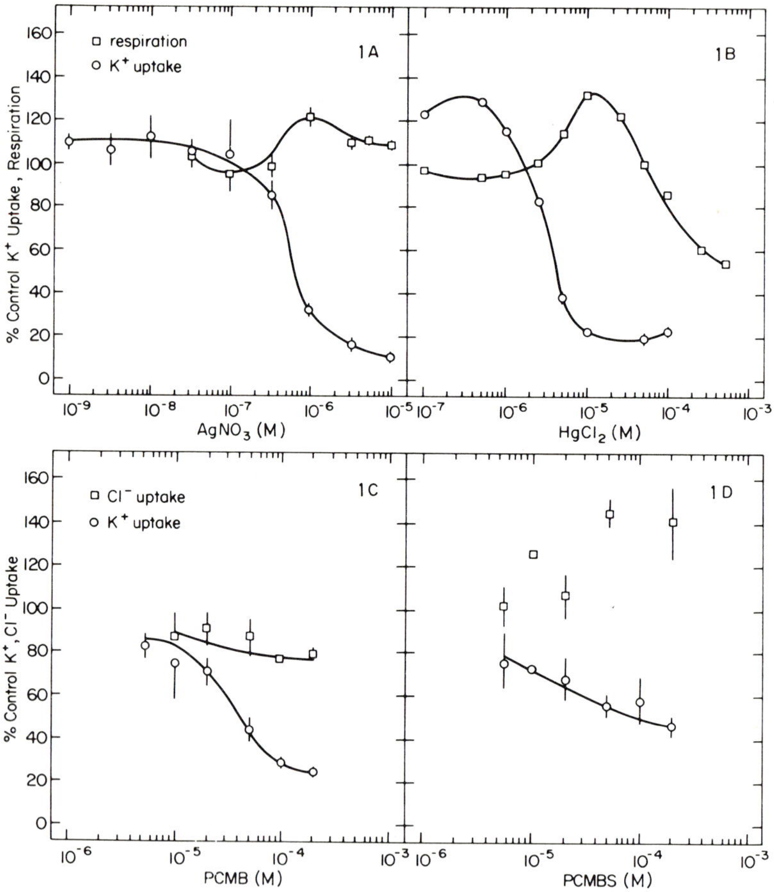

Fig. 1 Effects of Ag^+, Hg^{2+}, PCMB and PCMBS on ion uptake and respiration in etiolated pea stem segments. Tissue was treated for 3.0 h in all measurements. Points represent the means of three or four replicates and bars, where shown, stand for standard deviations.

Thallium (Tl^+) was utilized to see what effect a heavy metal would have which did not form the extremely insoluble mercaptides formed by mercury and silver ions. Thallium, which is known to be an inhibitor of cell division and growth in higher plants (HEWITT and NICHOLAS, 1963), appears to stimulate K^+ uptake at low concentrations and have little effect on respiration in etiolated pea stem segments.

II. Kinetic Analysis of Ag^+ Inhibition

To further test the hypothesis of a heavy metal reaction at or near cation uptake sites, a Lineweaver-Burk plot was prepared from the inhibition of K^+ uptake by Ag^+ (Fig. 2D). In several replications of this experiment it was found that the data most closely approach the type of inhibition termed competitive.

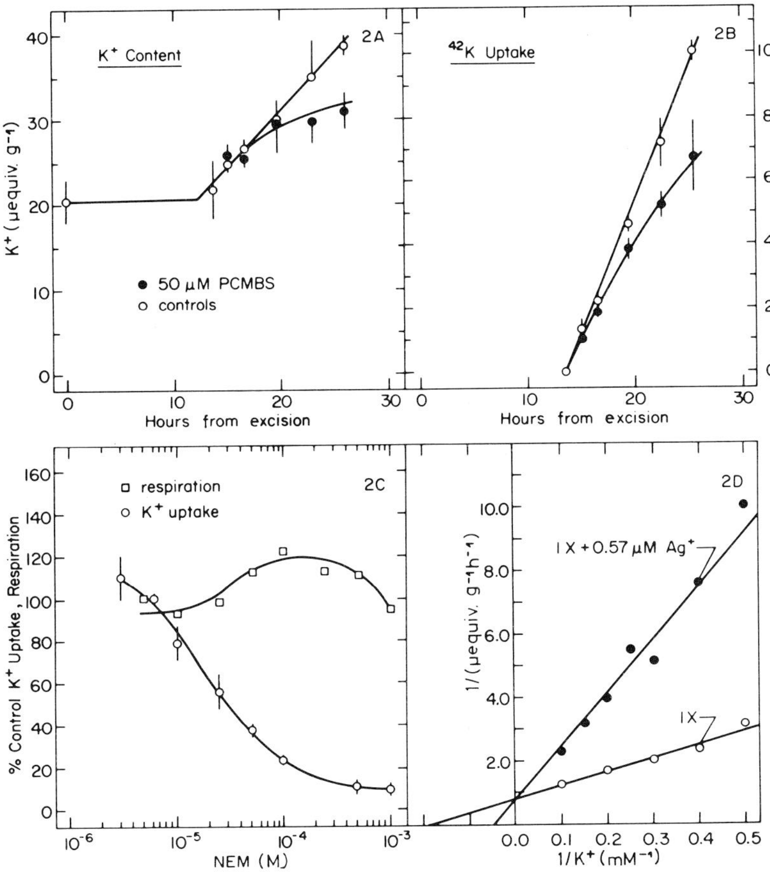

Fig. 2 A., B. Effects of PCMBS on the time course of K^+ content and K^+ uptake in pea stem. C. The effect of NEM on respiration and K^+ uptake. D. A Lineweaver-Burk plot of K^+ influx versus Ag^+. Pea stem cells were treated for 3.0 h. Points represent the means of four replicates and bars, where shown, indicate standard deviations.

III. Cell Electropotential Measurements

These reagents had almost no effect on cell electropotential when investigated by the techniques described elsewhere (PITMAN et al., 1971). They definitely do not produce sudden or marked changes in cell electropotential similar to those induced by CN^-, but two main problems hindered our evaluation of the data: (1) these reagents take a matter of hours to act (Figs. 2A, 2B) while respiratory inhibitors act relatively quickly (HIGINBOTHAM et al., 1970), and (2) silver, in particular, can plug the tips of KCl microelectrodes with AgCl. To overcome these difficulties, as many cells as possible were measured over a period of several hours and silver was only added to a cell previously impaled so that the tip of the electrode would not come in contact with added silver. Electrodes filled with KNO_3 were also tried in silver experiments (MULLINS and MOORE, 1960). In spite of the problems involved, it seems clear that these reagents have a much more pronounced effect on K^+ uptake than on the transvacuolar electropotential in pea stem cells.

D. Discussion and Conclusions

Current theories about ion uptake in higher plant cells ascribe to the mitochondrion the high energy compounds which move to the plasmalemma where they are utilized by ion pumping mechanisms. There are many ways of investigating this hypothesis, one of the more common methods involving poisoning the mitochondria and thus eliminating the energy source for the ion 'pump'. What we propose here is an alternate approach involving materials which block the ion recognition (or pump) sites in the plasma membrane. It should be emphasized that although specific pathways for K^+ movement may be required this need not require energy for transport.

If used at certain concentrations, sulphhydryl reagents appear to be able to cause such a blockage, as witnessed by their blockage of K^+ influx while not disturbing respiration. The fact that K^+ absorption was competitively inhibited by Ag^+ seems to indicate that the sulphhydryl groups affected are at or near the K^+ uptake site. However, sulphhydryl groups are so ubiquitous in living systems that data involving intact cells must be viewed with caution.

It is interesting to note that the K^+- Mg^{2+}-stimulated ATPase from the plasmalemma of oat roots which has recently been implicated in ion uptake (LEONARD and HODGES, 1973; HODGES, 1973) is also extremely sensitive to Hg^{2+}, Ag^+ and PCMBS but is quite insensitive to NEM (HENDRIX and HODGES, unpublished). This enzyme, however, appears to show uncompetitive kinetics with respect to K^+ concentration when poisoned with either Hg^{2+} or Ag^+.

Finally, this selective blockage at the plasmalemma may allow one to distinguish between those ions which are and those which are not pumped by an electrogenic mechanism. Since K^+ uptake is depressed in these cells by PCMB and PCMBS at concentrations which do not affect cell electropotential or Cl^- uptake, it is postulated that K^+ movement is not electrogenic in pea stem tissue; however, we do not rule out the possibility of a neutral metabolic K^+ mechanism, particularly since K^+ influx in pea stem is strongly depressed by metabolic inhibitors (HIGINBOTHAM et al., 1970).

E. Acknowledgements

This research was supported by the National Science Foundation under Grant GB 19201 to N.H. and a fellowship from the National Defense Education Act, Title IV to D.L.H.

F. References

BOYER, P.D., J. Am. Chem. Soc. **76**, 4331–4337 (1954)
ETHERTON, B., Plant Physiol. **38**, 581–585 (1963)
ETHERTON, B., HIGINBOTHAM, N., Science **131**, 409–410 (1960)
HENDRIX, D.L., HIGINBOTHAM, N., Plant Physiol. **52**, 93–97 (1973)
HEWITT, E.J., NICHOLAS, D.J.D., in Metabolic Inhibitors: A Comprehensive Treatise (eds. R.M. Hochster, J.H. Quastel), Vol. 2, pp. 311–436, Academic Press, London and New York 1963
HIGINBOTHAM, N., Bot. Rev. **39**, 15–69 (1973a)
HIGINBOTHAM, N., Ann. Rev. Plant Physiol. **24**, 25–46 (1973b)
HIGINBOTHAM, N., GRAVES, J.S., DAVIS, R.F., J. Membrane Biol. **3**, 210–222 (1970)
HODGES, T.K., Advan. Agron. **25**, 163–207 (1973)
LEONARD, R.T., HODGES, T.K., Plant Physiol. **52**, 6–12 (1973)
MADSEN, N.B., in Metabolic Inhibitors: A Comprehensive Treatise (eds. R.M. Hochster, J.H. Quastel), Vol. 2, pp. 119–143, Academic Press, London and New York 1963

MULLINS, L.J., MOORE, R.D., J. Gen. Physiol. **43**, 759–773 (1960)

PIERCE, W.S., Ion Accumulation, Charge Balance and Cell Electropotential in Pea Epicotyls and Roots. Ph.D. Thesis, Washington State University, Pullman (1971)

PITMAN, M.G., MERTZ, S.M., JR., GRAVES, J.S., PIERCE, W.S., HIGINBOTHAM, N., Plant Physiol. **47**, 76–80 (1971)

ROTHSTEIN, A., in Current Topics in Membranes and Ion Transport (eds. F. Bonner, A. Kleinzeller), Vol. 1, pp. 135–176, Academic Press, London and New York 1970

WEBB, J.L., Enzyme and Metabolic Inhibitors, Vol. 2, Chapters 4.7, Academic Press, London and New York 1966a

WEBB, J.L., Enzyme and Metabolic Inhibitors, Vol. 3, Chapter 3. Academic Press, London and New York 1966b

Effect of Gas Partial Pressure on the Electrogenic Mechanism of Plant Cell Membranes

K. Ichino, K. Katou, and H. Okamoto

A. Introduction

The distribution of surface electric potential difference (PD) along the hypocotyl of a seedling is correlated with the distribution of elongation rate (OKAMOTO, 1955). An electric potential minimum corresponds with the elongating zone, and it migrates acropetally with the growth of the axial organ (OKAMOTO, 1955; ICHIMURA and OKAMOTO, 1958) (Fig. 1A).

A hypothesis explaining this correlation has been worked out and reported elsewhere (OKAMOTO, 1955).

Further experiments produced two important results.

1) When a hypocotyl was cut into several pieces (each 10 mm in length), each segment maintained a stable electric PD between its cut ends for up to several hours. Moreover, the over-all potential distribution pattern could be reconstructed by combining each PD in series. This pattern was essentially the same as that observed on an intact hypocotyl (KATOU and OKAMOTO, 1970). (This means that the distribution of surface PD on an intact hypocotyl is mainly determined by the axial PD in the organ, and is not greatly dependent on the differences in radial PD in each part.)

2) The hydrogen ion in xylem sap is the only ion whose activity distribution had a close relationship with the axial distribution of electric potential (see curve A in Fig. 1B) (KATOU and OKAMOTO, 1970).

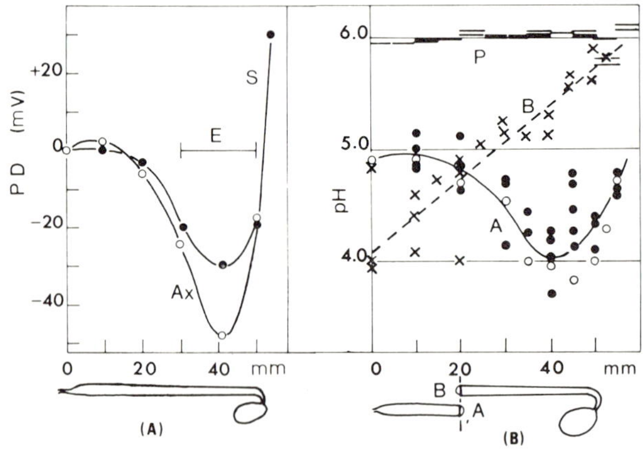

Fig. 1 (A) Distribution of electric potential on the surface (S) of the hypocotyl of *Vigna sesquipedalis*. E = elongating zone; Ax = axial PD.
(B) Distribution of pH along hypocotyl; A = acropetal bleeding sap: ● from KATOU and OKAMOTO (1970), ○ by the late M. Hayashi; B = basipetal bleeding sap (x); P = tissue homogenate.

These results gave rise to the following considerations.
1) Is the axial PD in a hypocotyl segment a result of an unequal distribution of ions in the segment, which in turn results from the ion transport activity of the intact embryo as a whole? Or does each segment have its own intrinsic electrogenic activity? If the latter is the case, what is the cellular basis of this phenomenon?
2) There is a pH difference of 2 units between xylem sap and parenchyma in the most rapidly elongating zone (curves A and B in Fig. 1B). This region can be regarded as the 'absorption centre' of this organ, for the most rapid entry of K^+ and water takes place here (KATOU and OKAMOTO, 1970). What is the meaning of this sharp pH difference in this particular region and what is the mechanism that maintains such a pH difference?

One possible answer to these problems is that there exists in the elongating zone an active H^+ transport from parenchyma into xylem, along which an H^+ concentration gradient is generated. This H^+ concentration difference may then generate an axial PD.

To examine the validity of this hypothesis, the responses of both axial and radial PD's of hypocotyl segments to changes in the partial pressures of O_2 and/or CO_2 in the surrounding atmosphere were examined and these results were compared with similar responses of the membrane potential of both leaf cells and cells in bean hypocotyl.

B. Materials and Methods

The hypocotyl segments were cut from seedlings of ***Vigna sesquipedalis*** 3 days after germination at 25°C. Leaf tissue of mature plants of ***Antirrhinum*** as well as plumule and hypocotyl slices of ***Vigna sinensis*** were used as the material in the experiments on cell membrane potential.

The measurement and recording of axial PD across a hypocotyl segment under various gas partial pressures were carried out with a system described elsewhere (ICHINO et al., 1973). The radial PD was measured in an Acrylite chamber containing two electrodes. One electrode made contact with the cut end of the segment, and the other electrode made contact with the outer surface of the segment, near the first electrode. In both chambers the gas mixture could replace all the air in a few seconds.

The membrane potential of a cell in a tissue was measured by the ordinary microelectrode technique. The tissue was mounted on the end of a glass rod with silicone grease, and was covered with a tiny droplet of working solution in which the tip of a reference electrode was dipped. The mounted tissue was enclosed in a Perspex chamber in which the gas phase could be quickly replaced.

The composition of the standard working solution was KCl, 0.67 mM; $CaCl_2$, 2.0 mM; mannitol 0.4 M; KH_2PO_4, 0.33 mM; $MgSO_4$, 0.3 mM; pH 5.6.

C. Results and Discussion

I. The Typical Response of the Axial and Radial PD's of a Hypocotyl Segment to Changes in Gas Partial Pressure

1) With a segment excised from the zone of a hypocotyl just below the absorption centre, the observed response to anoxia is shown in Fig. 2. After the air in the Acrylite chamber had been quickly replaced by N_2 or He, there was always a definite lag period (τ) before a sudden PD drop took place. The PD disappeared nearly completely within a short time. τ could be shortened by raising the chamber temperature. The apparent activation energy of the process characterized by $1/\tau$ was 18 kcal/mole between 14 and 30°C. Recovery from anoxia started immediately after the start of re-aeration and was complete in a few minutes at room temperature. The relation between the maximum rate of recovery of the PD and the

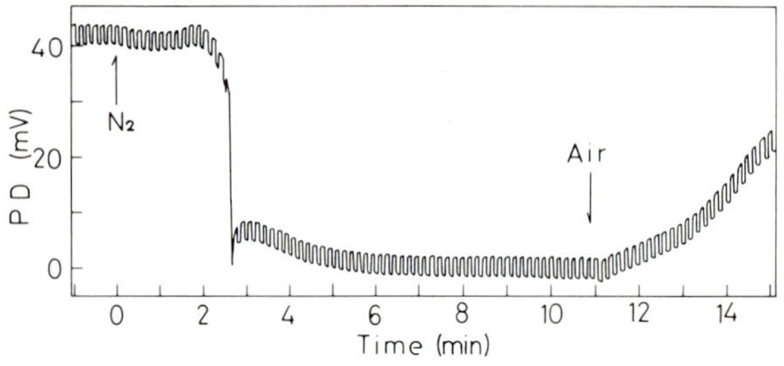

Fig. 2 Effect of anoxia on the axial PD of a hypocotyl segment. Rectangular current pulses of $2 \cdot 10^{-8}$ amp, 10 s duration, were applied through the segment.

oxygen concentration in the re-aeration gas mixture was of a Michaelis-Menten type, for which the apparent K_m was $2.1 \cdot 10^{-5}$ M O_2 (ICHINO et al., 1973), being of the same order as that of cytochrome oxidase in higher plants. Therefore a quite low partial pressure of O_2 (e.g. 2 %) is enough to maintain the normal PD level. The axial resistance did not change during the rapid change in potential. We can conclude that each segment has its intrinsic potential-generating activity which is dependent on aerobic metabolism.

2) Typical responses of PD to changes in the CO_2 partial pressure are shown in Fig. 3. The axial PD across a segment fell immediately after the increase of CO_2 pressure in spite of the presence of a normal supply of oxygen. The decrease showed an overshoot and a small, slow recovery. Simultaneously, the radial PD, which was normally quite small, increased (in this case by up to 60 mV, outside positive).

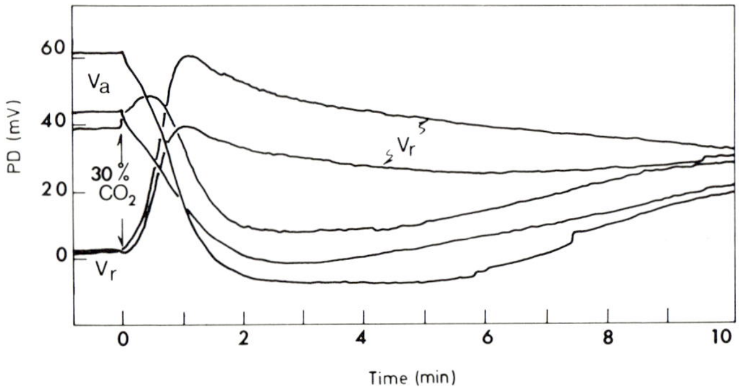

Fig. 3 Effect of high concentration of CO_2 (30 % in air) on both axial (V_a; 3 experiments) and radial (V_r; 2 experiments) PD of a hypocotyl segment. In this case part of the air was replaced by CO_2; similar results were obtained when O_2 tension was kept constant (e.g. 80 % CO_2 + 20 % O_2).

II. Typical Responses of Membrane Potential of Cells in Leaf Tissue to the Change in Gas Partial Pressure

Most experiments were conducted in the dark so as to eliminate the influence of photosynthesis. The plumule of *V. sinensis* gave results similar to those obtained with the mature leaf of *Antirrhinum*.

1) Usually a lag period, similar to the above, appeared before the potential drop due to anoxia took place. In 1 mM KCl working solution there always remained a 'residual potential' of about 60 mV. However, if the external KCl concentration was raised beforehand to 80 mM, equal to the overall K^+ concentration in the leaf tissue, almost all the resting potential disappeared under anoxia. The recovery from the effect of anoxia always took place immediately after re-aeration started in both cases.

Thus, the 'dark potential' of our leaf cell membrane comprises two fractions; one is an 'active fraction' dependent on aerobic metabolism and the other is a 'passive fraction' dependent on passive diffusion (mainly of K^+).

Although a hypocotyl segment responded to anoxia in much the same way as a leaf cell, the easily accessible hypocotyl parenchyma cell did not repond to anoxia. There thus seems to be a possibility that a hypocotyl consists of two different groups of cells: a quite special anoxia-sensitive cell group (possibly some of the parenchyma cells which are responsible for the radial active transport of H^+ into the xylem) and the majority of the cells which do not respond.

2) The responses of the membrane potential of leaf cells to high concentrations of CO_2 were rather more complex than the responses to anoxia. Two main types of reactions were observed (Figs. 4A and B), but sometimes intermediate types were observed. Where the

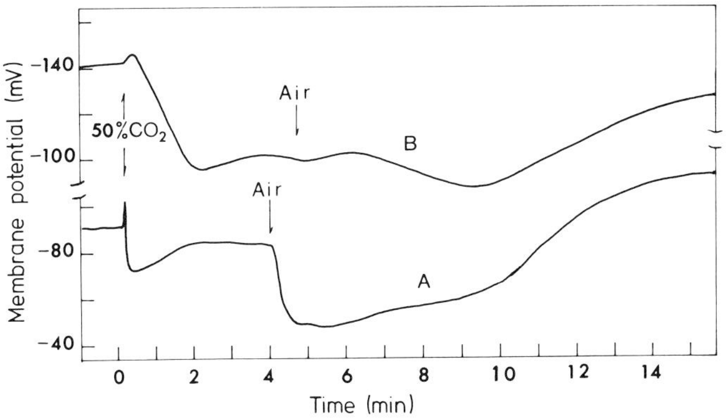

Fig. 4 Effect of high concentration of CO_2 (50 % in air) on the membrane potential of leaf cells in a mature plant of ***Antirrhinum***. K^+ concentration was 1 mM. (For explanation of curves A and B see text.)

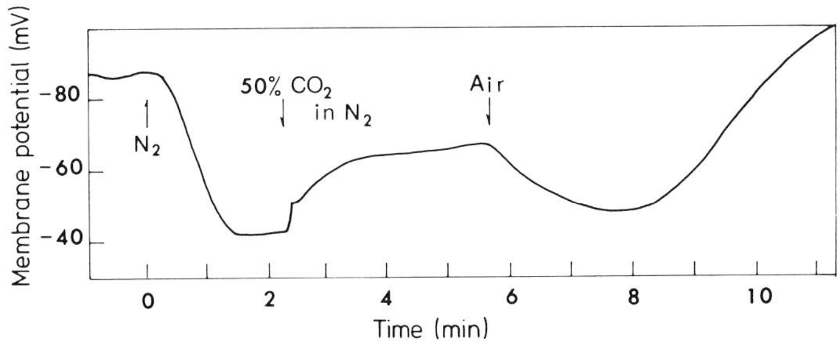

Fig. 5 Effect of high concentration of CO_2 and anoxia on the membrane potential of leaf cells in a mature plant of ***Antirrhinum***. KCl concentration was 1 mM. (Compare with Fig. 4.)

membrane potential did not at first drop very much (type A) a further drop took place when the CO_2 pressure was restored to normal. On the other hand, where the membrane potential showed a larger decrease in CO_2, no further significant drop occurred when the CO_2 was removed (type B). In both cases the total potential drop was approximately the same as that caused by anoxia. In a working solution of 80 mM KCl, CO_2 brought about an ordinary response of type A. Under anoxia, the response to CO_2 was a hyperpolarization (Fig. 5).

In those parenchyma cells of bean hypocotyl which lacked any response to anoxia, CO_2 brought about a quite simple reversible hyperpolarization reaction, which was similar to the response of the leaf cell to CO_2 under anoxia. This CO_2 effect at the cellular level may help us to understand the CO_2 effect on the radial PD of the hypocotyl segment (PD becoming outside positive) which may involve a decrease of radial transport of H^+ from the parenchyma into the xylem.

D. Acknowledgements

The authors are very grateful both to Dr. J. Thain of the University of East Anglia and to Professor T. Mori of Nagoya University for their valuable discussions. They are also much obliged to Dr. K. Oda for his helpful advice and to Mr. P. Linstead for his skillful technical assistance. The mechanical workshop in both Universities and the electronic workshop of the University of East Anglia afforded them indispensable aid in constructing the apparatus for these experiments.

E. References

ICHIMURA, K., OKAMOTO, H., Bot. Mag. Tokyo, **71**, 201–207 (1958)
ICHINO, K., KATOU, K., OKAMOTO, H., Plant Cell Physiol. **14**, 127–137 (1973)
KATOU, K., OKAMOTO, H., Plant Cell Physiol. **11**, 385–402 (1970)
OKAMOTO, H., Bot. Mag. Tokyo **68**, 1–13 (1955)

Round Table Discussion 8

Chairman: N. Higinbotham

The first part of this discussion was about plasmodesmata. LÜTTGE pointed out that they are very typical plant structures which, in a sense, arise because it is impossible for the plasma membranes of two adjacent plant cells to come very close together: they are separated by cell walls. He contrasted plant cells with animal cells, which have a different mechanism; they make 'tight junctions' when they want ready communication. The electrophysiological evidence, his own and that of others, shows good electrical contact between plant cells when plasmodesmata are present and there can thus be no closed membrane barriers across the plasmodesmata. WALKER agreed that the electrical contact was quite good, but he referred to an apparent discrepancy between the speed of diffusion of Cl^- through Characean plasmodesmata, as measured in Sydney and Toronto, and the electrical conductance as measured by Spanswick. He thought the conductance might be as much as 50 times smaller than it ought to be from the Cl^- experiments. He did not know what the answer was to this problem; it may be that there are mechanisms other than diffusion in the plasmodesmata. However, despite this discrepancy, there is still good electrical contact between cells.

BENTRUP pointed out that there seem to be two different types of plant cells and tissues with respect to their transmission of electrical signals. In cells such as ***Chara*** and ***Acetabularia***, lighting one part of a cell just produces an electronic spread to the rest of the cell, whereas in leaves of the moss, ***Hookeria***, and of LÜTTGE's experimental plant, and in his own single rhizoid cells, the spread is not purely electronic; it is as if there was something moving faster than any ion could diffuse and he raised the question of the nature of this signal relay. RAVEN partly referred to this question when he pointed out the local apparent 'action–at–a–distance', effects of light irradiation of parts of the cells of ***Mougeotia***, on the one hand, and ***Chara*** or ***Nitella*** on the other hand. In both these cases there is transmission of something, locally, between a patch of membrane and a chloroplast. But the times involved in these local transmissions are unknown.

Both ZIEGLER and HIGINBOTHAM asked questions about the effect of plasmolysis and deplasmolysis on plasmodesmata, the former on the question of the transmission of electrical signals and the latter on their subsequent reconstitution, if any. Apparently the answers to these questions are not known.

LÜTTGE again spoke of Brinckmann's experiments on transmission of electrical signals between green and white cells in the leaves of the variegated photosynthetic mutant of ***Oenothera***. He said the cycloheximide, while not inhibiting photosynthesis, stopped this electrical signal and he wondered whether, in the light of VREDENBERG's chloroplast potential measurements, cycloheximide stopped the signal from getting out of the chloroplasts in the green cells. HANSON asked whether cycloheximide abolished most of the membrane potential in the leaf cells, as it did in his maize root cells. It did not affect the −140 mV potential in the ***Oenothera*** leaf cells, according to LÜTTGE.

KUIPER then asked about the effect of hydrostatic pressure difference across plasmodesmata. He referred to the fact that if you puncture one of a pair of ***Chara*** internodal cells the other one does not lose hydrostatic pressure. He also raised the question of the relevance of the pressure theory of flow across sieve plates. ZIEGLER pointed out that one can get pressure flow of solution through plasmodesmata, for the pores in the 'sieve plates' of ***Laminaria*** are really plasmodesmata and here exudation from a cut ***Laminaria*** can go on for half an hour or more. Thus the instant occlusion of the plasmodesmata in ***Chara*** does not always occur.

HERTEL next spoke about auxin transport in coleoptiles: whether it is via the symplasm or from cell membrane to cell membrane through the apoplast. He quoted experiments, by Goldsmith and Ray, using coleoptiles in which the protoplasm had been moved either to the basal or to the apical ends of cells by centrifugation. These experiments and also the use of inhibitors in 'mid-stream' and some experiments of his own on rice coleoptiles all seemed to be difficult to reconcile with transport of auxin in the symplasm. But he did not doubt the reality of symplasmic transport in ARISZ's experiments on ***Vallisneria***.

WALKER asked the electron microscopists whether they had found any association between the plasmodesmata and the endoplasmic reticulum and, if so, how would such an association survive in an actively streaming cell? ZIEGLER replied that some, but not all, plasmodesmata seem to be connected with elements of the endoplasmic reticulum. He emphasized the difficulties of getting a good interpretation because the endoplasmic reticulum is such a dynamic structure. LÄUCHLI added that he looked for such an association in his chloride localization work and was able to find some cases, in barley roots, in which the ER does extend through the plasmodesmata and furthermore chloride does precipitate in both structures. He agreed with WALKER, however, that protoplasmic streaming would tend to disrupt the connections between ER and plasmodesmata.

CRAM then spoke a little about (hormonal) feed back control in higher plants. He warned about putting arrows in models and calling them feed back and appealed for some quantitative data about this 'feed back business'. LÜTTGE thanked him and then PITMAN spoke about some work he participated in on salt glands in a mangrove tree. In nature the activity of the salt glands shows a strong diurnal rhythm, but the glands on an excised leaf do not. He wondered here whether the water stress, which also showed a strong diurnal rhythm, gave the signal to the glands. Such a signal would be extremely rapid.

Finally HIGINBOTHAM referred to the observations of Scott and others on waves of potential on coleoptiles and their possible association with auxins. OKAMOTO pointed out that now we know that auxin stimulates H^+ pumps in elongating cells we are closer to understanding the correlations between auxin movement and the voltage waves.

Session 9

Regulating Factors in Membrane Transport

Chairman: A. Läuchli

The Photomorphogenic Pigment Phytochrome: A Membrane Effector?

M. Weisenseel and W. Haupt

A. Introduction

When investigating the action of external or internal factors controlling plant growth and development, one is always faced with the problem of whether the primary effect is, roughly speaking, biochemical or biophysical in nature, i.e. either a regulation of enzyme synthesis or activity, or a change in cell structure and ultrastructural cell organization. In both hormone and light physiology, there is good evidence for biochemical as well as for biophysical processes as the primary effects. A very preliminary and rough separation can be derived from the kinetics of a response: whenever the response appears within minutes or even seconds, a purely biochemical effect is improbable. But premature generalization should be avoided.

I. Properties of Phytochrome

There are so many different phytochrome responses in plants that listing them would exceed the limits of this paper (cf. BORTHWICK, 1972). As only one example, de-etiolation of seedlings of dicotyledonous plants may be mentioned; in complete darkness the internodes grow very long, and this growth is inhibited by light; in contrast, there is very little growth of the leaves in the dark, and a light signal is needed for normal leaf growth. Besides these and other morphological effects, increases or decreases of the activities of several enzymes can be observed as an effect of light; and all these effects are mediated by absorption of light by the phytochrome system.

Phytochrome is known to be a chromoprotein, the chromophore of which is a tetrapyrrol derivative (MOHR, 1972). The pigment can exist in two forms with absorption maxima near 660 nm (red) and 730 nm (far-red); they are called, therefore, P_r and P_{fr} (or P_{660} and P_{730}). By light absorption either form is photoconverted to the other one, according to the spectral conversion cross section. Consequently, in saturating light conditions, we always find a photostationary state of P_{fr}/P_{total}. The fraction of phytochrome in the P_{fr} form depends on the wavelength. For example, in saturating red light (660 nm) the P_{fr}/P_{total} ratio is about 80 %, whereas in saturating far-red light this ratio can be as low as a few percent, decreasing from about 5 % (710 nm) to about 1 % (730 nm). These properties of phytochrome are reflected in the physiological responses. P_{fr} is the physiologically active form, whereas P_r is considered to be inactive. Hence, whenever a physiological process is under the control of phytochrome in the classical sense (e.g. de-etiolation), the effect can be induced by a brief pulse of red light, which produces active P_{fr} from inactive P_r; if, then, this red pulse is followed by far-red, the active form P_{fr} is photoconverted back to P_r thus cancelling the induction effect. This back and forth by red and far-red can be repeated as long as the response is still under phytochrome control.

The question now arises as to what is the primary effect of a phytochrome photoconversion. For instance can phytochrome act as a membrane effector? Different ways have been sought to tackle this problem.

II. Evidence for Phytochrome being Bound to Cytoplasmic Membranes

1) Analysis of dose-response relationships gives some clues to the interaction of phytochrome with structural sites. For instance certain responses show a threshold effect. In mustard seedlings the activity of the enzyme lipoxygenase is switched off if P_{fr} exceeds a certain level, and it is switched on if the P_{fr} level is sufficiently lowered. This response seems to be almost 'all—or—none' with a very low critical P_{fr} level (Fig. 1).

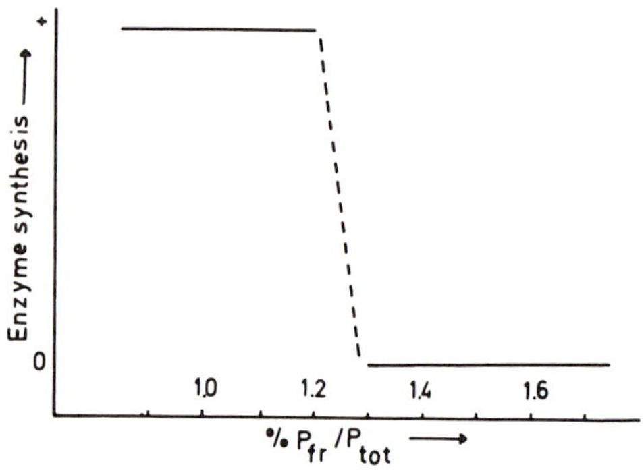

Fig. 1 Synthesis of the enzyme lipoxygenase in seedlings of ***Sinapis alba*** under the control of phytochrome. P_{tot} denotes the total concentration of phytochrome at the beginning of the experiment (36 h germination in the dark); then a certain fraction of phytochrome has been photoconverted to P_{fr} and kept continuously in that form (abscissa) (MOHR and OELZE-KAROW, 1973).

OELZE-KAROW et al. (1970) have demonstrated that this system is able to discriminate between P_{fr} levels which differ by less than 10 %, which can be interpreted as a cooperativity effect between P_{fr} molecules, the cooperativity factor being greater than 10. Such high cooperativity, in turn, requires a highly ordered structural matrix for the primary action of P_{fr}. It seems very likely that this matrix is a membranous structure.

2) Recently, methods of isolating phytochrome from cells have been much improved. It is now possible to detect phytochrome in well-defined cell fractions. In such experiments, part of the phytochrome is found in particle fractions (RUBINSTEIN et al., 1969) which probably consist of fragments of cytoplasmic membranes.

3) It is also possible to localize phytochrome ***in vivo*** by means of physiological responses. This has been achieved with the germ tube of a fern (ETZOLD, 1965) and with the green alga ***Mougeotia*** (HAUPT, 1970), by making use of the light oriented growth and the movement of the chloroplast in the cell, respectively. The most detailed information has been obtained from the ***Mougeotia*** system.

In earlier studies, small parts of the ***Mougeotia*** cell were irradiated with a microbeam in order to establish a strictly localized gradient of P_{fr} (Fig. 2A). Such a gradient can be recognized by a localized movement response, which shows that the effect of the light absorption does not diffuse. Furthermore, we were able to conclude that the photoreceptor pigment phytochrome is not localized in the chloroplast but in the surrounding cytoplasm; the chloroplast turns in a gradient of P_{fr} in such a way as to avoid the regions of highest P_{fr} concentration (Fig. 2B). Schematically this is shown in Fig. 3; if we start with P_r all around the cell, local irradiation with red light increases the P_{fr} level there, and hence the response is induced (Fig. 3A). If far-red follows red at the same region, P_{fr} is lowered again, and the response is cancelled (Fig. 3B). Thus the local response always indicates a local P_{fr} gradient in the cytoplasm.

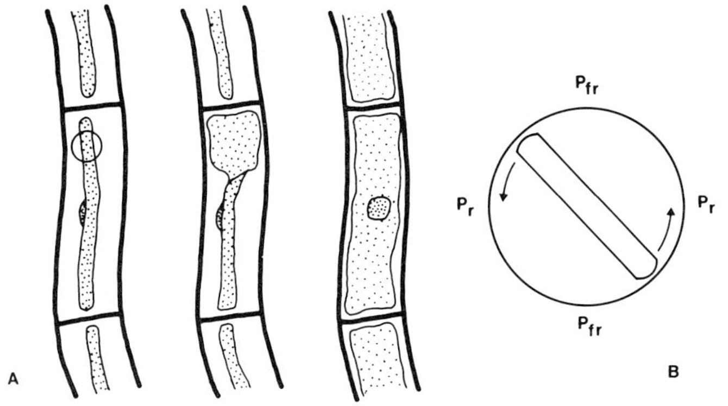

Fig. 2 A ***Mougeotia*** cells with the chloroplast orienting to the light. Left: profile (starting) position; right: face position after total irradiation with red light; middle: partial response of the chloroplast after partial irradiation with a microbeam (circle, left).
B Schematic cross section through a ***Mougeotia*** cell, showing the chloroplast movement in a gradient of P_{fr}. After HAUPT (1970, 1972).

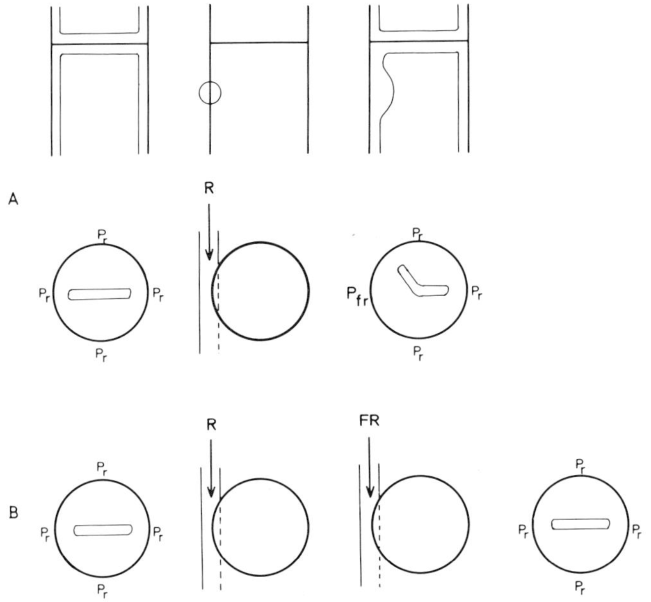

Fig. 3 P_{fr} gradient and chloroplast orientation in ***Mougeotia*** as a result of partial irradiations.
A Microbeam of red light. Schematic surface view (upper row) and cross section (lower row). Left: starting position; middle: positioning of the microbeam (chloroplast omitted); right: result of the irradiation.
B Same experiment, but red followed by far-red microbeam: no gradient, no response.

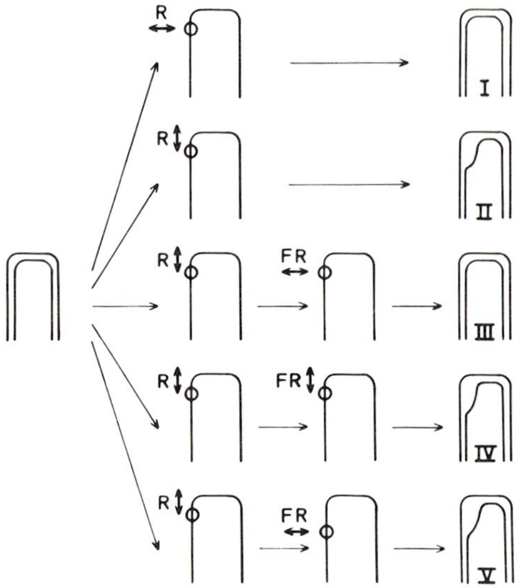

Fig. 4 Partial response of the ***Mougeotia*** chloroplast as a result of microbeam irradiations with polarized red and far-red light according to Fig. 3. The double arrows denote the direction of the electrical vector. The lower row (V) shows a control where R and FR were placed at different sites, hence no reversion of R effect by FR (HAUPT, 1972).

In this system, important additional information has been obtained by using plane-polarized red and far-red light. Such experiments have resulted in important conclusions about dichroism of phytochrome (Fig. 4).

1) Red light is effective in inducing the response only if its electrical vector is oriented parallel to the cell surface (compare in Fig. 4, I with II). Hence, P_r is dichroic with its double bonds parallel to the surface.

2) The induction is effectively reversed by far-red only if its electrical vector is oriented normal to the surface (compare in Fig. 4, III with IV). Hence, P_{fr} is also dichroic, but with its double bonds normal to the surface.

These results, and also those of the reverse experiments starting with P_{fr} all around the cell, lead to a picture of the position and orientations of the phytochrome which is schematically shown in Fig. 5, where the dashes denote the orientation of the transition moments, i.e. of the double bonds. This figure also shows that P_r is not only oriented parallel to the cell surface, but in a diagonal pattern.

We thus conclude that, in ***Mougeotia***, the phytochrome molecules are arranged in a very well defined order, which is possible only if they are strongly associated with a very stable cell structure. There seems to be no reasonable candidate for this localization other than the outer cytoplasmic cell membrane, i.e. the plasmalemma. The changes in dichroism during photoconversion demonstrate that the photoreceptor-membrane system remains highly ordered even after drastic changes in the photoreceptor orientation have occured.

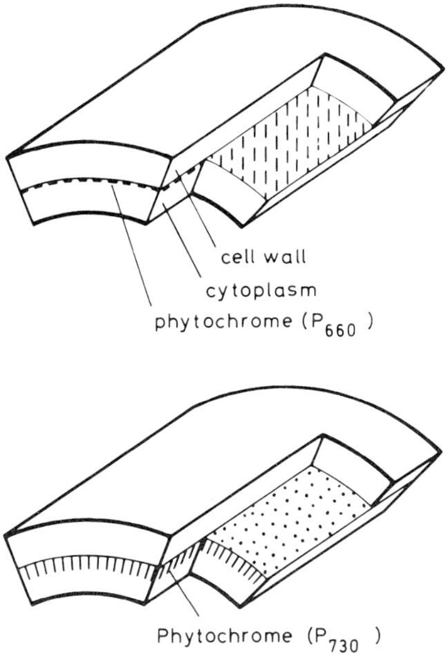

Fig. 5 Schematic diagram of part of a cylindrical ***Mougeotia*** cell with the dichroic absorption of phytochrome (dashes) parallel ($P_{660} = P_r$) or normal to the surface ($P_{730} = P_{fr}$) (HAUPT, 1970).

III. Membrane Effects of Phytochrome

The *Mougeotia* experiments have shown that phytochrome is localized in or at the membrane; but chloroplast movement cannot yet be related to membrane effects. Hence, the membrane as the primary effector of phytochrome photoconversion in this case cannot be asserted. However, there are responses in other systems which seem to be closely related to membrane effects and which are best explained, because of their kinetics, as direct effects of phytochrome on the membrane. Only three examples will be given.

1) The leaves of *Mimosa* and *Albizzia* open and close in response to light and dark. This response is controlled by phytochrome. The movement is at least partly the result of changes in membrane permeability and/or transport of K^+ in the motor cells of the pulvini. Since the response to phytochrome is seen after a few minutes, a direct effect of phytochrome on the membrane is likely (FONDEVILLE et al., 1966; SATTER and GALSTON, 1971a, b).

2) Root tips of barley or beans, suspended in a specific solution, stick to the negatively charged bottom of a glass vessel after red irradiation and are released by far-red. This is no doubt related to properties of the surface of the organ, possibly caused by biophysical changes in the membrane. The effect can be observed within less than 1 min (TANADA, 1968; JAFFE, 1970; RACUSEN and MILLER, 1972).

3) Biopotentials, measured in root tips and in coleoptiles, are reversibly changed by red and far-red irradiations, the tips becoming more positive in red light. These changes occur within minutes or even less (JAFFE, 1968; NEWMAN and BRIGGS, 1972). It seems difficult to explain these responses via a reaction chain containing biochemical steps. A direct effect of phytochrome on the membrane seems to be most reasonable.

Convincing as these experiments are, there is nevertheless a serious gap; in none of these latter systems has the localization of phytochrome at the membrane been proven. Therefore we have looked for direct effects of phytochrome upon the membrane in the alga *Mougeotia* where we believe it to be associated with the plasmalemma. These most recent investigations will be referred to in more detail.

WAGNER (1974) has measured the fluxes of K^+ and Cl^- across the membranes of *Mougeotia*. He reports that the efflux of neither ion responds to phytochrome photoconversion. It thus seems unlikely that phytochrome alters the permeability to K^+ and Cl^-. However this is not a disproof of membrane effects of phytochrome in *Mougeotia*, for the photoconversion of phytochrome may change the state of the membrane in such a way that it can be demonstrated by some other parameter.

IV. Kinetics of Plasmolysis and Deplasmolysis in *Mougeotia* as Influenced by Phytochrome

Investigating the effects of phytochrome on water permeability seemed worthwhile, for in general water does not traverse a membrane *via* the same channels as ions do (DAINTY, 1969). It is not difficult to envisage conformational changes in phytochrome-proteins affecting the ease with which water would permeate a membrane.

Such considerations (WEISENSEEL and SMEIBIDL, 1973) have led us to study the kinetics of plasmolysis and deplasmolysis in *Mougeotia* as a function of the state of phytochrome. Some of our results are summarized in Table 1 and Fig. 6.

Table 1

Influence of phytochrome on plasmolysis and deplasmolysis in *Mougeotia*.
Left: time lag (s) for the onset of plasmolysis in mannitol after irradiation with different light regimes. The osmotica were applied after the irradiations. The average times (± S.E.) are those which elapsed between arrival of the osmoticum at the *Mougeotia* cells under observation and retraction of the protoplasts from the cross walls. Each value is the average of about 40–60 cells. Right: onset and time (s) for completion of deplasmolysis in culture medium after a total of 25 min in 0.5 M mannitol + ENS. Initial length of the plasmolysed protoplast given as percent of the cell length. Abbreviations: R = red light (664 nm), FR = far-red light (724 nm), GS = green safelight, ENS = culture medium (ca. 10 mosm/kg). After WEISENSEEL and SMEIBIDL (1973).

	Plasmolysis Time lag for onset (s)		Deplasmolysis		
Pre-irradiation	0.5 M mannitol + ENS	1.0 M mannitol + ENS	Initial degree of plasmolysis (% of cell length)	Time lag for onset (s)	Rate of de-plasmolysis (s^{-1})
5 min R	13.7±0.64	3.5±0.14	67.9±0.83	19.7±0.79	0.50
5 min R + 10 min GS	13.8±1.02		72.3±0.90	22.5±0.48	0.45
5 min R + 10 min FR	27.7±1.16	7.3±0.22	70.7±0.91	28.7±0.87	0.36
5 min FR	33.9±1.72		68.5±0.81	30.0±0.36	0.34
5 min GS	32.4±2.16	8.6±0.40	72.0±0.72	32.0±0.78	0.32

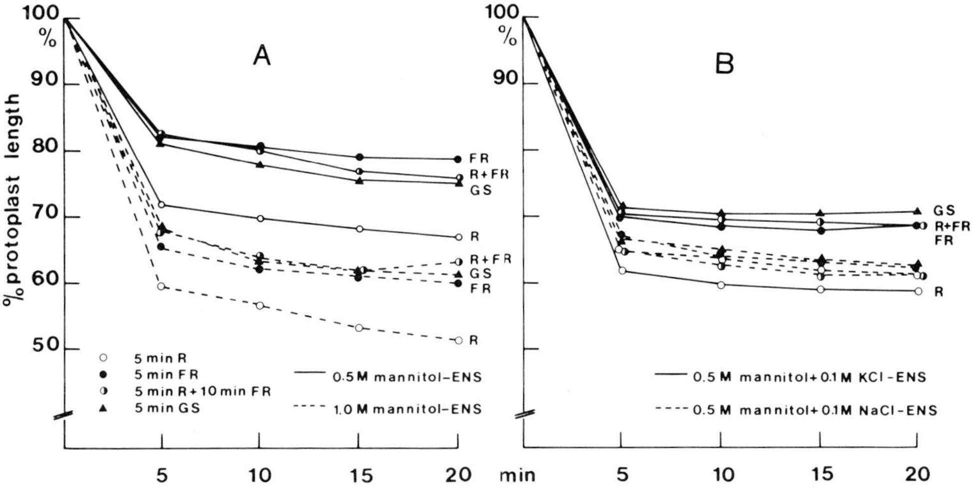

Fig. 6 Plasmolysis in ***Mougeotia***: time course of protoplast length contraction after irradiation with different light regimes (see insets) prior to application of the osmotica. A. Mannitol as osmoticum added to the culture medium. B. Mannitol supplemented with inorganic ions. The standard errors are small (cf. Table 1) and have been omitted for clarity. R = red light; FR = far-red; GS = green safelight; ENS = culture medium (0.01 osmol/kg). After WEISENSEEL and SMEIBIDL (1973).

The time lag between application of the osmoticum and onset of plasmolysis in culture medium (ENS) supplemented with 0.5 M or 1.0 M mannitol is shorter after red light than after far-red light. Photoreversibility shows that this phenomenon is a typical phytochrome response. Addition of 0.1 M KCl or 0.1 M NaCl to the osmoticum (data not shown in Table 1) shortens the time lag after far-red irradiation. The time lag for deplasmolysis after the cells have been plasmolysed for 25 min in 0.5 M mannitol culture medium is also decreased by a high level of P_{fr}. The initial rates of plasmolysis and the rates of deplasmolysis are controlled by phytochrome (Fig. 6A, B for plasmolysis). High levels of P_{fr} increase the rates of both plasmolysis and deplasmolysis. Addition of 0.1 M NaCl to the osmoticum abolishes the differences between the rates of plasmolysis. 0.1 M KCl has no such effect. Some preliminary results from electrical measurements show no change in the membrane potential.

From these observations the following conclusions might be drawn. The rates of plasmolysis and deplasmolysis are governed by the equation $dV/dt = AL_p\Delta\Psi$ where A = area, L_p = hydraulic conductivity, $\Delta\Psi$ = water potential difference.

It is extremely unlikely that P_{fr} can affect the driving force, $\Delta\Psi$, thus it must be affecting L_p. The differing degrees of plasmolysis (see Fig. 6) are puzzling. Calculations indicate that it would need an improbably high permeability to mannitol for mannitol entry to influence these results; and 'unstirred layers' cannot be a factor. Additional experiments are clearly needed.

In spite of these limitations, our results seem to demonstrate a control of fundamental membrane properties by phytochrome. We are thus encouraged to continue along this line for we must test whether we really are dealing with the primary action of phytochrome in the cell.

B. Acknowledgements

This work was supported by the Deutsche Forschungsgemeinschaft.

C. References

BORTHWICK, H.A., in Phytochrome (eds. K. Mitrakos, W. Shropshire Jr.), pp. 25–44, Academic Press, London and New York 1972

DAINTY, J., in The Physiology of Plant Growth and Development (ed. M.G. Wilkins), McGraw Hill, 1969

ETZOLD, H., Planta **64**, 254–280 (1965)

FONDEVILLE, J.C., BORTHWICK, H.A., HENDRICKS, S.B., Planta **69**, 357–364 (1966)

HAUPT, W., Physiol. Veg. **8**, 551–563 (1970)

HAUPT, W., in Phytochrome (eds. K. Mitrakos, W. Shropshire jr.), pp. 554–569, Academic Press, London and New York 1972

JAFFE, M.J., Science **162**, 1016–1017 (1968)

JAFFE, M.J., Plant Physiol. **46**, 768–777 (1970)

MOHR, H., Lectures on Photomorphogenesis. Springer-Verlag, Berlin–Heidelberg–New York, 1972

MOHR, H., OELZE-KAROW, H., Biologie in unserer Zeit **3**, 137–147 (1973)

NEWMAN, I.A., BRIGGS, W.R., Plant Physiol. **50**, 687–693 (1972)

OELZE-KAROW, H., SCHOPFER, P., MOHR, H., Proc. Nat. Acad. Sci. **65**, 51–57 (1970)

RACUSEN, R., MILLER, K., Plant Physiol. **49**, 654–655 (1972)

RUBINSTEIN, B., DRURY, K.S., PARK, R.B., Plant Physiol. **44**, 105–109 (1969)

SATTER, R.L., GALSTON, A.W., Science **174**, 518–520 (1971a)

SATTER, R.L., GALSTON, A.W., Plant Physiol. **48**, 740–746 (1971b)

TANADA, T., Plant Physiol. **43**, 2070–2071 (1968)

WAGNER, G., in Membrane Transport in Plants (eds. U. Zimmermann, J. Dainty), Springer-Verlag, Berlin–Heidelberg–New York 1974

WEISENSEEL, M.H., SMEIBIDL, E., Z. Pflanzenphysiol. **70**, 420–431 (1973)

Evidence for Binding of Phytochrome to Membranes

E. Schäfer

A. Introduction

Whether phytochrome can act as a membrane effector in photomorphogenesis is still controversial. Recently phytochrome-mediated threshold responses have been observed which imply significant cooperativity of the primary reaction, $P_{fr} + X \rightarrow P_{fr}X$ (OELZE and MOHR, 1973; SCHOPFER and OELZE-KAROW, 1971). But it has also been found that, in the same organ (mustard cotyledons), the primary reaction of P_{fr} does not involve any significant cooperativity in the case of phytochrome-mediated anthocyanin synthesis (DRUMM and MOHR, 1974). This indicates that the concept of a common initial reaction of P_{fr} can hardly be maintained. It was furthermore concluded by DRUMM and MOHR (1974) that in phytochrome-mediated anthocyanin synthesis a phytochrome (P_{fr}) induced change in the reaction matrix occurs. However the actual threshold value was not obviously affected by different P_{fr} concentrations (OELZE and MOHR, 1973). Against the background of these physiological data new results on phytochrome membrane interactions will be discussed.

B. Results

I. Characterization of Phytochrome Containing Membrane Fractions

Analysis of the binding properties of phytochrome to particulate fractions from cell homogenates has been tried for some years. It was recently shown that the cell fractions to which phytochrome molecules bind are very probably vesicles originating from cytoplasmic membranes (MARME et al., 1974b; QUAIL et al., 1973). Electron microscopy of negatively-stained preparations shows that, with 3 mM Mg^{2+}, distinct membranous vesicles (40 to 60 nm in diameter) appear (MARME et al., 1974b). The formation and aggregation of these vesicles depend strongly on the divalent ion concentrations. The apparent density of these phytochrome-containing vesicles, when prepared from dark-grown squash hooks following sedimentation on sucrose gradients, corresponds to 15 % sucrose (w/w) with 0.1 mM Mg^{2+} and to 40 % sucrose with 10 mM Mg^{2+} (MARME et al., 1974b). These differences in apparent density were explained on the basis of solubilisation and aggregation of the membrane vesicles. Most of the vesicles formed a pellet when centrifuged for 30 min at 20 000 X g after a precentrifugation at 500 X g for 10 min, if the Mg^{2+} concentration in the buffer medium was 5 to 10 mM (QUAIL et al., 1973). The amount of pelletable phytochrome as defined by the ratio of P_{tot} in the 20 000 X g pellet to P_{tot} in the 500 X g supernatant is a function of light pretreatment and pH (QUAIL et al., 1973; QUAIL and SCHÄFER, 1974) as well as being dependent on the concentration of divalent ions (MARME et al., 1974b; QUAIL and SCHÄFER, 1974).

II. Analysis of the Binding- and State-Function of the Phytochrome Accepting Membranes

1. *In vivo* Irradiations

The fact that the capacity of vesicles to bind phytochrome depends on light pretreatment seems to be the most interesting effect for an understanding of the physiological data.

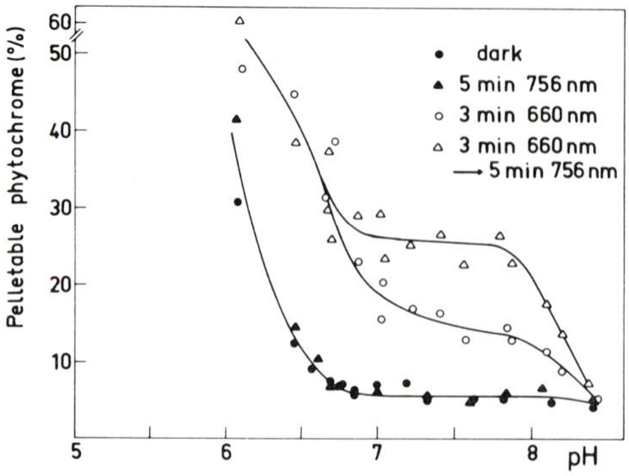

Fig. 1 Percent pelletable (bound) phytochrome from maize as a function of pre-irradiation at 25°C *in vivo* and pH in the final homogenate. The pH of the extraction medium was adjusted to provide a range of pH's in the final homogenate. MOPS was used as the buffering agent below pH 7.5 and TRIS above pH 7.5.

The percentage of bound phytochrome (defined by the ratio of phytochrome in the 20 000 X g pellet to the total amount of phytochrome in this pellet and in the 20 000 X g supernatant) is very low in etiolated seedlings (Fig. 1, ●), which are known to contain only the physiologically-inactive phytochrome species P_r.

This supports a previous suggestion (MOHR, 1972) that the equilibrium binding constant of P_r to the membrane is very low. In contrast, after an irradiation with red light which establishes a photo-stationary state P_{fr}/P_{tot} of approximately 80 %, the percentage of bound phytochrome increases over the whole pH range up to pH 8, indicating that the equilibrium binding constant of P_{fr} to membranes is high (Fig. 1, ○). In contrast to most of the physiological effects, this P_{fr}-induced increase of the binding constant for P_r is not reversible with far-red light (756 nm) (Fig. 1, △), which transforms nearly all P_{fr} back to P_r. However, this in fact enhanced P_r binding is reversed to the P_{fr} binding level by a second red irradiation. This indicates an induced transition of a dark-adapted receptor site (X) after red light induced P_{fr} binding ($P_{fr}X \rightarrow P_{fr}X'$). This reaction occurs within 1min. P_r and P_{fr} bound to

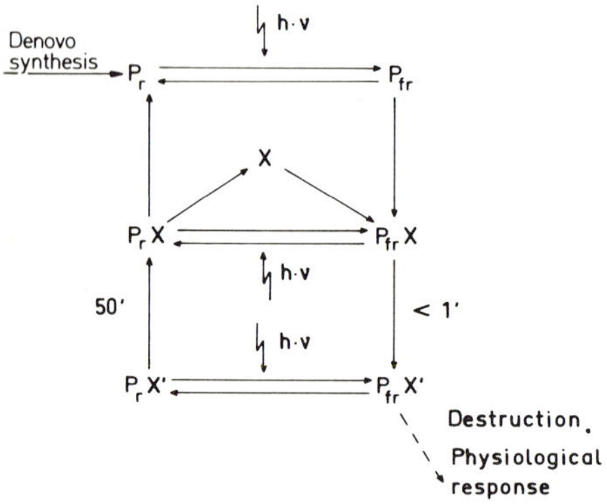

Fig. 2 Reaction scheme for the interaction of phytochrome with its binding sites. P_r and P_{fr} are the two photoreversible forms of phytochrome, X and X' are the two reversible forms of the receptor sites (modified after QUAIL and SCHÄFER, 1974).

X′ are still photoconvertible by light. The binding of P_r to X′, induced by red followed by far-red light, is not stable. It decays with a half-time of about 50 min (QUAIL et al., 1973). These reactions are summarized in the scheme shown in Fig. 2.

This special behaviour of the phytochrome membrane receptor system offers an easy approach to the binding as well as the state function of this system. The binding function describes the relation between bound and free ligand concentration. It can therefore be analysed by measuring the percent pelletable phytochrome as P_{fr} as a function of the P_{fr} concentration. The state function describes the relation between the amount of transformed receptor sites and the ligand concentration. In this case dark adapted and transformed receptor sites can easily be distinguished by their different affinity to P_r. To analyse the state-function (CHANGEAUX and THIERY, 1968) (X′ as a function of ligand concentration), corn coleoptiles were irradiated for 3 min with saturating doses of monochromatic light between 660 and 720 nm in order to establish different photostationary states, P_{fr}/P_{tot}, between 80 and 0.25 %. After this pre-irradiation nearly all the P_{fr} was transformed back to P_r by irradiation with far-red (756 nm) light. This is a prerequisite for the study of P_r binding to X′ as a function of the initial P_{fr} concentration. Fig. 3 demonstrates that the state-function is not only a function of the photostationary state but also a function of the light irradiance (QUAIL and SCHÄFER, 1974).

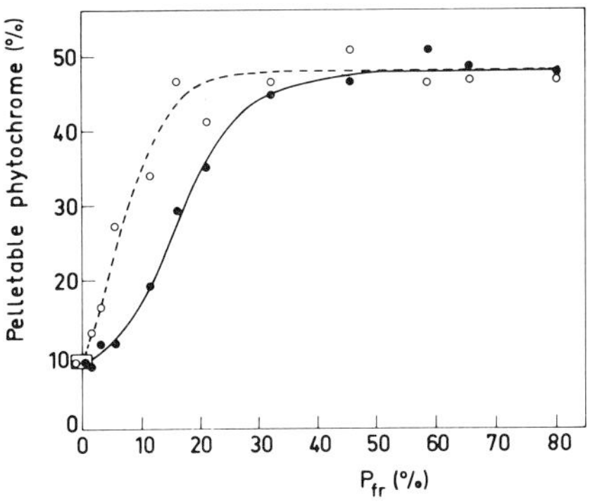

Fig. 3 Percent pelletable phytochrome in extracts from maize coleoptile tips following irradiations at 25°C *in vivo* of 3 min with wavelengths between 660 and 730 nm followed immediately in all cases with 5 min 756 nm before extraction. Percent P_{fr} refers to the photostationary state level of P_{fr} established *in vivo*, during the initial 3 min irradiation. (●—●) irradiations of approximately 500 erg cm^{-2} s^{-1}; (○—○) irradiations of approximately 3000 erg cm^{-2} s^{-1} (modified after QUAIL and SCHÄFER, 1974).

The binding-function ($P_{fr}X$ as a function of P_{fr}) and the state-function (X′ as a function of P_{fr}) can both be analysed without the problems of irradiance dependence, if different P_{fr} concentrations are established without any significant cycling of the pigment system ($P_r \rightleftharpoons P_{fr}$). The binding-function was studied by irradiating the plant tissue with short pulses of monochromatic red light (< 1 ms). Such a light pulse is much shorter than the dark reaction between the excited state of P_r, produced by the photon absorption, and eventual P_{fr} formation (LINSCHITZ et al., 1966).

The binding curve (P_{fr} in the 20 000 X g pellet as a function of total P_{fr}) in preparations of corn coleoptiles has a typical sigmoidal shape indicative of cooperativity (Fig. 4). The Hill-coefficient (CHANGEAUX and THIERY, 1968) of the binding-function is 1.6.

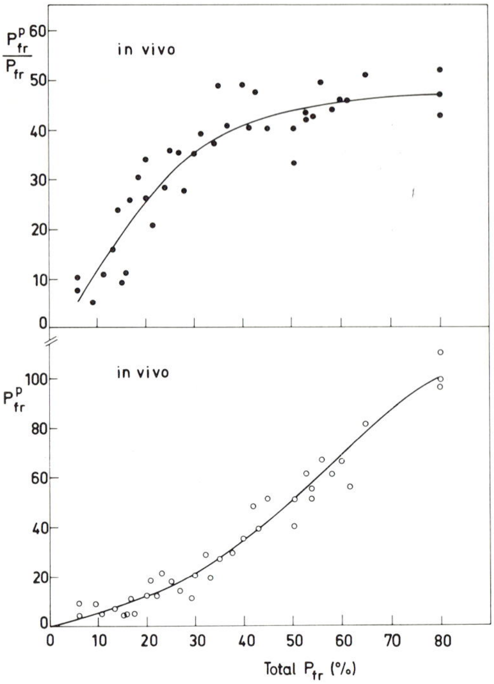

Fig. 4 The ratio of percent pelletable P_{fr} ($P_{fr}{}^{p}$) to total P_{fr} (upper graph) and percent pelletable P_{fr} (lower graph) plotted as a function of total P_{fr} in extracts from maize coleoptile tips following irradiations at 25°C *in vivo*. Short light flashes (< 1 ms) of monochromatic red light (655 nm) of different irradiations were used to establish different P_{fr} concentrations.

The state-function can be analysed by measuring the percent of bound P_r after pre-irradiation with red light pulses of different irradiance followed by 5 min far-red (756 nm) light. The Hill-coefficient of the state-function is 2.1 and thus significantly higher than that of the binding function.

2. *In vitro* Irradiations

Phytochrome membrane interactions can be studied not only after treatments performed *in vivo* but also *in vitro*. Brief irradiations of buffer extracts (20 000 X g supernatant) of dark-grown squash seedlings with red light result in a binding of P_{fr} to the membrane vesicles. With low concentrations of Mg^{2+} (0.1 mM) there is a partial reversion by far-red; higher concentrations (10 mM) yield enhanced binding following the far-red irradiation. The binding is also strongly pH dependent and decreases with higher divalent ion concentrations; it is not affected by monovalent ions (MARME et al., 1974a).

The binding-function could also be studied *in vitro* by irradiating the 20 000 X g supernatant of dark-grown seedlings in the same manner as *in vivo* and analysing the binding to the 50 000 X g pellet. The shape and the Hill-coefficient of the binding curves *in vivo* and *in vitro* are the same in the corn coleoptile system (SCHÄFER, 1974). These observations indicate that the extraction and centrifugation procedure does not alter significantly the binding properties of the membrane vesicles.

The degree of cooperativity, expressed by the Hill-coefficient (N^H), varies from plant system to plant system (squash hooks, N^H = 3.5; ***Avena*** coleoptile, N^H = 2.5; corn coleoptile, N^H = 1.6). The highest cooperativity was found in a hook system, which is in good agreement with the high cooperativity suggested for the primary reaction of P_{fr} in the phytochrome-mediated suppression of apparent lipoxygenase synthesis in the mustard cotyledon by hook phytochrome (OELZE and MOHR, 1973, 1974).

C. Discussion

I. A Photoconvertible Ligand-Membrane-System

The main difference between the phytochrome membrane systems described here and the acetylcholine membrane system (CHANGEAUX and THIERY, 1968) appears to be, beside differences at the membrane level, that in the former the ligand exists in two photoconvertible forms (P_r, P_{fr}). P_r has a low and P_{fr} a high equilibrium binding constant to the dark-adapted receptor site X (Figs. 1, 2), indicating stereo-specific binding. P_{fr} seems to induce a receptor transition (QUAIL et al., 1973). The data thus appear to be compatible with the following aspects of CHANGEAUX's (1969) concepts: (a) the capacity of the receptor site to exist in two reversible conformational states; (b) the capacity of the receptor sites for stereospecific recognition of the ligand; (c) the change of the affinity of the receptor site towards the corresponding ligand following the transition from one conformational state to another (QUAIL et al., 1973); and (d) the cooperativity of the binding process. This system does not appear to be compatible with the 'selective' aspect of CHANGEAUX's (1969) model i.e. the preexistence of two conformational states of the receptors (X,X'), because the $X \rightarrow X'$ state transition is not rapidly reversible upon removal of the P_{fr} effector (Fig. 2 and QUAIL et al., 1973). The data are better explained by an 'induced fit' approach (KOSHLAND and NEET, 1968), where the $X \rightarrow X'$ transition is induced by P_{fr} subsequent to binding. Removal of P_{fr} would then result in a slow relaxation of the induced state X' back to X, as was observed (Fig. 2, and QUAIL et al., 1973).

II. Physiological Relevance of the Phytochrome Membrane System

This phytochrome membrane system seems to enable plants to measure P_{fr} concentrations established by short light pulses by recognition of the amount of X'. They can also measure the irradiance in continuous irradiations, because light- and dark-reactions interact in this system in such a way that, up to the saturation level, the $P_{fr}X$ concentration is a function of irradiance and wavelength.

This has been shown for irradiation ***in vivo*** (QUAIL and SCHÄFER, 1974) and ***in vitro*** (QUAIL, 1974). Therefore the model presented in Fig. 2 might provide a 'molecular' basis for explaining the 'high irradiance response' (HARTMANN, 1966). HARTMANN (1966) has obtained convincing experimental evidence, using dichromatic irradiation, that the far-red peak of the action spectrum is due in some way to phytochrome. However, in spite of considerable effort the manner in which phytochrome is involved in the 'high irradiance response' is not clear. The major problem is that any explanation of the action spectrum of the 'high irradiance response' in terms of phytochrome requires the assumption that the optimum effectiveness of phytochrome under continuous irradiation is already reached at low $P_{fr}/(P_r + P_{fr})$ ratios and that the effectiveness decreases with increasing $P_{fr}/(P_r + P_{fr})$ ratios.

It has recently been shown by mathematical calculations on the basis of the model presented in Fig. 2 that the $P_{fr}X$ concentration is irradiance and wavelength dependent and that the optimum concentration occurs under continuous far-red irradiation (SCHÄFER, 1974).

D. References

CHANGEAUX, J.P., in Symmetry and Function of Biological Systems at the Molecular Level (eds. A. Engström, B. Strandberg), Nobel Symp., Vol. 11, pp. 235–256, Wiley, Interscience Division, New York 1969
CHANGEAUX, J.P., THIERY, J., in Regulatory Functions of Biological Membranes (ed. J. Järnefelt), BBA Libary, Vol. 11, pp. 116–138, Elsevier, Amsterdam 1968
DRUMM, H., MOHR, H., Photochem. Photobiol., in press (1974)
HARTMANN, K.M., Photochem. Photobiol. **5**, 349–366 (1966)
KOSHLAND, D.E., NEET, K.E., Ann. Rev. Biochem. **37**, 359–410 (1968)
LINSCHITZ, H., KASCHE, V., BUTLER, W.L., SIEGELMANN, H.W., J. Biol. Chem. **241**, 3395–3403 (1966)
MARME, D., BOISARD, J., BRIGGS, W.R., Proc. Nat. Acad. Sci., in press (1974a)
MARME, D., MACKENZIE, J.M., BOISARD, J., BRIGGS, W.R., Plant Physiol., in press (1974b)
MOHR, H., Lectures on Photomorphogenesis. Springer-Verlag, Berlin–Heidelberg–New York 1972
OELZE, H., MOHR, H., Photochem. Photobiol. **18**, 319–330 (1973)
OELZE, H., MOHR, H., Photochem. Photobiol., in press (1974)
QUAIL, P.H., Planta, in press (1974)
QUAIL, P.H., MARME, D., SCHÄFER, E., Nature New Biology **245**, 189–190 (1973)
QUAIL, P.H., SCHÄFER, E., J. Membrane Biol. **15**, 393–404 (1974)
SCHÄFER, E., Photochem. Photobiol., in press (1974)
SCHOPFER, P., OELZE-KAROW, H., Planta **100**, 167–180 (1971)

Hormonal Control of Ion and Water Transport in the Excised Maize Root

J.C. Collins

A. Introduction

Hormonal control of ion and water fluxes in the plant has received scant attention, and it is not surprising that there is little evidence on whether *in vivo* hormonal control of ion and water fluxes exists in the plant. There have been several reports of auxin increasing the water permeability of plant tissues but there have been few studies with the gibberellins. Most recent studies have concerned two of the later discovered plant hormones, cytokinins and abscisic acid (ABA). There are several reports that applied ABA increases the water permeability of excised root systems (TAL and IMBER, 1971; GLINKA, 1973; COLLINS and KERRIGAN, 1974); the first and last of these reports also deal with kinetin and show that application of this artificial cytokinin causes a drastic reduction in the water permeability of the root. There is also evidence that kinetin can alter ionic fluxes; ILAN (1971) noted an increase in K^+/Na^+ selectivity of *Helianthus* leaf tissue on addition of kinetin, due to both an increase in K^+ accumulation and a decrease in Na^+ accumulation. ITAI (1967) has shown that endogenous cytokinins are intimately involved with the events that occur immediately after subjecting a plant to water stress, whether it be a high evaporative demand to the leaves or immersion of the root system into a saline solution.

The evidence so far gained on hormonal interaction with root transport indicates that ABA increases the water permeability and that there is a corresponding increase in fluxes of K^+ and Cl^-. Kinetin causes a decrease in the water permeability and of the ionic fluxes (Table 1). The levels at which these hormones are operative is sufficiently low (10^{-6} M and less) to suggest an *in vivo* role; also their action is reversible (COLLINS and KERRIGAN, 1974).

Table 1

Effect of kinetin and ABA on excised maize roots

	Control	Kinetin	ABA
L_p (cm s^{-1} atm^{-1})	$4.0 \cdot 10^{-7}$	$9.4 \cdot 10^{-8}$	$8.4 \cdot 10^{-7}$
ϕ_{K^+} (pmoles cm^{-2} h^{-1})	39.6	6.0	54.9

Both hormones were added at 10^{-6} M. All solutions contained $2.5 \cdot 10^{-3}$ % v/v DMSO, K^+ flux into the exudate measured 24–25 h after excision.

The question where do these hormones have their effect on the root, has been given some tentative answers. CRAM and PITMAN (1972) showed that ABA increases accumulation of tracer Rb^+ and Cl^-, but inhibits the transport of these ions from barley roots; on this evidence they argued that ABA is operative at the step involving secretion into the xylem.

COLLINS and KERRIGAN (1974) found no evidence to support this contention in maize roots, although they did find a similar situation for the effect of kinetin.

The site of action of ABA and kinetin on water fluxes in the excised maize root has been investigated further using cortical preparations, the 'cortical sleeves' of GINZBURG and GINZBURG (1970), from young maize roots.

B. Materials and Methods

Cortical sleeves were pulled from the primary roots of 3-day old maize plants. The final sleeve length was 4–5 cm and this corresponded to the 1–6 cm segment from the root tip. The sleeve was sealed into a microcap capillary with a molten wax-resin mixture and the stelar space washed out with 50 mM KCl + 0.1 mM $CaCl_2$ saline; this removed ruptured cell contents. When washed three times the sleeve – microcap assembly was filled with saline, care being taken to exclude all air bubbles. The sleeve tip was sealed by immersing it in a drop of molten wax-resin and sucking at the capillary end; the mixture solidifies immediately on contact with the saline. The sleeve microcap was immersed in 1.0 mM KCl + 0.1 mM $CaCl_2$ solution and rate of movement of the meniscus observed microscopically. All solutions were kept at 25 ± 1°C. The sleeves used in these experiments were pretreated (before pulling) for 4 h in 1.0 mM KCl + 0.1 mM $CaCl_2$ saline with or without 10^{-6} M ABA or kinetin; all solutions contained 0.01 % v/v DMSO.

C. Results

The information from these experiments (Table 2) yields values for σL_p of the isolated cortex. The cortical tissue pretreated in 10^{-6} M ABA is significantly more permeable to water than either control or 10^{-6} M kinetin treated tissue. This corroborates the earlier conclusion that ABA has its effect on the cortex whilst kinetin would effect the xylem secretion step.

Table 2

Effect of kinetin and ABA on osmotic flow across the isolated cortex

	Control	Kinetin	ABA
σL_p (cm s^{-1} atm^{-1})	$5.9 \cdot 10^{-7}$	$7.6 \cdot 10^{-7}$	$11.3 \cdot 10^{-7}$

To determine, separately, σ and L_p similar experiments were performed, except that flow was due to a hydrostatic pressure gradient rather than an osmotic gradient. Sleeves were pulled and sealed into a microcap as before. The sleeve-microcap was filled with 1.0 mM KCl + 0.1 mM $CaCl_2$ saline and then sealed into a measuring capillary attached to a water manometer (up to 100 cm water). Pressure was applied from a screw-driven syringe. Sleeves used in these experiments were the ABA and control sleeves earlier described. The external solution bathing the sleeve could be easily changed and in one experiment L_p and σ for various solutes could be measured. Measurements were made at room temperature, 18 ± 3°C. A full description of both methods will appear elsewhere.

The data from these last experiments is as yet rudimentary, but there is no evidence that ABA alters σ_{KCl} of the root cortex. The value for σ_{KCl} for the cortex is 0.05, this value was confirmed from a comparison of σL_p, from osmotic flow, and L_p from hydrostatic pressure flow measurements.

D. Discussion

The increase in water permeability of excised maize roots caused by ABA treatment is now found in the isolated cortical system as well, but there is no decrease in permeability on kinetin treatment of cortical tissue as is observed in excised roots. Taking the evidence from this paper with the earlier evidence (COLLINS and KERRIGAN, 1974) we conclude that ABA exerts its control in the cortex, and probably also on the secretion from the cortex step as well, whilst kinetin is only operative at the secretion from the cortex step.

The mechanism of hormonal control of water permeability is unknown. The contention that ABA acts on L_p, rather than on σ, for the cortex gives little insight into mechanism without further evidence. Water movement through the cortex follows two pathways (GINZBURG and GINZBURG, 1970) and it could be that ABA affects movement in the symplasm; it is more difficult to envisage a hormonal effect on the cell wall pathway. The very low value for σ_{KCl} of 0.05 is at variance with the σ_{NaCl} value quoted by GINZBURG and GINZBURG (1970) of approximately unity. Their value was derived from flux measurements; however, their root material was not grown in culture solution (as here) but in moist air which could cause cutinization of cell surfaces in the cortex. This is an important point that needs further clarification.

The biochemistry of hormonal control of transport is as yet unexplored for plant systems. Hormonal effects may well be mediated through cyclic AMP or act via protein synthesis (SHARP, 1972). GINETZINSKY (1961) found anti-diuretic hormone to increase hyaluronidase secretion in the nephron, and this increased permeability by eating away the inter-cellular cement of the basement membrane. WHITTEMBURY (1962) showed that anti diuretic hormone acted upon pore size in isolated toad skin, and a recent report (HAYS et al., 1971) has shown that vasopressin has its effect on the permeability of the toad bladder not by increasing pore size, but rather by increasing the number of pores. The last reports are not red herrings, I hope, but helpful suggestions.

E. Acknowledgements

It is a pleasure to acknowledge helpful discussions with Dr. C.R. House.

F. References

COLLINS, J.C., KERRIGAN, A.P., New Phytol., in press (1974)
CRAM, W.J., PITMAN, M.G., Aust. J. Biol. Sci. **25**, 1125–1132 (1972)
GINETZINSKY, A.G., Nature **182**, 1218–1219 (1958)
GINZBURG, H., GINZBURG, B.Z., J. Exp. Bot. **21**, 580–592 (1970)
GLINKA, Z., Plant Physiol. **51**, 217–219 (1973)
HAYS, R.M., FRANKI, N., SOBERMAN, R., J. Clin. Invest. **50**, 1016–1018 (1971)
ILAN, I., Physiol. Plant., **25**, 230–233 (1971)
ITAI, C., Shoot and Root Interactions under Different Water Regimes. Ph.D. Thesis. Hebrew University Jerusalem (1967)
SHARP, G.W.G., in Transport and Accumulation in Biological Systems (ed. E.J. Harris), pp. 147–192, Butterworths, London 1972
TAL, M., IMBER, D., Plant Physiol. **47**, 849–850 (1971)
WHITTEMBURY, G., J. Gen. Physiol. **46**, 117–130 (1962)

Modification of Phosphate Transport in *Vicia faba* by Boron Deficiency, Growth Regulators, and Metabolic Inhibitors

G.A. Robertson and B.C. Loughman

A. Introduction

The absorption of both phosphate and rubidium is reduced in boron deficient *Vicia faba* roots (ROBERTSON and LOUGHMAN, 1973, 1974). The reduction in phosphate absorption can be reversed by resupplying boron to the deficient roots 1–2 h prior to the absorption period (LOUGHMAN and ROBERTSON, 1973). The reduced ion absorption in boron deficient plants could result from effects on membrane permeability or from general and secondary effects on cell metabolism.

In an attempt to characterise the changes in phosphate absorption as being due to primary or secondary effects of boron we examined further aspects of absorption and transport of phosphate in *V. faba* seedlings and followed the effects associated with varying boron status. We also compared some responses to boron deficiency with those occuring after growth was altered by other means.

B. Methods

The methods used to grow boron deficient *V. faba* seedlings have been described elsewhere (ROBERTSON and LOUGHMAN, 1973). Each replicate in whole plant experiments usually had 10–12 plants per 900 ml of labelled solution. After the prescribed period, the roots were washed for a total of 2 min in 3 changes of deionised water, the portion being examined was excised, digested in acid and counted using Cerenkov methods in a liquid scintillation counter (LÄUCHLI, 1969).

Where absorption by the 0.5 cm root tip was examined, tips were excised immediately prior to the absorption period, and 8–10 root tips placed in 50 ml conical flasks containing 4 ml of the treatment solution. At the end of the absorption period the root tips were poured onto sintered glass funnels, washed with 150 ml deionised water, digested and counted as described. In experiments designed to examine transport of ^{32}P and boron out of the cotyledons, a small cavity was drilled in the upper cotyledon and either 0.1 μCi of ^{32}P (carrier-free), or 10 μl of 0.1 M H_3BO_3 were placed in the cavity. Transport of ^{32}P was determined by digesting and assaying the activity in various organs of the seedling. All experiments were replicated at least twice.

C. Results

V. faba seedlings are very sensitive to boron deficiency, and within 48 h of transferring to boron-free solutions, root elongation virtually ceased. Most absorption experiments were conducted between 44–48 h after transfer to −B solutions when root elongation had slowed, but not stopped completely. The growth rate of the shoot, however, was not affected for at least 3 weeks. This could have been due to a differing sensitivity to boron deficiency in the two organs, or to a preferential transport of the seed reserves of boron to the shoot. When every second seedling in −B solution cultures received 10 μl of 0.1 M H_3BO_3 in small cavities in the cotyledon, the treated seedlings continued to grow normally,

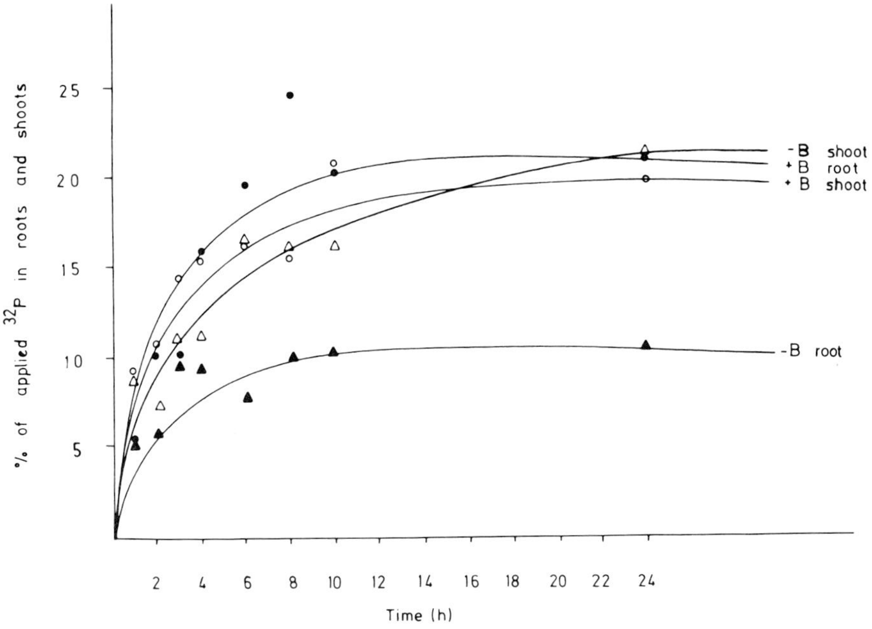

Fig. 1 Rate of arrival of ^{32}P in roots and shoots of boron deficient and non-deficient 7 day old *V. faba* plants. 0.1 μCi of ^{32}P was placed in a small cavity in the upper cotyledon. —B boron deficient root or shoot; +B non-deficient root or shoot.

whereas the non-treated seedlings exhibited reduced root elongation after 2 days. Thus transport of boron from the cotyledons to the roots is possible, at least when excess boron is present in the cotyledon.

V. faba seedlings transported less than 2 % of the phosphate absorbed over 24 h to the shoot while they were still dependent on seed reserves. This almost total reliance of the shoot on nutrients from the seed was not affected by variations in day length or light intensity, nor by phosphate concentration in the medium before or during the experiment. Removal of the cotyledons retarded root and shoot growth and reduced the ability to absorb phosphate. Transport to the shoot, however, can be turned on within 2 days of removal of the seed and, over 24 h, up to 40 % of total phosphate absorbed can be transported to the shoots. These are similar levels to those previously shown in barley seedlings of a comparable age (CROSSETT and LOUGHMAN, 1966). The high rate of phosphate transport from the seed to both the root and the shoot was demonstrated by placing 0.1 μCi of ^{32}P in 0.01 ml H_2O in a small cavity in the upper cotyledon and examining its translocation to the shoots and the roots. The results are rather erratic as seed size, as well as the exact position of the cavity in relation to the transporting vessels would have considerable effect, but the trends are obvious (Fig. 1). Initial rates of transport of ^{32}P from the cotyledons were high but declined with time, presumably as the specific activity of the inorganic phosphate pool declines as ^{32}P becomes incorporated into phosphate esters. Transport of phosphate to the shoots was similar in boron deficient and non-deficient plants but the roots of deficient plants received less than 50 % of that of roots of non-deficient plants. Thus cotyledons on boron deficient plants showed a reduced translocation of ^{32}P, the reduction being accounted for by the reduced effectiveness of the boron deficient root as a nutrient sink.

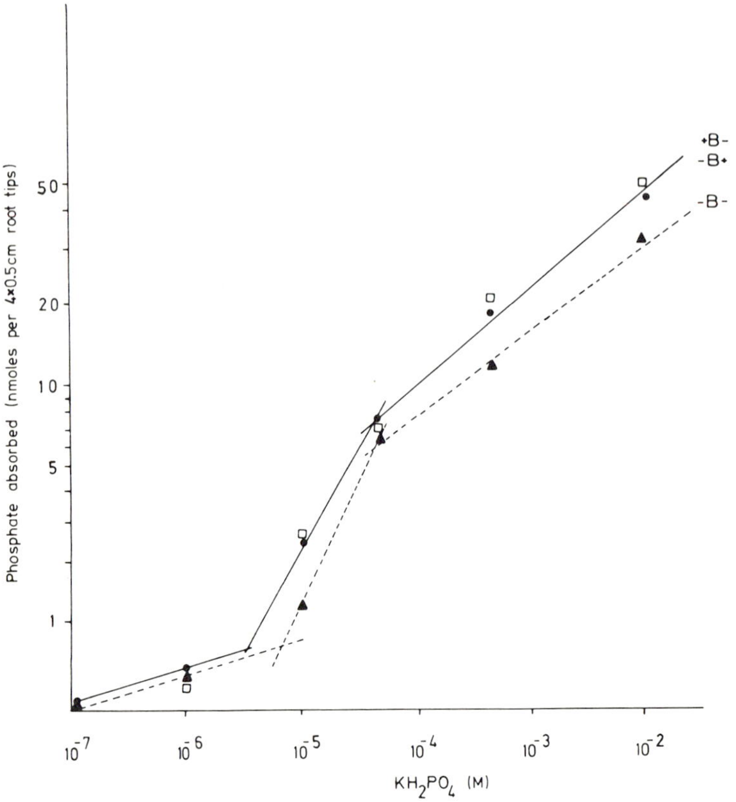

Fig. 2 The absorption of phosphate over 1 h by five 0.5 cm root tips of 7 day old *V. faba* from various concentrations of KH_2PO_4 in 10^{-3} M $Ca(NO_3)_2$ with ^{32}P at 0.05 μCi/ml. −B root tips excised from boron deficient plants. −B+ root tips from boron deficient plants resupplied with 10 μM H_3BO_3 1 h before excision. +B root tips from non-deficient plants. Results plotted on a log vs. log scale.

The ability to absorb phosphate over a wide range of concentration was examined with 0.5 cm root tips excised from deficient seedlings, from deficient seedlings resupplied with boron for 1 h prior to excision, and from non-deficient seedlings. The resulting isotherms were all triphasic and tips from boron deficient seedlings showed reduced absorption of phosphate at all concentrations from 10^{-7} M to 10^{-2} M KH_2PO_4 (Fig. 2). Addition of boron to deficient plants reversed this reduction at all concentrations except at 10^{-6} M and 10^{-7} M where it had little effect. The reduced ability of boron deficient tissue to absorb phosphate at all concentrations raises several interesting points with respect to possible mechanisms. The possibility that an increased efflux could account for the observed response was examined. Roots of 20 whole plants were allowed to absorb labelled phosphate for 3 h, rinsed for 1 min in 3 changes of deionised water and returned to aerated containers with 900 ml of non-radioactive 10^{-5} M KH_2PO_4 in 10^{-3} M $Ca(NO_3)_2$. 2 ml samples were taken at various intervals over the following 4 h and ^{32}P efflux estimated. The efflux curves were indeed different, non-deficient roots showing a much greater loss of previously absorbed phosphate. However, as previously demonstrated with Rb^+ (ROBERTSON and LOUGHMAN, 1973), the efflux curves for the deficient and non-deficient roots were similar in shape and the rate of loss of ^{32}P was closely related to the phosphate absorbed. Total efflux of the boron deficient roots after 4 h was 9.08 % of previously absorbed phosphate, compared to

9.14 % for non-deficient roots and assuming similar partitioning of absorbed phosphate in the root, the membranes appeared to be similarly permeable. Obviously, this is a measure of average efflux over the whole root surface and any differences in the root tip, the region most affected in boron deficiency, could be masked.

In whole roots of *V. faba* and ***Phaseolus aureus*** we were unable to demonstrate differences in the proportion of absorbed phosphate transported to the shoot, even under conditions of severe boron deficiency. However, using plastic racks that permitted feeding only the 0.5 cm of root tip with ^{32}P, clear differences between boron deficient and non-deficient roots were demonstrated. Transport was low, and over the 6 h experimental period, little of the absorbed ^{32}P reached the shoot. After 6 h absorption, the boron deficient root tips had exported only 8.4 % of absorbed phosphate compared to 17.5 % in the non-deficient roots. The increased thickness of the deficient root tip compared to the non-deficient tip (ROBERTSON and LOUGHMAN, 1973), would increase the time taken for transport to the conducting tissue and the anatomical changes accompanying boron deficiency in *V. faba* would also play a role.

As the deficiency develops, the endodermis in the elongating zone of the root tip loses its distinctive characteristics and, together with cells of the pericycle, begins to elongate and divide in a horizontal rather than a vertical manner. Thus the stele increases in both diameter and cell number, at the same time compressing the cortical cells. The xylem tissue is also crushed and disorientated and this also is likely to affect xylem transport. Considerable time has been spent studying morphological changes associated with boron deficiency and attempting to produce homologous responses by treatment with chemicals. Although the development of boron deficiency has been linked to a lesion in meristematic division (WHITTINGTON, 1959), mitotic poisons such as colchicine, dinitramine, and trifluralin (trade names, Cobex, Treflan respectively) did not produce similar anatomical responses to those of boron deficiency. They caused swelling near the root tip, due almost completely to increases in the cortical cell size. However, excess auxin type hormones (10^{-5} to 10^{-7} M) in the culture solution produced a homologous swelling and also produced reductions in phosphate absorption similar to those shown by boron deficient root tips (Table 1). In contrast the mitotic inhibitor studies demonstrate a stimulatory effect on phosphate

Table 1

Phosphate absorbed in 1 h by five 0.5 cm root tips of 7 day old *V. faba* seedlings. The tips were excised after 2 days growth in the various treatment solutions. Absorption at 25°C was from 10^{-5} M KH_2PO_4 in 10^{-3} M $Ca(NO_3)_2$ labelled with ^{32}P at 0.05 μCi/ml. Results, expressed as mean ± S.E., are drawn from several experiments where the controls (+B) differed by less than 4 %.

2 days pretreatment 10^{-3} M $Ca(NO_3)_2$	Phosphate absorbed by five 0.5 cm tips (nmoles/h)
+B	2.23 ± 0.08
–B	1.31 ± 0.06
+B + 2,4D 10^{-5} M	1.20 ± 0.18
+B + Colchicine 10^{-4} M	4.29 ± 0.15
+B + Colchicine 10^{-5} M	3.70 ± 0.84
+B + dinitramine 10^{-5} M	2.88 ± 0.26

Table 2

The effect of auxin type plant hormones and mitotic inhibitors on the absorption of phosphate by 5 excised 0.5 cm root tips of *V. faba*. Absorption was from 4 ml 10^{-5} M KH_2PO_4 in 10^{-3} M $Ca(NO_3)_2$ over 1 h with ^{32}P at 0.05 μCi/ml. Results as mean ± S.E. The chemical treatment was added at the same time as ^{32}P.

Treatment	Phosphate absorbed by five 0.5 cm tips (nmoles/h)	% control
Control	1.531 ± 0.015	100
Control + 10^{-6} M IAA	0.993 ± 0.039	64.6
Control + 10^{-6} M 2,4D	0.853 ± 0.055	58.8
Control + 10^{-6} M dinitramine	1.275 ± 0.239	83.2
Control + 10^{-5} M colchicine	1.505 ± 0.277	98.3
Control + 10^{-6} M kinetin	1.338 ± 0.172	87.3

absorption quantitatively related to their effect in increasing root tip thickness (Table 1). The presence of excess auxin during a 1 h absorption period with previously non-treated, excised root tips, also reduced phosphate absorption. A similar short term treatment with either mitotic inhibitors, or kinetin, had little effect on phosphate absorption (Table 2).

D. Discussion

Boron deficiency is usually first detected in the root rather than the shoot (WHITTINGTON, 1957). In our experiments there was no effect on the shoot for up to 3 weeks in boron deficient media suggesting either that boron requirements of the two organs differ substantially or that there is preferential translocation of boron from the cotyledons to the shoots. Addition of boron to the cotyledons overcame boron deficiency in the root tip indicating the presence of a pathway for boron transport in that direction and that shoots may be less susceptable to boron deficiency than roots.

In control plants the proportion of applied phosphate translocated from the cotyledons to both the root and the shoot was found to be similar (Fig. 1), implying that both organs presented similar sink capacities for phosphate. However, in boron deficiency, the translocation of phosphate to the root was markedly reduced suggesting a reduction in this capacity. As boron deficiency also reduces sugar translocation from leaves to roots (GAUCH, 1972) it appears likely that boron deficiency affects sink capacity rather than some specific transport step.

Boron status did not affect the shapes of either the absorption isotherms or the efflux curves indicating that boron is probably not involved in a specific transport step. As the relative efflux was similar in both treatments it also appears membrane permeability is little affected in boron deficiency, although we have no direct evidence for the root tip region. At all concentrations of phosphate up to 10^{-2} M, absorption by root tips was reduced in boron deficiency and the absorption isotherms showed no signs of saturation up to 10^{-2} M phosphate, concentrations 1000 times those of normal soil levels (BIELESKI, 1973). Rates of ion absorption can be controlled by internal ionic concentrations over a wide range of external concentrations (LEIGH and WYN JONES, 1973; SMITH, 1973). These reductions in net influx were not associated with enhanced efflux and it was suggested that plant hormones are involved in the control system (LEIGH and WYN JONES, 1973).

One specific difference between boron deficient and non-deficient roots was the reduced rate of transport out of deficient root tips. It is possible that metabolic pathways prior to translocation could be affected, but on average, in the whole root, boron deficiency does not alter transport of phosphate (ROBERTSON and LOUGHMAN, 1974). The increased thickness of the boron deficient root would have some effect, but the severe changes in polarity and structure of cells in the stele would be important.

The disruption within the stele in boron deficiency was similar to that produced by excess auxin type hormones and both treatments have similar effects on ion absorption (Tables 1, 2), echoing the suggestion already made by COKE and WHITTINGTON (1968) that many of the effects of boron deficiency could be mediated via excess endogenous auxins.

The hypothesis that boron deficiency causes, or leads to, an excess of endogenous auxins could be invoked to explain the greater sensitivity of the roots rather than the shoots to boron deficiency. Shoots are unaffected by internal concentrations of auxins up to 1000 times those normally associated with roots. Thus if boron deficiency does increase internal auxin concentrations it is likely that the root would initially be more affected than the shoot. Changes in hormone balances can also have large effects on sink-source relationships in plants (SETH and WAREING, 1967), possibly explaining reduced translocation to the root tip, after the deficiency has developed.

The absorption of phosphate, or indeed any ion, can be used as a tool in examining disruptions of cell metabolism. In this case boron deficiency reduces phosphate absorption, but it is unlikely that boron has a specific role in phosphate absorption. It is more probable that the hormonal balance of the root is affected in boron deficient plants and this in turn affects transport of phosphate. It is possible that hormones have a role in balancing the demand and supply factors in ion absorption and that in boron deficient plants this interaction is affected. However, excess hormones could cause disruptions in metabolic processes especially concerned in energy provision and thereby alter absorption rates of ions.

E. References

BIELESKI, R.L., Ann. Rev. Plant Physiol. **24**, 225–252 (1973)
COKE, L., WHITTINGTON, W.J., J. Exp. Bot. **19**, 295–308 (1968)
CROSSETT, R.N., LOUGHMAN, B.C., New Phytol. **65**, 459–468 (1966)
GAUCH, H.G., Inorganic Plant Nutrition, pp. 239–259, Dowden, Hutchinson and Ross Inc., Stroudsburg, Pa. 1972
LÄUCHLI, A., Int. J. Appl. Rad. Isotop. **20**, 265–270 (1969)
LEIGH, R.A., WYN JONES, R.G., J. Exp. Bot. **24**, 787–795 (1973)
LOUGHMAN, B.C., ROBERTSON, G.A., Atti. del. IX Symposio Internazionale di Agrochem. Planta Ala. 1972
ROBERTSON, G.A., LOUGHMAN, B.C., J. Exp. Bot. **24**, 123–129 (1973)
ROBERTSON, G.A., LOUGHMAN, B.C., New Phytol. **73**, 291–298 (1974)
SETH, A.K., WAREING, P.F., J. Exp. Bot. **18**, 65–77 (1967)
SMITH, F.A., New Phytol. **72**, 769–782 (1973)
WHITTINGTON, W.J., J. Exp. Bot. **8**, 353–367 (1957)
WHITTINGTON, W.J., J. Exp. Bot. **10**, 93–103 (1959)

Hormonal Regulation of Ion Transport in Parenchyma Tissue

R.F.M. Van Steveninck

A. Introduction

In a short space of time an impressive body of evidence has accumulated indicating that plant hormones play a regulatory role in salt transport (for a review see VAN STEVENINCK, 1975). It has been shown that ABA promotes net uptake of K^+, Na^+, and Cl^- in aged beet slices and also changes the K^+–Na^+ selectivity to favour Na^+ uptake (VAN STEVENINCK, 1972a). On the other hand kinins (kinetin and benzyladenine) prevent the development of a capacity to absorb K^+ and Na^+ in fresh beet slices, but a distinct short term promotion of cation uptake occurs (VAN STEVENINCK, 1972b). The long term effects of kinins could be interpreted as a condition of salt saturation (influx = efflux) at relatively high external concentrations, especially for K^+. The latter observation is of immediate interest in relation to PITMAN's proposal that hormones play a key role in the regulation of ion transport from the root to the shoot (PITMAN and CRAM, 1973; PITMAN et al., 1974).

IAA has been shown to stimulate a H^+ pump in *Avena* coleoptiles (HAGER et al., 1971). Its operation through K^+/H^+ exchange (HASCHKE and LÜTTGE, 1973) and the fact that ABA decreases the IAA induced H^+ secretion (RAYLE and JOHNSON, 1973) indicates that IAA should be included in the spectrum of hormones which play a regulatory role in salt transport.

In the following work with ABA, kinetin, benzyladenine and IAA, $^{42}K^+$, $^{24}Na^+$, $^{86}Rb^+$, and $^{36}Cl^-$ were used to measure the effects of the above hormones on the apparent influxes and on K^+–Na^+ selectivity in beetroot parenchyma tissue. The experiments were designed to make a clear distinction between the regulatory effects of hormone treatment in the development of the capacity for net salt uptake leading to new equilibrium levels, and the effects of hormones on aged tissue already capable of net ion uptake.

B. Materials and Methods

Slices of beetroot (*Beta vulgaris* L.) storage tissue, 1 mm thick and 15 mm in diameter, were cut with a hand microtome and used after 3 rinses in deionised water. Zero hour for fresh slices is taken at this time and slices were then either lightly blotted and distributed in lots of 50 slices to treatment flasks each containing 250 ml 0.5 mM KCl + 0.5 mM NaCl with or without hormone added at a concentration of $3.8 \cdot 10^{-5}$ M or, alternatively, allowed to age in aerated deionised water (about 500 discs to 1 litre) for 92 h prior to hormone treatment. All flasks were aerated with CO_2-free air at a rate of approximately 150 ml per min. At prescribed intervals after the commencement of hormone treatment discs were exposed for 1 or 2 h to small volumes of tracer solutions (4 ml 0.5 mM NaCl + 0.5 mM KCl with or without $3.8 \cdot 10^{-5}$ M hormone solution in 25 ml conical flasks placed in a shaking water bath at 23°C). The slices were subsequently rinsed for three 5 min periods in 3 changes of 5 ml deionised water at 0°C to remove free space ions, then lightly blotted and stuck to planchettes for counting with an end window counter. Slices were also analysed for tracer and ionic content.

C. Results

I. Fresh Beetroot Slices

1. Effect of Kinetin

All external solutions contained 0.5 mM KCl + 0.5 mM NaCl at zero hour and it is shown in Fig. 1 that the presence of kinetin prevents the development of net uptake from these solutions. A condition of salt saturation is implied and the concentrations of the external solution at which this equilibrium is reached are relatively high, especially for K^+ (compare VAN STEVENINCK, 1972b). Slices were sampled at points I, II, III, and IV for exposure to tracer solutions. It can be seen in Table 1 that the apparent $^{36}Cl^-$ influx is inhibited by kinetin from 24 h onwards, while $^{24}Na^+$ influx is inhibited appreciably only after 71 h. It appears that an increased efflux of Na^+ at the earlier stages of aging may be the reason for the diminished net uptake of Na^+. The addition of 0.5 mM KCl to the tracer solution resulted in an approximately 30 % inhibition of Na^+ influx which was not affected by kinetin treatment.

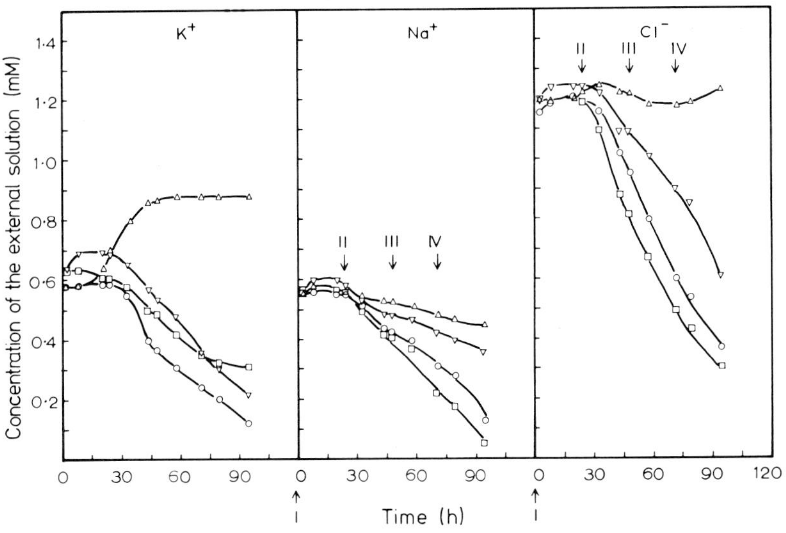

Fig. 1 The effects of hormone treatments on net ion movement of K^+, Na^+, and Cl^- during the aging of beet slices. Ordinate: concentration of the external solution in mM. ○—○, Control; △—△, 3.8 · 10^{-5} M kinetin; □—□, 3.8 · 10^{-5} M ABA; ▽—▽, 3.8 · 10^{-5} M IAA. All solutions contained 0.5 mM KCl + 0.5 mM NaCl at the start of the experiment; ↑ and ↓ I, II, III, and IV sampling of slices at 0, 24, 48, and 71 h respectively in order to measure $^{24}Na^+$ and $^{36}Cl^-$ influx.

In a similar experiment the apparent influx of $^{42}K^+$ and $^{36}Cl^-$ was measured at 4 intervals during aging (Table 2). Again at 24 h $^{36}Cl^-$ influx is strongly inhibited but in contrast with previous results this inhibition is followed by a promotion of $^{36}Cl^-$ influx. Inhibition of $^{42}K^+$ influx at 24 h and 48 h is also followed by a slight promotion at 70 h. These results show that the effects of kinetin can be quite inconsistent.

The percentage inhibition of $^{42}K^+$ influx by Na^+ is generally less than the inhibition of $^{24}Na^+$ influx by K^+ (approximately 20 % versus approximately 40 %); again kinetin does not

Table 1

The effects of kinetin ($3.8 \cdot 10^{-5}$ M), ABA ($3.8 \cdot 10^{-5}$ M) and IAA ($3.8 \cdot 10^{-5}$ M) on the apparent influx of $^{24}Na^+$ and $^{36}Cl^-$ measured over a period of 2 h at various intervals (I, II, III, and IV) during the aging of beet slices (means of four determinations; values between brackets are percentages of controls).

	Treatment	I (0 h)	II (24 h)	III (48 h)	IV (71 h)
cpm for $^{36}Cl^-$	Control	145(100)	749(100)	1210(100)	2010(100)
	Kinetin	160(110)	234 (65)	790 (65)	1283 (64)
	ABA	171(118)	985(132)	2150(178)	3080(155)
	IAA	184(127)	288 (38)	1060 (88)	1373 (68)
10^{-3} x cpm for $^{24}Na^+$	Control	15.8(100)	42.9(100)	53.7(100)	66.8(100)
	Kinetin	16.9(107)	37.9 (89)	46.4 (87)	44.8 (67)
	ABA	13.7 (87)	49.2(115)	54.1(101)	80.8(121)
	IAA	17.2(109)	43.2(101)	43.4 (81)	44.8 (67)
Percentage inhibition	Control	18.4	30.6	40.4	32.0
of $^{24}Na^+$ uptake	Kinetin	13.6	28.0	31.5	33.5
due to the presence	ABA	17.8	28.8	39.4	12.1
of 0.5 mM K^+	IAA	26.8	34.5	49.5	37.1

Table 2

The effects of kinetin ($3.8 \cdot 10^{-5}$ M), ABA ($3.8 \cdot 10^{-5}$ M) and IAA ($3.8 \cdot 10^{-5}$ M) on the apparent influx of $^{42}K^+$ and $^{36}Cl^-$ measured over a period of 2 h at various intervals (I, II, III, and IV) during the aging of beet slices (means of four determinations; values between brackets are percentages of controls).

	Treatment	I (0 h)	II (24 h)	III (48 h)	IV (71 h)
cpm for $^{36}Cl^-$	Control	224(100)	275(100)	455(100)	1087(100)
	Kinetin	326(146)	108 (39)	701(154)	1680(155)
	ABA	259(116)	1010(404)	3080(682)	2830(261)
	IAA	240(107)	111 (40)	488(107)	598 (55)
10^{-3} x cpm for $^{42}K^+$	Control	25.8(100)	106(100)	94(100)	88(100)
	Kinetin	31.2(121)	62 (58)	76 (81)	93(106)
	ABA	30.1(117)	84 (79)	112(119)	89(101)
	IAA	28.4(110)	86 (81)	90 (96)	97(110)
Percentage inhibition	Control	–	4.7	19.2	20.5
of $^{42}K^+$ uptake	Kinetin	8.3	9.8	27.6	16.1
due to the presence	ABA	25.6	–	20.5	25.9
of 0.5 mM Na^+	IAA	–	11.6	–	6.2

seem to have a significant effect on this inhibition and it appears that the kinetin-induced higher equilibrium concentration of K^+ compared to Na^+ is caused by an enlarged efflux of K^+.

2. Effect of ABA
Fig. 1 shows that ABA promotes net Cl^- and Na^+ uptake, but not net K^+ uptake. ABA causes a consistent increase of $^{36}Cl^-$ influx, but $^{24}Na^+$ influx is significantly promoted only at 71 h (Table 1), and hence we must assume a reduction of Na^+ efflux to explain the increased net uptake in the presence of ABA during the period before 71 h. Only when ABA stimulates $^{24}Na^+$ influx at 71 h does a distinct decrease in K^+ inhibition of $^{24}Na^+$ influx occur (Table 1). This result confirms the ABA induced change of K^+–Na^+ selectivity in favour of Na^+ observed earlier (VAN STEVENINCK, 1972a). Table 2 confirms that there is a substantial stimulation of $^{36}Cl^-$ influx, while $^{42}K^+$ influx remains unaffected.

3. Effect of IAA
Contrary to COMMONER and MAZIA's (1942) finding in potato slices, it was shown that IAA inhibits net uptake of K^+, Na^+ and Cl^- in beetroot slices (Fig. 1). However, IAA induced leakage of K^+ and an absence of ion uptake stimulation in beet slices were recorded by VAN STEVENINCK (1965). Tables 1 and 2 show that the IAA effects on fluxes are somewhat erratic with inhibition of $^{36}Cl^-$ influx after 24 h, inhibition of $^{24}Na^+$ influx after 48 h, and inhibition of $^{42}K^+$ influx at 24 h becoming a promotion after 48 h. There is a general tendency for a greater percentage inhibition of $^{24}Na^+$ influx in the presence of K^+ when IAA is added (Table 1), while IAA causes a stimulation of $^{42}K^+$ influx when external Na^+ is present (Table 2). Hence, IAA induces a distinct selectivity towards K^+ uptake.

II. Aged Beetroot Slices

1. Effect of Benzyladenine (BA)
When slices have been aged until the external solutions are depleted (i.e. to very low equilibrium concentrations for K^+, Na^+, and Cl^-) benzyladenine does not affect net uptake until at least 9 h after its application (Table 3). The inhibition of $^{36}Cl^-$ influx by BA within 1 h was therefore somewhat surprising. The net uptake of K^+ was not affected by BA over the period tested (25 h), however $^{86}Rb^+$ influx was promoted after 4–5 h, again a rather surprising result which might be explained by a concomitant larger efflux of K^+. However, caution is necessary in using $^{86}Rb^+$ as a tracer for K^+ (MARSCHNER and SCHIMANSKY, 1971), especially when a possible hormone controlled regulation of ion selectivity may be involved.

2. Effect of ABA
ABA promoted both net uptake of Cl^- and $^{36}Cl^-$ influx within the 1st hour of its application and consistently over the remaining test period of 25 h. $^{36}Cl^-$ influx was stimulated to a much greater extent than net uptake of Cl^- and hence the effect of influx stimulation may partly be cancelled by a greater efflux (Table 3).

The stimulatory effect of ABA on $^{86}Rb^+$ influx becomes apparent after 2 h and is of about the same magnitude as the stimulation of net K^+ uptake which also starts 2 h after the ABA application (Table 3).

3. Effect of IAA
IAA seems to have little effect on net uptake and on influx in aged slices, some inhibition of net uptake of K^+ and Cl^- being apparent only towards the end of the 25 h test period (Table 3).

Table 3

The effect of hormone treatment (3.8 · 10^{-5} M) on the apparent influx of $^{36}Cl^-$ and $^{86}Rb^+$ and on the net uptake of Cl^- and K^+ in 92 h aged slices of beetroot. Influx (cpm · 10^{-3}) and uptake (mequiv kg^{-1} h^{-1}) measurements over 1 h periods at various intervals following the exposure of slices to hormone treatment (means of four determinations; values between brackets are percentages of controls 0–1 h period).

Treatment	0–1 h	2–3 h	4–5 h	8–9 h	12–13 h	24–25 h
$^{36}Cl^-$ influx						
Control	1.80(100)	1.85(103)	2.47(137)	2.26(125)	2.27(126)	2.63(146)
Relative values						
Control	100	100	100	100	100	100
ABA	122	132	131	162	215	202
BA	79	85	72	91	96	79
IAA	114	105	110	110	93	101
Net Cl^- uptake						
Control	0.53(100)	0.61(115)	0.68(128)	0.71(134)	0.72(136)	0.80(151)
Relative values						
Control	100	100	100	100	100	100
ABA	117	118	120	131	121	115
BA	96	95	94	101	79	54
IAA	91	97	103	96	93	84
$^{86}Rb^+$ influx						
Control	8.35(100)	8.35(100)	8.12 (97)	7.71 (93)	7.59 (91)	7.97 (96)
Relative values						
Control	100	100	100	100	100	100
ABA	90	112	112	139	154	151
BA	97	94	111	124	139	120
IAA	91	102	102	105	104	94
Net K^+ uptake						
Control	0.38(100)	0.53(140)	0.71(187)	1.00(254)	1.17(308)	1.25(329)
Relative values						
Control	100	100	100	100	100	100
ABA	95	187	138	122	130	141
BA	90	119	97	108	102	101
IAA	110	94	101	107	89	87

It should be noted that $^{86}Rb^+$ influx in the control treatment does not change with time, while net uptake of K^+ does (Table 3, see relative figures for control treatment). This phenomenon may be due to a progressive lowering of efflux which is rather high at the beginning after handling of the slices. This seems unavoidable however carefully the transfers are made.

D. Conclusions

The following are some of the main points which emerge from this work.

1) The observed stimulation of net Cl^- uptake by ABA can be ascribed to a stimulation of the apparent Cl^- influx, and it is likely that this ABA stimulated Cl^- influx operates against an electrochemical gradient (HIGINBOTHAM, 1973). ABA stimulation of cation uptake is probably mainly due to a reduction in efflux of these ions.

2) When Na^+ influx is stimulated by ABA the inhibiting effect of competing K^+ ions is much reduced. At present there is no real explanation for the ABA induced change in affinity of the uptake mechanism.

3) Kinetin causes a condition of flux equilibrium at relatively high external concentrations, especially for K^+. The effects on tracer influx are variable, both promotion and inhibition being observed, but in general the results imply that kinetin promotes efflux of ions and hence it appears that the role of kinetin is in assisting the passage of ions across the plasma membrane. This property may be of importance in facilitating long distance transport via the symplasm and therefore supports the proposal that kinins play a key role in the regulation of ion transport from the root to the shoot (PITMAN and CRAM, 1973).

4) IAA did not stimulate salt uptake in beet slices as observed by COMMONER and MAZIA (1942) in potato tissue. This result indicates that some interesting differences may exist which seem to divide various storage tissues into at least two groups with distinctly different responses: those which show H^+ extrusion and a Tris effect during aging (beetroot , carrot, parsnip) and those which do not (potato, swede, artichoke) (compare VAN STEVENINCK et al., 1973). IAA however did have an effect on ion selectivity favouring K^+ uptake over Na^+ uptake which is opposite to the effect of ABA.

5) It was apparent that IAA and kinetin had variable effects on ion influx. These variable results may arise from possible differences in hormone status of the root at the beginning of the experiment which in turn may depend on the growing conditions (VAN STEVENINCK, 1972a). The variability may also depend on ruling electrochemical gradients which could easily vary from one root to another during aging or with a change of concentration of the external solution.

E. Acknowledgements

The work was supported by the Australian Research Grants Committee.

F. References

COMMONER, B., MAZIA, D., Plant Physiol. **17**, 682–685 (1942)
HAGER, A., MENZEL, H,, KRAUSS, A., Planta **100**, 47–75 (1971)
HASCHKE, H.-P., LÜTTGE, U., Z. Naturforsch. **28c**, 555–558 (1973)
HIGINBOTHAM, N., Ann. Rev. Plant Physiol. **24**, 25–46 (1973)
MARSCHNER, H., SCHIMANSKY, C.H., Z. Pflanzenernaehr. Bodenk. **128**, 129–143 (1971)
PITMAN, M.G., CRAM, W.J., in Ion Transport in Plants (ed. W.P. Anderson), pp. 465–481, Academic Press, London and New York 1973
PITMAN, M.G., LÜTTGE, U., LÄUCHLI, A., BALL, E., J. Exp. Bot.**25**, 147–155 (1974)
RAYLE, D.L., JOHNSON, K.D., Plant Physiol. **51**, Suppl., 2 (1973)
VAN STEVENINCK, R.F.M., Nature **205**, 83–84 (1965)
VAN STEVENINCK, R.F.M., Z. Pflanzenphysiol. **67**, 282–286 (1972a)
VAN STEVENINCK, R.F.M., Physiol. Plant. **27**, 43–47 (1972b)
VAN STEVENINCK, R.F.M., in Encyclopedia of Plant Physiology New Series (eds. U. Lüttge, M.G. Pitman), Springer-Verlag, Berlin–Heidelberg–New York 1975 (in press)
VAN STEVENINCK, R.F.M., MITTELHEUSER, C.J., VAN STEVENINCK, M.E., in Ion Transport in Plants (ed. W.P. Anderson), pp. 251–269, Academic Press, London and New York 1973

Auxin Transport and *in vitro* Auxin Binding

R. Hertel

A. Introduction

The plant hormone 3-indole acetic acid (IAA) is actively transported through coleoptile tissue (GOLDSMITH, 1969) and at the same time plays a regulatory role in at least two transport systems: proton secretion is promoted by hormone application (RAYLE, 1973) and auxin transport itself can also be modified by its substrate (HERTEL and FLORY, 1968). This regulatory function as well as the question of auxin control of cell elongation and tropisms will find a satisfactory explanation only by a model system which is made up of biochemical elements such as proteins, whose affinities towards auxins and other factors, conformational changes, enzymic activities, etc. are characterized and whose localization in the cell is understood; these elements and their interactions should then allow predictions of e.g. the early effects of the hormone, and the kinetic properties of the active transport system.

To this end the auxin problem must be attacked from two sides: the primary receptor sites should be studied biochemically and, on the other hand, all features of the early physiological events, including apparent paradoxes, have to be analyzed in detail.

B. Results and Discussion

I. *In vitro* Binding of Auxin and Inhibitors of Auxin Transport

The binding site for 1-naphthylphthalamic acid (NPA), an inhibitor of auxin transport, can be found in particulate fractions from corn coleoptiles using a simple pelleting technique (LEMBI et al., 1971; THOMSON et al., 1973). The method used is very sensitive since the particulate binding sites are concentrated by high speed centrifugation into the small volume of the pellet.

Homogenates or suspensions of particulate cell fractions such as membrane vesicles are mixed at 4°C with low concentrations of the labelled, physically interesting substance. To part of the sample, unlabelled material is added at a concentration higher than the presumed saturation of the sites. In the pellet radioactivity is specifically associated with the sedimented particles at low ligand concentration, while at the high concentration of unlabelled substance only unspecific retention — water volume of the pellet and/or partitioning — can be seen.

The binding sites for the transport inhibitor NPA are — unexpectedly — not competitively occupied by auxins, but showed strong affinity towards morphactins (THOMSON and LEOPOLD, 1974).

Most, if not all of the NPA binding sites are localized in membrane cell fractions that preferentially stain after a procedure which *in situ* seems to preferentially contrast the plasmalemma (LEMBI et al., 1971).

This intracellular localization is further supported by the fact that the density of the fractions containing the NPA-receptors agrees exactly with the density of the material that displays an alkali-stimulated ATPase (BALKE et al., 1974); in addition, the very quick and strong action of NPA on the auxin transport system indirectly indicates the cell surface. The NPA receptor has a dissociation constant $K_D = 2 \cdot 10^{-8}$ M, and its concentration in the tissue amounts to ca. $5 \cdot 10^{-8}$ M.

Subsequently, specific auxin binding was found (HERTEL et al., 1972). When low concentrations (e.g. 10^{-7} M) of labelled ^{14}C-IAA or 1-naphthaleneacetic acid (^{14}C-NAA) were added *in vitro* to homogenates of corn coleoptiles, radioactivity was reversibly bound to pelletable particles. The binding was auxin-specific since among many compounds tested, only auxins and auxin analogues that are known to interact directly with auxin transport and/or auxin-promoted growth were found to interfere with this binding (HERTEL et al., 1972). The following experiment of THOMSON (1973) further illustrates this specificity and the binding test in general. The auxin analogues of the trichlorophenoxyacetic acids (T) display very different auxin activity and similarly different transportability depending on the aromatic substitution (JACOBS, 1968). 2, 3, 6 – T, 2, 3, 4 – T, and 2, 4, 5 – T were tested for their ability to displace ^{14}C-NAA from its binding site, as shown in Fig. 1. The tested substances are competing corresponding to their physiological activity: NAA > 2, 4, 5 –T > 2, 3, 4 – T > 2, 3, 6 – T.

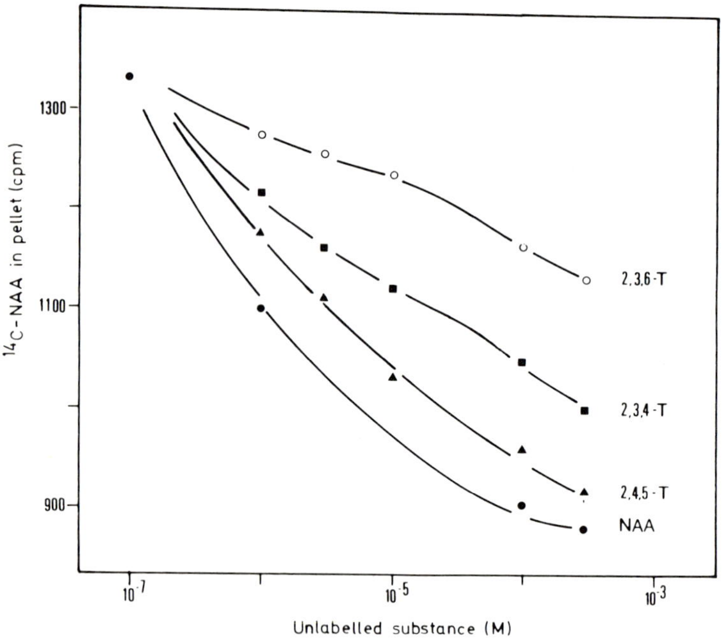

Fig. 1 Binding of ^{14}C-NAA in the presence of different concentrations of unlabelled trichlorophenoxyacetic acids and of NAA.
Corn coleoptiles (5 days old) were homogenized as described by HERTEL et al. (1972; medium Ia). ^{14}C-NAA ($1.5 \cdot 10^{-7}$ M; specific activity 54 mCi/mM) was added to the total homogenate at pH 5.5. 2 ml samples were provided with the concentration of unlabelled substance indicated at the abscissa. Each point is based on 4 parallel tests.

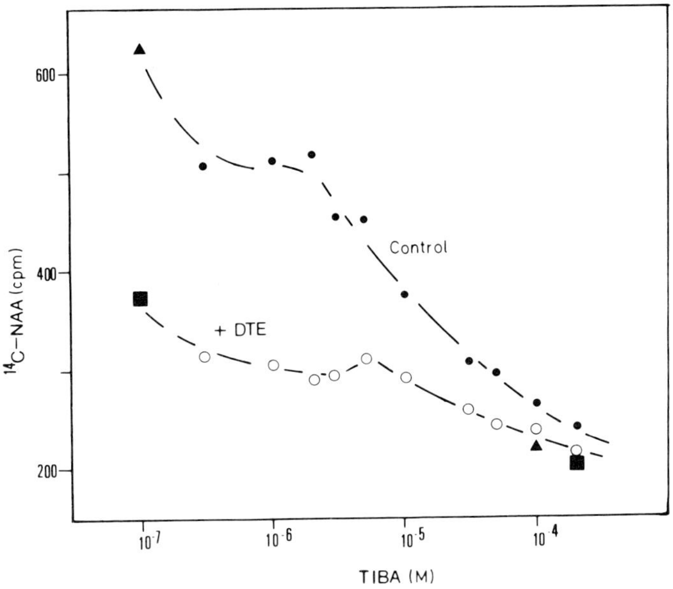

Fig. 2 Effect of different TIBA-concentrations and of 10^{-3} M DTE (= CLELAND's reagent, dithioerythritol) on ^{14}C-NAA binding.
The test was similar to that in Fig. 1 with the following modifications: no reducing agent was added during extraction; the particulate material was pelleted (100 000 X g, 20 min) and resuspended in fresh medium before the test. ▲, ■ = NAA only.

The binding of antiauxins and inhibitors like TIBA (2, 3, 5 – triiodobenzoic acid) warrants further investigation. Fig. 2 shows TIBA competing for the auxin site. This experiment also documents that the reducing agent dithioerythritol (-SH) strongly decreases the binding.

Recently the binding test has been greatly improved by removing the particulate fractions from the supernatant by pelleting and resuspending (RAY and HERTEL, unpublished). A sample of the results is shown in Fig. 3, where affinity and number of binding sites are determined with NAA in the presence and absence of a supernatant factor which is presently further studied. Without the factor the K_D is $5 \cdot 10^{-7}$ M; the concentration of receptors is 32 pmoles/0.6 g of tissue ($5 \cdot 10^{-8}$ M).

The distribution of the auxin binding activity in isopycnic sucrose gradients (RAY, unpublished) suggests a cellular localization at the endoplasmatic reticulum; a minor part of the receptors may reside at the plasmalemma. The auxin binding fractions can be obtained free of mitochondrial cytochrome oxidase; nevertheless they contain flavoproteins and cytochromes.

II. Optimum Function and Adaptation in the Auxin Transport System

The analysis of early geotropic bending has revealed two surprising features: (a) the bending reaction as well as the lateral auxin transport start in the 'wrong' direction; (b) after 15–20 min this response pattern changes to the 'normal' downward transport and upward bending although the external stimulus remains constant (HILD and HERTEL, 1972).

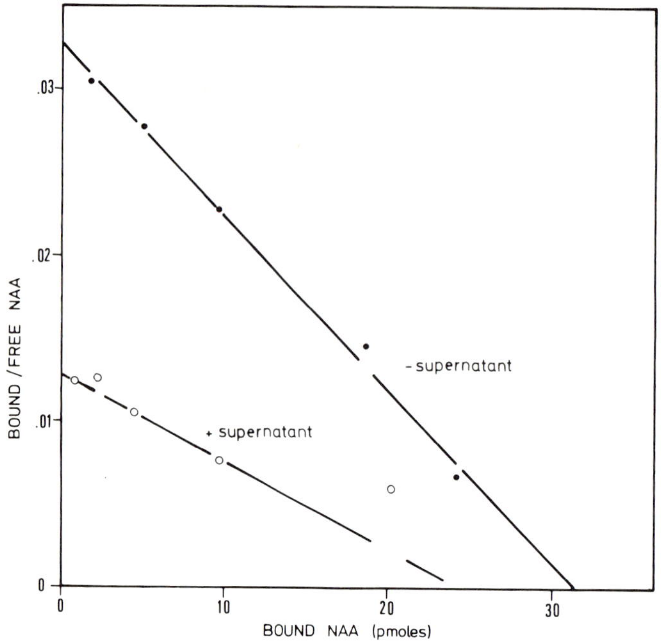

Fig. 3 Concentration dependence of NAA-binding in the presence and absence of 15 % supernatant (SCATCHARD-plot).
Coleoptiles were homogenized in 0.25 M sucrose, 50 mM Tris, 1 mM EDTA, 0.1 mM $MgCl_2$, pH 8. The homogenate was precleared at 2500 X g, 10 min; particulate material was then obtained at 100 000 X g, 20 min as a pellet which was washed in 0.25 M sucrose, 10 mM Na-citrate, 0.5 mM $MgCl_2$, pH 6 and repelleted again. The final resuspension for the test was in 0.25 M sucrose, 10 mM Na-citrate, 5 mM $MgSO_4$, pH 5.5 with or without 15 % of the supernatant. Test and evaluation were similar to Fig. 2 in HERTEL et al. (1972). Each point is based on 3 replicate determinations.

Using different g-forces and the amylomaize mutant containing smaller statoliths an optimum curve can be derived for the initial geotropic response; a high stimulus leads initially to an inhibition of auxin secretion while a medium size stimulus promotes. Symmetrical, alternating geotropic prestimulation resulted in a localized desensitisation (= adaptation) to the subsequent gravity stimulus (HILD and HERTEL, 1972).

Illumination with blue light and also the phytochrome system rather directly interfere with the geotropic adaptational level; this observation suggests a common filter in these sensory pathways (HILD, 1973).

The hypothesis is proposed that the auxin transport pump itself is the site of the localized adaptation and of the optimum type transducer where high substrate concentration or high stimulus intensity inhibit the output. The data in Fig. 4 support this scheme. The 'wrong' initial geotropic response is more pronounced at high concentrations of applied auxin.

Formal models to explain such features are easy to construct and provide new test implications.

The non-linearity or 'cooperativity' of the auxin transport system has already been indicated by the finding that the strong auxin IAA promotes the transport of the weaker auxin 2, 4 – dichlorophenoxyacetic acid (HERTEL and FLORY, 1968).

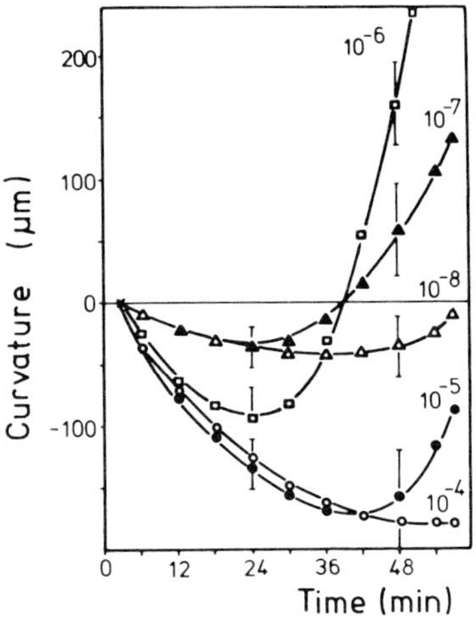

Fig. 4 Initial geotropic curvature of decapitated corn coleoptiles provided with different concentrations of IAA (data of HILD, 1973).
IAA (molar concentration as shown in the Fig.) was applied in agar blocklets to vertical coleoptiles at t = -120 min before horizontal exposure. Each value is the mean from 28 continuously monitored coleoptiles.

C. References

BALKE, N.E., SZE, H., LEONARD, R.T., HODGES, T.K., in Membrane Transport in Plants (eds. U. Zimmermann, J. Dainty), Springer Verlag, Berlin, Heidelberg, New York 1974

GOLDSMITH, M.H.H., in Physiology of Plant Growth and Development (ed. M.B. Wilkins), pp. 125–162, McGraw-Hill, London 1969

HERTEL, R., FLORY, R., Planta **82**, 123–144 (1968)

HERTEL, R., THOMSON, K.-S., RUSSO, V.E.A., Planta **107**, 325–340 (1972)

HILD, V., HERTEL, R., Planta **108**, 259–270 (1972)

HILD, V., Frühe geotropische Reaktionen: gemessen als Krümmung und seitlicher Auxintransport bei Koleoptilen von ***Zea mays*** L. Dissertation, Freiburg (1973)

JACOBS, W.P., Plant Physiol. **43**, 1480–1495 (1968)

LEMBI, C.A., MORRE, D.J., THOMSON, K.-S., HERTEL, R., Planta **99**, 37–45 (1971)

RAYLE, D.L., Planta **114**, 63–73 (1973)

THOMSON, K.-S., Versuche zur *in-vitro*-Bindung von Auxinen und Hemmstoffen des Auxintransportes an partikuläre Zellfraktionen aus Koleoptilen von ***Zea mays*** L. Dissertation, Freiburg (1973)

THOMSON, K.-S., HERTEL, R., MÜLLER, S., TAVARES, J.E., Planta **109**, 337–352 (1973)

THOMSON, K.-S., LEOPOLD, A.C., Planta **115**, 259–270 (1974)

Round Table Discussion 9

Chairmen: L. Bergmann and R.F.M. van Steveninck

VAN STEVENINCK opened the discussion by saying that, although we are still at the descriptive phase in the study of hormone action, there have been some exciting recent developments. With respect to the site and nature of primary auxin action, hormonologists have had their days of active water uptake, enzymology, pectinmethylesterase and calcium chelation hypotheses; finally the ball has come into our court with the excitement about hormone stimulated proton extrusion.

LÜTTGE then raised the question of the counter-ion involved in proton extrusion. One of his students, Haschke, has studied this problem using ***Avena*** coleoptiles and LÜTTGE described his experiments and results. IAA was used to stimulate growth and the experiments were carried out at various levels between 2 and 10 mM of potassium phosphate buffer. In every case there was a one—to—one H^+/K^+ exchange. ZIEGLER commented that since growth needs potassium, is it not more reasonable to assume that the primary effect of auxin is a K^+ uptake and that the H^+ efflux is secondary. MARRE pointed out, however, one can get auxin stimulated H^+ efflux for many hours in the absence of potassium in the external medium. And HERTEL said that it was reasonable to take H^+ efflux as the primary effect, because growth can only take place when the walls are made more 'stretchable'. As demonstrated by Cleland and others, H^+ does this job; and, although of course auxin needs metabolism for its action (presumably because the H^+ efflux pump needs energy), low pH will make cell walls more stretchable in the absence of metabolism. DAINTY pointed out, and HERTEL agreed, that growth needs more than the extrusion of H^+ to soften the wall; salt uptake must continue because the osmotic pressure, and hence the turgor pressure, of an expanding cell must be maintained more or less constant. BENTRUP then described some preliminary studies on the effect of auxin on K^+ and Cl^- fluxes in ***Petroselinum*** cell cultures. There was no effect on the K^+ fluxes but the Cl^- fluxes were affected in such a way as to suggest that Cl^- was the co-ion for the H^+ efflux; it is almost as if the cell is excreting HCl! MARRE commented that this could be a simple consequence of the fact that when auxin activates the H^+ pump, the cell electric potential becomes more negative, JAFFE suggested that one of the key consequences of auxin action may not be H^+ efflux but Ca^{2+} influx, for Ca^{2+} is essential for all (animal) secretion processes and wall growth is essentially a secretion process; he thought one could test this idea by using an inhibitor of Ca^{2+} movement such as lanthanum.

MARRE then described some of the work in his laboratory over the past two years using fusicoccin, a fungal toxin. Cell enlargement is stimulated, in stem sections by auxins and fusicoccin, in cotyledons by cytokinins and fusicoccin, and in leaf fragments by fusicoccin. Seed germination is promoted by gibberellic acid and by cytokinins and, usually much more efficiently, by fusicoccin. In general all these effects are more pronounced with fusicoccin than with the hormones. All the responses are accompanied, or rather preceded, by an increase in the capacity of the treated tissues to decrease the medium pH from near neutrality to values between 4.5 and 5.8; and the magnitude of the pH drop corresponds to that of the physiological response. Transmembrane electric potential difference changes were investigated, using fusicoccin on stem segments and cotyledons, and it was found that proton extrusion was accompanied by a consistent increase in the (negative) potential difference. Both the pH and PD responses were clearly detectable less than 5 min after the beginning of the fusicoccin treatment. Thus these results suggest that part of the physiological action of fusicoccin and of natural plant hormones is mediated by the activation of a metabolism dependent, electrogenic, proton extrusion pump.

Phytochrome was the next topic to be discussed. KUIPER felt frustrated by the plasmolysis results of HAUPT. He could not envisage how a loosely bound, as he understood it, molecule like phytochrome could affect water permeability. It also seemed to him that while differences could be seen between the effects of red and far-red light on plasmolysis, they were not so marked with deplasmolysis. And he pointed out that the more important effects of phytochrome were at the tissue level and thus cell-cell interactions, probably *via* the plasmodesmata, were the things to really be looking at; he cited the case of the collapse of the leaves of *Mimosa* as an example. HAUPT, in reply, first said that the *Mimosa* phenomenon is in fact at the cellular level because it depends on whether the few 'motor' cells are illuminated or not. With respect to the rate of deplasmolysis, he pointed out there is a distinct difference in rates; there are the time lags of 20 to 30 s but, when these are over, in red light deplasmolysis is complete in 2.8 s whereas in far-red light it takes 3.9 s. He also drew attention to the fact that there is not only a difference in the rates of plasmolysis but also in the extent; after red light plasmolysis is about 60 % in extent, but after far-red light it is 70 to 75 %. JAFFE thought, with respect to this latter point, that perhaps the permeability to mannitol had been changed by the red light. DAINTY felt much less frustrated than KUIPER about HAUPT's permeability results; he seemed to think that the results were rather exciting because here were controllable molecular events in the membrane influencing water permeability.

Although HAUPT has emphasized that in *Mougeotia* we had the first definite effect of phytochrome where we know that it is localized in the membrane, JAFFE suggested that the Tanada effect (the sticking of roots to glass surfaces) was also such an example, and an earlier one. Both BENTRUP and RAVEN thought it more likely that the Tanada effect was a cell wall (charge) effect and need not necessarily be membrane-mediated.

Questioned by STIEVE about his evidence for cooperativity in phytochrome action, SCHÄFER said that not only was the Hill coefficient 1.6 but also the behaviour of the amount of pelletable phytochrome as a function of ligand concentration was such as to rule out any simple adsorption process. And the Hill coefficient in the hook system was as high as 4.

Finally KYLIN talked about the time factor in ecological adaptation, i.e. the fact that plants may have a memory. He had two groups of pines, one group outside and the other group in the greenhouse. Some of the greenhouse pines were taken out into the open on 1st February and the cold hardiness of all these trees was studied through to 1st April. Of course the trees outside all the time were cold hardy and the trees inside all the time were rather sensitive to freezing; the group transferred adapted to some extent but never reached the degree of cold hardiness of the outside trees. He said 'they remember the treatment they had inside'. His experiments suggested that pines could keep their 'memory' for as long as 16 seasons whereas spruce, on the other hand, had no such memory. His point was that you must expect some such time factor of adaptation to operate when, say, plants with different histories are transferred from outside to the laboratory.

Subject Index

When the subject matter is a major topic discussed in a paper the page reference is given in **bold face** and the number then refers to the page on which this contribution begins